MACHINABILITY OF POWDER METALLURGY STEELS

MACHINABILITY OF POWDER METALLURGY STEELS

A. Šalak, M. Selecká and H. Danninger

Viva Books

New Delhi | Mumbai | Chennai | Kolkata | Bangalore | Hyderabad | Kochi

First Indian Edition 2008

VIVA BOOKS PRIVATE LIMITED

- 4737/23 Ansari Road, Daryaganj, New Delhi 110 002
  E-mail: vivadelhi@vivagroupindia.net, Tel. 42242200
- Plot No. 76, Service Industries, Shirvane, Sector 1, Nerul, Navi Mumbai 400 706
  E-mail: vivamumbai@vivagroupindia.net, Tel. 27721273, 27721274
- Jamals Fazal Chambers, 26 Greams Road, Chennai 600 006
  E-mail: vivachennai@vivagroupindia.net, Tel. 28290304, 28294241
- B-103, Jindal Towers, 21/1A/3 Darga Road, Kolkata 700 017
  E-mail: vivakolkata@vivagroupindia.net, Tel. 22836381, 22816713
- 7, Sovereign Park Apartments, 56-58, K. R. Road, Basavanagudi, Bangalore 560 004
  E-mail: vivabangalore@vivagroupindia.net, Tel. 26607409, 26607410
- 101-102 Mughal Marc Apartments, 3-4-637 to 641, Narayanguda, Hyderabad 500 029
  E-mail: vivahyderabad@vivagroupindia.net, Tel. 27564481, 27564482
- First Floor, Beevi Towers, SRM Road, Kaloor, Kochi 682 018, Kerala
  E-mail: vivakochi@vivagroupindia.net, Tel: 0484-2403055, 2403056

www.vivagroupindia.com

Published by arrangement with

Cambridge International Science Publishing
7, Meadow Walk, Great Abington
Cambridge CB1 6AZ, UK

ISBN-10: 81-309-0938-3
ISBN-13: 978-81-309-0938-7

Published by Vinod Vasishtha for Viva Books Private Limited, 4737/23, Ansari Road, Daryaganj, New Delhi-110 002. Printed and bound by Anand Sons, Delhi - 110 092.

Acknowledgements

The authors wish to express their gratitude to all persons and institutions that have contributed to this book in some or other way.

Our first thanks go to our families for their patience and understanding; there were many evenings and weekends we were not available to them.

Secondly, we want to thank the staff members of our respective institutes, i.e. the Institute of Materials Research, Slovak Academy of Sciences, Košice, Slovakia, and the Institute of Chemical Technologies and Analytics, Vienna University of Technology, Vienna, Austria. From both institutes came numerous contributions to this book such as micrographs, photographs, chemical analyses, data, process description and sample preparation. We thank also the colleagues from other universities and institutes who have helped us through information and advice; their names are listed in the references.

The authors are particularly grateful to the companies that have either directly contributed or granted permission to publish illustrations and data. Höganäs AB, Höganäs, Sweden; MIBA Sinter Austria GmbH, Vorchdorf, Austria; Chemetall GmbH, Arnoldstein, Austria; Boehlerit GmbH, Kapfenberg, Austria; Mashad Powder Metallurgy Co., Mashad, Iran, and Metalsint a.s., Dolný Kubín, Slovakia must be specially mentioned.

Finally, we want to thank our publisher, Cambridge International Science Publishing, for the excellent cooperation and the high technical quality of the book.

Andrej Šalak, Marcela Selecká
Košice, Slovakia, December 2004

Herbert Danninger
Vienna, Austria, December 2004

Contents

1 Introduction

Ferrous and nonferrous structural precision parts comprise about 80% in tonnage of powder metallurgy mass products. Of these parts, roughly 75% are used for transportation, primarily in the automotive industry, in which case particularly high requirements towards mechanical and functional properties, shape precision, and surface finish have to be met. In this area, powder metallurgy mass production is highly competitive to conventional metalworking techniques.

Here however it should be kept in mind that the production of components from wrought steels, cast iron, and various nonferrous alloys is done mostly by well established machining techniques with geometrically defined, in part also undefined, cutting edges. These machining processes have been revolutionized by the introduction of hardmetals, a special PM product, and of related, still harder cutting materials. This group of materials, for which PM is the only feasible manufacturing route, in production value even exceeds the PM precision parts, underlining the technical and economical importance of powder metallurgy products in general.

The competitiveness of PM precision components is enhanced by secondary operations which improve mechanical, functional, and geometrical properties. Among those operations, machining seems to be the most complicated and least understood and at the present time is not sufficiently managed. Machining of PM parts, employing different cutting methods, extends the range of shapes and thus also the range of applications without lowering the standard for static and especially dynamic mechanical properties, shape, geometrical precision, and surface finish. This latter precondition is an essential point when studying PM parts machining.

As mentioned above, the success of machining conventional materials was in part achieved by the use of new cutting tool materials supplied by powder metallurgy. The group of PM tool materials can be expected to boost also machining of PM precision parts, both with regard to increased productivity (e.g. through higher cutting speeds) and decrease of overhead cost.

Generally, the degree of machining and the amount of material this removed are drastically lower for PM parts compared to those from wrought steels which results in minimum solid and fluid waste generated and thus gives PM a clear advantage from the environmental viewpoint.

The reason for machining operations performed on PM precisions parts is that the standard PM shaping process by uniaxial die compaction, although very economical, cannot deliver all geometrical features; e.g. threads, cross-holes, undercuts, etc. cannot be pressed but have to be machined. Furthermore, it may be more economical or technically easier to introduce some geometrical features into sintered parts through machining rather than pressing even if the latter would be technically possible. Machining is increasingly also done to restore or improve the geometrical precision of components such as heat treated gears, powder forged heavy duty parts, and many others. This latter role of machining can be expected to become more and more important with increasingly higher requirements regarding both geometrical tolerances and mechanical properties (which frequently means heat treatment). It is estimated that at least for the European market about 40–50% of all PM ferrous parts undergo some machining process from these reasons [Beiss, P.: personal communication, 2004].

It is well known that machining of sintered steels is a difficult job, although the reasons for the problems encountered are not quite clear, and there are different explanations given. Many basic rules from machining of wrought steels can be transferred to machining of sintered steels; there are however peculiarities that are in part attributed to the porosity, in part to the specific microstructure of sintered steels, although the understanding of the various effects is still rather missing. It can be stated generally that the problems in machining of PM steel parts can be attributed to their microstructure when pores are regarded as microstructural constituents of zero Young's modulus; also in PM machining the result of a cutting process is determined by the interaction between the workpiece materials and the cutting tool under the given conditions.

Unfortunately the literature about machining of sintered steels is very scattered and incomplete, and due to the large variety of materials investigated and machining processes and conditions selected it is extremely difficult to identify general outlines and give general rules. Furthermore, the experience that has been collected by the PM parts manufacturers is commonly kept confidential, due to competition reasons. Therefore, quite a large proportion of the published literature and data about machinability comes from powder manufacturers rather than from those companies who actually do the machining, and there are hardly any data from the cutting tool manufacturers, quite in contrast to the extensive information given for conventional fully dense materials in the catalogues.

In this book it has been attempted to collect and structure as much as possible the information available from the viewpoint of the metallurgist. It was intended to describe all factors and parameters that are relevant for the formation of a solid PM material with defined physical and mechanical properties and with a characteristic microstructure since at the end all this affects the machinability. Thus the very important relationships between the 'history' of a PM component, which is affected by significantly more parameters than in case of wrought steel parts, and its machinability are described by giving an overview of powder metallurgy processes and materials, the principles of steel machining, cutting tools available, the various factors that influence – and the ways to improve - the machinability of PM steels, measures to improve machining of selected materials groups, and finally recommendations for machining sintered steel components.

Recommendations for the machining shop are already available, even as standards; in this book, it has been tried to link the machinability, in particular the optimum machining parameters – usually obtained empirically – , to the material, its properties, and its microstructure and thus ultimately to its manufacturing history. Due to the very scattered literature, the reader will find quite contradicting statements and also sizeable gaps in the knowledge base. In the following, the findings from the literature are given as published, with only some comments, in order to stimulate the reader to draw her/his own conclusions.

It is hoped that this book, by showing the deficiencies in current knowledge and practices, will also motivate the PM community to put more effort on machinability – at best in cooperation with cutting tool producers –, in particular on testing routines and standardization, in order to obtain a more reliable knowledge base about this frequently neglected but essential part of the PM production process. Machining should be regarded not as a mere 'secondary' operation but as an effective tool for enhancing the precision and extending the shape complexicity range of PM components. The aim of this book is to contribute to recognition of this potential.

2

Powder Metallurgy Processes and Materials

The powder metallurgy processes are more competitive than other fabrication methods for numerous materials and finished parts, in part due to their high flexibility regarding the manufacturing processes involved. PM offers chances for manufacturing of materials than cannot be obtained by classical metallurgical techniques, such as cemented carbides or very high alloy tool steels with an isotropic microstructure. In production of structural parts, PM offers higher precision, eliminating most of the machining operations required for castings and wrought semi-products. It is thus more cost effective and environmentally friendly, resulting in high material utilization. Figure 2.1 shows the basic steps of the powder metallugy process as used for manufacturing of structural parts and semi-finished products which may be subjected to supplementary machining. Each of these main steps contains some particularities and introduces specific physical, chemical and technological factors that affect the processing and final properties of a sintered material and by this also of each structural component.

2.1 METAL POWDER PRODUCTION

At present, there are two main methods for production of iron and steel powders. Together, the production of sponge iron powders by reduction of iron oxides and by water atomisation of iron and of low alloyed steel powders accounts for more than 90% of iron and steel powders produced around the world. (The production of powders by gas atomisation is limited to high alloy and some special powders). The main physical, chemical and technological properties of these two groups of iron-based powders are extensively documented by manufacturers and in the literature.

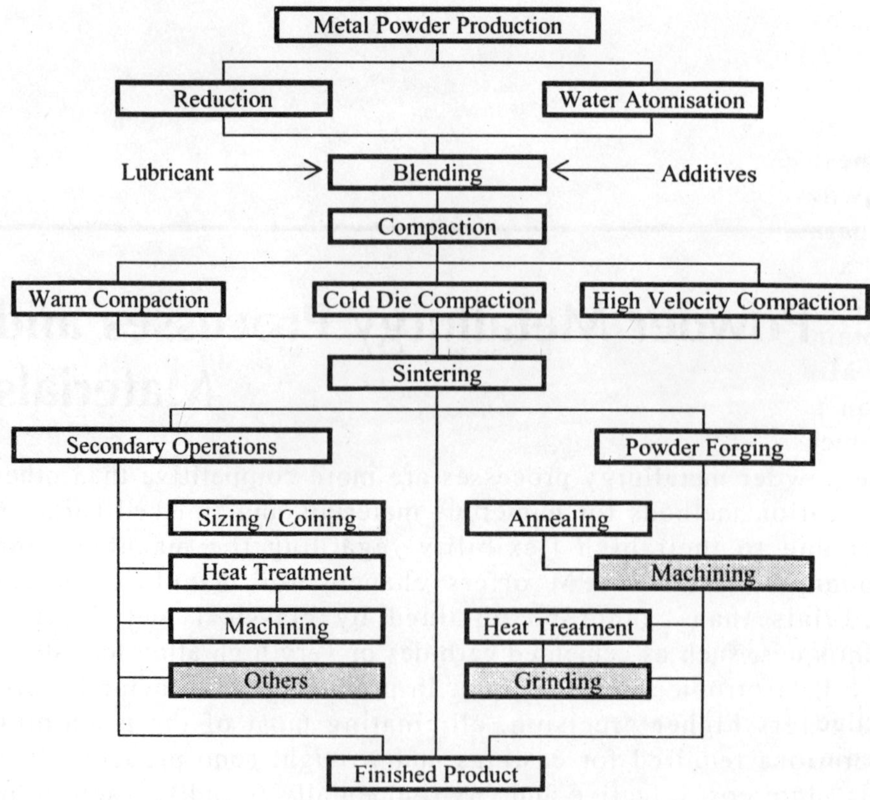

Fig.2.1 Basic steps of the powder metallurgy process.

Most of the chemical, physical, and technological properties of iron powder grades resulting from raw stock, manufacturing, and preparation methods have a pronounced impact on the green and final properties of sintered parts. Their effects are not fully suppressed by the following processing steps to the final sintered material. Therefore, the selection of the starting powder(s) is a critical step in PM parts production.

2.1.1 Reduction of iron oxides

The production of sponge iron powders by reduction of iron oxides is basically a chemical process. By using different iron oxide grades and processing parameters, it is possible to produce sponge iron powders with different physical and technological properties. A typical example of this production method is the Höganäs process, which has been the first mass production route for plain iron powder grades. The process is based on the use of quite pure magnetite (Fe_3O_4) ores. The milled and magnetically separated iron ore powder is reduced with a carbonaceous material in a tunnel kiln furnace (typically indirect reduction with coke with addition of limestone to bind any sulphur). After crushing and milling the powder cake, the resulting powder is carefully magnetically separated. The characteristic feature of the reduced iron powder is a spongy morphology with a higher specific surface area compared to smooth surface and full dense powder particles obtained by atomisation (see below).

Mill scale is another iron oxide suitable for production of sponge iron powder. 'Pyron iron powder' is produced from mill scale by reduction with hydrogen at moderate temperatures. The porosity of these iron powder particles is finer than that of powders produced from iron ore. The reduction of low-carbon steel mill scale following the Höganäs process produced a sponge iron powder with an apparent density ~2.1 g/cm^3 and a larger specific area. By additional milling it was possible to increase the apparent density of this powder to ~4.0 g/cm^3 for use in production of welding electrodes [1]. In another process, the very fine mill scale was reduced by combined reducing agents. Soot was added to the milled scale, and the admixture was compacted to briquettes, which were then reduced in converted natural gas [2].

The required low carbon and oxygen content of the reduced iron powders is obtained by subsequent annealing steps in a hydrogen-containing atmosphere. The final particle shape, particle size distribution, apparent and tap density and flowability are adjusted by separate preparation steps to attain the optimum specific properties of powders for the given application.

The microstructure of plain sponge iron powder particles is ferritic with different grain sizes, depending on the production method and final annealing temperature. In some cases, the reduced iron particles are very coarse-grained, it means one particle = one metallographic grain. The characteristic microstructure of sponge iron powder particles is shown in Figs.2.2 and 2.3.

The spongy character of the reduced powder is more pronounced in the iron powder reduced from mill scale as shown in Fig.2.4.

The internal pores are more evident in iron powder NC100.24 compared to SC100.26 iron powder. The microhardness of particles is one of the characteristic properties of the base powders; for these iron powder grades, it was in the range of 110–123 HV 0.02.

Figure 2.5 shows the morphology of a sponge iron powder

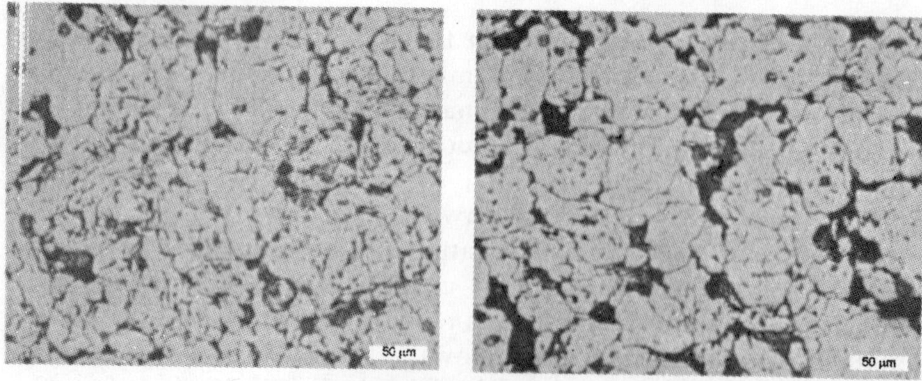

Fig.2.2 Microstructure of a green sample from sponge iron powder NC100.24 (Höganäs) compacted at 400 MPa. Optical micrograph. Nital etched.
Fig.2.3 (right) Microstructure of a green sample from sponge iron powder SC100.26 (Höganäs) compacted at 400 MPa. Optical micrograph. Nital etched.

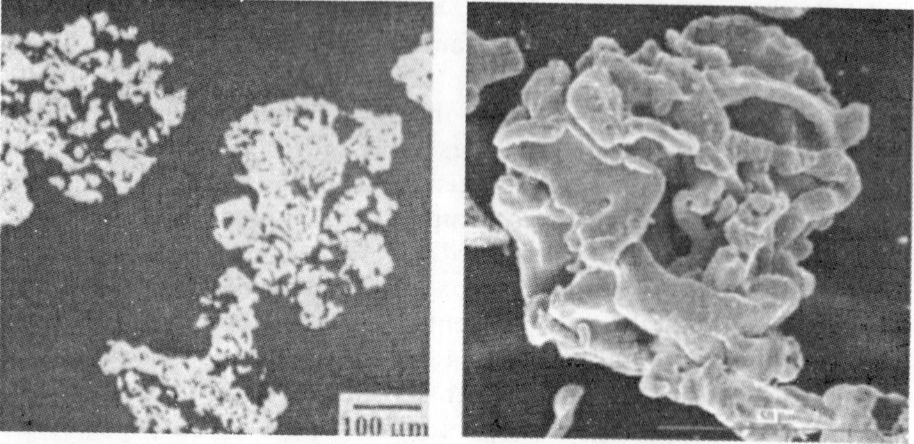

Fig.2.4 Cross-section of loose iron powder particles reduced from mill scale. Optical micrograph. Unetched.
Fig.2.5 Morphology of one particle of sponge iron powder (NC100.24, Höganäs). SEM.

particle in the as-received condition, *e.g.* after the standard milling, annealing, and screening operations. Comparison of Figs.2.5 and 2.6 shows large differences between the sponge iron powder grades which can affect the properties of intered iron.

The morphology of the iron powder particles, characterised by the specific surface area, is a physical property which clearly demonstrates the difference in the surface character caused by the starting raw material type and by the production method used. Compressibility, green strength, and interparticle contact formation

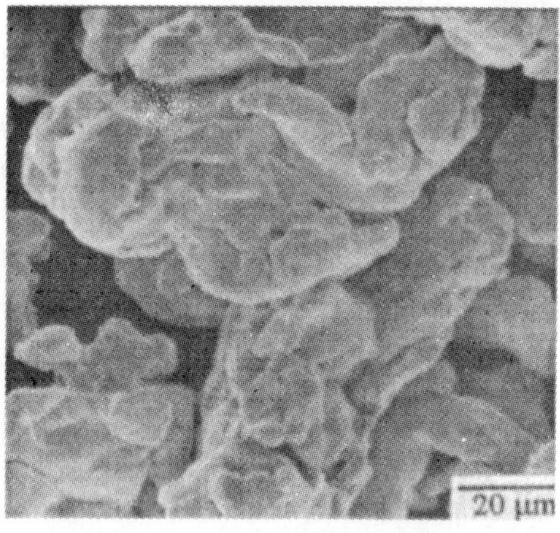

Fig.2.6 Morphology of a sponge iron powder particle (SC100.26, Höganäs), SEM.

during the initial sintering stages are affected by the particle morphology. Another characteristic of the reduced iron powders is the absence of segregation effects, *i.e.* nonreducible (slag) impurities remain in the powder (unless they can be removed by magnetic separation) and, therefore, also in the sintered product.

2.1.2 Water atomisation of metal powders

Atomisation of molten metal, low in Al and Si, by high-pressure water jets has become one of the main methods of producing iron and some low alloyed (prealloyed) powders. The main advantage of water atomisation is the possibility of producing chemically much cleaner metal powders. The water used for atomisation should be soft, without nitrates. Water atomisation is the standard production route for plain iron powder, some low alloyed prealloy powders and some special powders, such as stainless steel or tool steel powders.

The starting material used for iron powder production by atomisation is carefuly selected low carbon steel scrap, as best deep drawing steel sheet scrap or a pure and stable ilmenite ore body which afford the necessary high chemical cleanliness. Crushed reduced sponge iron billets and oversize iron particles are also suitable 'scrap'. The solidified droplets are superficially oxidised and can be hard with the non-equlibrium structure of the particles due to high cooling rate when the carbon for deoxidation was added or when the scrap with a higher carbon content was used. The cooling rate increases with the reduction of the droplet size, with the corresponding effect on the microstructure especially of alloyed materials. This effect is very sensitive in water atomised HSS powders.

The base particle size depends on the water pressure and, therefore, on the water jet velocity. The bulk in the iron powder production is targeted towards an average particle size near 150 μm. After drying, the powder is magnetically separated from slag particles, screened to various particle size fractions and homogenized, *i.e.* from the fractions a 'synthetic' powder with a given particle size distribution is mixed in order to ensure the necessary consistency of the powder properties.

After drying, screening and mixing, the final chemical and processing properties of water atomised iron powders are also established by low-temperature reducing annealing usually done at temperatures <900°C in a reducing hydrogen-containing atmosphere. The goal of this treatment is to remove oxide surface layers generated during atomisation and to decrease the carbon content to ~0.01% and lower. Carbon in water atomised powders contributes to the reduction of iron oxides to result in a minimum oxygen and carbon content. The typical oxygen content in the as-annealed powders is <0.15%.

The annealing temperature eliminates the non-equilibrium state of the particles and finally affects the matrix grain size and the type and amount of lattice defects in the powder particles. This is relevant especially in atomised powders because the cooling rate of solidified particles in water depends strongly on the particle size, and, therefore the as-atomised substructure properties can be different and this can also be retained partly after annealing (substructural properties).

Compared to sponge iron powder particles, water atomised powders are characterised by a more regular shape without internal pores and thus with a lower specific surface area, as shown in Fig.2.7. The microstructure of water atomised ASC100.29 iron powder particles seems to be be clean, without internal pores and without visible non-metallic inclusions, *i.e.* the particles are fully dense. The morphology of water atomised iron powder particles is shown in Fig.2.8.

In general, the iron powders differ in some characteristics, irrespective of the manufacturing method. The starting micro- and substructure characteristics of iron powder particles are usually not fully characterised, although they affect to a great extent the sintering and mainly the alloying behaviour of the base iron powder matrix.

In addition to the microstructure and morphology of iron powder particles, their substructure properties play an important role in

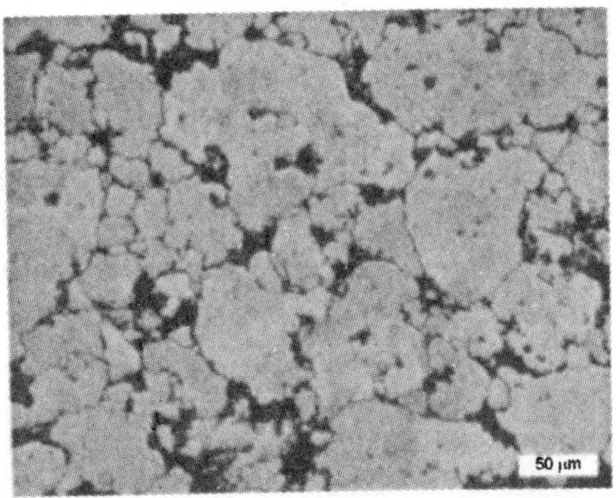

Fig.2.7 Microstructure of a green sample from atomised iron powder (ASC100.29, Höganäs) compacted at 400 MPa. Optical micrograph. Nital etched.
Fig.2.8 (right) Morphology of water atomised iron powder particles (ASC100.29, Höganäs). SEM.

sintering. X-ray diffraction analysis performed by the 'size–strain' method demonstrated some differences in the microstructure and substructure defects between the iron powders in relation to the manufacturing method (NC100.24, SC100.26, ASC100.29). These powders in term of crystals (clusters of crystals) are anisotropic. *i.e.* the crystal size and lattice distortion can differ in particular directions. The atomised iron powder grade analysed exhibited higher dislocation density in comparison to both sponge iron powder grades. Furthermore, in green samples compacted at 800 MPa, the lattice defect density was on the level of wrought cold rolled and annealed deep drawing steel.

These substructure characteristics in the starting powders affect in the first stage the formation of interparticle bonds during heating and the as-sintered properties of a material. The grain size and lattice defects of iron powder particles affect the sintering behaviour and the alloying process of iron matrix from the early stages of sintering process, *i.e.* during heating in some cases still in the α-region. As an example, Fig.2.9 shows the tensile strength of Fe–4Mn–0.3C steel sintered at 875°C based on different iron powder grades. A small difference in tensile strength between atomised and two sponge iron powder grades and a high increase in Hametag iron powder compacts was demonstrated. It was the proof of different sintering and alloying activities of iron powder grades in the compacts towards alloying by manganese vapours at this extremely low sintering temperature. The 'lower' sintering and alloying activity of sponge iron powder compared to atomised one

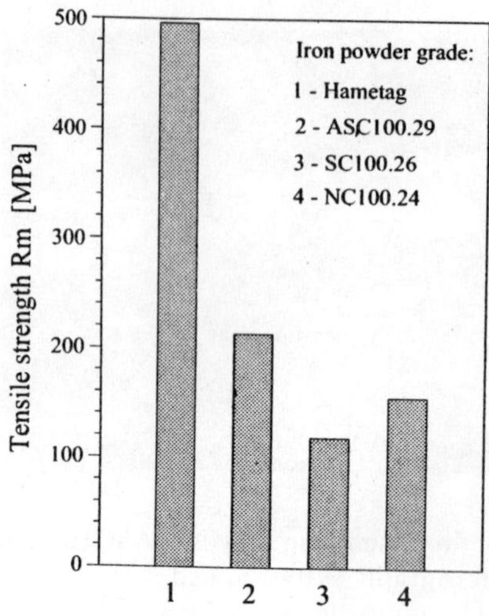

Fig.2.9 Tensile strength of Fe–4Mn–0.3C steel sintered at 875°C for 30 min in cracked ammonia, dew point –30°C. Compaction 600 MPa. Iron powder grade: 1 – Hametag (eddy milled, no more in production), 2 – water atomised (ASC100.29), 3 – sponge (SC100.26), 4 – sponge (NC100.24) [5].

can be attributed to the oxygen content originating from the internal pores in the powder particles which are closed during compaction, thus preventing easy oxygen removal, and partly to the specific surface area. The 'high' sintering and alloying activity of Hametag iron powder resulting in an as-sintered density increase by ~0.15 g/cm^3 compared with the atomised powder was caused by the fine-grained microstructure and higher dislocation density [3]. Sintering of Mn-steel in the α-region could be used as a test method for comparing the substructural activity of various iron powder grades [4].

2.1.3 Water atomisation of prealloyed powders

The production of water atomised prealloyed powders for structural application in alloying is limited by the (solid solution) hardening effect of the alloy element(s) on ferrite with the resulting adverse effect on the cold compressibility of the powders, as shown in Fig.2.10. The hardness of powders increases almost proportionally to the concentration and the hardening effect of the alloying element in ferritic iron.

The difference in the effectiveness of an additive in iron as a strengthening and hardening agent is traced to many factors, including atomic structure, atomic size, and chemical reactions in the steel. The reasons for the possibility of producing the water-atomised low alloyed Cr-, Mo- and/or Ni powders follow from this graph.

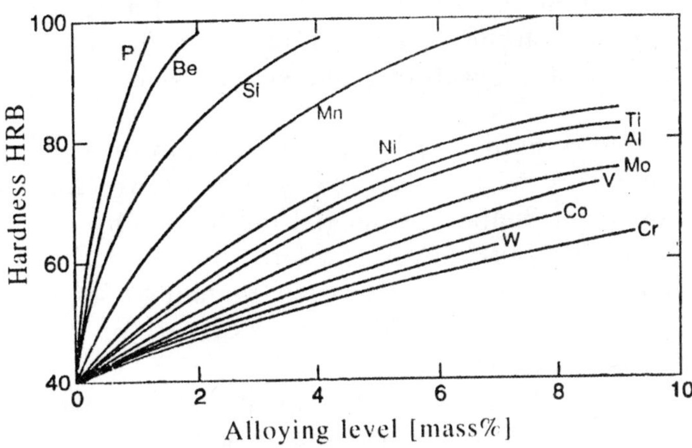

Fig.2.10 Hardness as a function of alloying level for several additions to ferritic iron [6].

Water atomised prealloyed powders can be divided into two groups. The first one includes powders containing up to 2% nickel as a base alloying element with small additions of Mo, and Mn and Cr (<0.3%). The newer second group of prealloyed powders contains powders alloyed with chromium as the base element (up to 3%) with the addition mostly of molybdenum (<0.5%) and the third group are those alloyed by molybdenum only (0.85–3.5%). A separate group are water atomised stainless and tool steel powders. The water atomised prealloyed powders are characterised by the uniform distribution of the alloying elements in the solid solution. The compressibility of these powders is slightly reduced because they are harder as a result of solid solution strengthening by the alloying elements.

The production of low-chromium prealloyed steel powders, usually with 0.3 or 0.5% addition of molybdenum, is of considerable technical and economical interest, mainly due to the high hardenability when combined with carbon. As follows from Fig.2.10 in terms of compressibility of a prealloyed powder, chromium is the most suitable alloying element. On the other side, chromium belongs to the elements characterised by a high oxygen affinity, which is a problem from the thermodynamic view for annealing of as-atomised powders and must also be considered when sintering parts from such powders.

The reduction-anneal treatment of chromium-containing pre-alloyed powders is done either in vacuum, in which case the carbon contained by the powder particles is used as a reducing agent, or in a hydrogen-containing atmosphere at a suitably high temperature. For these powders, carbon is a less effective reducing agent than hydrogen at lower temperatures, but becomes more effective at higher ones [5].

A contribution to the more effective production of low chromium prealloyed powders with a low oxygen content could be the atomisation by water containing special inhibitors. By this method, atomised Fe–1Cr–0.7Mn–0.5Mo powders annealed at 1100°C for 30 min in hydrogen exhibited an oxygen content of ~0.02%. When compacting the powder at 600–700 MPa, a density of the compacts of ~7.0 g/cm^3 was attained [7].

The final chemical and technological properties also of water atomised prealloyed powders are the result of the preparation methods used and are specific for the respective powder producer.

2.1.4 Oil atomisation

Despite the fact that oil atomisation is not conducted any more, the powders (iron and prealloyed) produced by this method, mainly Cr-prealloyed powders, exhibited the highest chemical cleanliness so far attained, including low oxygen content and high compressibility [8]. The powder forged parts based on oil-atomised Fe–1Cr–0.7Mn–0.2Mo powder exhibited higher fracture toughness and other mechanical properties properties than wrought bearing steels.

It is possible to presume that the atomisation of a melt alloyed also with chromium to a powder with a low oxygen content using a diesel oil jet perfected by nitrogen through a special injection nozzle to prevent inflammation may be a further contribution to the development of the liquid atomisation technique [9,10].

2.1.5 Inert gas atomisation

Atomisation by an inert gas (nitrogen, argon) is used for the production of high-alloy steel powders, in most cases stainless steels, tool steels, superalloys, and other special alloys, or powders of high chemical purity from easily oxidisable metals. The nitrogen concentration in nitrogen atomised powders, especially in those alloyed with nitride-forming elements is higher compared to argon atomised powders; the latter, however, are more prone to contain entrapped gas. The gas atomised metal powders are spherical in shape and usually hard and, therefore, not cold compactible in a rigid die, also resulting in extremely poor green strength. Gas atomised powders are not used for the production of structural components.

2.2 CHEMICAL, PHYSICAL AND TECHNOLOGICAL CHARACTERISTICS OF METAL POWDERS

2.2.1 Basic characteristics of some iron and prealloyed powders

The data regarding the mentioned properties of base metal powder grades used, and their differences are those which form a complex of the specific properties. Tables 2.1, 2.2 and 2.3 list the data for the given sponge and atomised iron powder grades.

Tab.2.1 Particle size, apparent density, flowability, and chemical composition of base sponge and water atomised iron powders [11]

	Powder grade	Particle size [μm]	Apparent density [g/cm³]	Flow rate [s/50g]	C [%]	H_2-loss [%]
Sponge	NC100.24	~<150	2.45	30	<0.01	0.21
	SC100.26	~<150	2.65	28	<0.01	0.12
	MH80.23	~150-45	2.95	33	0.08	0.35
Atomised	AHC100.29	~150	2.95	25	<0.01	0.10
	ASC100.29	~<150	2.95	24	<0.01	0.08
	ABC100.30	<200	3.00	24	0.002	0.05

Typical particle size distribution data of main iron powder grades (which may vary in a small range from one batch to another without changing the base processing properties) are listed in Tab.2.2. Table 2.3 contains further characteristics and chemical data of the given iron powders.

Tab.2.2 Particle size distribution of common sponge and atomised iron powder grades (determined by the authors)

Iron powder grade	Particle size [μm]					
	>160	100	71	63	45	<45
NC100.24	1.6	38.7	22.3	13.1	9.3	15.0
SC100.26	1.7	16.7	26.1	22.7	14.3	18.5
ASC100.29	6.4	20.9	17.2	19.1	15.8	20.6

Remark: *matter insoluble in HCl, HV 0.01 = average microhardness measured in the core of iron powder particles

Water atomised powders are characterised by higher apparent density and faster flowability due to smoother surface and lack in internal pores and due to lower specific surface area.

Tab.2.3 Chemical and physical properties of common atomised (ASC100.29) and sponge (SC100.26, NC100.24) iron powder grades (determined by the authors)

Iron powder grade	Apparent density [g/cm^3]	Tap density [g/cm^3]	Flow rate [50 g/s]	Specific surface area [m^2/g]	C [%]	H_2-loss [%]	*HCl [%]	HV 0.01
NC100.24	2.5	3.2	40	0.06	0.004	0.22	0.32	132
SC100.26	2.7	3.5	38	0.04	0.003	0.12	0.23	142
ASC100.29	2.8	3.5	32	0.02	0.002	0.13	0.20	138

Minimum carbon (<0.01 %) and oxygen content (<0.2%) are the condition for optimum compressibility of a powder. Minimum difference between the sponge and atomised as-delivered iron powder particles in microhardness values was determined.The particle size distribution in the powder is a result of the production method and/or preparation operations. As described above, for as-delivered powders it is established by "synthetically" mixing powder fraction following a given formula; thus the slight variations between the atomising runs can be eliminated. The mixing formula is defined with the aim to reach an economical and technical compromise between the flow behaviour, bulk density, and compressibility of a powder, and thus also the green strength, on one side and high sintering activity on the other side to attain the highest possible mechanical and/or toughness properties of a sintered body. For some applications, iron powders with specific particle size distribution and other technological properties (high compressibility) are produced, as, *e.g.* atomised iron powder ABC100.30 (0.002% C, 0.05% H_2-loss, density 7.3 g/cm^3 at 600 MPa) for soft magnetic application.

The basic characteristics of typical prealloyed powders from which some have a stable position in PM parts production, *e.g.* Astaloy A and B and some Cr- and Mo-alloyed are intensively investigated and partly yet applied in manufacturing regarding their metallurgical and cost advantages, are listed in Tab.2.4.

The composition including chemical cleanliness of some powder grades and materials which are commonly used in powder metallurgy is expressed also in their theoretical pore-free density measured by pycnometry as listed in Tab.2.5.

For comparison, some properties of pure iron are: room-temperature density = 7.87 g/cm^3, thermal conductivity = 80.2 Ω/m/°C, thermal expansion coefficient = $11.76 \cdot 10^{-6}$/°C, yield strength = 133 MPa, ultimate tensile strength = 266 MPa, elastic (Young) modulus = 210 GPa, elongation to fracture = 45%, hardness = 21 HRB, electrical resistivity = 9.7 μΩ·cm [6].

Tab.2.4 Chemical compostition and base processing properties of water atomised prealloyed powders (all powder grades - particle size <150 μm)

Powder grade	Chemical composition [mass %]							Apparent density [g/cm³]	Flow rate [s/50 g]
	C	Ni	Mo	Cr	Mn	V	O_{tot}		
Astaloy A		1.90	0.55	-	0.20	-	0.13	3.1	24
Astaloy B		0.45	0.60	-	0.30	-	0.13	3.1	24
Astaloy CrM	<0.01	-	0.50	3.00	-	-	<0.25	2.85	25
Astaloy CrL	<0.01	-	0.50	1.00	-	-	0.15	2.85	
Astaloy Mo	<0.01		1.50	-	-	-	0.1	3.1	24
Astaloy 85Mo	<0.01	-	0.85	-	-	-	0.1	3.1	25
*KIP 103V	0.01	-	0.30	1.00	0.30	0.3	0.15	2.8	25
*KIP 30CRV	0.01	-	0.30	3.00	0.1	0.3	0.24	2.8	25

Supplier: *Kawasaki – vacuum annealed, other Höganäs

Tab.2.5 Approximate pore-free density (theoretical density ρ and theoretical specific volume 1/ρ) of some standard powder grades and additives [12]

Material	ρ [g/cm³]	1/ρ [cm³/g]	Material	ρ [g/cm³]	1/ρ [cm³/g]	Material	ρ [g/cm³]	1/ρ [cm³/g]
NC100.24	7.75	0.129	Distaloy AB	7.86	0.127	Graphite F10	2.25	0.444
SC100.26	7.76	0.129	Astaloy Mo	7.85	0.128	*LUBE	1.03	0.971
ASC100.29	7.83	0.128	Nickel	8.84	0.113	MnS	3.86	0.259
ABC100.30	7.83	0.128	Copper	8.90	0.112	-	-	-
Distaloy AE	7.88	0.127	Graphite UF4	2.28	0.439	-	-	-

Remark: *binder/lubricant used in a Densmix powder. Graphite: UF4 natural, F10 artificial

2.2.2 Impurities

All types of impurities of the microstructure play a deteriorating role in the processing of metal powders and in the properties of sintered parts. Impurities in a sintered material are evaluated therefore also in connection to machining.

According to their origin, the chemical composition and the effect on the processing and properties of sintered materials, impurities in powder metallurgy materials can be divided into the following groups:

– non-reducible, known as non-metallic inclusions,
– reducible by hydrogen and/or carbon monoxide
coming from:
– raw materials used for production of metal powders (oxides, scrap),
– oxides formed during powder atomisation and not fully re-

duced during the subsequent anneal,

- formed on the particle surfaces firstly during cooling in a low-temperature post reduction furnace and during handling and preparation at ambient temperatures (oxide films),
- formed during sintering in insufficiently pure atmospheres (especially in the presence of oxygen-sensitive alloying elements),
- formed during cooling from sintering temperature (decreasing temperature) on open pore surfaces in the compact through the reaction of oxygen in the sintering atmosphere with the alloying elements of high oxygen affinity (oxide films). In this stage, the atmosphere is not in equilibrium conditions with the material system.

Generally, impurities in metals include extrinsic metallic (unsoluble phases), non-metallic, and gas matters. The impurities in metal powders cannot be completely removed under economical conditions, it is possible only to limit their amount and type by technical means in relation to the required properties of the parts.

Contrary to wrought (ingot) materials, in powder metallurgy each powder particle can be a carrier of non-reducible impurities (in reality they are not all), and, therefore, they are relatively homogeneously distributed in the sintered material without formation of clusters. This minimises their deteriorating effect. Typical examples are powder forged steels exhibiting higher toughness properties than wrought steels of corresponding composition, despite a higher total oxygen content compared with wrought steel. Usually, the effect of porosity (>5%) prevails over the effect of impurities on the properties of sintered materials when compared with properties of fully dense materials.

The reducible impurities coming from raw materials are reduced in the final annealing step of the powder preparation route. The oxide films formed on the powder particles during storage and handling are reduced during heating to sintering temperature in a reducing atmosphere (or also in an inert atmosphere, if carbon is present). In terms of thickness, these oxide layers are very thin (several atomic layers) but in terms of processing they are very important and in some cases they play a very complicated role. The oxide layers show interdiffusion between the adjacent metal particles at sintering, therefore, they should be reduced before the initial stage of interparticle neck formation.

The oxide films formed on open pore surfaces in the compacts during cooling are persistent and may have an effect on machining.

This problem becomes relevant especially when sintering materials alloyed with elements of high oxygen affinity. The effect of these surface oxides on the mechanical properties of the sintered compacts is not sufficiently documentated in particular alloys.

In terms of reducible impurities, the state of powders is characterised by the reducible oxygen content, which for iron powders is determined as H_2-loss (ISO 4491), *i.e.* the mass loss measured when annealing the powder in a hydrogen atmosphere. The total oxygen content in alloyed steels (reducible + non-reducible oxygen) is determined by vacuum hot extraction, *i.e.* reaction of the molten specimen with carbon at very high temperatures and infrared detection of the CO generated.

The non-reducible impurities in iron powder are commonly measured chemically as matter insoluble in hydrochloric acid (ISO 4496). This method of determining non-metallic inclusions is, however, not reliable for alloyed steels because some carbides are insoluble in HCl.

These acid-insoluble impurities are, *e.g.* non-metallic inclusions, which adversely affect the as-sintered properties (while the reducible oxygen is commonly removed during sintering). Sponge iron powder reduced from mill scale contains less non-metallic inclusions compared with sponge iron powder grades reduced from iron ore. Atomised iron powders contain still lower amounts of acid insolubles since slag inclusions in them are at least in part removed during melt processing of low carbon steel scrap.

In addition to the the total amount of non-metallic inclusions, another important factor is their size. Larger non-metallic inclusions have a stronger deteriorating effect on the mechanical properties than finer ones, especially regarding the fatigue behaviour [13].

Non-metallic inclusions are formed mainly by oxides, such as SiO_2, Cr_2O_3, Al_2O_3 CaO, MgO, MnO, ZnO. They are not reducible by hydrogen or carbon monoxide at standard temperatures [10]. They form nuclei of dimple facet failure in a (ductile) sintered material, *e.g.* at tensile or bending load up to fracture, and can also affect the machining process.

The non-metallic inclusions impair the strength and mainly the toughness properties of the material. Figure 2.11 shows fracture surfaces of sintered iron specimens after tensile testing. Non-metallic inclusions in this case initiated the formation of larger dimple facets. The fracture of the failured sample without non-metallic inclusions is characterised by dimple and line facets, Fig.2.12.

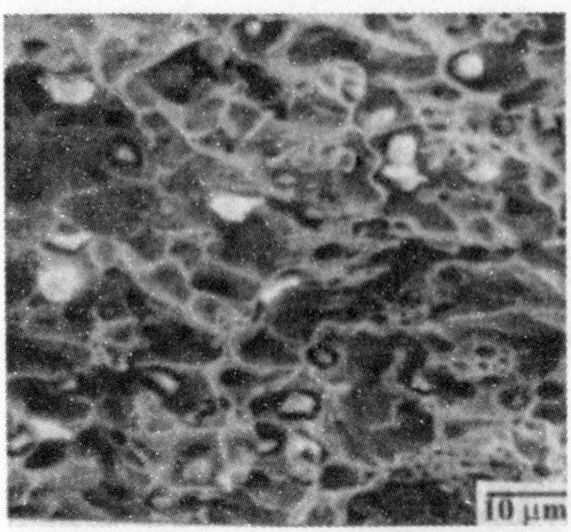

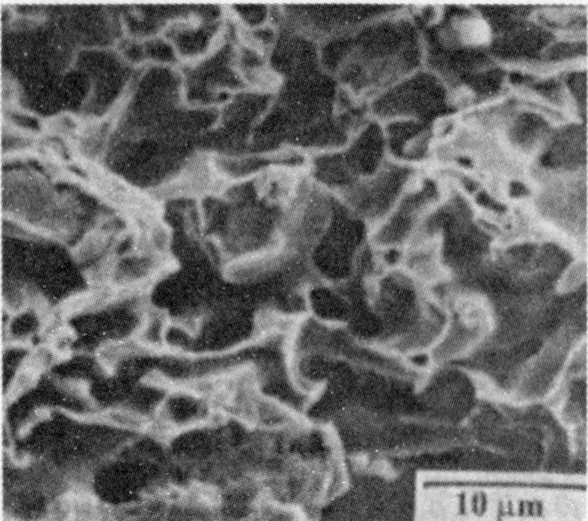

Fig.2.11 Tensile fracture surface of sintered iron specimen with non-metallic inclusions. Sponge iron powder MH140. Compacting at 600 MPa. Rm = 219 MPa, SEM.
Fig.2.12 (right) Tensile fracture surface of sintered iron specimen without non-metallic inclusions. Sponge iron powder SC100.26. Compacting at 800 MPa. Rm = 248 MPa. SEM.

In terms of machinability, it would be necessary to study this general evaluated deteriorating effect of non-metallic inclusions on machinability in more detail by special tests with regard to their rather low amount in the materials. Considering existing production and preparation methods, the presence of some impurities belongs to the base characteristics of metal powders.

2.3 MIXING AND COMPACTION

In the PM parts production route, compaction is the shaping step in which the usually very complex geometry of the component is defined while the mechanical strength is still low.

One goal for compaction of iron and alloyed powders is to reach the highest possible density with the lowest compaction force and to attain the green strength required for safe handling of the compacts regarding also the requirement for green machining. The green density attained depends on the compacting pressure as well as on the physical and technological properties of base iron and/or other powder particles, type and amount of lubricant and the friction between the die wall and the powder. Green strength is affected mainly by the morphology (specific surface area) of the powder particles. Spherical particles with a relatively low specific surface area result in the lowest green strength. The admixed pressing lubricants also lower green strength.

Prior to compaction, the powders are mixed with lubricants that lower the friction between the powder particles during compaction and between the compact and the die walls, improving compactibility and drastically reducing die wear. Lubricants used are stearates, stearic acid, polytetrafluorethylene, polyvinyl fluoride, and even waxes. The latter are used to the largest extent in PM steel structural parts production although for complex shapes Zn-stearate containing lubricants are still necessary. The undesirable residue after decomposition of a lubricant is 'ash'. The stearates have the highest amount of ash (zinc stearate ~14%) and waxes have the lowest (parrafin wax ~1%) [6]. The addition of a lubricant should be as small as possible, typically 0.5 to 0.8 mass% being added. Currently, trials with direct spray lubrication of the compacting die ('die wall lubrication') are performed which enables significant reduction of the amount of admixed lubricant. In addition to the lubricant, other powders are frequently, added such as metallic alloy powders and graphite; in practice, stepwise mixing is done with the metallic powder – *e.g.* Cu and Ni – being added first, then after some mixing graphite is introduced and, after further mixing, the lubricant with subsequent final mixing. Thus, segregation and cluster formation during mixing can be drastically reduced. To minimise the segregation of powder particles during handling and filling of the die cavity, low-cost binders are used to glue the mostly fine additive powder particles to the base ferrous powder.

In production of structural parts, standard cold pressing, repressing (double pressing), and warm compaction methods are used, and new methods are permanently being developed for increasing efficiently the density of parts at the lowest pressure.

2.3.1 Cold die compaction

Cold uniaxial die pressing of metal powders, used for compaction of structural parts, is a multistage densification process in which the individual stages are controlled by different mechanisms and can overlap. Die compaction enables manufacturing of parts of complicated shapes [10,14]. An increase of compacting pressure increases the extent of interparticle contacts (size and number) as a result of starting and continuing plastic deformation. Densification occurs by: 1) rearrangement of the particles, and 2) plastic deformation of powder particles (the final stage of the pressing process) in the regions of initial contacts (after filling the die) and those newly formed during densification.

Cold work hardening of iron powder particles during cold die pressing manifests itself in an increase of microhardness, Fig.2.13. A minimum compacting pressure of 50 MPa ensured the

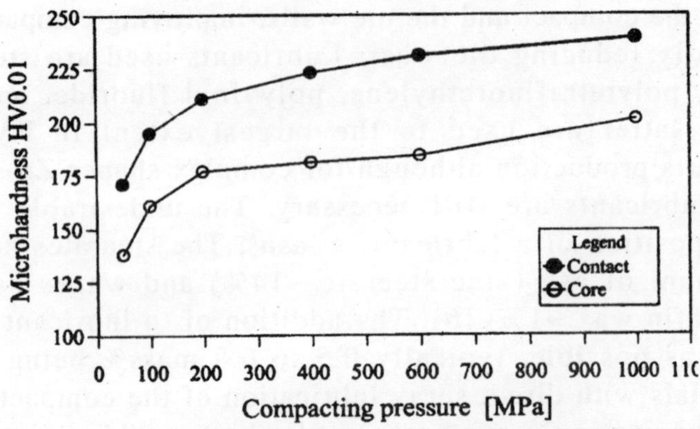

Fig.2.13 Microhardness in sponge iron powder particle (0.10–0.16 mm in size) cores and in interparticle contact areas in a green compact as a function of the compacting pressure [10,14].

consistence of the green compact. This shows that the work hardening of the particles already takes place when compacting the powder at this lowest pressure, *i.e.* already in the starting stage of compacting. An increase of compacting pressure caused an increase of the particle core microhardness by 80% and in the interparticle contact areas by 115% in relation to the starting (annealed) state (90–120 HV 0.01).

According to X-ray diffraction analysis, compacting at 50 MPa already increases microstresses by 80%, and by compacting at standard 600 MPa by 460%, as indicated by changes in the diffraction line width (β), which was ~10–15 mm for powder annealed in hydrogen at 650°C/1h. The microstresses are released at heating to sintering temperature at about 700°C, but in the interparticle regions work-hardened during compaction a new fine grained microstructure with higher dislocation density is formed. The dislocation density seems to be a factor that greatly accelerates the initial sintering stages, which means the formation of interparticle metal bonds. It is the stage of the formation of a continuous metallic body from the powder particles which affects all physical and mechanical properties and, therefore, also the machining behaviour.

Compacting is controlled by the metal powder particle properties, mainly by hardness and particle shape and particle size distribution. The compacting stages are characterized by both a gradual reduction of the possibility of rearrangement and displacement of the particles and an increase of work hardening. Due to the latter mechanism, the known compressibility curves, differing depending on the powder grade, show clearly that it is not possible to reach the full density of powder compacts by cold compaction following common techniques, *i.e.* at practically feasible compacting pressures (below 1000 MPa). It is the area for new compaction methods. On the other side, powders of high compressibility are more active in sintering compared to lower compressibility powders.

In industrial practice, the compacting pressures for compacting powder mixes (plain iron, mixed-alloy powders and prealloyed powders with additives) are mostly in the range of 600 to 700 MPa. A rigid tool set is required, usually consisting of a hardmetal die and cold work tool steel punches and core rods. Ejection of the compacted (green) body from the die is also a critical process. The powder mix properties, compacting pressure used and the coefficient of friction between the powder and the die affect the ejection force as does the elastic relaxation of the die and of the green body, resulting in 'spring back' (growth of the compact compared to the die cavity) and in some re-densification.

Repressing (or double pressing) is a cold die compaction operation used for increasing the density of pre-sintered parts (by 5 to 20%) before final sintering in a double pressing and sintering technique. The green parts, usually compacted at pressures lower than for final compaction, are soft annealed in order to relieve the work hardening of the iron particles. This enables a second compaction of the part by common compacting pressures with a further increase of density. For sintered ferrous materials, the density can be increased from typically 7.0 to 7.1 g/cm^3 for single pressing to 7.3 to 7.4 g/cm^3 for the double pressing technique.

2.3.2 Warm compaction

The main role of the warm compaction process is in lowering the pressure required for attaining green densities of the compacts above 7.0–7.2 g/cm^3 in dependence on the properties of the powder used. These value cannot be reached by common cold compaction. During warm compaction, simultaneous heating of powder and the die is perfected, resulting in an isothermal process leading to excellent dimensional stability. Both powder and tooling are heated to about 125–150°C giving a slight softening of the iron which produces increased green density and green strength when compacting usually at 700 MPa. The process is based on the knowledge that the yield point of iron at 125°C is lower than that at ambient temperature. A slightly smaller decrease of the yield point was also determined for Fe–1.5Mo [12]. The powder is coated with polymers (various stearates or other lubricants, including teflon), facilitating rearrangement of the particles during compaction. Close temperature control is necessary, since the product uniformly suffers if the polymer is too hot. After cooling the warm compacted powder part to room temperature, the green density usually increases by 0.15 to 0.20 g/cm^3 compared to standard cold compaction. The warm compaction process is more costly than the typical cold die compaction, therefore it is used mainly for high strength steels, *e.g.* diffusion alloyed powders or Cr- and Mo-prealloy variants. If warm compaction is performed at higher temperatures, polymer degradation then leads to inferior sintered properties, as confirmed by experiments. After sintering, the steel warm compacted at 130°C was the strongest, followed by that compacted at room temperature, while the steel warm compacted at 150°C exhibited the lowest strength [6].

2.3.3 New and developing compaction processes

The direct pressing of cross holes in the green part is desirable because it omits additional mechanical handling after sintering (reduction of production capacity) and eliminates the secondary operation – machining. For a long time, however, cross hole formation was considered technically impossible, because of difficulties during pressing as well as removal of the compact. This newest development results show that the cross hole formation in a green part is possible by a different method.

2.3.3.1 Cold die pressing with cross holes

As shown in Fig.2.14, conventional fully automatic hydraulic cold pressing (CNC) enables cold pressing of cross holes with the possibility of tapping in the sintered state in a ring-shaped part with four cross holes.

A similar feature of cross hole formation in a compact by cold pressing in a CNC press is shown in Fig.2.15.

2.3.3.2 Warm flow compaction

The warm flow compaction process (WFC) is a variant of powder metallurgy shaping that combines features of powder pressing, warm compaction, and metal injection moulding. It is a compaction process that enables in part pressing of parts, *e.g.* with undercuts, through holes, and tapped holes and by this minimizes or eliminates machining which would otherwise be necessary. This implies that

Fig.2.14 The specimens – rings cold compacted in one stroke in a rigid closed die from mixed alloyed powder with four cross through holes. (Courtesy of Institute of Materials Research of SAS, Košice).

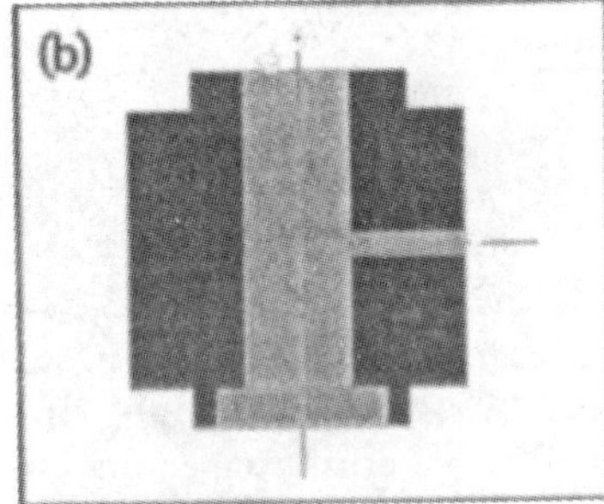

Fig.2.15 a) Ferrous compact with cross hole made by using CNC press with two upper and three lower punches and a one-side cross pin, b) cross-sectional diagram of the part [15].

no additional steps should come into the process, and tool setup must not become too complex at a marked density increase. In WFC, radial powder movement in the die was ensured by the use of a binder (~3%) forming a liquid during compacting at elevated temperatures (80 to 130°C). Thus, in order to increase the packing density, the powder mixture (also Distaloy type), containing about 10 to 20% of fine powder (<22 μm in size), has the characteristics of a very viscous liquid [15]. The principle of the warm flow compaction is applicable to virtually all powders. The balance between sinter shrinkage and tolerances on parts must be permanently regarded [16].

As example of a part formed by this method is 'T-shape' and a 'cross' shown in Fig.2.16. It indicates that the WFC method enables powders (WFC-powder blend) to flow perpendicular to the pressing direction and, therefore, form parts with higher geometrical complexity compared to cold and warm compaction, although the parts shown were compacted only from top and bottom. Figure 2.17 shows a PM steel part with blind threaded hole produced by WFC.

Figure 2.18 depicts an experimental part with side branches

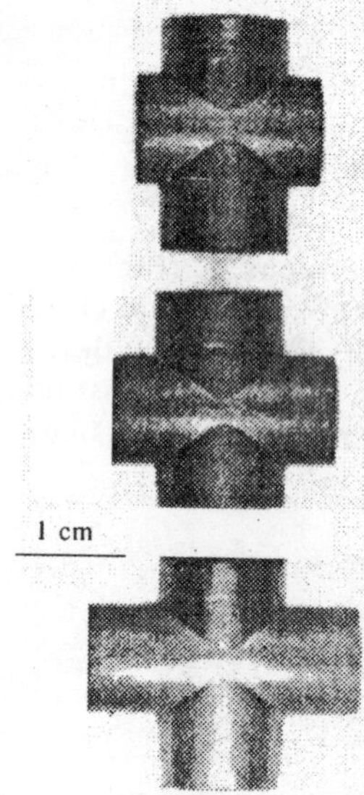

Fig.2.16 Green parts produced by WFC from a cross-shaped die [16].

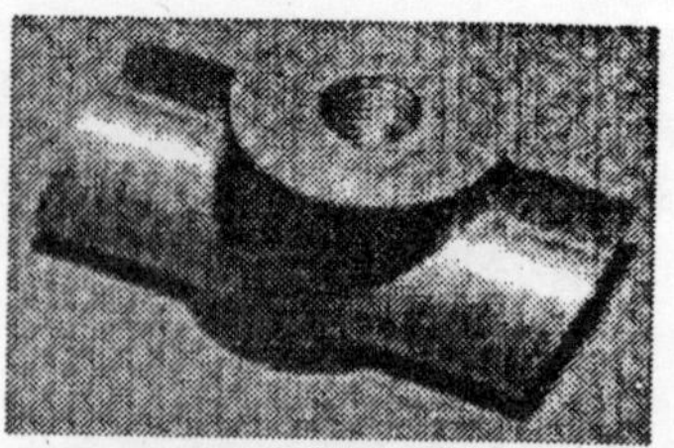

Fig.2.17 Threaded blind hole (M6) in cross-flow PM steel green part produced by WFC [16,17].

Fig.2.18 Trial part with side branches made of stainless steel 316L (left: green part, right: sintered) [17].

produced by WFC which underlines the possibility to use also stainless steel powders for this method.

With regard to machining of PM parts, it can be stated that the WFC process can be used for producing parts with cross holes and threaded cross holes, Fig 2.19, and undercuting is possible by uniaxial die pressing, thus avoiding the need for machining operations.

2.3.3.3 High energy compaction

High energy (velocity, HVC) rate (~10 m/s) powder consolidation methods were investigated and used mainly for compacting powders with low compressibility. High velocity compaction (dynamic compaction) is a new method for compactimg also high compress-

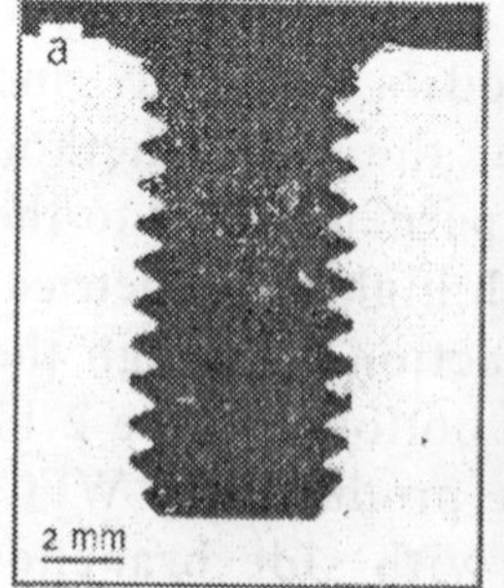

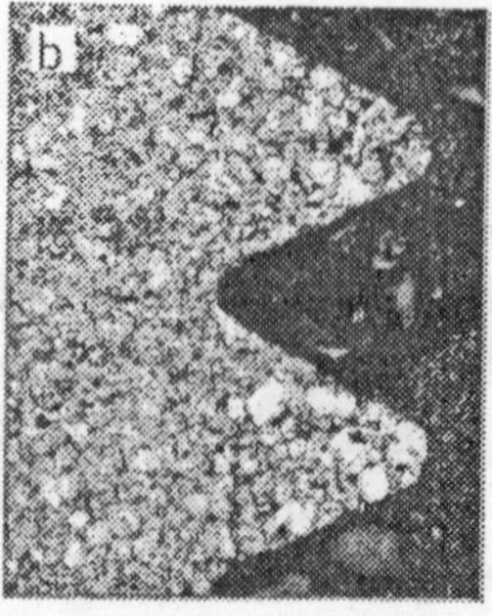

Fig.2.19 a – and b – Fine detail of a threaded hole formed by WFC [17].

ibility metal powders offering high densification. High velocity compaction seems to prevent particle rearrangement, leading to lower stresses than when compacting very spongy particles in conventional die compaction (CDC). This is achieved by shock waves, created by a hydraulically operated hammer, that transfer the compaction energy through the compaction tool (punches) to the powder. The use of controlled energy die compaction (CEDC) adjustable instead of force or displacement, enables densification in a time shorter than 0.01 s, shorter than the conventional process (0.5 s). Dynamic powder compaction differs from such high speed techniques as Petro forge or Dynapark in that the latter processes are closely related to crank presses and drop forges.

Parts production by HVC consists of the same processing steps as conventional compaction. The advantage of the process was proved by pressing some commercial press-ready mixes (*e.g.* Distaloy HP + 0.4% C, NC 100.24 and Distaloy AE powder) to a density of 7.4 g/cm^3, and 'full' as-sintered densities (~7.5 g/cm^3) were achieved. The sintered compacts (Distaloy HP1 + C) showed a tensile strength of 920 MPa. For prealloyed powder grade CrM + C compacted by HVC to 7.5 g/cm^3 and sinter hardened, the tensile strength was almost 1400 MPa and the yield strength was >1200 MPa. Potential structural parts manufactured by HVC are those requiring high wear resistance, fatigue endurance and high rigidity, and also very large parts [18,19], although currently only single-level parts such as gears have been produced by HVC. The products of this compaction method present new tasks for PM machining.

2.3.4 Powder forging and cold forging

The purpose of forging, whether performed hot or cold, is to improve part performance. Densification is an essential part of the forging process, and deformation has been shown to aid densification and part performance. In respect to the rate of densification, forging is nearer to high energy compaction, *i.e* the densification of the preform occurs in a few tenth of a second. The most important factor for powder forging (success or failure) is therefore proper preform design. The lateral material flow in contrast to upsetting or repressing is a condition for forged parts with the required high density without cracks. With lateral constraint, the rate of densification due to an increment of punch displacement is greater than that for upsetting, as one would anticipate. The contact and by this also the friction of the preform being forged with the die walls at the lateral flow is minimised to the final stage of the densification process (>95% relative density).

Forging includes the production of preforms by conventional compaction and sintering followed by the selected forging process. The density of preforms for cold forging is higher (~7.0 g/cm^3) and for powder forging lower (~6.0–6.8 g/cm^3).

2.3.4.1 Powder forging

Powder forging (PF), also called 'sinter forging', is an economically advantageous process for manufacturing 'fully' dense parts, especially for high mainly dynamic loading conditions. Powder forging enlarges the possibility of powder metallurgy production of parts in the region with the properties higher than those of wrought steel. The process is characterised by the simultaneous effect of dynamic load, high velocity and heat on the porous preform. Basically, the elimination of pores diminishes their deteriorating effect on the properties of parts. In such a case, high toughness, fatigue behaviour and structural homogeneity are mandatory properties of the parts. 'Fully' dense powder forged parts should have a maximum porosity of ~1.0–0.5%. The so-called residual porosity in powder forged parts is concentrated in the subsurface layer only what is advantageous for machining (short chips). Large parts for agricultural machines were powder forged with a porosity of up to 5%. Direct powder forging of parts for full density greatly shortens the lifetime of the forging dies, impairs the final dimensions of the forgings and increases the forging force. The density of the preforms must also be matched with their dimensions. To prevent the formation of a flash and achieve uniform density, the mass tolerance of the preforms should be ±0.5%. Powder forging can be realised in two ways, *i.e.* direct forging the preforms from the sintering temperature, or forging sintered preforms after new heating to forging temperature.

Powder forging is an additional technological step requiring special equipment and instrumentation, and, therefore, seems to be effective only for the production of parts with the properties which cannot be attained by conventional methods used in powder metallurgy for the densification of green compacts. An optimum solution is to install automatic forging lines.

The main difference in methods for the preparation of 'fully' dense parts by conventional powder metallurgy methods (liquid phase sintering, rapid consolidation, warm flow compaction, etc.) and by powder forging lies in the deformation process of the powder preform in a closed die. During powder forging of preforms preheated to ~1100°C, always in a die with a temperature of 150–250°C, the lateral flow of sintered preforms (~10%) (shear stresses) should be ensured to achieve 'full' density also for larger but mainly for highly dynamically loaded parts. Therefore, closure of pores in a preform occurs at uniform pressure under the combined effect of shear stresses in lateral yielding of the material during forging. This state cannot be produced by the previously mentioned conventional cold and warm pressing methods.

In general, the powders used for forging could not, at least not as yet, be made as exact copies of the wrought steels they are supposed to replace. It is advantageous and feasible to prepare the powders with the composition which makes it possible to achieve properties of the parts on the highest technical and economical level. The mixed and prealloyed powders are used for powder forging. It depends on the individual technical and economical decision. Recent positive experiences with the preparation of water atomised low alloyed chromium–molybdenum and molybdenum steel powders form a perspective for a broader use of powders of the similar composition to the wrought steel. The oxygen content in atomised powders used for forging is <0.20% and, in some cases, <0.15–0.10%. This is a good starting point for further processing. Structural homogeneity of powder forged parts based on prealloyed powders is a source of their higher mechanical and mainly toughness and fatigue properties, *e.g.* for contact fatigue endurance, fracture toughness, etc.

Suitable examples of successful manufacturing of large-scale powder forged parts can be pinion gears [20], connecting rods [21] and rolling bearing rings [22,23] as heavy duty and some other parts.

2.3.4.2 Cold forging

In contrast to selective surface densification, the cold forging process results in the formation of the final shape and dimensions of a part and its bulk densification from a simpler preform. This is advantageous for parts for which full density is not critical. Cold forging can be realised as back extrusion, forward extrusion or as cold die forming regarding the properties and the shape of the product. The positive effect of the lateral material flow on densification is shown in Fig.2.20, where a higher density at the given forging pressure was achieved for the cold forming process that includes metal flow, as opposed to the strict repressing process.

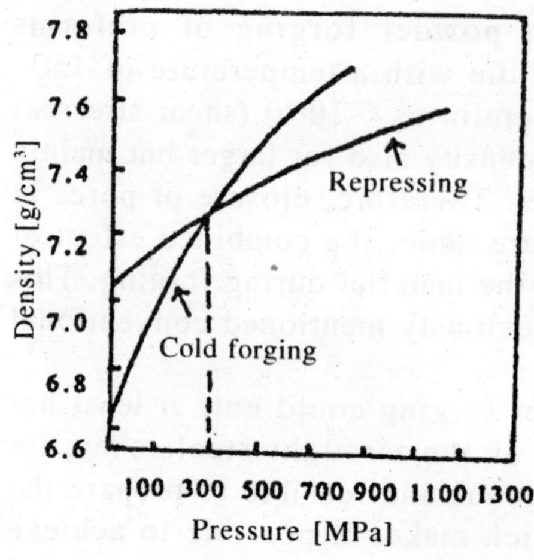

Fig.2.20 Comparison of the rate of densification for cold forging and conventional repressing of porous preforms [24].

In cold forging, preform design must take into account the poor workability of the porous body, the beneficial effect of the lateral flow on densification and performance, and tool loads. Intermediate cold forming technology involves localised densification. Deformation loads are lower than in powder forging and the properties are between those of powder forged parts and press-sinter parts.

2.4 SINTERING

Sintering is an active thermal process during which transforms, in our case the dispersed metal body – consisting of pressed-together powder particles – into a mechanically strong metal body of defined properties. In contrast to other sintered materials such as, for example, hardmetals and ceramics, sintered steel components shrink only marginally during sintering, the porosity thus being present also in the as-sintered components.

The macroscopic manifestation of the sintering of a compact is its densification which is characterised as a length, volume and porosity decrease or as a density increase. Due to simple measurement, the linear dimension change is preferred as the method for the determination and characterisation of the course of densification. Dilatometric measurements are used to express the rate of densification as a function of time and temperature. Generally, densification is an isothermal process. There have been indications of the non-isothermal sintering stage also in metallic materials, mainly in multi-components systems (intensive shrinkage stage). In a green metal compact, there are many easily diffusing

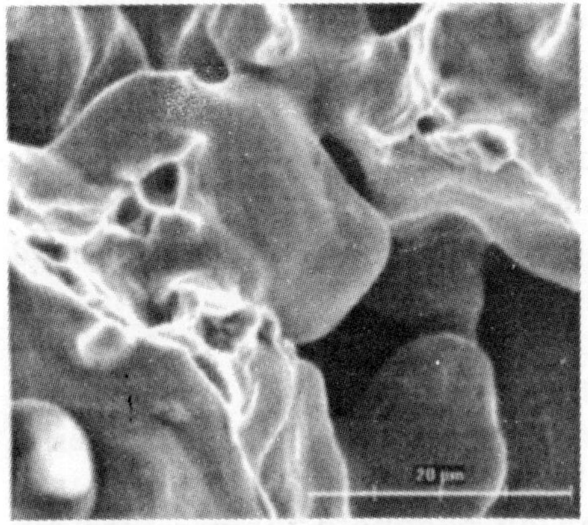

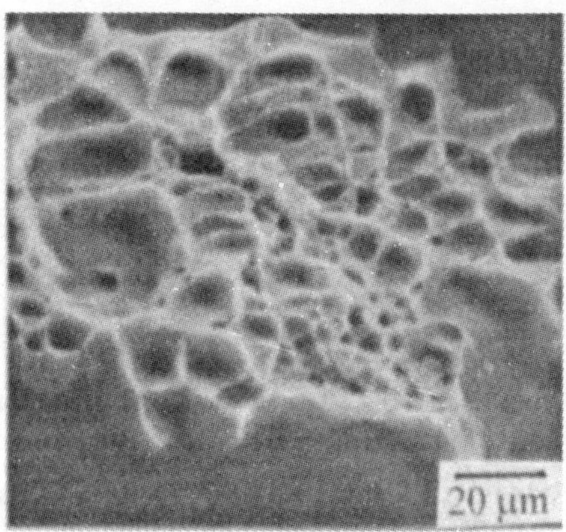

Fig.2.21 Interparticle necks formed during sintering of a Fe–0.3C sample compacted at 600 MPa. SC100.26 iron powder, sintering at 1120°C, 30 min. SEM.
Fig2.22 (right) Fracture (tensile load) of an iron powder sample compacted at 400 MPa and sintered at 875°C for 1 h in hydrogen. Sponge iron powder reduced from mill scale. SEM.

atoms at sintering with a small energy in the crystal structure. Consequently, the existing surface roughness of powders increased by dislocations (including all lattice defects) formed in compaction (plastic deformation) forms the structural and energetic state of a compact entering the sintering process [25].

Formation and growth of interparticle necks and alloying of the matrix take place during sintering. Figure 2.21 shows in detail interparticle necks formed during sintering of a compact, which demonstrates the most important role of sintering in ferrous precision parts – formation and growth of interparticle necks, not only densification, resulting in mechanical and physical properties and in densification. Active formation of interparticle necks in iron compacts occurs during heating yet in the α-phase shown in the form of dimples on the fracture surface of a sintered iron compact after the tensile, Fig.2.22.

Sintering as a complex and special heat treatment process requires a special set of sintering conditions, because driving forces and material transport mechanisms associated with the process in particular sintering stages are different. Various mechanisms operatons taking place in sintering in this case include lubricant removal/debinding, reduction of surface oxides, and surface, volume, grain boundary and pipe diffusion, evaporation and condensation and plastic deformation.

In the final stage, sintering is responsible for the formation of a new microstructure of sintered products with defined physical and mechanical properties and dimensions.

The determining process variables involved in sintering of a loose powder and/or of a powder compact are temperature, time, and atmosphere. The cooling rate plays an important role in the final microstructure formation of the sintered materials, especially alloyed ones.

2.4.1 Sintering temperature and time

The first stages of sintering and alloying of a metal powder material occur during heating considering the high diffusivity in the α-iron range. All sintering stages require an adequate temperature for successful formation of a metal body. Basically, with increasing temperature the increased atomic motion progressively leads firstly to the formation of interparticle necks from pressing contacts, followed by their growth at higher temperatures in all material systems, and secondly, in mixed or hybrid alloy powder systems, to alloying of base iron and/or prealloyed matrices through solid state diffusion including initially the carbon originating from graphite.

Main reduction–degassing processes take place during heating. For Fe–C and for Fe–Mo–C prepared from water atomized powders, mass loss occurs predominately in the range 700°C and 950–1100°C, together with the formation of CO and some CO_2 indicating carbothermic reduction of the oxides present in the starting powders. Sponge iron powders, which initially contain more oxygen, exhibit a much larger mass loss in the temperature range >900°C. With Cr-prealloyed powder steels, the mass loss at the lower temperature is insignificant, most of degassing occurs at temperatures >1000°C. Gas formation processes are strongly dependent on temperature but virtually independent of time [26,27].

The common sintering temperatures are in the range 1120 to 1300°C. They depend on the material systems and the preparation method, especially on alloying and on the required properties. Sintering time is closely related to sintering temperature and varies between 30 and 60 min. There is a trend to shorten sintering time and increase sintering temperature, due to economical reasons.

When high density parts (>approx. 93% relative density), with increasing sintering temperature, at high sintering temperatures (HTS) >1150°C mostly at 1280°C, in addition to the alloying of the matrix and the formation of interparticle bonds also the tendency to formation of closed, isolated pores becomes pronounced because the shape of pores depends more on temperature than on sintering time. In this respect, sintering time can be reduced by 60% when sintering at 1280°C instead of conventional 1120°C. High temperature sintering is also attractive for this reason, enabling higher furnace throughput. However, HTS is not always beneficial. Diffusion-alloyed systems, Distaloy type, show little change or even worse properties after HTS, the reason being that the higher homogeneity resulting from HTS gives rise to fewer nickel-rich zones in the microstructure, reducing the mechanical, especially endurance, properties. In this particular case, alloy design relies on conventional sintering to deliberately produce microstructures with pronounced chemical heterogeneity for optimum properties [28]. This example shows the very complex relationships between composition, alloying technique, manufacturing parameters, and properties that are typical for sintered steels.

High temperature sintering at, for example, ~1250°C, is preferred for the production of high strength steels but its effect on possible grain coarsening of some alloys, as mentioned, *e.g.* diffusion alloy steels, must also be considered. Conventional and new alloys are processed at 1280°C in high volumes and in complex shapes, with acceptable dimensional stability, *e.g.* high performance heat treatable steels containing chromium, molybdenum and, most recently, also manganese.

2.4.2 Sintering atmosphere

The sintering atmosphere is a factor that affects the course of all sintering process stages involving delubing, reduction and diffusion, thus strongly influencing the final properties and by this also affecting the machining of parts. In general, the sintering atmosphere plays a more important role for the properties of a sintered material than might be assumed.

Basically, the sintering atmosphere should be reducing. In this respect it is also essential to consider the reducing effect of carbon for material systems which enables sintering also in inert atmospheres as, for example, nitrogen. The furnace atmosphere provides three functions when sintering metal powders:

- removal of the lubricant or binder,
- preventing oxidation,
- reduction of surface oxide films to form metallic surface (*in statu nascendi*).

Some major problems encountered with the furnace atmosphere can be:

- ineffective lubricant removal,

- poor control of the carbon content,
- oxidation of open pore surfaces during cooling, especially in a compact alloyed with an element with high oxygen affinity.

Thermochemical reactions are important when selecting the atmosphere for sintering ferrous PM alloys. The dominant reactions are oxidation–reduction and carburization–decarburization. The reactions involving iron (Fe) are basic and well understood reactions. Equilibrium between oxygen and an oxide such as FeO can be written in the reaction form as follows:

$$Fe_{(s)} + \frac{1}{2}O_{2(g)} \Leftrightarrow FeO_{(s)}.$$

A high concentration of iron and oxygen leads to iron oxide formation. Alternatively, a high concentration of iron oxide favours reduction and the formation of oxygen and pure iron.

The iron oxides are easy to reduce (direct reduction) using hydrogen. The reaction is described as follows:

$$FeO_{(s)} + H_{2(g)} \Leftrightarrow Fe_{(s)} + H_2O_{(g)}.$$

If iron (Fe) is heated in the presence of steam (water vapour, H_2O), the iron oxide (FeO) and hydrogen (H_2) form. Alternatively, if iron oxide and hydrogen are heated together, the reaction produces iron and water vapour. A constant supply of hydrogen is required for total oxide reduction, *i.e.* the water vapour product must be constantly replaced with fresh hydrogen. The higher sintering temperatures make reduction easier. An adequate oxide reduction must be ensured during sintering in order to encourage interparticle necks formation and by this develop the required properties of the material. Also, the more stable oxides will require a higher temperature to be possibly reduced. Thus, atmosphere purity and its thermodynamically optimum composition is a concern in sintering.

The diagram, Fig.2.23, shows that oxides such as iron, nickel, molybdenum and copper oxides are reducible during heating in an atmosphere of a common dew point <–30°C. For oxides of Cr and Mn, lower dew points and/or higher temperatures are necessary to ensure complete reduction according this diagram. In the case of manganese its high vapour pressure must be considered [29].

It follows from the Ellingham–Richardson diagram that oxides of iron, nickel, molybdenum, and copper (elements with low oxygen

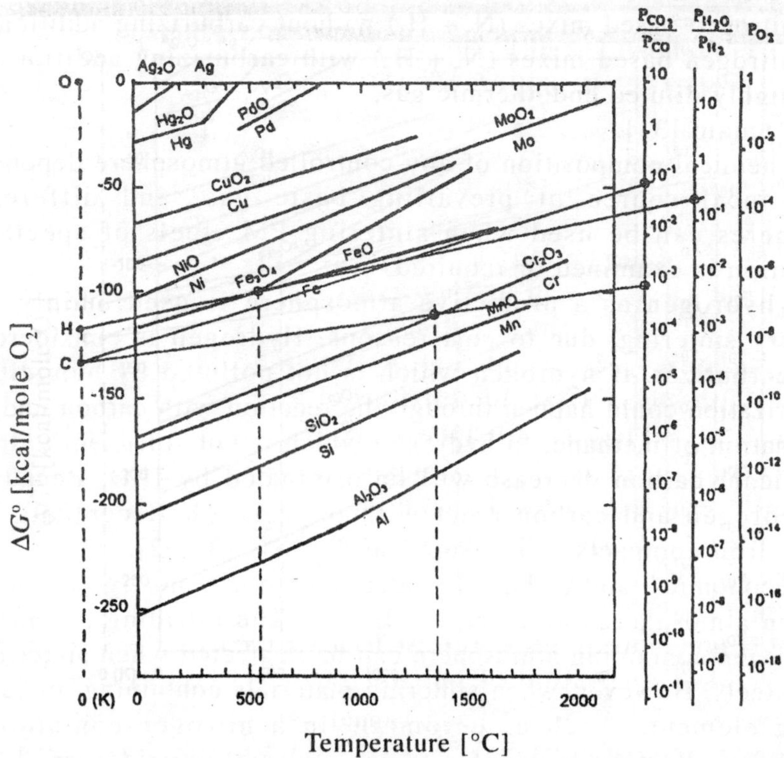

Fig.2.23 Graphical determination of equilibrium temperatures for Fe and Cr in and H_2O/H_2 – and in a CO_2/CO atmosphere [30].

affinity) are reducible during heating of a powder (compact) in an atmosphere of relatively low purity even when the chemical activity of the metal is taken as unity [10]. The real material systems usually do not correspond to unit activity, commonly a <1. High purity atmospheres and/or higher temperatures are necessary for reducing the oxides of Cr, Mn, V, and Si. Also, for this reason it is not necessary to note that the purity of the sintering atmosphere is a concern in sintering.

Considering the sintering of carbon-containing steels, the possible atmospheres which also depend also on the type of alloy additions, may be:

- endothermic gas (endogas) from gaseous hydrocarbons (usually methane or propane),
- synthetic endothermic gas (mixtures of methanol and nitrogen),
- pure hydrogen,

- cracked ammonia,
- nitrogen based mixes ($N_2 + H_2$) without carburizing additions,
- nitrogen based mixes ($N_2 + H_2$) with carburizing additions,
- highly diluted endothermic gas,
- vacuum [31].

The chemical composition of any controlled atmosphere depends on the local source of prevailing basic gas, and different atmospheres can be used when sintering PM steels of specific composition is examined or required.

Pure hydrogen as a protective atmosphere is used mainly in laboratory sintering, due to cost reasons. Hydrogen is considered as a decarburizer. If hydrogen which is not polluted by humidity, decarburization could happen through the reaction with carbon under the formation of methane. In hydrogen which is not sufficiently dry, a significant carbon decrease will be originated by H_2O. Because both hydrogen and carbon react with oxygen. Their contents in sintered iron compacts is measured as H_2 loss [31,32].

The economical and technical tendency is to use mostly nitrogen-hydrogen atmospheres up to pure nitrogen. The nitriding potential in a nitrogen containing atmosphere can be neglected when sintering carbon steels. However, when sintering materials containing nitride-forming elements, such as chromium in a nitrogen-containing atmosphere, the formation of nitrides and carbonitrides at slow cooling rates is a consequence. Molybdenum contributes to the formation of high-dispersion complex nitrides. The microhardness of these compounds is above 1200 HV, with a very negative effect on machinability. Nitrides form mainly in chromium- and molybdenum-alloyed structural steels and, especially, in stainless steels.

Neglecting any nitriding effect, dry nitrogen is neutral when sintering carbon steels. Nevertheless, a significant carbon decrease is observed because added carbon removes the oxygen bonded to starting iron powders (decrease ~0.11%). The resulting carbon loss has to be considered in order to result in a sufficient final (combined) carbon content to achieve the required strengthening or hardening effect. The carbon loss is compensated either by adding some extra graphite or by sintering in a slightly carburizing atmosphere, *e.g.* nitrogen with a well defined hydrocarbon content.

Molecular nitrogen in a sintering atmosphere cannot diffuse into the iron matrix to form nitrides. The penetration of nitrogen (gaseous phase) into the solid phase is a result of the surface reaction:

$$N_2 \rightarrow 2N_{ads}$$

(thermal dissociation of molecular nitrogen) and in the presence of hydrogen:

$$N_2 + H_2 \rightarrow 2NH_{ads}.$$

Two reactions occur at the interphase boundary, namely chemisorption involving dissociation of nitrogen on the surface and dissolution of nitrogen in the iron matrix from 'adsorption sites' [33,34]. The rate of all reactions that proceed during sintering of the porous PM materials increases due to the large surface area of open pores and the high affinity of chromium to nitrogen. For this reason, nitrides form in the entire volume of the compacts. This results in significant strengthening.

Cracked ammonia is a relatively low cost-reducing atmosphere to sinter many metallic materials. Residual ammonia is typically below 250 ppm, and as long as the moisture content is low the atmosphere is nearly neutral with respect to carbon [6].

In nitrogen-based atmospheres, some additions of a reducing agent are quite common. Therefore, in most cases, hydrogen is present in quantities between 5 and 12%, in some cases up to 30%. A small amount of CH_4 is added when carburizing action is required.

When sintering is done in an endothermic gas from gaseous hydrocarbons, both hydrogen and carbon monoxide are reducing, the latter can be more or less strongly carburizing. As a consequence, it is necessary to consider the relations between temperature, carbon concentration and alloy type and additions that could be present in the material [32].

The carbon activity can be changed by some alloying elements, Fig.2.24. For example, Si and Ni, increase the a_c value, whereas other alloy elements, like Cu, Mn and Cr, decrease the carbon activity. This means that the presence of Ni and Si originates a decrease of carbon content of saturated austenite and in pearlite as well. When using an endothermic gas to sinter carbon steels at high temperatures, a drop in carbon content is to be accepted, and to compensate this drop, *e.g.* soaking time at a suitable lower temperature is needed.

A new type of atmosphere which can be seen as a strongly diluted endothermic gas, with a low dew point is the *Alnat I* atmosphere. Therefore, the continuously changing conditions during heating – which are peculiar of any endothermic atmosphere – have

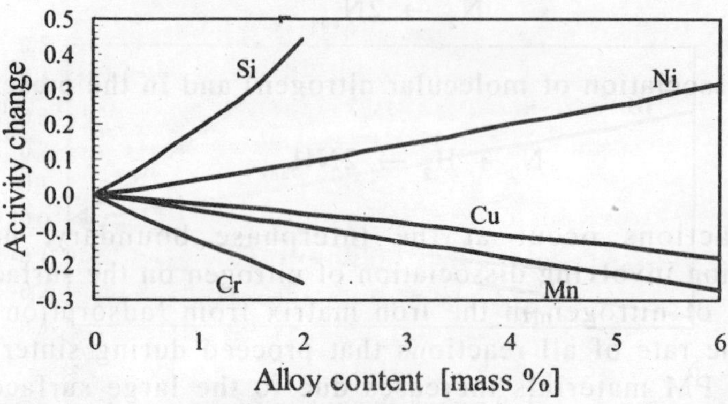

Fig.2.24 Changes of carbon activity (a_c) due to presence of alloy elements evenly distributed in austenite [32].

reduced effects, if any. The H_2 content is high enough to offset – at least partially – the carbon decrease due to the oxygen content of the powder. When using consistent raw powders, the carbon content decrease upon sintering remains within narrow ranges. In relation to carbon and especially to high-strength high-temperature sintered steels, the acceptable carbon levels and uniform carbon distribution have to be provided by eliminating decarburizing effects. Higher hydrogen (~10%) concentrations in a nitrogen-based atmosphere were found in some cases to be decarburizing by the formation of water vapour and did not produce uniform carbon distribution even with methane additions. Low levels of H_2 and CO were found to provide a uniform carbon profile. Nitrogen–methanol–methane atmospheres are considered as viable alternatives to N_2–H_2–CO and N_2–H_2–CH_4 atmospheres for high-temperature sintering of iron-based materials [35]. Successful sintering of iron–carbon material, *e.g.* elimination of non-controlled decarburizing, depends on the furnace atmosphere (carbon potential). At present, sufficient sintering conditions can be established for sintering iron–carbon steel.

The carbon content in a final sintered part, as follows from the thermodynamics and practical experiences, is the result of carburization and decarburization processes occuring during sintering. As the first example, the general reaction between iron (Fe) and its carbide (Fe_3C) is considered in terms of carbon monoxide (CO) or methane (CH_4) reactions [6]:

$$3\,Fe_{(s)} + 2\,CO_{(g)} \Leftrightarrow (Fe_3C) + CO_{2(g)}$$
$$3\,Fe_{(s)} + CH_{4\,(g)} \Leftrightarrow (Fe_3C) + 2\,H_{2(g)}.$$

Consider also the reaction involving the carbide in steel with water vapour (steam) during sintering:

$$Fe_3C_{(s)} + H_2O_{(g)} \Leftrightarrow 3\ Fe_{(s)} + CO_{(g)} + H_{2(g)}.$$

The same reaction occurs in the case of carbon dissolved in austenite which is the usual state at T>750°C. Understandably, the reaction is drastically more pronounced in this case both due to kinetic and thermodynamical kinetic reasons, H_2O becoming less stable at higher temperature while CO becomes more so. Therefore, decarburization through humidity in the atmosphere is most pronounced at high temperatures.

As the water content in the atmosphere increases, the steel is more strongly decarburized. Thus, atmospheres with low water vapour levels (dry or low dew points) are most effective in controlling carbon during sintering. Furthermore, decarburization is significantly faster at higher temperatures, therefore dry atmospheres are necessary especially for high temperature sintering. The relation between the dew point (water content) of the atmosphere and the final carbon content level for endothermic atmosphere sintering is shown in Fig.2.25.

The common purity of the sintering atmosphere, characterised by the dew point, is in the range of –20 to –60°C, except for endothermic gases. The cleanest atmospheres without additional cleaning are those based on cryogenic gases. The dew point of the atmospheres depends mainly on the correct process conditions at the supplier and could be –60°C and lower (nitrogen from a

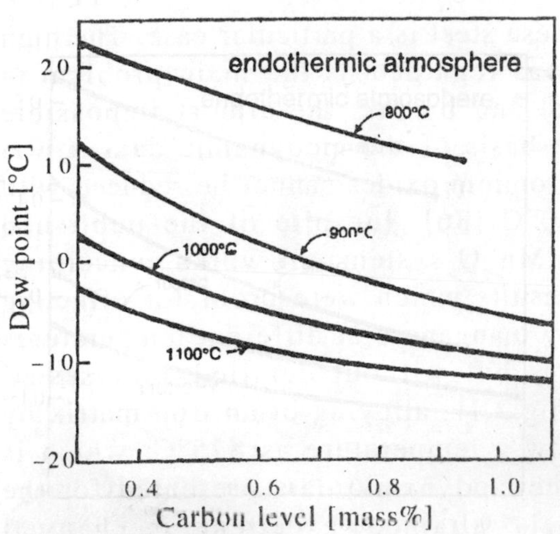

Fig.2.25 Effect of dew point and temperature on the stabilisation of various carbon level during sintering in an endothermic atmosphere [6].

cryogenic plant has an initial dew point of –196°C). In connection with furnace atmospheres, the possible proper control of flow rate and flow direction in continuous sintering furnaces, which usually consist of four zones, enables to sinter the parts with the required properties. In terms of different functions of each of the zones (burn-off, hot, carbon restoring, cooling), each requiring combination of temperature, time and atmosphere composition, the resulting effect of the applied atmosphere is an optimum compromise in relation to the properties of the parts and technical possibilities.

In continuously working sintering furnaces, the gas counterflow in regard to the transport of the being sintered parts in the furnace occurs. The inlet of the purest atmosphere applied is on the exit side of the furnace and the outlet of the atmosphere, which is contaminated by the gas reaction products formed during foregoing sintering stages (H_2O, CO_2) is at the entry of the parts into the furnace. Higher oxygen content existing in a real atmosphere in the heating zone is necessary for delubing. The first step of interparticle bonding, of graphite (carbon) diffusion into the iron particles and the reduction of oxides occurs in reality in the atmosphere with substantially higher dew point as in the atmosphere at the inlet. The thermodynamics of the reduction processes in the processed materials during sintering are commonly calculated for the nominal purity of the atmosphere entering into the furnace. The atmosphere really in contact with the solid surface is however quite different, especially within the pores of the compact. This state is not sufficiently analysed regarding its effects on sintering and alloying, *e.g.* with chromium and silicon or to sintering of chromium prealloyed powders.

Sintering of mixed manganese steel is a particular case. The high affinity of Mn for oxygen was regarded as the main problem in sintering of manganese steel and by this the almost impossible reducibility of MnO. On the basis of thermodynamic data it was stated that manganese and chromium oxides cannot be reduced with hydrogen or carbon at 1120°C [36]. In spite of the published thermodynamic data for the Mn–O system, all works concerning manganese steel published results which were proof for effective alloying of the iron matrix by manganese at different temperatures and in different atmospheres but without detailed analyses of reasons for it. As shown in Fig.2.11, alloying of an iron matrix by manganese occurred at as low a temperature as 875°C, which is not in agreement with the thermodynamic data presented for the Mn–O system. These steels attained excellent mechanical properties in spite of the mentioned problems with reducibility of manganese oxides.

In reality, the sublimation of manganese during sintering due its high vapour pressure enables sintering of mixed manganese steels in an atmosphere of common industrial purity with ~–30 to –50°C dew point, not corresponding to the thermodynamic data for MnO reduction. The reduction of $MnO_{(s)}$ is possible only with $C_{(s)}$ at above 1280°C. In reality, alloying of the iron matrix by manganese occurs in solid (iron)–gas (manganese vapour) phase. Manganese vapour partly escaping from the compacts through the open pores actually reacts with the oxygen in the sintering atmosphere under formation of volatile MnO which is transported by the atmosphere away from the furnace. This forms *a self-cleaning (getter)-effect* for the sintering atmosphere formed by manganese for equilibrium conditions [19,29]. The proof of this cleaning-effect is the formation of green in colour MnO oxide on inner sides of steel box at sintering manganese steel. This oxide was not reducible at 1200°C in N/H atmospheres with the dew point –60°C. The existence of MnO on the surface of initial milled manganese powder (ferromanganese, electrolytic manganese) used in iron powder-manganese mixes independently on milling in air or in nitrogen was not proved. Other manganese oxides brought into the system via the manganese powder are reducible or decompose during the heating period.

Initial stages of sintering in formation of interparticle necks and dissolution of graphite and alloying of chromium-containing and of Astaloy Mo–C materials also begin with the first deoxidation maximum (= CO–) at about 1000°C [27].

Further, information concerning the interactions of the atmosphere constituents with the constituents of the processed parts in the partly limited space in boxes at limited gas flow is completely lacking. In reality, the steady state differs from a true equilibrium here in that only a partial or local, equilibrium is established at the metal–vapour interface. In systems with a continuously flowing atmosphere the steady state condition is the one most frequently achieved. This is done by supplying atmosphere at a rate sufficient to overpower any compositional changes that might occur by reaction of the atmosphere with the metal surfaces [31].

The atmosphere plays also a role of a cooling medium in the cooling zone of the furnace. For example, hydrogen in comparison with nitrogen exhibits higher thermal conductivity and, therefore, increases the cooling rate of the parts compared to nitrogen.

The size, shape, and mass can also affect the final properties of a part of a defined chemical composition. During cooling the phase transformations in the sintered material occur in dependence on cooling rate, measured in the furnace as the temperature of the flowing atmosphere. As an example, the differences in the hardness of the parts (ring-shaped) of different dimensions and mass measured in perpendicular and in pressing direction placed in the boxes each on another are listed in Tab.2.6.

In both cases, significantly higher hardness was attained for the smaller parts with a larger specific surface area. It means that the cooling of these parts in the critical temperature zone (600–400°C) proceeded faster due to the larger surface in relation to the mass (volume) and by this due to the more effective contact with the flowing cooling atmosphere. For this reason, the hardness measured perpendicular to the pressing direction (circumference of the rings) was markedly increased compared with the hardness in the pressing direction, due to the more active flow of the atmosphere in contact with this surface of the parts (gas–part interface). Regarding the machinability, the differences in hardness caused by the size of the part of the same chemical composition cooled with the same 'furnace' rate may play a role, larger parts being usually softer.

2.4.3 Carbon control

The effect of carbon in powder materials, except some pure metal, is directly connected as mentioned with sintering as reduction matter and with hardening all PM steels.

For the production of sintered structural parts, carbon is added to the iron or prealloyed powders as elemental carbon – natural or artificial graphite – or as combined carbon in carbides or in other master alloys containing carbon in higher amounts. The amount of graphite added to all PM steel mixtures is sligthly higher than the required combined carbon content for alloying for enhancement of

Tab.2.6 Hardness (HV 10), density (ρ), mass (G) and specific surface area (F) of ring shaped parts (roller) from Distaloy SE and Fe–3.5Mn–0.5Mo powder (Ø 22/12.2 × 8.5 mm, G = 14 g and Ø 40/18.3 × 14.5 mm, G = 105 g). Sintering in a pusher furnace at 1180°C for 40 min, $75N_2/25H_2$ atmosphere (inlet dew point –55°C) in different sintering cycles, furnace cooling rate of ~10°C/min

Alloy	Roller Ø 22 mm				Roller Ø 40 mm			
	ρ [g/cm³]	HV 10 ↓	HV 10 →	F [cm²/10 g]	ρ [g/cm³]	HV 10 ↓	HV 10 →	F [cm²/10 g]
Dist. SE-0.6 C	7.04	234	250	10	6.99	206	215	4.4
Fe-3.5Mn-0.5Mo-0.9C	6.74	252	273		6.66	216	231	

Remark: Distaloy SE: Fe–4Ni–1.5Cu–0.5Mo; ↓ pressing direction, → perpendicular to the pressing direction

the reduction activity of the atmosphere. Especially for Cr prealloy steels, even if sintering in a hydrogen-containing atmosphere, the actually reducing agent is carbon and not hydrogen [37]. As shown in Fig.2.26, graphite dissolution becomes noticeable at about 750°C but occurs mainly in the temperature range 850 to 950°C, *i.e.* significantly above the eutectoid temperature at which dissolution becomes thermodynamically possible.

At about 850°C only about half of graphite has been dissolved in the matrix. It indicates that carbon reacts rather sluggish at low to moderate temperatures but quite fast at higher ones. It seems also that graphite dissolution in N_2 occurs at slightly lower temperatures than in H_2. In the latter case, the maximum of carbon dissolution is found in the range of 950°C at which almost all the graphite has disappeared in N_2. No substantial difference in the solubility of purified natural and high quality artificial graphite grades (chemically cleaner) was observed [38].

Above the α–γ iron phase transformation temperature for plain iron, in the range 910–1100°C, carbon dissolves relatively rapidly (solubility in austenite 2.1%) in iron. In an unalloyed Fe–C system practically only 1.5% C is in solid solution and the rest remains as graphite.

Dissolution of graphite in iron in dependence on sintering temperature was also analysed by the measurement of magnetic polarisation on 5 mm samples in diameter as shown in Fig.2.27. With decreasing free carbon and by this with increasing combined carbon content magnetic polarisation decreases.

After completing the process of carbon dissolution, shrinkage of the dimensions takes place during holding at sintering temperature, and extent of shrinkage is almost proportional to the carbon content.

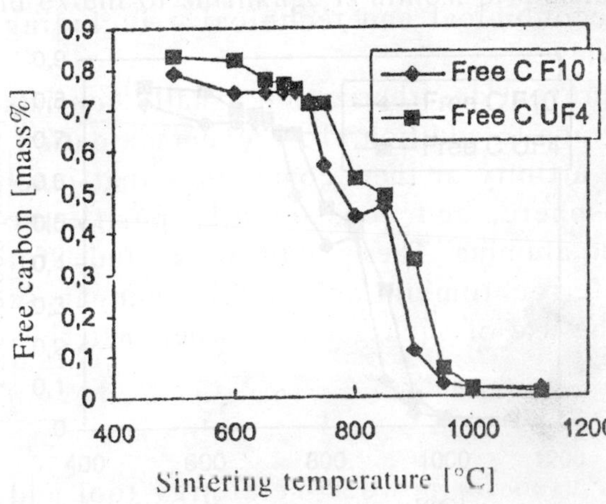

Fig.2.26 Free carbon in Fe–0.8C material prepared from atomised iron powder. Compaction at 600 MPa, sintering 60 min in nitrogen [38]. Graphite: UF4 natural, F10 artificial.

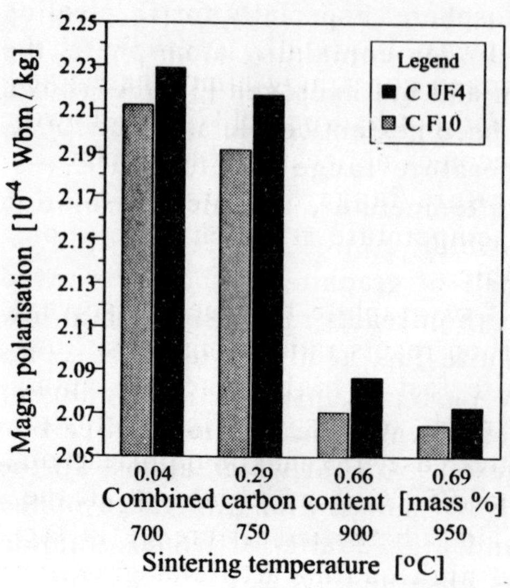

Fig.2.27 Magnetic polarisation as a function of dissolved carbon content and sintering temperature. Graphite: UF4 natural, F10 artificial. Magnetic polarisation of the sample without graphite = 2.249 [10^{-4} Wbm/kg].

2.4.4 Sintering under getter

The aim of using getter is to either prevent oxidation of sintered materials because of their high oxygen affinity (if a sufficiently pure furnace atmosphere is not available, especially for the cooling stage) or to eliminate the decarburisation and to secure an uniform concentration of carbon in the compact. Sintering under a getter is carried out more frequently under laboratory conditions and is less common in practice. The parts are either covered with the proper getter powder (which however might result in unwelcome reactions through interdiffusion) or are sintered in double boxes with the getter in the gap between the outer and the inner box. Sintering of some materials in double steel boxes under getter must be very carefully evaluated from economical and technical – including environmental – points of view.

The getter employed to prevent decarburization usually consists of alumina with 1 to 5% graphite addition. To prevent oxidation, elements with high oxygen affinity or their compounds that react with oxygen in the atmosphere, and consequently purify the atmosphere, are added to the alumina. These additions include, for example, ferroaluminium, ferrochromium and/or ferromanganese powder. As a side effect, the use of such getter powder materials prevents also decarburisation.

2.4.5 Vacuum sintering

Vacuum sintering is mostly used for powder metallurgy tool and stainless steels and in a limited range also of low-alloyed high strength steels mainly under laboratory conditions in stationary furnaces. Some extra graphite is added to the mixes for reduction of oxides brought into the systems by powders.

2.5 ALLOYING METHODS AND ALLOYING ELEMENTS

2.5.1 Alloying methods

There are four main methods for the preparation of sintered alloy steels, each of them with specific characteristics and with particular effects on the properties of sintered parts. These are:

- from mixed powders; admixture of iron powder and of alloying elements,
- from diffusion alloyed (partially diffusion bonded) powders; mixed powders after additional annealing of the admixture to obtain partial bonding between the iron powder particles and alloy elements particles, but without significant formation of solid solution,
- from prealloyed powders,
- from hybrid powders; admixture of a prealloyed powder and of an additional alloying element, and others:
- alloying through coating iron powder particles, *e.g.* with copper through cementation, electrolysis,
- infiltration of a solid skeleton by a molten alloy metal.

Alloying elements can be introduced into the metal powder system (powder admixtures) also by a number of other techniques. The alloying elements are added to the base iron powder in form of:

- elemental powders,
- master alloys: special for powder metallurgy prepared low-melting alloys or ingot metallurgy ferroalloys such as ferrochromium, ferromanganese, ferrosilicon, ferromolybdenum, ferroboron,
- carbides: molybdenum carbide, boron carbide, special carbides containing Cr, Mn, Mo and V or Si and Mn (*e.g.* the MCM, MVM, SM alloys), and others.

All carriers of alloying elements containing carbon are hard and abrasive and lower the compressibility of the powder admixture and can shorten the life of the compacting die.

The alloying methods for the preparation of sintered steels indicate the wide range of possibilities of PM to prepare various alloyed systems and, as regards the alloying elements, concentrations in the systems which cannot be prepared by ingot metallurgy because of solubility. Each method has its specific and general application and is used in cases in which are the most suitable technical, and, in particular, economical conditions formed for the application.

2.5.1.1 Mixed alloy steels

The preparation of alloyed steels based on mixed powders, including diffusion alloyed powders, is most convenient, because it can be easily applied and can be used to produce the required alloy for singular applications, even in small batches. Such a powder mix retains the high compressibility of base iron powder, which is also a technical advantage; from this reason carbon is added exclusively through the mixing route, as fine graphite. The metallurgical reason for this method is that there are also no limits of mutual solubility of the components, which is a base advantage of powder metallurgy. The alloying of the iron powder matrix in the compact takes place mostly by solid state diffusion of the alloying elements at corresponding sintering conditions regarding specific properties of iron powder grade, including graphite and lubricant used. In some cases, transient liquid phase can be used to accelerate homogenisation of the alloy metals.

For ensuring technically possible homogeneous distribution of alloy particles in the compact it is desirable to attain as many contacts between alloy particles and iron particles in the green compact as possible. The smaller the additive particle size, the more uniformly dispersed they are in the green compact structure.

The elemental alloy powders added to iron powder are very fine in size depending on the production method (particle size <1 up to 45 μm, mean size ~10–15 μm; also in case when some of them are milled and industrially sieved. It is difficult to avoid the segregation and dusting of the alloying elements and prepare a mix with fully uniform alloy particle distribution. During a production run this might lead to variations in the chemical homogeneity of a sintered part microstructure and in variations of the chemical composition from part to part. These variations will influence the dimensional tolerances and the mechanical properties of sintered parts which are taken into account and minimised.

The main characteristic of materials from mixed powders in general is the microstructure heterogeneity. It is a manifestation of a difference in concentration between the margin area and the core of initial powder particles and new formed matrix grains in dependence on mutual diffusivity of the base powder and alloy element. The heterogeneity of the microstructure of a sintered compact can be partly affected also by segregation of particular powder additions during handling the admixtures beside those caused also by different local and total densities of the mix components.

In general, it is necessary to consider the microstructure heterogeneity of mixed-powder systems which in most alloys provides also higher mechanical and toughness properties compared, *e.g.* with prealloyed powder material of equal composition. On the other hand, the presence of different phases in the microstructure with various microhardness can play a negative role for parts in which the surface is exposed to load (fatigue, friction) or to machining.

In terms of microstructure formation and, by this way, of the properties of a sintered metal part, sintering of mixed-powder compacts can be regarded as most important and complicated. Many processes interact simultaneously in the solid state. Compared to mixed-powder systems, the microstructure and properties of sintered iron, iron–carbon and of prealloyed powders can be more exactly predicted in relation to the starting powders and processing conditions. Alloying of the iron matrix with the additive elements depends on diffusivity of the alloying elements mainly in the γ-region, but begins to some extent during heating below 900°C. The transfer of the interacting components with limited or complete solubility takes place by heterodiffusion the driving force for which is the concentration gradient generated by contact formation between the interacting phases. Since atomic motion is faster at higher temperatures, sintering of mixed-powder compacts at higher temperatures in general results in better sintered properties.

When sintered mixed steels are evaluated, except in iron–carbon material the characteristic microstructural heterogeneity is determined by the thermodynamic and diffusion characteristics of the starting components at the given sintering conditions and finally during cooling. Usually the problematic achievement of full homogeneity of an alloying element in the iron matrix depends on its physical–metallurgical properties and on common sintering conditions (temperature, time), especially on diffusion activity in austenite of iron or prealloyed matrix and added alloy elements, especially if more elements are involved. In contrast to the expectation the full homogeneity of the microstructure nor of copper alloyed steel after sintering at 1250°C is achieved.

Some peculiarities in microstructure formation of mixed steels and in porosity are responsible for specific behaviour of a sintered material also with regard to machining. The grade of the homogeneity of the microstructure affects the interval of the maximum stability of the austenite regarding the base character of the alloying elements (ferrite or austenite stabilizer).

For the preparation of mixed alloy steels, in addition to carbon further alloying elements such as nickel, copper, molybdenum and phosphorus are used. The preparation of mixed manganese steels is intensively and permanently investigated [29,39]. Chromium and/or molybdenum alloyed steels are at present in most cases manufactured from prealloyed powders and not by mixed ones.

The distribution of the alloying elements in mixed systems (Cr, Mo) in the iron matrix occurs more homogeneous with increasing sintering temperature. The homogenization of admixed alloy elements Mo and Cr is drastically enhanced if transient liquid phase is formed during sintering, which for Mo occurs above about 1240°C at 0.7%C [40]. The insufficient quality (stable oxide content) of iron powder and purity (dew point) of sintering atmosphere can lower the homogenization of the microstructure [41].

For the preparation of mixed manganese alloy steels the high manganese vapour pressure and its sublimation during sintering must be taken into account, the vapour transport mechanism (condensation of manganese vapour on the whole surface of iron particles) resulting in similarly enhanced homogenization as does transient liquid phase but without the pronounced temperature threshold typical for the latter mechanism. This alloying mechanism of Fe–Mn system occurs also with addition of other elements and in prealloyed powders [10,29].

2.5.1.2 Diffusion alloyed steels

The development in production of powder metallurgy steels is oriented markedly on high performance steels, with all consequences for machinability. Diffusion alloying – occasionally also dubbed partially prealloying – as a special case of mixed powders is considered to make a large contribution to it. The goal of this method was to eliminate or at least reduce the variations in distribution of alloying elements caused by demixing and segregation of fine admixed powders. These unwelcome effects can amount to ~2 to 10% in the mix during handling and filling the compacting die cavity. By diffusion bonding, the homogeneity of all properties of the sintered parts can be improved due to minimum segregation of the additives. The alloying powder particles are bonded to the surface of base iron powder particles during a heat treatment in a reducing atmosphere at the temperatures <900°C.

During this treatment the oxide layers on the particle surfaces and in open pores are reduced, and the mostly fine alloying element particles sinter to the iron powder particles but without significant interdiffusion, thus avoiding solid solution strengthening effects.

These powders thus retain virtually the compressibility of the base iron powder. As a result of reduced segregation of the alloying elements during handling of the mixture – the fine particles which are prone to demixing being bonded to the larger ferrous particles – the dimensions of the sintered products vary only slightly and their microstructure homogeneity is higher than that of the parts produced from the mixed powders only, which is exactly the aim of the process. As a result of more uniform alloying the diffusion alloyed powders also have higher strength and other properties, of course also depending on carbon content and sintering conditions.

With the method of diffusion alloying of powders, the elements Ni (1.75–4%), Cu (1.5, 2.0%), Mo (0.5–1.5%) (total amount of alloying elements of 3.5 to ~8%) in base mixes are mostly used. These powders are based on plain sponge or atomised iron powder or on molybdenum prealloyed ones. By a further addition of copper or nickel in the amount of 2% to some diffusion alloyed powders the special properties of sintered parts (dimensional accuracy, hardenability, etc.) are achieved.

The Ni–Cu–Mo diffusion alloyed powders, usually mixed with 0.5%C, are used for a broad range of sintered precision parts, typical ones being the synchronizer hubs in automotive transmissions. These materials are singular since they result in better mechanical properties when sintered at standard 1120°C than at higher temperatures. The reason is the very heterogeneous microstructure, esp. with regard to Ni, distribution, typical for Ni containing mixed steels. This microstructure which from ferrite and pearlite to martensite and retained austenite contains virtually any microstructural type found in steels, gives favourable mechanical properties but on the other hand makes the material tricky to machine, since the occurrence of hard and soft areas in the material within short distances (20–50 μm) results in rapidly varying load and friction on the cutting edge. Sintered steel components from diffusion bonded powders thus generally offer a true challenge for machining.

2.5.1.3 Prealloying

Prealloying is a special powder metallurgy term for the production of alloyed powders, usually from a melt by water or gas atomisation, the melt typically eshibiting the same composition as the emerging powder particles (with the notable exception of the oxygen content). Prealloyed powders for structural component application are produced by water atomisation. Prealloying augments the possibility for preparation of new varieties in production of PM parts with high mechanical and some special properties.

Alloying elements are uniformly distributed in the solid solution in all prealloyed powder particles or are present as separate phases in high alloy powders. In prealloyed powders used for the production of structural steels, the first case applies. For the production of these prealloyed powders alloying elements having low hardening effect in ferrite are chosen to ensure their compressibility, Fig.2.10.

The main advantage of the prealloyed powder steels is the fully homogeneous matrix microstructure, also with regard to carbon in the as-sintered state. The compressibility of these powders is slightly lower compared to the compressibility of plain iron and mixed alloy powders, the difference depending on the hardening effect of the alloy elements contained. The production of prealloyed powders if high compressibility is required is concentrated on material alloyed with elements with low strengthening effect in ferrite. Prealloyed high speed steel and stainless steel powders exhbit with lower compressibility. To obtain sufficient compressibility, the carbon content in the prealloyed powder must be very low, and carbon is added as admixed graphite. Only in the case of tool steel powders, also carbon is added through prealloying.

The recent trend is to use some prealloyed powders to a larger extent to attain the highest possible mechanical and toughness properties in a more simple way. The use of prealloyed powders extends the range of powder metallurgy routes, not as replacement of mixed powders but as a supplement.

As listed in Tab.2.4, one group of prealloyed steel powders is formed by powders low alloyed mainly with nickel (0.5, 2)%, molybdenum (~0.5%) and in some cases to a minor extent by manganese (<0.3%) (Astaloy type powders). These powders are admixed also with copper and are used often for powder forging. The second group of prealloyed powders, used mainly for the production of high strength structural parts, is at present formed by the (0.85, 1.5, 3.5)% Mo powders and the (1.0, 1.5, 3.0)% Cr powders with 0.2–0.5% Mo addition and in one case with 0.3% V

[11,42]. The problem in the production of chromium prealloyed powders is the reduction of chromium oxides that are formed during atomization on the surface of the droplets in contact with the oxidising atomising medium – water – and in some cases also in the interior of the powders during the solidification (inclusions). Therefore, the requirements for the final oxygen content in the powders are high, typical levels being <0.2% O. The chemical activity of chromium as element with high affinity for oxygen is reduced by its relatively low content in the prealloyed powder compared with data calculated for an unit activity of the system Cr–O–H–C, according to the Richardson–Ellingham diagram. The remaining oxygen is removed during the early sintering stages, when attaining about 1000°C, by the admixed carbon, about 0.2% C being sufficient to result in sufficient deoxidation [43] although complete oxygen removal requires temperatures of at least 1250°C [44].

The molybdenum prealloyed powders are also considered as a further base (also as hybrid) for the preparation of a new generation of high strength steels. Molybdenum reduces the γ-region. The Fe–3.5% Mo powder is fully ferritic, and for carbon-free compacts sintering at 1250–1280°C occurs in the range with its significantly higher self-diffusivity and resulting fast sintering that results in densification from, for example, 7.0 g/cm^3 green density to >7.5 g/cm^3 after sintering. These steels are rather soft after sintering and, for reasonable mechanical strength, are carburized or carbonitrided [45].

In most cases, however, the mentioned prealloyed powders (containing carbide formers) are mixed at first with carbon to achieve the highest possible combination of strength and toughness properties of the sintered steels for a broader application. Both molybdenum and chromium prealloyed powders do not suffer from some disadvantages of the sintered mixed steels alloyed with these elements, *e.g.* the need for high sintering temperatures to obtain reasonable microstructural homogeneity, although also the prealloy steels, especially the Cr alloyed ones, exhibit markedly better resuts when sintering at 1250°C and above [46].

2.5.1.4 Hybrid systems

Hybrid systems are formed by a base prealloyed powder to which one or more other alloy elements in powder form are admixed or diffusion bonded. The microstructure of such material is affected by affinity of additional element for the base prealloyed element under given sintering conditions. The microstructure homogeneity of a hybrid material system is much higher due to the homogeneous microstructure of base matrix alloyed with one or more elements compared with the mixed system including diffusion alloyed materials. The microstructural heterogeneity of hybrid alloy steels is caused by additional alloying.

Hybrid powder systems offer many effective combinations in chemical composition and in final properties of the sintered materials. Generally, alloy elements that do not strongly affect the compressibility of powders, *e.g.* Mo and Cr, are introduced through the prealloy route while those that would result in strong solid solution strengthening, such as Ni, Cu, and Mn, are either admixed or diffusion bonded. Typical commercial systems are Fe–1.5Mo + 2% Ni, Fe–1.5Mo + 2% Cu, and Fe–1.5Mo + 4% Ni+ 2% Cu, which are available as diffusion bonded grades.

The hybrid system Fe–Cr–Mn–Mo–C based on prealloyed FeCrMo(V) and FeMo powders with addition of manganese (Cr and Mn the cheapest alloying elements) constitute a technically and economically very promising material systems for the production of complex high strengh and high toughness steels with a total amount of alloying element not exceeding ~4–5% as predicted [47,48]. These systems could replace to some extent higher nickel and copper alloyed steels. For example, the Fe–3Cr–1.5Mn–0.3Mo–0.3V–0.15C steel (690 MPa, 1200°C, 60 min) attained tensile strength of 929 MPa, transverse rupture strength of 1207 MPa and impact energy of 18.9 J [49]. The optimal properties of both particular powder systems by this are exploited. Since both chromium and molybdenum are carbide forming elements that also increase the hardenability of steels, the advantageous and high mechanical properties of these materials will require special efforts in machining.

2.5.1.5 Coating and infiltration

The other mentioned alloying systems, *e.g.* metal coatings of powders [50] and infiltration of porous ferrous skeletons [51], are used to a limited extent in production of structural steels, mostly for special purposes. Copper infiltration (electrolytic powder or Cu wire pieces, today also through pastes) is used for local or total strengthening of a part or for joining two or more simple parts.

Coating as alloying method is used only sporadically probably because the cost of the coating operation is relatively too high. Copper coating of ferrous powders through cementation being the easiest way which is employed in, for example, diamond–metal cutting segments and has been proposed for two-phase valve seat inserts, for example [52].

2.5.2 Alloying elements

The alloying elements used in powder metalurgy for sintered structural steels can be divided in two main groups, according to the requirements towards the sintering process. The first one is formed by elements with low affinity for oxygen, such as nickel, copper, molybdenum, and phosphorus, and by this without special requirements for the cleanliness of the furnace atmosphere. The second group is formed by elements with high affinity to oxygen such as manganese, chromium, vanadium, and silicon, with requirements for high purity of the sintering atmosphere. The use of each alloying element in appropriate concentration and/or combination enables manufacturing the materials with the physical–mechanical and service properties required for a given application.

2.5.2.1 Alloying elements with low oxygen affinity

Nickel, copper, molybdenum, and phosphorus have been common elements of choice for alloying PM steels. These elements form easily reducible oxides and are readily processed in mesh belt furnaces at sintering temperatures up to 1150°C, with the dew point of the atsmosphere below −10°C. Their use, therefore, has been determined by processing conditions rather than by metallurgical reasons. As a consequence, the PM industry has resorted to relatively high alloy contents to produce high strength properties [31].

With regard to machining, at standard sintering conditions the substitutional elements phosphorus, nickel, molybdenum, *etc.* form a higher alloyed zone around the central part of the iron particle during sintering due to the slow diffusion rate of these alloy elements in the iron lattice compared to carbon. This results in a heterogeneous core–rim structure of the former ferrous powder particles with in part widely varying hardness, and as a consequence, machinability is decreased. Interstitial elements like carbon, in contrast, diffuse rapidly in the iron lattice. The formation of pearlite also in the central part of the starting iron particles is a consequence. The material with this homogeneous microstructure are 'easy' to machine.

Copper

Copper is one of the most widely used alloying elements in powder metallurgy. Copper is used because of its strengthening effect (substitutional solid solution and precipitation hardening) and for forming a – usually transient – liquid phase (melting point at 1085°C) to activate sintering. Liquid copper rapidly fills the pores of a compact, from where it diffuses relatively easily into the iron particles; however, secondary pores remain at the sites at least of larger Cu particles [53]. Copper is soluble in γ-iron (austenite) up to approximately 9%, but only up to 4% in α-iron (ferrite) (mutual solubility of iron and copper).

Iron–copper alloys can be precipitation hardened by low-temperature annealing after sintering, and they actually do so to a certain extent anyway, when passing the cooling zone of the sintering furnace [54].

Copper causes growth of the dimensions of a part by liquid phase penetration of the pressing contacts ('copper swelling'), depending on green density and copper content. Generally, the iron powder grade does not have any great effect on the mechanical properties of parts alloyed with copper because of the high diffusivity of copper. There are mixed sintered materials alloyed with 1.5–8% Cu and up to 20% by Cu-infiltration. Copper is added to plain iron and prealloyed powder in the form of electrolytic copper; also coating of iron powder particles by copper, *e.g.* through cementation, is performed. Copper belongs to the main alloying elements of diffusion alloyed powders. The strengthening effect of copper depends on the particle size. Finer copper particles contribute to higher strength and hardness of sintered alloy, which is valid also for diffusion alloy steels.

Mixed alloys with 1.5–2.0% Cu have enabled considerable advances to be made in the production of sintered parts. Sintered alloy steels with even larger amounts of copper are usually prepared mainly by infiltration, to a small extent by electrolytic coating or cementation of iron powders. The microstructure of sintered copper steels is relatively homogeneous. Small differences in the homogeneity of the microstructure – *e.g.* a slight core–rim structure – are caused by the copper concentration changes (single or two phase microstructure).

Carbon in the ternary Fe–Cu–C system has a hardening effect and reduces the extent of the increase of the dimensions of the parts. Large carbon additions can almost completely suppress the increase of the dimensions caused by copper. Alloying with copper results in a tensile strength of 450–600 MPa with an elongation of up to 6%, depending on copper content, higher in combination with carbon. The Fe–Cu–C alloys are therefore often used for the production of parts. In this connection the commonly known better machinability of copper alloyed steels compared to other sinteerd steels is a very important factor for practice.

Nickel

Nickel is the second mostly used alloying element in powder metallurgy. Nickel belongs to those elements whose behaviour during sintering differs from that of copper. Nickel causes shrinkage of the dimensions of the parts and with iron forms at higher temperatures substitution solid solution. Nickel in iron extends the γ-area and by this as an alloying element is similar like manganese.

The rate of diffusion of iron into nickel is higher than that of nickel into iron. For this reason, the inhomogeneities in sintered mixed nickel alloy steels cannot be completely removed under acceptable sintering conditions. The concentration of nickel in regions of interparticle necks is higher compared to the cores. The microstructure inside the grains is ferritic if the nickel addition is too low to stabilise martensite. This microstructural heterogeneity, which is the largest in sintered steels, is detrimental for machining. Nickel is added to the iron powder in the form of the finest powder (carbonyl), and the material should be sintered minimum at 1100°C, better at 1200°C.

In Fig.2.28 the diffusion rate of nickel atoms spreading into iron is shown and in Fig.2.29 the nickel concentration profile in an alloy of Fe–2Ni after sintering at three temperatures for 1 h. In the first figure the sudden decrease in the diffusion rate at α- and γ-iron transformation temperature and its recovering until a sintering temperature of 1120°C. In the second figure it is seen that the homogenisation requires a temperature near 1200°C. The microstructure of mixed iron–carbon materials alloyed, *e.g.* with nickel, copper and nickel and and copper molybdenum are typically heteregeneous which is not possible to define by equilibrium phase diagrams.

Phosphorus

Phosphorus is regarded detrimental in wrought steel but it is a common sintering alloying element in powder metallurgy steels. It is usually admixed to the iron powder as ferrophosphorus consisting of Fe_3P (16% P) and Fe_2P (22% P), respectively. For the production of structural parts the Fe–(0.3–0.45–0.6)% P–C steels

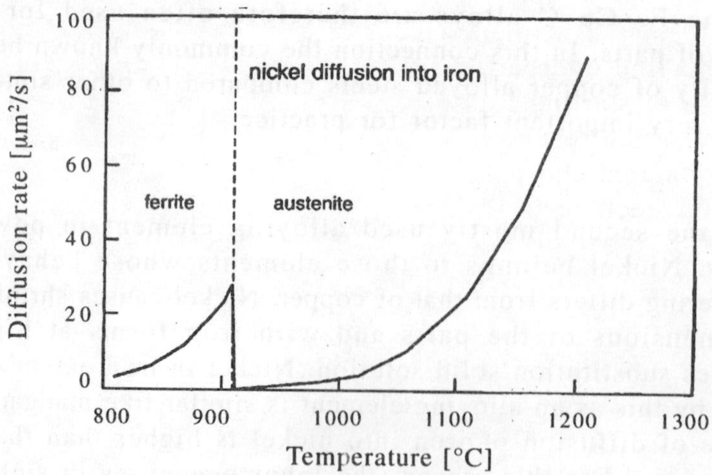

Fig.2.28 Diffusion rate of nickel atoms into α- and γ-iron [6].

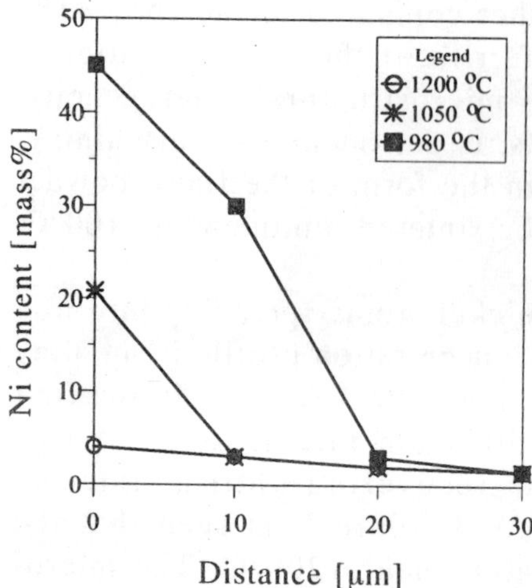

Fig.2.29 Nickel concentration profile in an alloy of Fe–2Ni after sintering at three temperatures for 1 h [6].

based on sponge or atomised iron powder, in part also with addition of copper, are used. Increase in strength properties and hardness of Fe–(0.3, 0.45)% P is achieved with addition of copper in amounts of 2.0 to 4.0%. The mechanism of alloying of iron with phosphorus during sintering is characterised by the formation of a transient binary phosphide eutectic melt followed by its dissolution in the matrix and diffusion of P into iron with which phosphorus forms solid solution. The transient liquid phase accelerates the sintering process. Phosphorus, as a strong ferrite stabilizer, also enhances the sintering process through increasing the self-diffusion rate of iron. Due to low diffusivity of P in Fe a higher concentration of phosphorus remains in the surface area of the iron particles and mostly in the fine interparticle necks formed during sintering. By this the cores of starting iron particles are softer and the necks are strengthened which results in high strength at contemporary higher ductility of sintered phosphorus containing steels compared to some other sintered steels or to wrought steels with the equal phosphorus content [55]. Phosphorus-alloyed grades with 0.6% P based on sponge and atomised iron powders (PNC60, PASC60) also with addition of copper complete this alloy steel group [11]. Binary Fe–P is frequently used for soft magnetic components, the high density and rounded porosity attained through the effect of P resulting in higher magnetic saturation and lower coercive force than with similarly processed plain iron [56].

Molybdenum

Molybdenum is a strong carbide forming element with fairly low thermodynamic stability of the oxides. It is therefore the most commonly used carbide-forming alloy metal in sintered steels. Molybdenum increases the stability of the steel at the tempering and by this lowers the temper brittleness. Molybdenum in small amounts (0.15–0.25%) is addition to many wrought structural case-hardening and nitriding steels what is also the case in sintered Mo- containing steels. Mo is either admixed to iron powder in the form of elemental molybdenum or ferromolybdenum, commonly in amounts of ~0.5% mostly in combination with copper and nickel beside carbon, and also in formation of complex alloyed carbides. For wear-resistant sintered steels, *e.g.* used in valve seat inserts, 2–3% Mo are added as an elemental powder. The other route is prealloying in which case the very low solution strengthening of Mo in ferrite is helpful. Molybdenum lowers the carbon content in pearlite, iron–molybdenum carbides of cementite type – $(Fe, Mo)_3C$ are formed if austenite containing dissolved Mo is cooled. By suitable sintering conditions, dissolution of the Mo particles can be suppressed in favour of carbide formation, M_6C and, at higher carbon levels, Mo_2C being obtained which gives attractive wear resistance. The diffusion coefficient of molybdenum in γ-solid solution is very low, but higher than of chromium. Mo homogenisation is very much accelerated by sintering with transient liquid phase which however necessitates sintering temperatures >1200°C, the critical temperature threshold depending on the carbon content [41].

The ternary mixed Fe–Mo–C systems are not often used for structural precision parts in spite of a broader investigation [40], having been superseded by the introduction of prealloyed Fe–Mo powders which yield microstructure homogeneity also at moderate sintering temperatures. However, the mixed Mo steels – in part also containing Co and Ni – are largely employed for valve seat inserts, in this case a heterogeneous austempered microstructure reinforced with Mo_2C being aimed at [57]. Not surprisingly, the machinability of these wear-resistant composite materials is rather tricky.

2.5.2.2 Alloying elements with high oxygen affinity

The group of elemer*s with high oxygen affinity used for alloying of PM steels is formed by chromium, manganese, vanadium, boron and silicon ranged in order with increasing oxygen affinity. Titanium investigated in structural microalloyed steels completes this group of elements.

Chromium

Chromium is added into mixed alloy steels as elemental chromium, ferrochromium, chromium carbides, sigma phase and/or complex carbides (Cr, Mn, Mo, V). When a low carbon ferrochromium powder is used, usually the surface oxides formed at milling are present, which are from thermodynamic point of view not reducible below 1000°C, and high temperature sintering at ~1250°C is required which is however recommendable anyhow to obtain reasonably homogeneous Cr distribution.

Chromium as carbide forming alloy element is, when combined with carbon, very effective in hardening of the steel and relatively cheap and among the mentioned elements has a low hardening effect in ferrite as shown in Fig.2.10. When austenitising of chromium steel is performed the dissolution of chromium carbides proceeds. For sintering the mixed chromium alloyed steels high sintering temperatures are required in an atmosphere with low dewpoint to achieve sufficient dissolution of chromium in iron.

The addition of chromium to iron powder in form of a nitrogen atomised low melting masteralloy Fe–33Cr–5C–1.60Mn–1.60Ni–1.60Mo–2.60P powder (melting temperature at 1130–1140°C) enabled the sintering of compacts at 1150°C in the atmosphere of common purity in spite of high affinity of chromium for oxygen. The higher hardness, transvesre rupture strength and more homogeneous microstructure in comparison when using chromium carbide (Cr_2C_3) and/or ferrochromium as chromium carrier were attained [58]. The main problem with carbon-containing masteralloy powders is the abrasive effect of these powders on the tools during compaction.

The main approach to manufacture Cr alloy sintered steels is however through prealloy powders. While the first attempts to manufacture Cr–Mn–Mo prealloy powders through oil atomisation [59,60] were technically successful but not so economically, a water atomisation process with subsequent vacuum reduction anneal resulted in industrial production of Cr–Mn–Mo and Cr–Mo–V prealloy powder grades, though at limited scale [61]. The recent introduction of fairly cheap Cr–Mo alloy powder grades with low oxygen content [11] boosted interest in this class of sintered steels; it could be shown that the more stable oxide layers covering these particles can be effectively reduced at sufficiently high temperatures and clean atmospheres. Actual results confirm the statement that chromium and molybdenum as alloying elements seem to be obvious choices also for developing hardeability [28].

Studies of the reduction behaviour of Fe–3Cr–0.5Mo prealloy steels [62] have shown that there are two degassing maxima indicated by mass loss and CO formation in the ranges 1000°C and 1250°C, respectively, while for plain Fe–C or for Fe–Mo–C maxima at about 700°C and 1000°C have been found, Fig.2.30. These maxima have been attributed to reduction of surface and internal oxides, respectively [63], and it has been shown metallographically, Fig.2.31, that in Cr–Mo prealloy steel powders as well as in steels sintered at 1120°C, oxide inclusions are found inside the particles. After sintering at 1250°C these particles are no more present [64]. Such fine internal oxides may have considerable effect on the machinability, and drilling tests on green (warm compacted) specimens have in fact revealed a significantly more abrasive effect of the Cr–Mo steel compacts than in other types of PM steels [65]. In this case, high temperature sintered materials should be better machinable than those sintered at standard temperatures.

In the case of sintering of Cr-prealloyed powder materials for the reduction of chromium oxide the indirect carbothermal reduction reaction, based on CO/CO_2 mass transport mechanism for transferring carbon to the surface of the oxide, would appear as the fastest and more likely reduction process [66].

Manganese

In spite of the high hardening effect in iron and low price, manganese is used as alloying element in powder metallurgy structural parts production only in a very limited range. As a solid solution strengthener, manganese is approximately four times more effective than nickel. Manganese alloy steels can be prepared only

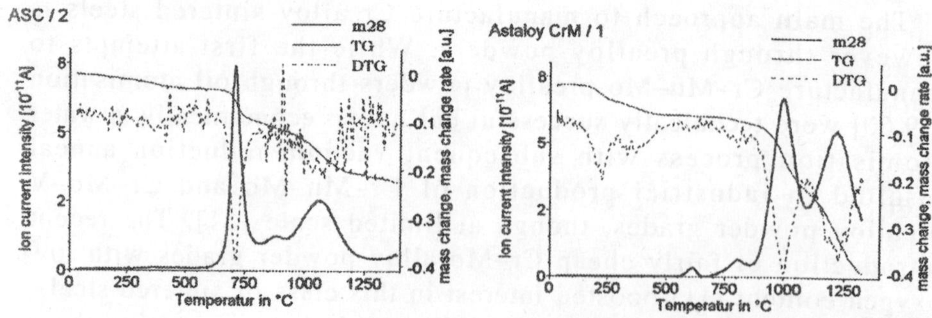

Fig.2.30 Mass loss (DTG) and mass spectra (m28=CO) for a – Fe–1C (ASC/2) and b – Fe–3Cr–0.5Mo–0.1C (Astaloy CrM/1) powder compacts [63].

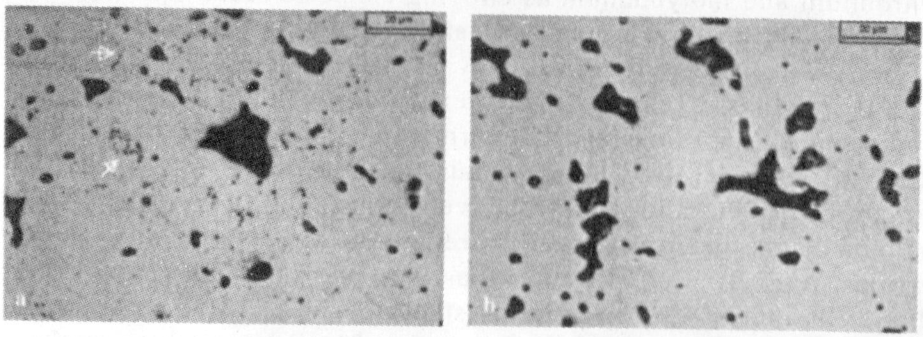

Fig.2.31 Sections of Fe–3Cr–0.5Mo prealloy powder steels with admixed graphite sintered at 1100 and 1240°C, respectively. a – in H_2 at 1100°C centre, b – in H_2 at 1240°C centre [66], unetched, oxides are arrowed on the micrographs.

through the mixing route due to high hardening effect in ferrite which very much deteriorates its compressibility. Manganese can be added to iron powder mixtures as elemental electrolytic manganese, ferromanganese or some special high manganese–carbon or in combination with other elements manganese containing masteralloys. Manganese addition slightly lowers the compressibility of the iron–manganese powder mix regarding also the lower density of manganese measured in relative sintered density of the compacts [39].

Contrary to other powder alloying systems, the alloying of base iron powder by manganese in a compact occurs in solid phase (iron powder)–gas phase (manganese vapour) system. The detailed original thermodynamic and physical–chemical analysis confirmed this process [29,67]. The sintering of manganese steel with high mechanical properties under industrial conditions without special respect to thermodynamics of the Mn–O system occured [68,69].

At alloying the manganese vapour formed fills the pore space of the compact and condenses uniformly on the whole surface of the iron particles (surface diffusion) and hence diffuses into these particles (volume and grain boundary diffusion), Fig.2.32. It means, that the diffusion alloying of iron particle cores depends also on their structure and substructure properties. Figure 2.33 shows the section of Fe–4.5Mn–0.5Mo–0.5C alloy based on Hametag iron powder chracterized by higher sintering activity [70]. In this case the matrix is fully alloyed by manganese and ferromolybdenum particles added to the base iron powder formed a poreless joint what contributed to the density increase. Dissolution of molybdenum in the Fe–Mn–C matrix followed at higher temperatures.

To the given data it is possible to note that, while conventional PM alloys are of course well proven for many traditional applications, their mechanical and cost limitations become increasingly apparent as PM moves up to markets with more demanding applications. It is known that copper and nickel would not be obvious choices when designing an alloy for maximum dynamic endurance. Ferrite solid solution strengthening is more effective by additions of manganese, chromium and molybdenum in combination with carbon. The advantage of chromium and molybdenum over nickel for hardenability is clear. In terms of cost effectiveness, the advantage of manganese is even greater.

As mentioned, the synergistic effect of alloying with Cr, Mn and

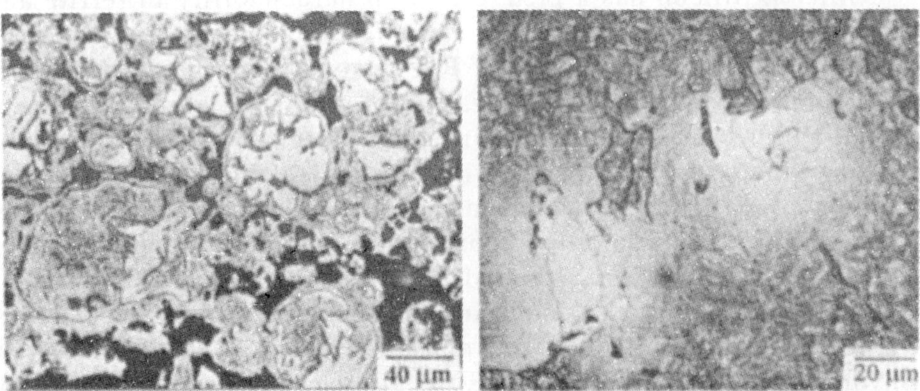

Fig.2.32 Microstructure of Fe–4Mn–0.4C sintered compact. RZ iron powder (air atomised), density 6.6 g/cm^3, sintering 1100°C, 10 min. Optical micrograph. Nital etched.

Fig.2.33 (right) Microstructure of Fe–4.5Mn–0.5Mo–0.5C sintered compact. Iron powder Hametag (eddy milled). Compaction at 600 MPa, sintering at 1120°C, 60 min, cracked ammonia, density 7.2 g/cm^3, Rm = 690 MPa [71]. Optical micrograph. Nital etched.

Mo, coupled with well documented benefits of high temperature sintering on interparticle bonding quality and pore morphology, may result in components with a significantly superior balance of mechanical properties than is possible with conventional PM processing and alloying concepts [28,47]. Making separate additions of each alloying element enables precise tailoring of compositions. Alloy design therefore becomes a powerful tool in optimising product performance with considerations also for their machining.

2.5.2.3 Vanadium, silicon and boron

Vanadium. Vanadium is ued as a precipitation hardening element in non-tempered wrought steel. The addition of 0.3% V in prealloyed Fe–3Cr–0.3Mo steel (sintering at 1150°C, 60 min nitrogen) lowered hardenability and extended a range for pearlite structure compared with the steel without V addition. Consequently, the addition of vanadium should extend the range of cooling rate after sintering (sinter hardening) within which high strength can be obtained. Also higher rotating bending fatigue strength of 310 MPa compared to 280 MPa for Fe–1Cr–0.3Mo steel was attained. Increase in strength by adding vanadium was caused by the precipitation of vanadium carbonitrides [72]. Vanadium in heat threated steels forms V_4C_3 carbide which can contain also a small amount of molybdenum and chromium. Vanadium can be used also as microalloying element in structural steels [73].

Silicon. Silicon has a great effect on hardenability in ferrite and therefore was investigated as a suitable replacement for expensive nickel and copper in alloying of PM steels. Silicon has a stabilising effect on α-iron. A great disadvatage of silicon as alloying element is high sintering activity casusing high shrinkage of the compacts. In combination with addition with manganese this disadvatage was supressed. Silicon was added to the iron powder as ferrosilicon or later in form of masteralloy consisting mainy of intermetallics (Fe, $Mn)_5Si_3$ and (Fe, $Mn)_3Si$ (called SM) which is able to form also a transient liquid phase. At sintering the specimens (Fe–1.4Si–3.2Mn–0.4C) compacted at 600 MPa and sintered at 1180°C for 1 h in hydrogen with a dewpoint of –30°C attained the tensile strength of 900 MPa, hardness 305 HV 20 and elongation 2% [74].

Silicon also belongs to the base elements for the production of soft magnetic materials. At sintering in vacuum at 1200°C of the powder compacts Fe–(2.5–6.0)Si (Si in form of Fe45Si) compacted at 300–800 MPa a relative density up to 94% was attained. The start of alloying of iron matrix by silicon at about 900°C was observed [75].

Boron. Boron in powder metallurgy is used as sintering enhancer by the formation of the liquid phase with iron. The materials containing intermatelic eutectics (Fe_2B) on the grain boundaries with the microhardness up to ~1400 HV without affecting the microhardness of iron matrix can detrimentally decrease the machining of such materials. In contrast to the Fe–B system it was found that in some sintered alloyed materials, *e.g.* Fe–Mo or Fe–Cr, boron diffuses from the eutectics into the matrix without the eutectics in microstructure. By this the density of the samples and the microhardness of the Fe–3Cr–0.3Mo–0.3V–0.2C matrix and the apparent hardness up to ~300 HV 10 was increased [76]. Boron diffused into the matrix and formed borides with some alloying elements, as in the case of Cr and Mo. Sintered alloyed steels with a small boron addition without presence of eutectics in microstructure is necessary to take into acount also in term of machining.

2.6 SECONDARY OPERATIONS

The goal of a secondary operation is to modify the structure of the surface or the substrate of a PM part, irrespective of its density or hardness, in order to obtain some special technical property, such as:

- dry or fluid adhesive or abrasive wear resistance under low or high compression loading,
- resistance to scuffing and seizure,
- fatigue in general, especially rolling contact fatigue, resistance to case cracking (surface collapse),
- resistance to corrosion.

Secondary operations can be divided into three main groups:

- mechanical,
- heat treatment and surface hardening,
- hardfacing.

2.6.1 Mechanical operations

The base secondary operations in production of PM parts are both sizing and coining. The hardness of parts to be sized or coined should not exceed 180 HV after sintering. The external forms should be sized before the holes, to prevent cracking. At both operations the parts are subjected to elastic and plastic deformation. The sizing or coning load required is dependent upon pressing area and final density of the part. This load is usually higher especially in coining than those used for compaction. An important factor in sizing and coining is the lubrication of the surface of the part and/or of the die.

Sizing is used to obtain high dimensional accuracy, thus compensating for warpage or other dimensional defects occurring in the sintering operation. Only a slight plastic deformation is necessary, and the forces required for the sizing are normally quite higher in relation to the density of a part being be sized. An increase in density nor in hardness of deformed layer is not intended. Sizing of holes can be performed as ball sizing.

Coining improves the dimensional accuracy, as in sizing, and the surface configuration of the part. High presures required for coining results also in density increase of the part, mainly in surface area, not in the bulk as in repressing. Due to considerable strain-hardening occurring in the coining operation, tensile strength and hardness increase while the elongation decreases. This increase in mechanical properties may be in many cases very important that soft, unalloyed sintered parts often gain sufficient strength for use under quite severe conditions with a positive effect on machinability.

Repressing means increasing the overall density of a compact by subjecting it to a second pressing treatment after eliminating the cold work from the first compacting step by intermediate annealing. For standard ferrous compacts, density levels can be increased from typically 7.1 g/cm^3 maximum to 7.3 to 7.4 g/cm^3. Since the tools for the second pressing can be made with simpler geometry, no powder transfer being necessary, higher pressures can be applied. In the case of carbon containing compacts it is essential that the annealing treatment is carried out at temperatures that do not result in significant carbon dissolution, *i.e.* about 800°C maximum. It is possible to expect that the repressed parts with the green strength in bending over 20 MPa could be subjected to 'green' machining.

Local surface densification by the controlled plastic deformation of the surface layer on sintered (porous) materials may greatly improve their fatigue properties which are strongly dependent on the density. By locally increasing the density where the stresses are concentrated on a component, the mechanical properties can be improved. The plastically deformed surface of a component may be successfully surface hardened by various method required for service properties of a part.

Cold surface plastic deformation of a porous component results in:

– work hardening of the deformed area,

- formation of residual compressive stresses in the region,
- increase in density up to full density in the region and in its vicinity,
- formation of a smooth surface, depending on the method of deformation and material and its starting porosity [10].

Surface (selective) densification forming a poreless subsurface area of a PM part is essential *e.g.* for parts subjected especially to contact fatigue loading and/or to wear under different conditions. The investigation of surface-densified PM transmission gear shows one of the regions for the application of the method [77,78,79]. Density requirements for different automotive parts are shown in Fig.2.34.

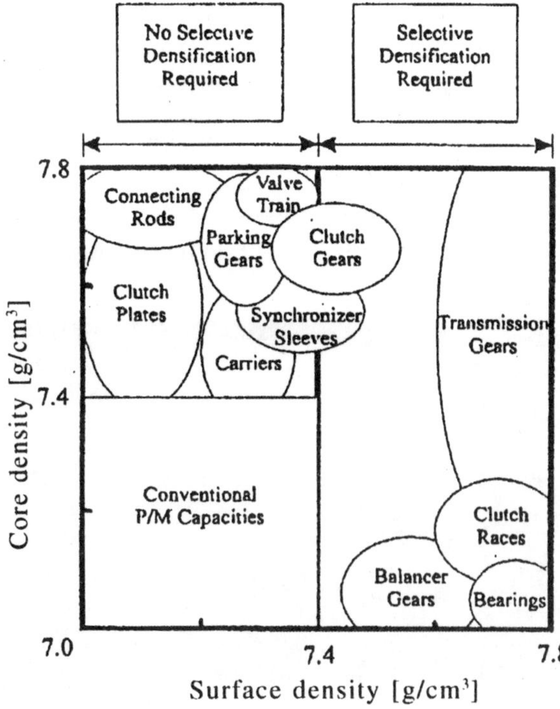

Fig.2.34 Density requirements for PM automotive parts [80].

Surface densification may by realised by 'coining', rolling, roll burnishing, shot peening, and blasting [10].

By a 'coining' method based on a new concept of the coining die it was possible to form the selectively densified surface layer in one stroke with the different thickness required for adequate properties of a part, *e.g.* for contact fatigue endurance with application of all chemical–thermal hardening processes. In regard to contact fatigue it can be noted that at rolling line contact fatigue test the defects initiating crack formation are situated at a depth of 70 to 80 μm below the surface in which the maximum shear stresses under applied load are formed [81]. The lodading conditions of a such part determine the minimum depth of the densified layer needed. Table 2.7, for example, shows the hardness of the surface layer ~0.3 mm thick densified by a special coining method for 6 alloys compared to the initial hardness of the samples. The sections of densified sample layers are shown in Figs.2.35, 2.36.

Tab.2.7 Hardness of selectively densified surface layer by 'coining' and as-sintered hardness and density of tested alloys. Specimen: disc 30 mm in diameter and 10 mm in height, sintering 1120 °C, 30 min, cracked ammonia. C – graphite

Alloy	Density [g/cm^3]	Hardness HV 10	
		As-sintered	Densified layer
Fe-0.5C (ASC100.29)	7.04	81	145
Fe-2Cu-0.5C (ASC100.29)	7.07	102	183
Fe-1Cr-0.3Mo-0.3V-0.5C (KIP103V)	7.02	114	205
Fe-1.75-1.5Cu-0.5Mo-0.5C (Dist. SA)	6.90	151	199
Fe-4Ni-1.5Cu-0.5Mo-0.5C (Dist. SE)	6.93	159	245
Fe-0.85Mo-0.5C (Ast. 85Mo)	6.97	118	210

Remark: as-sintered – in pressing direction, densified layer – perpendicular to the pressing direction

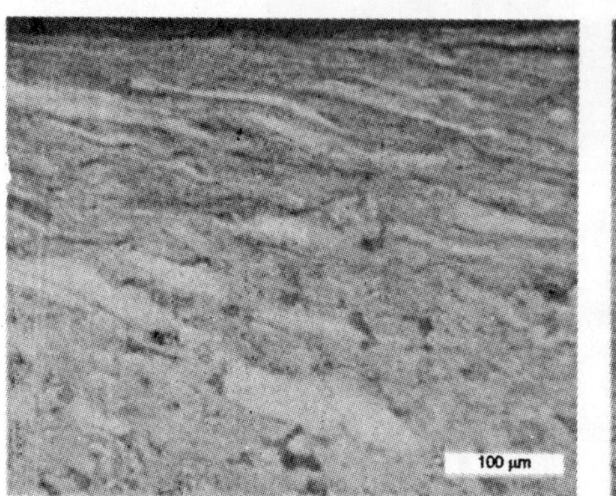

Fig.2.35 Section from a surface densified sample Fe–2Cu–0.5C (alloy No. 2, Tab.2.7). Optical micrograph. Nital etched.
Fig.2.36 (right) Section from a surface densified sample Fe–1.75Ni–1.5Cu–0.5Mo–0.5C (alloy No. 4, Tab. 2.7). Optical micrograph. Nital etched.

Because machining operations such as drilling should be applied after these operations, problems with machining such parts, esp. with tool life, can be expected, in particular if the machining parameters are adjusted for the non-densified base material. As shown above, surface densification results in different material properties and thus also different machinability within the volume of a part.

Parts can be smoothed or formed with rolls with a defined rolling force or with a given geometry. Rolling to the final dimensions of the parts is less used procedure than sizing.

Rolling trials on flat specimens designed, for example, for bend tests and specimens for rolling fatigue tests can be carried out.

Ball burnishing is used in certain cases to modify the dimensions of bore because this method is cheaper than sizing.

One of the methods to improve machinability without changing the composition can be, for instance, to repress or otherwise densify the area to be machined.

2.6.2 Sintered gears from manganese steel for high dynamic loading

In contrast to surface densification of the gears, the gears for hydrogenerators from as-sintered manganese steels were prepared and tested. The gears, Fig.2.37, were prepared from Fe–3Mn–0.5Mo–0.35C alloy by compaction at 600 MPa and sintered at 1120°C for 140 min in cracked ammonia. The density of the sintered gears was 6.96 g/cm^3 and hardness 178 HV 10. The final tolerances of the functional dimensions of the sintered gears were attained by grinding and the allowance in as-sintered dimensions for this operation was 0.10–0.15 mm. The groove for the tongue was slotted under the cutting conditions used for wrought steel (AISI 6150). The broaching seemed to be less advantageous for this process.

Regarding the operational pressure in a hydrogenerator of 1.5 MPa, the sintered Fe–Mn–Mo–C gears were tested by the standard method for these parts with the oil pressure in the range of 0.7–5.0 MPa at 2800 rev./min. The testing varied between constant pressure during some time period and cyclic pressure

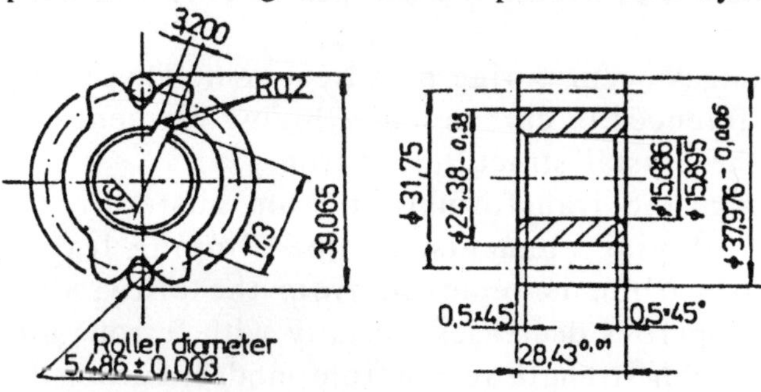

Fig.2.37 Dimensions in *mm* of a tested gear for a hydrogenerator [82].

changing all 5 s between zero and the 'nominal' test pressure of 3 or 5 MPa. The dimensions of the gears were controlled after 600 h test, then after 1000 h test. The dimensions after the tests were in the tolerances. Also a successful test for 4000 h with the pressure of 3 MPa was performed without any change in tolerances dimensions required or in material behaviour [82].

2.6.3 Heat treatment and surface hardening

It is necessary to stress in advance that all heat treated and/or surface hardened parts offer problems for machining, except by grinding. The solution is the machining of the parts before heat treatment, *e.g.* as-sintered, considering the dimensional changes that are linked to these processes.

The aim of heat treatment and surface hardening processes is to establish special improved properties of the porous or fully dense powder metal parts adjusted to the particular requirements of the application. By these processes, the as-sintered character of the microstructure in volume or in surface regions is changed in the direction of the formation of harder constituents (martensite, bainite). Hardness and tensile strength and some special properties increase as a result of these processes. Porosity and microstructural heterogeneity of the sintered part will exhibit an effect on the properties of processed part.

Heat treatment and surface hardening processes used are mainly:

- through hardening,
- surface hardening – gas carburizing, low-pressure carburizing, plasma carburizing, carbonitriding, nitrocarburizing, gas nitriding, plasma nitriding.

The use of some of these processes depends on the required properties of a part and on costs. In term of machining of a sintered part, the machining before the heat treatment – or other processes which substantially increase the hardness – is regarded the optimum solution.

Major changes occur during heat treatment. These changes result from the differences in carbon and alloying element contents with changes in the crystal structure for iron.

On cooling, the transformation from austenite to ferrite is impeded by alloying elements. As the alloying level decreases, especially as carbon is removed from the alloy, the ability to manipulate properties decreases. Usually with ferrous alloys, a high hardness and high strength result from modest carbon levels in the range of 0.4–0.8%. Other alloying ingredients such as nickel, molybdenum, chromium, and manganese affect the heat treatment response, even at low concentrations.

In relation to the heat treatment response for each alloying element addition, there are three key changes. The first relates to the relative effects on carbon dissolution in the various phases. Most metals stabilise either ferrite or austenite, but some form carbides. Secondly, the alloying addition changes the transformation temperature. Thus, they can enhance or retard the formation of desirable phases in heat treatment. Finally, the transformation rate is affected, and much interest exists in alloying additions that slow the change from austenite to the diffusion controlled on cooling. The slower the transformation, the easier it is to form high-strength sintered components in heat treatment.

In respect to heat treatment, a critical carbon content is at the eutectoid compositions, which is 0.77% carbon for simple steel. In highly alloyed steels, the carbon level corresponding to the eutectoid composition shifts to lower carbon contents for alloying ingredients such as nickel, manganese, chromium, silicon, tungsten, molybdenum, and titanium. Figure 2.38 shows the carbon content at the eutectoid composition and the eutectoid temperature versus alloying level for several alloy element additions to Fe–C.

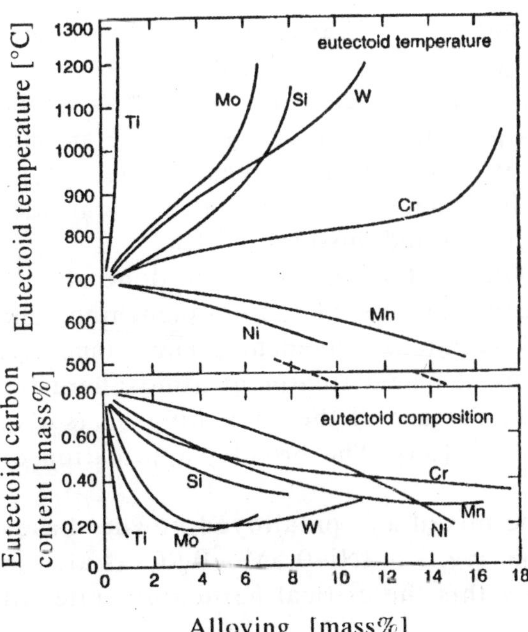

Fig.2.38 Plots of the carbon content at the eutectoid composition and the eutectoid temperature *vs* alloying level for several additions [6].

2.6.3.1 Through hardening

The through hardening process (austenitizing, quenching, and tempering) is applied in cases when high strength and hardness of the parts are required. The effect of the process is affected by porosity, carbon content, and alloying and processing conditions (austenitising and tempering temperatures, quench medium).

Because of the porosity, the austenitizing temperature for carbon and medium-alloyed steels can exceed the Ac_3 temperature by 50–80°C because they are not sensitive to overheating but it is not ncecessary to use for heat treatment. The carbon content needed for attaining the required tensile strength and hardness depends also on alloying. Through hardening is linked to dimensional changes due to prevailing martensite in the microstructure, and there is some tendency to distortion.

The porosity has also a strong effect on the beginning of the martensite transformation. The density increase of Cr, Ni, Mo alloyed powder steels at continuos cooling causes the decrease in the critical rate of the transformation of austenite and by this the decrease in the temperature of the beginning of the martensite transformation.

The consequence of it is a decrease in hardness. The austenite grain growth for the porous sintered steels is much less or none compared to wrought steels, the pores pinning the grain boundaries.

Optimum quenching temperature increases with increasing amount of alloying elements and the austenitizing time is longer. In some cases, therefore, the hardness of heat treated steels with increasing amount of alloying elements is not higher than of 40–50 HRC. The hardly disoluble primary carbides in Cr–Mo steels, *e.g.* $Me_{23}C_6$ and Me_7C_3 bind a part of carbon an by this the amount of carbon in solid solution is lower thus decreasing the carbon content in martensite and other phases. The presence of carbides retards the homogenisation of retained austenite [41,83].

Through hardening in powder metallurgy could be done also in modified sealed sintering/quench furnace, including quenching. The process eliminates a separate operations – quenching and tempering. The work load is protected by the heat treatment atmosphere for the entire treatment cycle. The surface of the quenching oil is also protected by the gas atmosphere [84]. The broader application of this furnace is not known.

The difference between the mixed and prealloyed powder based steel of the equal composition, *e.g.* Fe–2Ni–0.5Mo–0.9C, is also in the critical hardening rate. By this the critical hardening rate of the mixed steel was 19°C/s and of prealloyed powder 2.3°C/s as the consequence of the different stage of the homogenisation of the solid solution [41]. According to this, in the heterogeneous microstructure the soft zones could occur if cooling is not sufficiently fast.

2.6.3.2 Sinter hardening

Sinter hardening as a special PM production process enables the production of PM parts with high strength and hardness more effectively than conventional heat treatment. The entire heat treatment process of sintered steels is very much simplified by self-hardening achieved by faster gas cooling only of the parts from sintering temperature, without separate heat treatment installation.

The low alloy powder steel developed for sinter hardening must provide sufficient hardenability to make the martensite transformation possible during the cooling portion of the sintering cycle.

In order to predict the required as-sinter hardened final properties, the critical factors are the powder mix composition, *i.e.* carbon content, the level of alloying elements, processing variables, cooling rate after sintering, and density. One of the important parameters in designing sinter hardening parts and in optimising the sintering process is to determine the cooling rates in the martensite transformation temperature range that can be achieved in a given furnace operating under specified conditions [85].

In general, up to present mainly the Fe–Ni–Cu–Mo–C steels have been regarded suitable for this process. The range of sinter hardening steels is enlarged by chromium and manganese additions in different amounts. Powders with lower hardenability can show good sinter hardened properties if the cooling rate is fast enough to avoid formation of pearlite and minimize the formation of bainite during cooling. Hardness up to 40 HRC for Ni-alloyed steel (Fe–0.45Mn–1.0Mo–0.9Ni–0.45Cr; and Fe–1.75Ni–1.5Cu–0.5Mo) with (0.9 resp.0.6)% C at either 0.4 or 1.5°C/s was reached. The highest cooling rates in modern industrial sinter hardening furnaces can exceed ~1°C/s. It was also indicated that Cu and (Ni)–free Fe–Cr–Mo and Fe–Mn–Cr–Mo steel grades, *i.e.* Fe–2.5Mn–1Cr–0.45Mo–0.7C, are capable of sinter hardening [86] in part even in conventional PM processing furnaces [87].

The apparent hardness of sinter-hardened steel depends on cooling rate and also on density of the part, Fig.2.39. At relatively low cooling rates, apparent hardness increases rapidly with increasing cooling rate to reach a platea for cooling rates larger

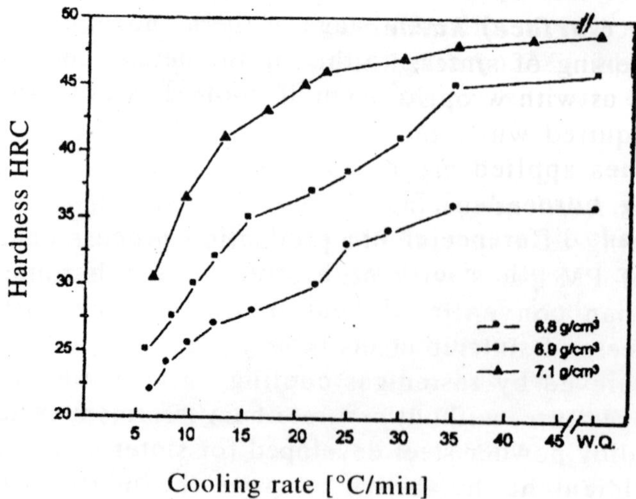

Fig.2.39 Apparent hardness of Fe–1.80Ni–0.73Mo–0.50Mn–0.80C steel (MP55) as a function of cooling rate over the temperature range 550°C to 350°C for three sintered densities [85].

than 30–35°C/min. The highest hardness was achieved for the material with highest density. A further factor affecting the cooling rate at sinter hardening is mass of the part, Fig.2.40. The cooling at shown temperature range is critical for martensite transformation. According to this graph, when the mass of sintered compact was less than 200 g, the cooling rate was greater than 1.5°C/s. The cooling rate was measured by thermocouple buried into the center of singular green compacts. An other state occurs at large-scale sintering of the parts. An air hardenable PM steel may be prepared through infiltration of a porous steel skeleton with copper–manganese alloys (~30% Cu–70% Mn), resulting in high hardness without quenching [89].

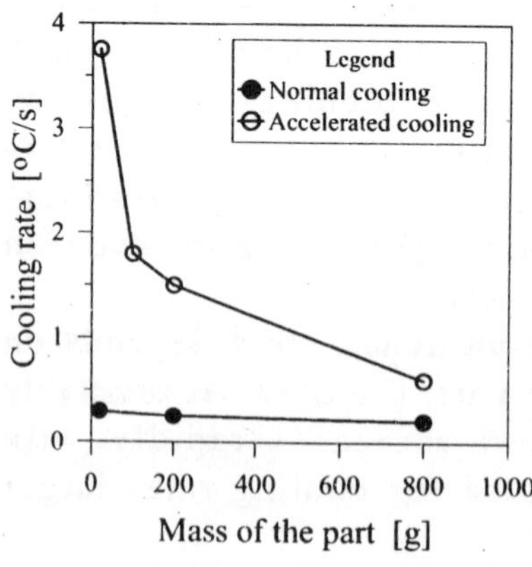

Fig.2.40 Correlation between mass of sintered compacts and cooling rates over the temperature range of 500 to 300°C [88].

2.6.3.3 Case (surface) hardening

Surface hardening of sintered steel components is done from the same reasons as with wrought steels, *i.e.* a hard and wear resistant surface is required while the core should be sufficiently ductile. The techniques applied are also relatively similar, carburizing, carbonitriding, nitrocarburizing, and nitriding being employed.

A significant difference compared to wrought steels is found in the penetration depth, esp. with components of average to high porosity in which case the pores are mostly open and interconnected. This results in fairly deep penetration of the active reagents – CO, NH_3, etc. – in part in formation of thick carbide layers [90] and in part even in through hardening effects. To prevent these unwelcome results, the parameters must be adapted, for example by lowering the treatment temperature, superficial pore closure or, for example, by shot peening, and by plasma treatment in which case the reactive component – for plasma nitriding nascent nitrogen – is generated only above the part surface but is quickly deactivated by recombining within the pores. Thus, chemical reactions are restricted to the external surfaces. Also low pressure carburizing techniques are increasingly being used for porous sintered parts [91].

Surface hardening without changing the chemistry can be obtained by induction hardening at a sufficiently high frequency, making use of the skin effect. This technique is used quite frequently for sintered parts, *e.g* [92,93]. However, it must be considered that quenching – which is in part done with water – results in comparatively high stresses in the material, and there is the danger of crack formation due to the low ductility of the porous sintered steels.

2.6.4 Hardfacing

Hardfacing is a special treatment for the formation a surface layer of very high hardness (1100–2000 HV) mainly on high speed steel tool components, and for high abrasive wear resistance also on structural steel components. Appropriate variants are manganising, chromising, boriding [94,95] laser beam treatment, and CVD and PVD processes [10].

In any case the machining of the parts with the modified high surface hardness must be performed before surface processing.

2.6.5 Steam treatment and coating

Steam treatment increases the resistance to corrosion, to

compression stresses, and considerably increases the hardness. There is a negative effect on ductility and toughness, which are lowered, particularly in parts with low relative density. For Fe–Cu and Fe–Cu–C materials, steam treatment is always combined with some age-hardening. The process involves treatment of the sintered parts with overheated water vapour in inert atmosphere, absence of oxygen being an essential requirement. All exposed surfaces, interior and exterior, are coated with a hard dark blue to black magnetite layer – iron oxide Fe_3O_4. The process temperature is in the range of 450 to 600°C, typically 510–540°C, higher temperatures resulting in thicker and darker layers. Steam treatment is also applied for HSS drills. Machining of the parts is usually done before steam treatment regarding the dimensional change by the process.

Coating. Powder metallurgy components can be electroplated with the same metals as cast and wrought components. However, the PM components should have their porosity sealed by resin impregnation to avoid entrapment of plating solutions in the pores, which would cause corrosion later on. Acid copper plating generally gives the best result on PM components because of its excellent throwing power. Deposition of copper on iron powder from cyanide bathes is mostly used. Nickel, applied either alone or over a copper base, is used to increase wear resistance. However, this is the currentless nickeling which >10% P contains and consists mainly from phosphides to provide a bright corrosion resistant surface. Chromium plating is done for either decorative reasons or to obtain a wear resistant surface.

2.7 POROSITY AND MECHANICAL PROPERTIES OF SINTERED IRON AND STEEL

A permanent effort in studying machining is to find a relationship between the machinability of a sintered alloy and some easier to define mechanical or other characteristics. The main characteristic property of a sintered material parameter is the porosity. The relationships between the mechanical properties of sintered iron and, in a limited range, of some alloyed steels and the porosity are used to investigate if the porosity can be accepted as a general characteristic also for the machining chracterisation in combination with mechanical properties.

Mechanical and physical properties are related to the porosity also in research and practice in machining of PM materials. Porosity is considered to be the main factor deteriorating the machinability of a sintered material. From this reason, in the following some results are presented showing that the relationships between the porosity on one side and hardness or tensile strength values on the other side show a great scatter and therefore have a validity only for a material of a given composition and starting powders, prepared under very exactly defined processing conditions.

In general, pores lower the physical and mechanical properties of materials compared with wrought materials unless they represent the main service characteristic of such a material as in, for example, self-lubricating bearings or filters. The detrimental effect of pores on the properties of sintered materials is reflected in a reduction of the load-bearing cross-section and in the micronotch effect [96,97]. Porosity is a result of the complex effects of all material and processing factors determining the final properties of a sintered material. Therefore, the porosity cannot be used to unambiguously characterise the behaviour of the properties of a sintered material.

The term porosity includes the open pores, which are interconnected in the volume of the component and with the surface, and the closed pores. Strictly speaking, however, for the mechanical properties rather the pore connectivity is relevant, and therefore 'interconnected' and 'isolated' pores can be defined (open pores are always interconnected, but not the other way round; there may be interconnected pore networks that are not open to the surface [98]. This corresponds to the 'sponge' and the 'swiss cheese' microstructural types, in one case one single complex pore network existing, and in the other case numerous more or less spherical cavities are present in a metallic matrix. The fraction of open porosity is easily determined by infiltration (see standard ISO 2738) while the pore connectivity can only be determined reliably by fractographic investigations [97]. Their mutual ratio of open/closed porosity as a function of the total porosity depends on many factors, such as iron powder grade, particle size distribution and specific area of the powder, compacting pressure, also lubrication, sintering conditions, esp. formation of liquid phases, and alloying. The shape of pores (size, shape factor) is very helpful for better characterisation of the properties, especially failure at loading of a material.

2.7.1 Sintered iron

Figues 2.41 and 2.42 show the correlations between porosity and tensile strength and hardness for sintered iron compacts. The relations presented are based on some hundred literature data involving all iron powder types, sintering temperature of 900 to 1200°C, sintering time of 0.5 to 16 h and single and double pressing technique, *i.e.* all basic powder and processing variables affecting the properties of sintered iron [99,100]. When evaluation a limited range of variables a lower spread in the relationships between the mechanical properties and porosity of sintered iron is determined.

It follows from these relationships that, for instance, the compacts with 10% porosity may exhibit real tensile strength value in the wide range of 160 to 250 MPa, elongation of 7 to 19% and hardness of 55 to 90 HB. The total porosity, or density, as shown does not generally determine any property with sufficient accuracy because it is a result of the mentioned many factors which are in some cases not known and not easily definable. It is shown that the dependence of tensile strength is not linear. On the other hand the relation between the hardness ad porosity can be evaluated as linear.

It is necessary therefore to take into account a certain specific spread of the values of the evaluated property in relation to the total porosity. The previous relationships are accomplished with the relationship between hardness and tensile strength of sintered iron regarding the previous data which are often used for the characterisation of machinability of some material, is also not linear, Fig.2.43. The effect of open and closed porosity on the tensile

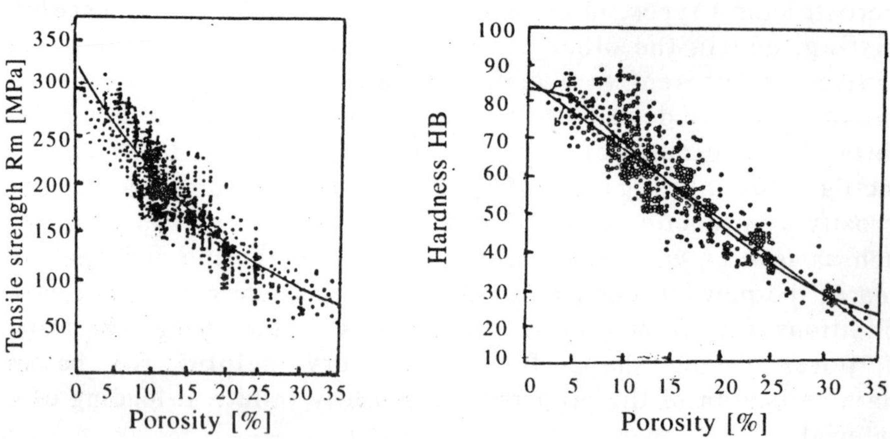

Fig.2.41. Ultimate tensile strength (Rm_s) vs. porosity for sintered iron powder compacts. Mean curve computed to eq. $Rm_s = Rm_c \exp - k \times P$ [MPa] = 32.4 $e^{-0.043}$ P [MPa]. P is porosity in %, Rm_c is the tensile strength of wrought plain iron [99].

Fig.2.42. (right) Hardness (H_s) vs. porosity for sintered iron powder compacts. Mean curve computed to eq. $H_s = 83.1 \times P^{0.127}\ e^{-0.049} \times P$ [99].

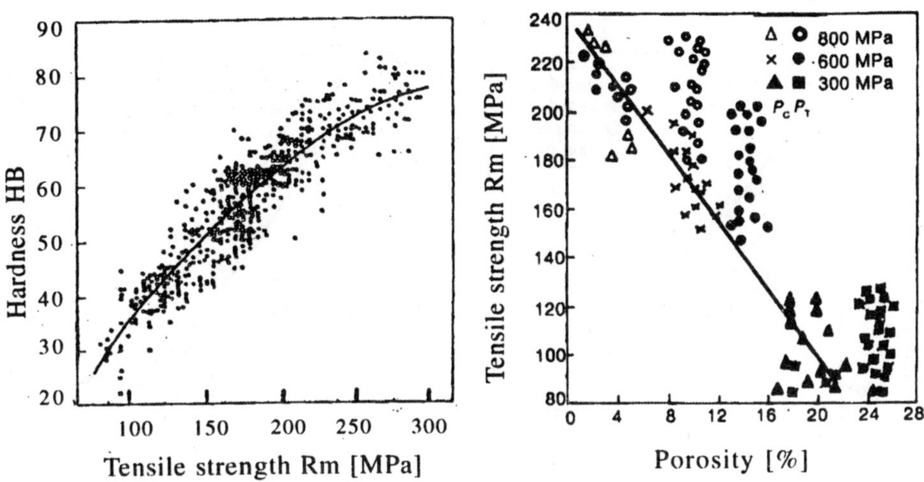

Fig.2.43. Relationship between hardness and tensile strength of sintered iron. Data as in Figs. 2.40, 2.41 [99,100].

Fig.2.44. Dependence of tensile strength of sponge iron powder specimens compacted at 300, 600 and 800 MPa and sintered at 930, 1000 and 1100°C for 1/4 to 16 h in dissociated ammonia on total – P_T and open – P_0 porosity [10].

strength of sintered iron compacts prepared from one iron powder grade, shown in Fig. 2.44, confirms the complexity of this problem.

It follows from mentioned data that the variation range of mechanical properties in relation to the total porosity depends in final outcome on the quality and quantity of interparticle bonds formed during sintering, but in dependence on many mentioned factors. This manifests oneself also by different character of facets of ductile failure and by their number [101].

2.7.2 Plain carbon steels, copper alloyed steels

Iron–carbon alloys are very often used in production of structural parts with additional machining. For the completion of previous data concerning sintered iron, Figs. 2.45 and 2.46 show dependences of tensile strength and hardness of Fe–C sintered material on porosity. Thereof that for the determination these relationships were used data regarding a smaller range of materials and processing conditions, the spread in the results is lower compared to sintered plain iron. For the design of these figures the data from Ref. 11 for sponge iron and atomised iron powders were used.

The curves for tensile strength and hardness designed for particular carbon content are not linear. A larger effect of carbon content on these properties is shown for decreasing porosity. The scatter in tensile strength and hardness values at given porosity and

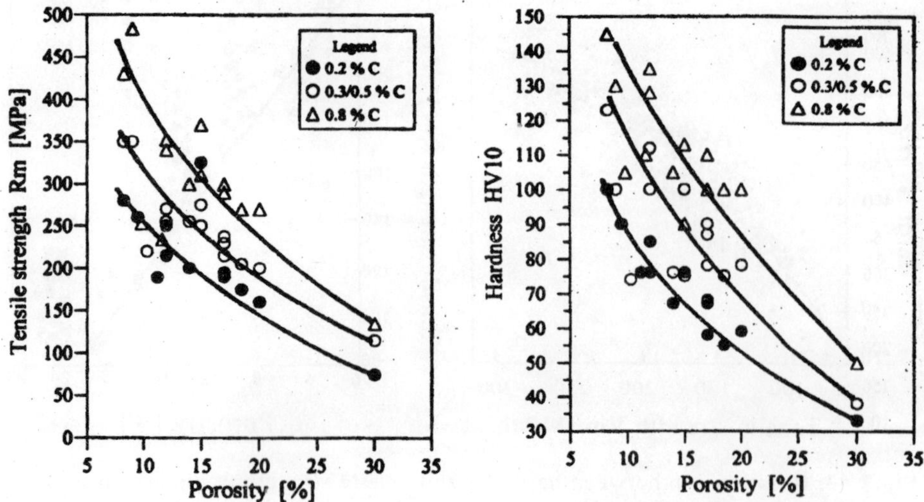

Fig.2.45 Tensile strength vs. porosity of sintered Fe–(0.2, 0.5, 0.8)C steel based on 3 sponge and 3 atomised iron powder grades. Sintering 1120°C, 30 min, endothermic atmosphere, density 5.3–7.2 g/cm^3 (acc. data in Ref. 11).
Fig. 2.46 (right) Hardness *vs* porosity of sintered Fe–(0.2, 0.5, 0.8)C steel as in Fig. 2.44.

at the equal carbon content caused by iron powder grades is in this case relatively small and not exceeds ~50 MPa. Hardness values show a larger scatter in relation to the evaluated factors compared with tensile strength values. The proof of this is, for instance, hardness of 100 HV 10 at the porosity of the materials in the range of 8.3 to 20%. It was possible, therefore, to design a curve only for 0.2% C materials.

Figures 2.47 and 2.48 show dependences of the tensile strength and hardness on the porosity of Fe–2Cu–C sintered steels.

The steels with higher carbon content did not exhibit larger differences between them at higher level of tensile strength values. As seen for 0.2% C steel at the porosity of 11–12% the scatter in tensile strength values was ~75 MPa and for 0.8% C steel ~185 MPa. Similarly a smaller scatter in hardness values (~20 HV 10) was determined for 0.2% C steels. Small differences in hardness values for the steels with higher carbon content in relation to the porosity were observed.

As shown the real value of the porosity may be related only to the proper property of a sintered material prepared under defined same conditions, including a specific base powder. The properties of someone material sintered under equal sintering conditions but, *e.g.* in other type of furnace or by other PM part producer can be different.

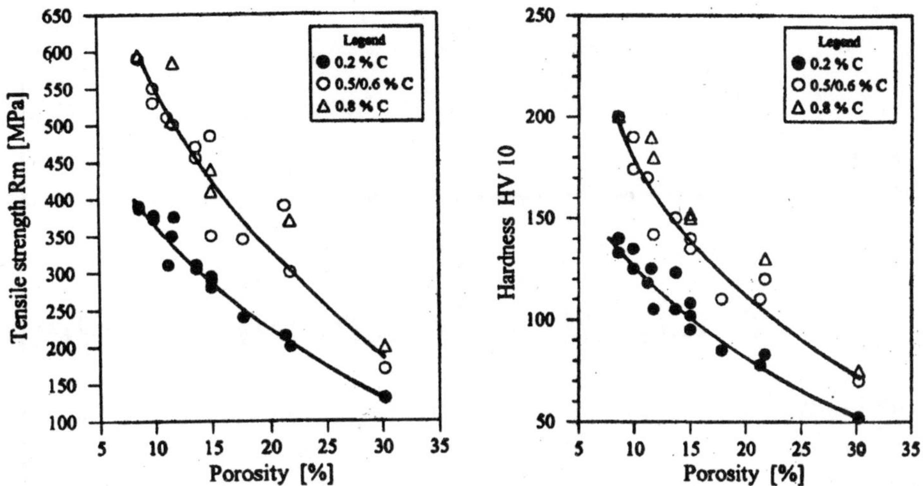

Fig.2.47 Tensile strength *vs* porosity of sintered Fe–2Cu–(0.2, 0.5, 0.6, 0.8)C steel based on 3 sponge and 3 atomised iron powder grades. Sintering 1120°C, 30 min, endothermic atmosphere, density 5.5–7.2 g/cm^3 (acc. to Ref.11).

Fig.2.48 (right) Hardness *vs* porosity of sintered Fe–2Cu–(0.2, 0.5, 0.8)C steel as in Fig.2.47.

3

Principles of Machining of Steel

The knowledge of the basic characteristics of machining for wrought steels has been gained during some decades and is therefore the basis and valid for studying the machinability of PM steels. There are, however, differences between wrought and PM steels in physical and chemical properties and thus in microstructure and mechanical properties, and this is the reason for the needed complex investigation of machinability of PM steels.

For determining the optimal machining conditions of a material it is necessary to define all factors affecting the machining process. The result of the interaction between them will be the PM product with requested shape attained by machining what is not a simple process.

Machining is a metal cutting and shaping method which includes a large collection of manufacturing processes designed to remove unwanted material, usually in the form of chips, from a workpiece. It is a purposeful form of failure of material under specified geometrical, mechanical, and friction conditions. All methods of mechanical cutting processing of conventional and sintered materials cause their elastic and plastic deformation, which is accompanied by friction and thermal effects, densification or failure of machined surface layer.

The basic characteristics of all factors participating in the metal cutting process are common for wrought and PM steels. General description and definitions of machining processes and tools may contribute to the more detailed view on the role of single factors for fulfilling the requirements for dimensional accuracy and surface quality of the product and for the technical and economical effectiveness of the machining process.

3.1 MACHINING PROCESSES

According to the cutting edge characteristic of the tool, the machining processes are divided into two groups, *e.g.* the processes with geometrically:

- defined cutting edge,
- undefined cutting edge.

The first group involves: drilling and boring, turning, tapping, threading, reaming, milling, shaping (in PM parts not necessary to perform), broaching, and ball sizing. The second group involves: grinding, honing, lapping, as well as special methods. Burnishing and polishing are not usual in PM machining. Tumbling for deburring sintered parts is an other mechanical process commonly performed in PM production.

Figure 3.1 shows schematic examples of common machining practices in metal cutting.

The main field of application for machining methods with geometrically defined cutting edges are the finishing processes, also in machining powder metallurgy steels. The machining parameters,

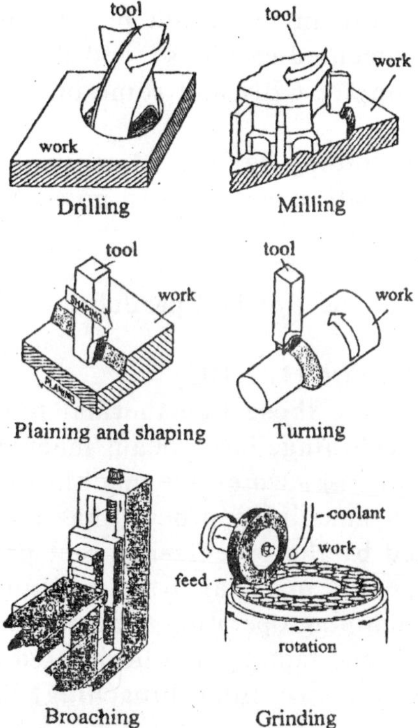

Fig.3.1 Examples of common machining processes by drilling milling, planing, shaping, turning, broaching and grinding the PM components [6] (Reprinting permitted by John Wiley and Sons).

tools and cutting materials employed basically correspond to those used in the machining of conventional materials. Nevertheless, it is necessary to adjust all processing parameters to PM materials.

The primary chip formation processes include more specific shaping steps such as:

- drilling (reaming, tapping, spot facing, counterboring, countersinking, shallow drilling),
- turning (boring, facing, cut off, taper turning, form cutting, chamfering, recessing, thread cutting),
- abrasive machining (grinding, honing, internal and surface lapping–broaching),
- shaping (planing, vertical shaping, *e.g.* vertical milling),
- milling (hobbing, thread milling),
- sawing (filing).

In the machining process there are independent and dependent variables:

Independent input variables are:

- tool material; the three most common cutting tool materials currently in use for production machining operations are high-speed steel (HSS), both in ingot and powder metallurgy (PM) form; hardmetals (cemented carbides), and the respective coated tools; a special group are CBN and some other tools,
- cutting parameters,
- tool geometry, work holding devices, clamping,
- cutting fluids (to a small extent used in powder metallurgy).

Dependent variables are:

- cutting force and power,
- size and properties of the finished product,
- surface finish,
- tool wear and tool failure [102,103].

The special methods are those like thermal removal processes, electrical discharge machining, laser beam machining, as well as laser beam assisted turning, water jet machining and ultrasonic assisted grinding. Each machining process has its special field of application, determined by the process-inherent properties and the specific features of the PM material to be machined [104].

The predominant machining operations common in PM are mainly drilling (~30%), turning and tapping and boring (each ~25%), thread cutting, grinding, and others (milling, broaching) in a minor extent [105].

3.1.1 Drilling

Drilling is a term covering all methods of making cylindrical holes in metal products. The term can be divided into two categories: short hole and deep hole drilling. The difference between them is not restricted to the relationship between depth and the diameter. Parameters such as completeness of chip evacuation and removal rate also form the basis for differentiating short and long hole drilling methods. The term 'drilling' covers thus also machining processes like reaming, counter boring, and various forms of finishing operations. For PM components most drilling involves short holes.

Drilling is a combination of two movements: a main rotating motion plus linear feed motion. With short hole drilling in conventional machines the most usual form of working is that the tool does both the rotating and feeding motions. In principle, this cutting operation remains the same, whether it is to be done on a drilling machine or on a lathe. At drilling and boring operations performed on a lathe (universal NC and CNCX controlled lathes) the workpiece is chucked in a three-jaw chuck and the drill does not rotate [106].

For counterboring bore-holes on a lathe, twist-type counterbores or shell drills are used. The quality of the inner surface of the counterboring mostly is somewhat finer than that achieved in boring. However, a high dimensional accuracy is not ensured in counterboring.

If high surface quality of drilled, bored or countersunk holes of a high dimensional accuracy are to be produced, they are finished by means of reamers. Machine reamers are used in the powder metallurgy machining. To ream the holes with breaks in the form of grooves etc., reamers with spiral flutes are used. This prevents the cutting edges from jamming and breaking.

3.1.2 Turning

The term turning denotes various cutting methods which are primarily used for machining workpieces with a circular section. Very much research has been done on turning, since it is one of the most commonly used machining operations, beside the drilling, and relatively easy to perform under controlled conditions since the cutting edge is fixed (and can thus be easily observed and controlled). Turned parts may be finished parts, parts which are directly assembled or used, and/or may be subjected to further machining operations. Furthermore, the surface quality achieved on a lathe may be still insufficient for certain purpose. In such cases the turned part is ground. Turning as machining process had as variables cutting speed, feed rate, and depth of cut [107]. Although turning is mostly considered as a non-interrupted cutting procedure, interrupted cut is also frequent, *e.g.* when turning a shaft with grooves. This considerably affects the choice of the tool, esp. the tool material, regarding the cyclic loading involved.

3.1.3 Milling

Milling is a metal cutting operation at which chips are detached by a series of cutting edges arranged on the circumference of rotary milling cutters. During the cutting operation, part of the edges are in contact with the workpiece while the others move freely; milling is therefore a typical machining operation with interrupted cut. This has considerable impact on the tool, especially regarding mechanical and thermal fatigue properties, similarly to turning with interrupted cut. Due to the high number of cutting edges, a large volume of material in the form of chips can be removed in one single revolution. Powder metallurgy workpieces – parts can also be finished in a single operation, Workpieces of a great variety of types and shapes can be milled by means of appropriately shaped milling cutters. Milling machines are horizontal and vertical ones.

The milling cutters used on horizontal milling machines are called hobs. The type of milling cutters which is most widely used on vertical machines is the type with cutting edges on the end face. Cutters of this type are used to produce flat surfaces. Further, vertical milling machines can be used to produce recesses of various types in workpieces. For example, shank-type cutters can be used to produce a necked-down portion of workpiece, whereas slotting end mills are used to produce grooves, and others as slotting and milling cutter, angular cutter, circular saw blade, formed milling cutter, shell and mill, side milling and plain milling cutter with spiral flutes [106].

3.1.4 Tapping

Tapping is one of the most difficult operations especially in PM machining. This would explain the frequent use of oil impregnation to assist the operation. Tapping is closely related to drilling. There is an obvious need for a hole to perform a tapping operation. The diameter of the hole is a very important factor in determining both thread quality and the useful life of a tap. A correct thread can only be made if the hole is round and straight. The reason for the problems associated with tapping are, for example, the large number of edges that cut simultaneously and the difficulty to remove the very numerous chips, and the cut in a hole surface layer of the material deformed and work hardened at drilling.

3.1.5 Grinding

Grinding is a machining process which is used to finish workpieces that must show a high surface quality, accuracy of shape and dimensions. Grinding is commonly the finishing operation because grinding permits the machining of a workpiece to the required size within such close tolerances as a few micrometers. A great advantage of grinding is the possibility to grind workpieces which have already been hardened. Besides the main grinding of tools, there are the following grinding processes: face grinding, cylindrical grinding of shafts and holes, grinding of threads and profile grinding of gear teeth, cutting-off by grinding. With regard to possible removal efficiency and attainable surface quality, grinding, particularly of hard (heat treated) brittle PM materials, is nowadays the most important method [108].

In grinding the sharp abrasive grains of the grinding wheel detach metal chips from the workpiece. The grinding wheel spindle usually rotates at high speeds. The grain size of the grinding wheel is determined by the desired surface quality and the amount of metal to be removed. Grinding wheels with coarse grains are used for rough-grinding, whereas grinding wheels with fine grains are required for finish-grinding.

A typical feature of grinding is that tool wear is essential for successful operation since only the removal of blunted edges – the respective grain breaking out of the grinding wheel – creates fresh, sharp edges that continue the machining operation. Carefully balanced bonding of the abrasvie grains is therefore of decisive importance; too weak bonding results in unnecessarily fast wear while too strong bonding retains also blunted grains and lowers the cutting effect. For grinding, the ratio between the removed volumes of workpiece and tool life is a dominant criterion. Figure 3.2 shows schematically the contact between a disc wheel and a PM part in grinding.

The grinding tools can be subdivided into grinding wheels made of conventional grinding materials and diamond or CBN grinding wheels. Conventional grinding wheels, normally SiC or Al_2O_3, are particularly suitable for the machining of standard low alloyed PM steels and can therefore be used for all short- or medium-cutting

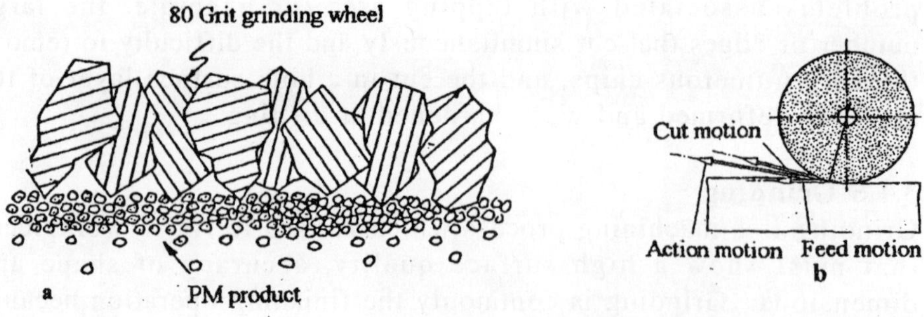

Fig.3.2 Left –Type of contact between a disc wheel and a PM product [109], (right) – Cutting feed and action motion in grinding [110].

alloys. For PM high speed steel they are not practical, given the hardness of the carbides and the extensive wear of the grinding wheels. CBN grinding wheels are used frequently for this task. Hardmetals are machined mostly using diamond grinding wheels. The basic requirement of the optimal application in grinding is that the topography of the grinding wheel must be adjusted to the grinding task. Conventional grinding wheels can be profiled and sharpened in one process step. Diamond and CBN grinding wheels are normally profiled and sharpened in different working processes with diamond profile and tracing rolls or at cylindrical profiles also by means of small cup wheels [108].

Ultrasonic grinding is another new promising method to machine tribologically important surfaces. In ultrasonic grinding, a workpiece oscillation is superimposed perpendicularly to the cutting movement of the grinding wheel.

3.1.6 Honing

Honing is, like grinding, a cutting process with abrasive grains. Compared to grinding, honing offers basic advantages such as longitudinal form tolerance, avoidance of short/wave concentricity tolerance, and higher bearing ratio of the honed surface. The motion of a honing tool with corresponding cutting conditions is shown in Fig.3.3.

The main field of application of honing is internal finishing of drilled holes, or in PM production of holes formed by standard die compaction. Accordingly, the rotating honing tools have radial honing stones which are either hydraulically or mechanically pressed on the machined surface during the honing process.

In general, the honing of powder metallurgy and cast and ingot metallurgy materials does not differ. However, the honing of porous

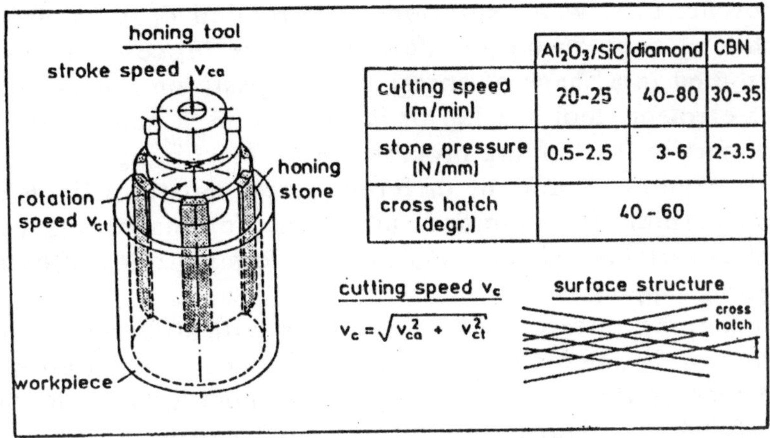

Fig.3.3 Motion of a honing tool [108].

PM materials proves to be problematic because, as in grinding, the pores are clogged by removed grinding wheel and workpiece particles. Thus a separate cleaning process is normally necessary, *e.g.* ultrasonic cleaning [108].

3.1.7 Lapping

Lapping is defined as machining with loose grain mixed into a paste or liquid (lapping mixture), which is fed to the lapping process on a shape lapping tool. With regard to porous PM materials, it is identical to grinding and honing. By means of lapping, almost all PM materials can be machined. High surface quality can be reached, Fig.3.4.

The most influential factors concerning material removal and achievable material removal rate are lapping wheel speed, lapping pressure, the material of the lapping tool and grain size of the lapping mixture as well as concentration and viscosity of the lapping liquid. Abrasive material after mixing with a carrier

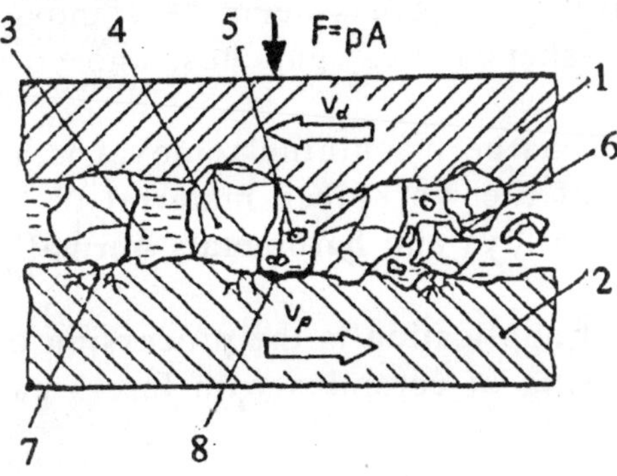

Fig.3.4 Diagram of the lapping process. 1 – lapping tool, 2 – workpiece, 3 – liquid, 4 – abrasive micrograin, 5 – chipped stook, 6 – cracking of an abrasive micrograin, 7 – microcracks, 8 – stock sliced away. *F* – pressure between a lap and a workpiece, *p* – unit pressure, A – nominal contact surface, v_d – lapping tool speed, v_p – workpiece speed [111].

(kerosene, oils, water solutions) is supplied drop by drop to the machining zone during the duration of the process. It can be also distributed in a shape of abrasive micropaste on the surface of the other element (tool – a lap or workpiece as a mating part) which takes part in the lapping process before machining.

For lapping, grains of different sizes, mostly silicon carbide, boron carbide and diamond, are used. For machining requiring polished surfaces, which is the case for forming tools of hardmetals, polishing lapping is used [108].

In lapping tests of PM Fe–C steels with different portions of pearlite in the microstructure, maximum linear capacities of lapping have been obtained by using a paste with micrograines of 99C green SiC (F500/13) whereas the lowest ones by using a 95A (F1200/3) electrocorundum paste.The sintered steels (hardness ~133 HV1) should be subjected to the preliminary lapping using 99C carbide micropaste whereas the softer materials (~84 HV1) by using 95A electrocorundum paste [111].

Another lapping method is ultrasonic lapping. In this case the grains, which are mixed into lapping paste or liquid, get their workability by the impulsion force of the forming tool vibrating in the ultrasonic field. An advantage of this method is that there is no restriction concerning geometry to be produced.

3.1.8 Other processes

Electrical discharge machining (EDM). In contrast to the previous processes, electricity-conducting PM materials can be machined quite cheaply by electrical discharge machining. For the machining of PM materials, thermal removal processes are a promising alternative compared to conventional cutting processes. Thermal removal processes offer the opportunity to produce complex three-dimensional shapes even with highly alloyed PM steels, hardmetals, and some ceramics, regardless of their hardness and strength. Especially for powder pressing tools, EDM has been the standard procedure for preparing complex-shaped dies, punches, and core rods.

A precondition of electrical discharge machining removal is an electrical minimum conductance of the workpiece material to be machined of 0.01 S/cm. PM metal alloys and hardmetals normally meet these requirements.

At the moment the main field of application is the processing of metallic materials, but the processing of ceramics with lasers has spread in the last years.

Laser beam machining. In thermal removal processes by means of a laser beam machine, light energy is generated in an optical resonator and transferred to the material in form of heat by absorption. In order to achieve a thermal removal, the material to be machined must be heated to the sublimation or evaporation temperature.

At the moment the main field of application is not the processing of metallic materials, but the processing of ceramics.

Laser assisted turning. Provided due account is taken of the strength behaviour of hard and brittle PM materials which is dependent on temperature, it is possible to machine sintered parts with a defined cutting edge. In the case of silicon nitride tools, there are amorphous areas at the grain boundary of the largely crystalline structure which results in softening of the material at a temperature of approximately 1100°C. This permits the heated material to be machined in a turning operation.

Ultrasonic assisted grinding. In ultrasonic grinding, a workpiece oscillation is superimposed perpendicularly to the cutting movement of the grinding wheel. By this, friction between grains and workpiece is reduced, since adhesive processes are diminished by the high frequency used. Ultrasonic grinding is promising concerning future use in some kinds of tribologically used PM materials [108].

3.2 CHARACTERISATION OF MACHINING PROCESSES

Machining is a change of the shape of a workpiece or part through the detachment of material particles by mechanical ways. The effectiveness of machining is preferentially determined by tool wedge and by operating of the processes in the zone of the chip formation.

During machining, the workpiece and the tool are in relative motions to each other which is formed mostly by a cut- and a feed motion, which together form a final cut motion. In relation to the cut the shape of a part, the cutting processes are divided into those as mentioned before, *e.g.* with a) the geometrically defined cutting edge, and b) the geometrically undefined cutting edge.

3.2.1 Motions in cutting of metals

The basis of the cutting process is that during the sequence of operations workpiece and tool perform certain motions relative to each other. The individual motions involved are called:

working motion which is either a reciprocating or rotary motion and may be performed either by the tool or the workpiece,

traverse motion, the motion that exposes new areas of the workpiece to the tool, broadening the cut,

feed motion, the motion that tends to deepen the cut.

The examples of the relative motions at some machining operations are illustrated in the following figures.

In drilling, the rotating drill detaches material within the reach of its diameter during the straight feed motion. The workpiece is stationary. There is no traverse motion; the width of cut is determined by the diameter of the drill as shown in Fig.3.5.

In longitudinal or straight turning, the tool set at a certain depth of cut is fed in the longitudinal direction of the workpiece. With the revolution of the workpiece, the turning tool is adjusted perpendicularly to the rotary motion (the tool is traversed) by a certain amount as shown in Fig.3.6.

In facing, the turning tool is moved perpendicularly to the axis of revolution of the work (the tool is traversed), removing material from the work face as shown in Fig.3.7.

In the milling process, the workpiece is traversed and fed while the cutting tool rotates as shown in Fig.3.8.

The base important planes and terms related to the workpiece are illustrated in Fig.3.9.

3.2.2 Cutting tool characteristics

The tool in the region of the cutting edge is heated by the energy generated in plastic deformation of the workpiece material near the

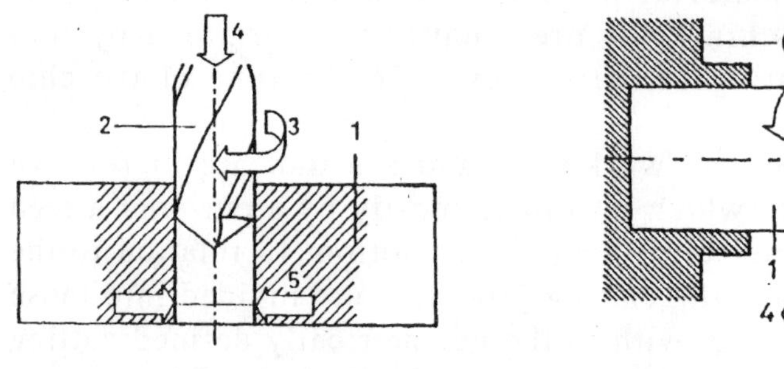

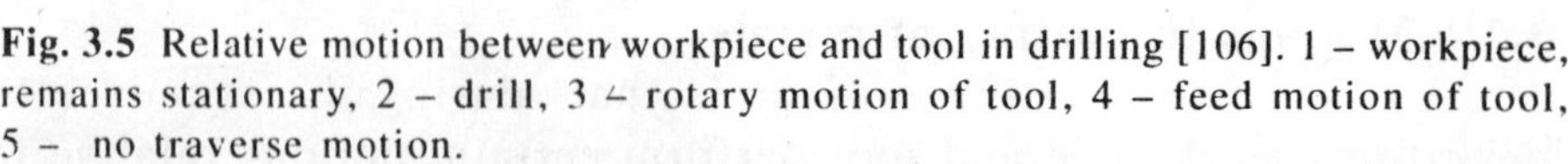

Fig. 3.5 Relative motion between workpiece and tool in drilling [106]. 1 – workpiece, remains stationary, 2 – drill, 3 – rotary motion of tool, 4 – feed motion of tool, 5 – no traverse motion.

Fig.3.6 (right) Relative motion between workpiece and tool in straight turning [106]. 1 – workpiece, 2 –turning tool, 3 – rotary motion of workpiece, 4 – traverse motion of tool, 5 – feed motion of tool, 6 – chuck.

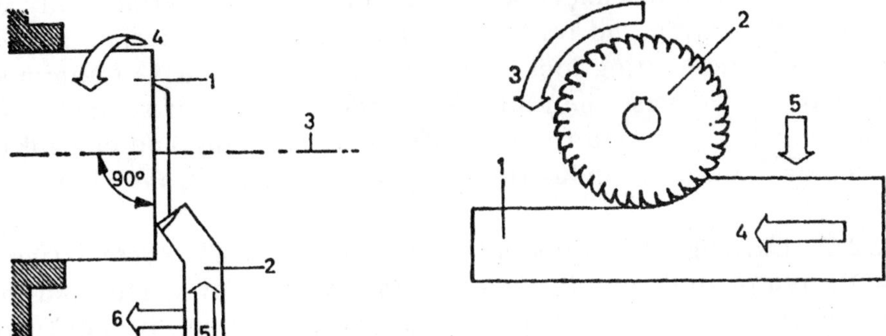

Fig.3.7 Relative motion between workpiece and tool in facing operation [106]. 1– workpiece, 2 – turning tool, 3 – axis of rotation, 4 – rotary motion of workpiece, 5 – traverse motion of tool, 6 – feed motion of tool.

Fig.3.8 (right) Relative motion between workpiece and tool in milling [106]: 1 – workpiece, 2 – milling cutter, 3 – rotary motion of tool, 4 –traverse motion of workpiece, 5 – effect of feed motion of workpiece.

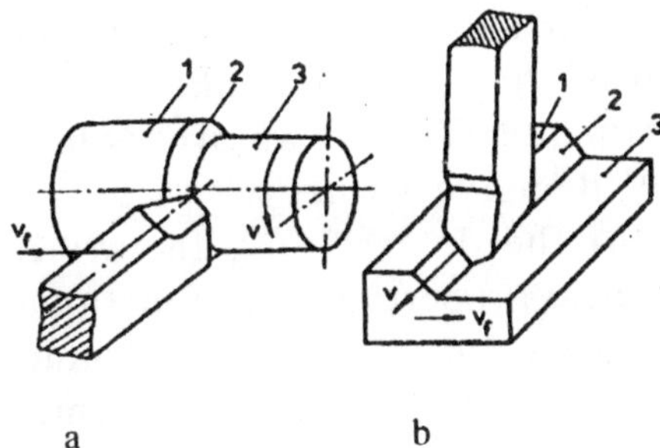

Fig.3.9 Important planes on the workpiece [112]. a – cylindrical work, b – angular work; 1 – machined face, 2 – cutting face, 3 – finished face.

tool/workpiece interface. Temperature and temperature distribution near the tool edge largely determine limits to the rate of metal removal, the properties required of the tool materials, and the tool wear mechanisms. Temperatures generated in bands of thermo-plastic shear have particular significance for machinability. In machining, the amount of plastic deformation near the tool/work interface is very high and the imposed compressive stress is also very high [113].

Most metal cutting tools are metallic. With these, very strong metallic bonds at the tool/workpiece interface would be expected, especially when cutting contact times are long, *e.g.* greater than 1 s (as, for example, for noninterrupted turning), and the speed is not very low, *e.g.* not less than 1 m/min (supposed as limits). When using hard non-metallic refractory materials such as alumina, sialon, diamond, and cubic boron nitride, the tool/work interface is likely to be much weaker than with metallic tools but not too weak to influence movement at the interface. Movement of workpiece material over the tool under seizure conditions can continue only by plastic deformation and fracture of the workpiece material. To this service state the cutting tool characteristics are adapted to minimise the wear and increase the tool life. Understanding of seizure is of fundamental importance for scientific understanding and control of the major features of industrial metal cutting operations [113].

3.2.2.1 *Cutting edge nomenclature and base characteristics*

The design of tools involves an immense variety of shapes and the full nomenclature and specifications are very complex. It is difficult to appreciate the action of many types of tools without actually observing or, preferably, using them. The performance of cutting tools is very dependent on their precise shape. In most cases there are critical features or dimensions which must be accurately formed for efficient cutting. This may be, for example, the clearance angle, the radius of the nose and its blending into the faces, or the sharpness of the cutting edge. The importance of precision in tool making, whether in the tooling room of the user, or in the factory of the tool maker, cannot be overestimated. This is an area where excellence in craftsmanship is still of decisive value.

The cutting tool is active as one of the determining factors in machining of materials with given specific properties. The functional element of a cutting tool is the cutting wedge. The part in contact with the workpiece is the cutting part of a tool which is formed by the cutting edge. The frontal and back face are composed from one, two or more cutting edges, as shown in Fig.3.10.

The angles of a cutting tool, elements of tool geometry, are

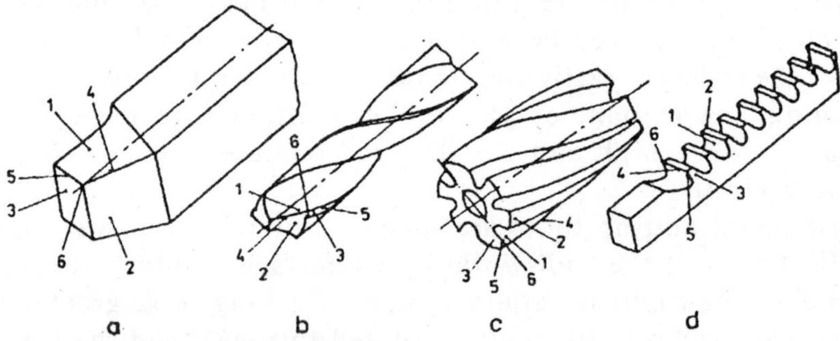

Fig.3.10 Important faces and edges of various cutting tools [112]. a – tool for turning, b – twist drill, c – cylindrical cutter, d – broaching tool: 1 – top edge, 2 – main back edge, land clearance for drill, 3 – auxiliary back edge, 4 – top cutting edge, 5 – auxiliary cutting edge, 6 – tool tip.

important for the definition and description of the cutting process. These angles are determined in two ordinate systems:

- tool ordinate system (static) is used for determination of the tool geometry at design, manufacturing, and control,
- tool cutting angles are used for determination of the tool geometry,
- working ordinate system (kinematic) is used for the determination of the tool geometry during the cutting process – working cutting angles.

By adjusting the tool angles it is possible to affect the cutting process to a great extent.

The angles have to be changed in dependence on mechanical and physical properties of the workpiece, the stiffness of the machine, the speed of singular motions, the depth of the cut, and other factors [112].

It follows that the wedge is the basic shape of any cutting tool, whereas the specific shape of the wedge is dependent on the intended purpose.

While the force applied to it remains constant, the cutting action of the wedge is dependent on several properties, firstly, on the value of the wedge angle, secondly, on the width of the wedge, force applied resolution and angles at the cutting edge are shown in Fig.3.11.

Experiences and studies of the cutting operation have enabled the determination of the most suitable cutting edges for the various purposes. There are certain angles at a cutting tool which determine the efficiency of the tool. The most important angles at a cutting

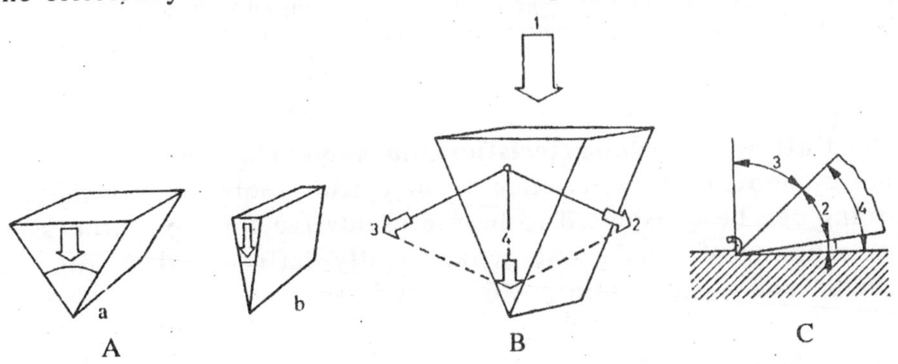

Fig.3.11 A – Large (a) and small (b) wedge angles in a turning tool. B – Resolution of forces setting on a wedge: 1 – force applied to the big end on the wedge, 2 – right face force, 3 – left face force, 4 – feed force (F). C – angles at the cutting edge: 1 – clearance angle α, 2 – wedge angle β, 3 – top rake angle γ, 4 – cutting angle δ [106].

tool are the following four angles which are denoted: α (alfa), β (beta), γ (gamma), δ (delta).

Due to the clearance angle α, only the cutting edge of the tool contacts the (freshly prepared) surface of the workpiece. Thus, friction is reduced and an additional rise in temperature in cutting is avoided. The wedge angle β determines the resisting force of the cutting edge. The tool life is in some relation to the wedge angle. Grossly speaking: The larger the wedge angle, the longer the tool life. This is the period for which a sharpened tool can be used without interruption until it is blunt. The wedge angle is limited by the top wedge, also known as the true rake, and the lower face of the wedge, known as the flank.

The mentioned four cutting tool angles are involved in any of the various cutting methods as shown for three cutting edges in Fig. 3.12. The force applied to the thick end of the wedge is resolved into the face forces. The resolution of the drilling force in cutting, action and feed motion in drilling is illustrated in Fig. 3.13.

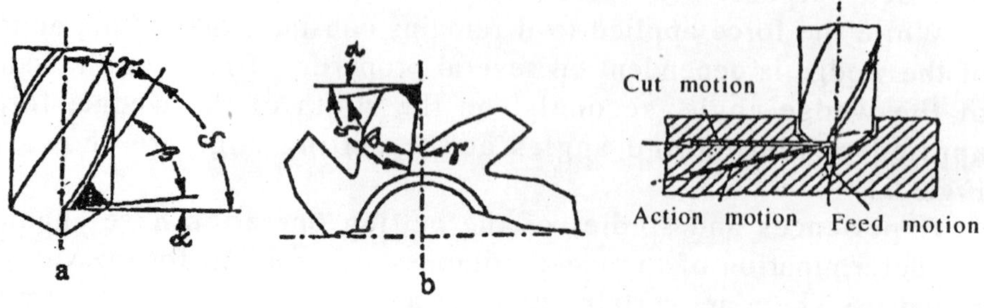

Fig.3.12 Angles at the cutting edge [106] a – in drilling, b – in turning.
Fig.3.13 (right) Resolution of cutting forces in drilling [110].

3.2.3 Cutting tool characteristics and nomenclature

The geometrical characteristics of the cutting tools for drilling and turning can be considered to be representative for most processes in machining of metals with geometrically defined cutting edge in general, and especially in powder metallurgy.

3.2.3.1 Drilling tools

For drilling metals, twist drills are primarily used. The typical feature of a twist drill are the twisted flutes. The flutes are designed to discharge the chips from the workpiece. The nomenclature and geometrical characteristics of a twist drill are

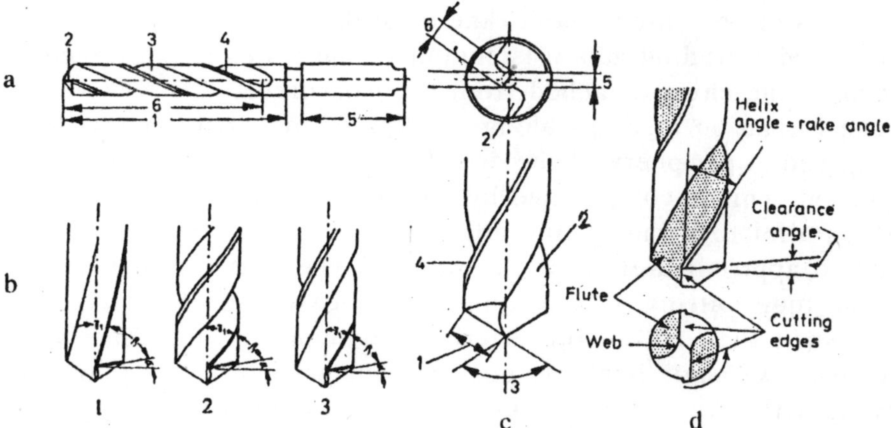

Fig.3.14 a – Twist drill nomenclature: 1 – flute length, 2 – lip edge, 3 – flute (twisted), 4 – margin, 5 – shank, 6 – drilling depth; b – rake of helix angle of the flute: 1 – for hard brittle materials, 2 – for soft materials, 3 – standard; c – terms used to denote the parts at the drill edge: 1 – cutting lip, 2 – flute (twisted), 3 – point angle, 4 – margin, 5 – core diameter, 6 – chisel edge; d – nomenclature of angles in the cutting edge of drill [106,114].

illustrated in Fig.3.14.

Explanation of the relevant nomenclature term for a drill is [115]:

- drill point – the cutting edge of the drill,
- flutes – the grooves that are formed in the body of the drill to allow fluids to reach the point and remove chips from the workpiece,
- land – the remainder of the outside of the drill body after the flutes are cut,
- helix angle – the angle that the leading edge of the land makes with the drill axis,
- thrusting (thrust) force – the force that arises from the cutting action along the lips and the extrusion and cutting forces acting at the chisel edge,
- web – the external portion of the drill body that connects the lands,
- lips – the angle between these denotes the angle of the drill,
- heel – the inside portion of the land facing the drill body.

A twist drill has two cutting regions, the cutting edges (or lips), which cut, and the chisel edge which partially cuts but also rub extrudes, and indents. Twist drills are made with diff angles which should be selected accordi properties to be drilled. Due to the twist drill design, however, the cutting and extruding action is complex. The flutes are designed to discharge the chips detached from the workpiece.

The formation of chips along the lips of the drill occurs along the lip and at periphery of the drill for which a positive rake angle is present. This is complicated by the fact that these lips are offset (*i.e.*. parallel to a radial line ahead of the centre) by an amount equal to approximately half the web thickness at the drill point. The oblique cutting action tends to increase the effective rake angle such that, at the intersection between lip and chisel, the effective rake angle is highly positive; geometrically this is true of a twist drill. An extremely complex metal removal mechanism occurs under the web. [116].

Generally, a large rake angle is required for soft materials such as light alloys, whereas a small rake angle is required for hard and brittle metals, Fig.3.14b. The drills with the appropriate helix angle should be selected when machining large number of the same workpiece in term of effectiveness. The cutting edge on the feed side, also known as chisel edge, of twist drills must be sharpened in such a way that high cutting capacity is ensured. The chisel edge angle ψ (psi) can be too large (>55°) or to small (<55°) and the chisel edge can be reduced to 1 mm, Fig.3.15. A chisel angle of 55° is considered as correct.

For the point angle, various values have been determined in practice to ensure economical drilling for any given purpose.

When grinding by means of the twist-drill grinder or by hand, the point angle (for steel from 116 to 118°) should be measured. Experience has shown that the lip relief angle α should be 6°.

If the chisel edge angle ψ is higher than 55°, the lip relief angle α will be higher than 6°. The backing-off clearance becomes too large. As a consequence, the drill will no longer cut but squeeze the material.

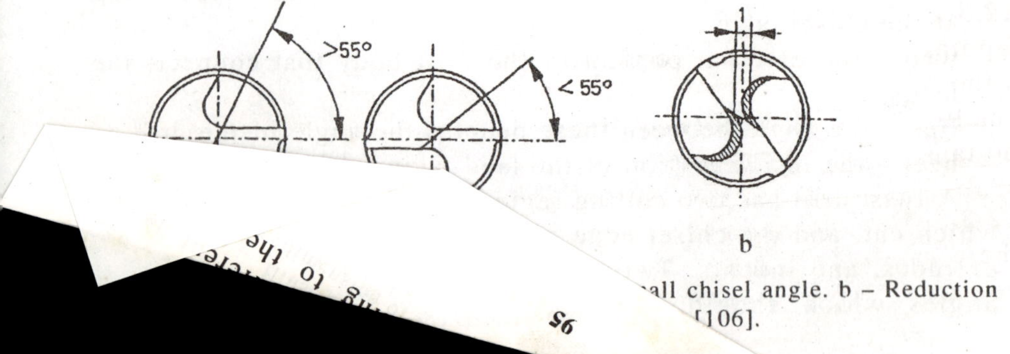

...all chisel angle. b – Reduction ... [106].

The most unfavourable cut of the chisel edge will be obtained in this case due to the fact that it is located at the centre of the drill. A large portion of the power of the feed motion is absorbed by the chisel edge. If the chisel edge width is reduced, this power requirements can be cut down to a reasonable limit.

At the same time the wedge angle β will become too small. The cutting lips may break out and the drill easily stick in the material.

The selection of the drill spindle speeds is governed by the general rule that soft materials, *e.g.* aluminium and its alloys, are to be drilled at higher speeds, whereas hard materials, *e.g.* steel, are to be drilled at lower speeds. The spindle speed is also governed by the material of which the drill is made. It is not permissible to operate drills of plain tool steel at the same high speeds as drills of high-speed steels since overheating and thermal softening will occur rapidly, causing blunting of the cutting edges.

An essential feature of drilling is the variation in cutting speed along the cutting edge. The speed is maximum at the periphery, which generates the cylindrical surface, and approaches zero near the centre-line of the drill, the *web,* where the cutting edge is blended to a chisel shape, and the action is no longer that of a cutting tool in the same sense. This variation in speed along the edge is responsible for many aspects of drilling which are peculiar to this operation [117].

It is necessary to stress that the deeper the drill penetrates into the material, the higher the friction at the wall of the hole will be.

Counterboring. The re-boring of pre-drilled holes, the spot-facing of plane for screw heads and similar operations are called counterboring or countersinking, Fig. 3.16. The speeds for counter-boring are little lower than those used for drilling. Counterborers usually have a larger number of cutting lips than drills. The counterborer must be carefully applied to avoid damage to the cutting edges.

3.2.3.2 Tapping

Tapping is very closely related to drilling. A characteristic feature is the large number of cutting edges in operation at the same time ('multi-point tool'). The nomenclature of the base characteristics of the taps is shown in Fig.3.17.

3.2.3.3 Turning

There are turning tools of different shapes which are a[illegible]
the various kinds of work. Although the shapes of [illegible]

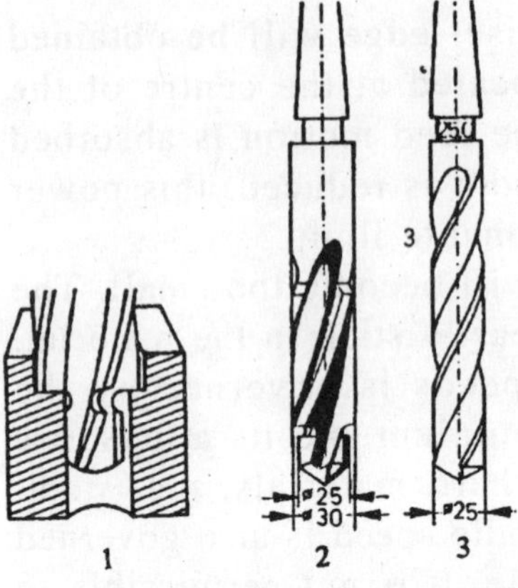

Fig.3.16 1 – Counterboring. 2 – Counterboring tool. 3 – Drilling tool [106].

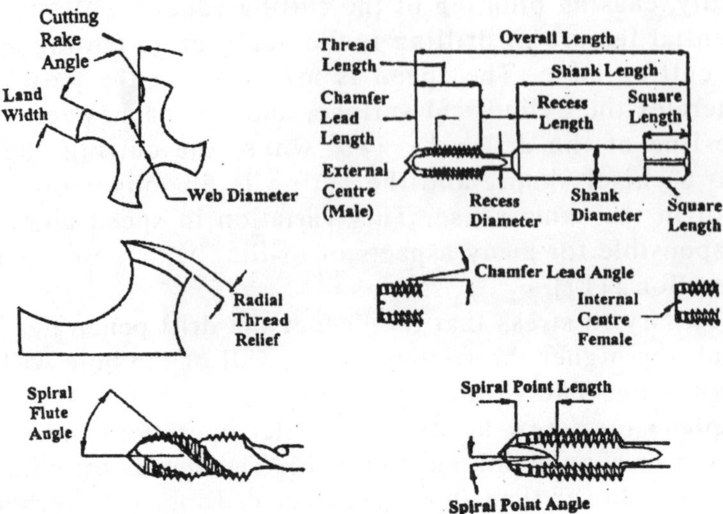

Fig.3.17 Defining characteristics of the taps [118].

all of them are subject to the same geometry of cutting edges. The base cutting tool nomenclature is shown in Fig.3.18.

The cutting properties of the tool are primarily determined by the clearance (lip relief) angle α, the lip angle or wedge angle β, the ...le γ and the cutting angle δ, Fig.3.19.

... at a turning tool should be just large ... friction between tool and workpiece. ... a size of anything between 5° ... between the clearance angle ... increases, the clearance

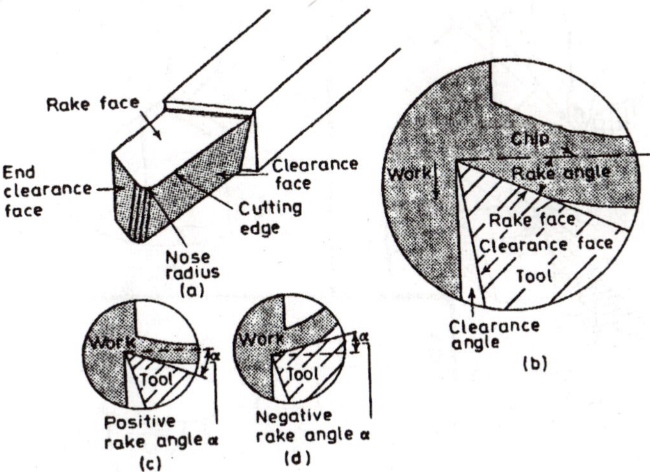

Fig.3.18 Nomenclature for cutting tool in turning [117].

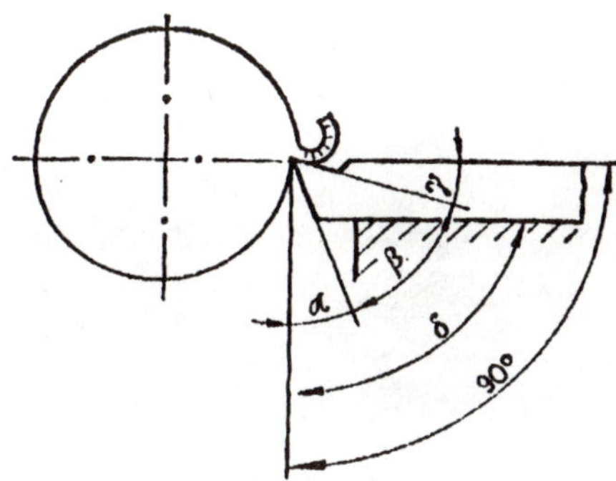

Fig.3.19 Angles at a turning tool [106].

angle decreases provided the lip angle is constant. As a rule, the rake angle γ should be made large enough to enable the chips to be peeled off easily. In general, the value of rake angles is within the range of 8° to 20°. The size of the lip angle β is dependent on the workpiece material.

Further angles which largely determine the flow of the chips, the life of the turning tool, and the power required in machining are the side angle λ (0, negative or positive), the nose angle ε (epsilon) and the side angle κ (kappa) (betwen the cutting tool edge and the generating line of the workpiece), Fig.3.20. The nose angle should be anything between 90° and 110°. The side rake angle γ which is formed by the horizontal plane and the cutting edge is equal 0° in thc case when the turning tool runs parallel to the horizontal plane.

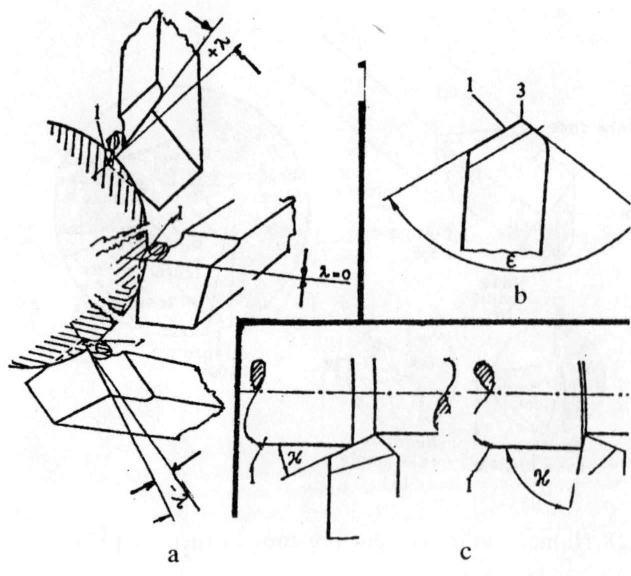

Fig.3.20 Characteristics of: a – λ, b – ε (1–leading cutting edge, 2–trailing cutting edge, 3–nose), c – κ angle of the turning tool [106].

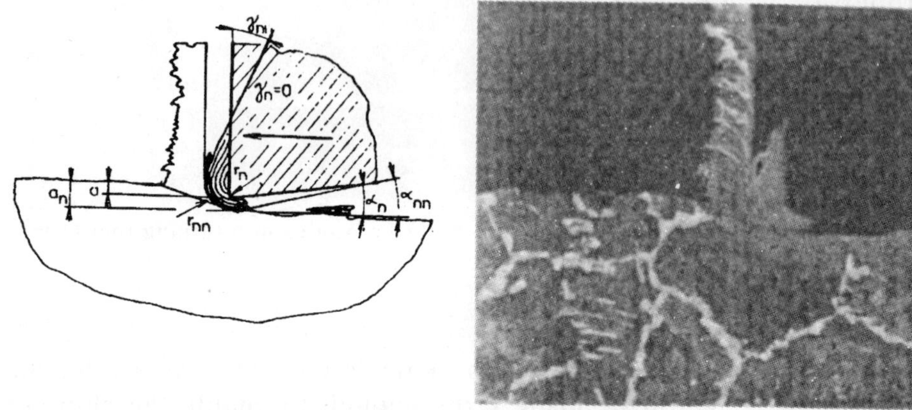

Fig.3.21 Schematic of cutting process with a built-up edge [112].
Fig.3.22 (right) Section of the base of a chip formed as in Fig.3.21 showing the beginning of the formation of the machined margin layer. Optical micrograph. Nital etched. (Material: wrought steel C45, ASTM 1045 – annealed, HM P20 tool, v_c = 40 m/min, feed 0.1 mm, γ_a = 0°) [112].

3.3 ANALYSIS OF THE MACHINING PROCESS

The cutting process is generally illustrated by considering metal flow toward and over a simple cutting tool in a turning operation which is prevailing in base analysis as shown in Figs.3.21 and 3.22.

During cutting, tool movement into the workpiece generates a complex stress field around the tool tip. These stresses initiate cracks in the workpiece ahead of the tool tip. The cracks grow and join to separate the metal to be removed from the bulk of the workpiece. The layer of metal deforms, flows across the tool face, and is removed from the cutting area. Ideally, the separate metal fractures so that it is removed easily from the cutting area as 'chips'. From this simplified description, it is apparent that the workpiece, cutting tool, and chips undergo high stresses and strain rates during the cutting process. These conditions, plus friction between the tool and chip, cause undesirable local heating. In extreme cases, the heat can cause welding of chips to the tool face, local plastic deformation, or even melting of the tool or workpiece surface.

In order to select the right tool and machining parameters, knowledge of the loads on the tool material and the properties of the tool material is necessary, together with an analysis of the wear mechanisms, and with knowledge of the workpiece material properties, which should be characterized in more detail for the machining process, and their interaction. Typically, the loads on the edge of a cutting tool are different at various locations. The primary input factors are feed rate (pressure) and cutting speed (velocity).

Understanding of the metal cutting process involves prediction of the behavior of various types of metals as they are formed into chips. Partly, this means predicting the effect of deformation, temperature, and mechanical forces, as these play a dominant role in the quality of machining operations. Temperature affects the cutting process, and high temperatures will negatively change the cutting material. Cutting forces affect the power needed to perfom the operation. Designing an accurate cutting edge means considering temperature, cutting forces, and chip formation under the given machining conditions. The effects of the process on tool life and stability of the edge are important factors in the design of a cutting geometry.

3.3.1 Cutting process

When cutting material to an edge as occurs, *e.g.*, during turning, the tool deforms some of the workpiece material which then separates as a chip. Large stresses build up as the layer of which is to become the chip approaches the cutting edge. Elastic and plastic deformation of the metal occurs as the cutting forces reach the yield strength of the material. The basis dynamics of the cutting process (*i.e.* chip formation) is illustrated in Fig.3.23.

The boundary between the chip and the workpiece, the zone which separates the deformed and undeformed metals, is called the shear plane (SP). This has an angle to the workpiece called the shear plane angle (∅). The metal to the right of the plane is the deformed chip, with thickness (h_2), and the metal to the left is the undeformed chip, thickness (h_1).

As the strength of the materials increases, the workpiece becomes more resistant to the shear deformation caused by the tool action and thus the friction force increases at the chip–tool interface. In the drilling thrust force test, the tool revolution and penetration rates are constant. Therefore an increment in shear and friction forces results in an increase of the torque and the thrust force applied to the workpiece. These forces are illustrated in Figs. 3.11 and 3.24, which isolates a cross-section of the drill face. The principal forces acting on the chip are the shear force causing the metal deformation and the friction force created at the chip–tool interface. Torque and thrust which are the forces through the drill affects the workpiece through the shear and wear forces.

3.3.2 Cutting forces

Metal cutting requires a lot of power to separate chips from the workpiece. An understanding of cutting forces will lead to well balanced cutting edges through positive cutting actions and good cutting edge strength.

There is a relationship between the power needed for the cutting process and the cutting forces involved. Seen from an orthogonal point of view, a state of equilibrium exists with the forces involved and in relation to the shear plane. The forces at work on the

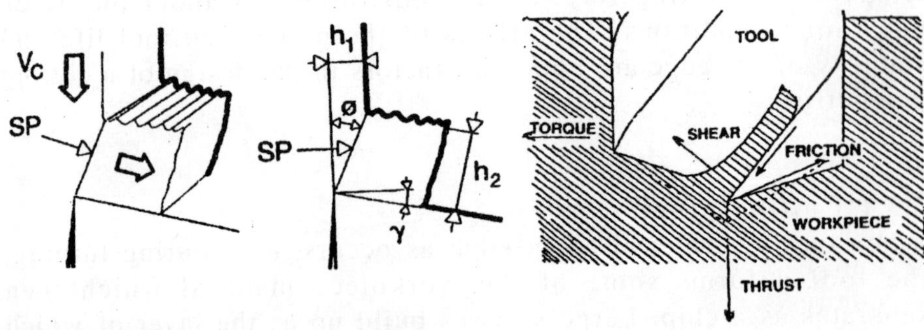

Fig.3.23 Fundamentals of cutting process: shear plane (SP) and chip-forming factors [118].

Fig.3.24 (right) Principal forces developed during the drilling operation [119].

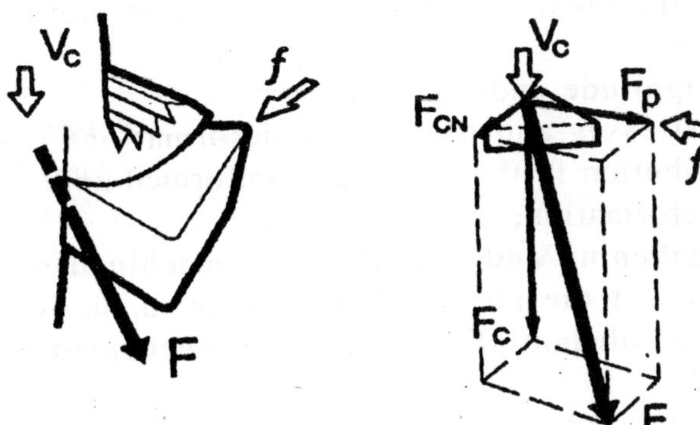

Fig.3.25 Forces at work to produce cutting during machine turning [118].

workpiece and chip, along the shear plane, and between the tool face and chip are, in principle, equal.

Seen in three dimensions, the cutting forces can be divided into three components: tangential or main force (F_C), radial or positive force (F_{CN}) and an axial or feed force (F_P) in a dynamic equilibrium as shown in Fig.3.25.

The main force is to a great extent dependent upon not only the contact and friction between the workpiece and tool, but also the condition of contact between chip and rake face of the cutting edge. The quality of the actual chip formation and breaking affects the main force considerably. There is also a direct relationship between the undeformed chip thickness (h_1) and the magnitude of this force, Fig.3.23. For most workpiece materials, increasing cutting speed leads to lower cutting forces (higher temperature, softer material). The decrease in forces varies with material and range of cutting speeds in question.

For determination of the cutting forces it is necessary to take into account also the friction forces at the front surface of the cutting tool. The friction forces are basically affected by the properties of the tribosystem involving the machined workpiece and the tool material under the cutting conditions applied [120].

3.3.3 Formation of a chip

The chip, which becomes a waste, is an unavoidable product of the machining of a material. The knowledge concerning the mechanism and force conditions of the chip formation can be useful for adequate machining of a material since the chip shape and its properties result from the workpiece material properties with their changes during the cutting and from the cutting conditions including the cutting tool material. The formation of a chip has a complicated and specific character of the load, at which in a short time interval on the deformed material act:

- stresses of various magnitude and direction,
- high level of acting stresses and of relative deformations and their nonuniform distribution in the plastically deformed zone,
- high velocity of the deformation,
- development of strengthening and failure of the machined material as a consequence of the intense heat release in the cutting zone, which has a substantial effect on the mechanism of deformation [121,122].

Chip deformation is related mainly to three factors: a) the thickness of the undeformed chip, b) the rake angle (γ) between the chip face and the workpiece surface, and c) the mechanical and structure properties of the workpiece material. These factors also affect the shear plane angle during the cutting process. Thus the principal cutting action occurs at the shear plane, determined to a large extent by the cutting ratio between the undeformed and deformed chip thickness which results in a longer chip what is not the common case in PM machining.

The electrical and magnetic properties, thermal conductivity, and other properties of the material are changed and by this the workpiece material becomes physico–chemically more active in the change of the properties in the cutting process. The machined material in this state comes into contact with the working surfaces of the cutting wedge of the tool.

From this reason the process of the chip formation at machining belongs to the complicated processes of heterogeneous deformation and destruction processes. The tool acts on the workpiece by the forces which can be characterized as the force of a normal pressure of the chip on the front of the tool F_a and the friction force F_t, which acts on the contact face of the tool with the generating chip, Fig.3.26.

The last one acts on the contact face of the tool with the chip being formed. The chips formed at cutting of plastic materials have a sectional character with various degrees of sectioning. This process surely is related to the heterogeneity of plastic deformation processes, which are characterized by the instability zones. The force relationships immediately depend on the mechanism of the chip formation.

The course of plastic deformation affects the thermoeffect at machining, contact length of the chip with the tool, and the course of tool wear. From the begining of the machining process, the

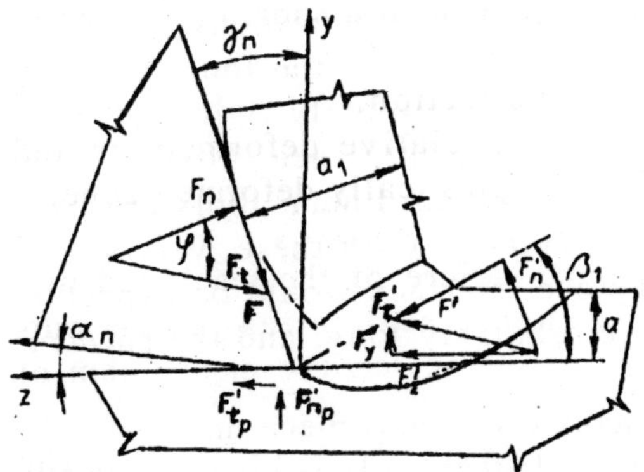

Fig.3.26 Relationships between the forces in the zone of chip formation [123].

structure of the machined material and its mechanical and physical properties are the leading fundamental factors with the consequences on the effects occuring in the cutting zone of the formation of the chip.

During cutting a stagnation zone follows the tip of the edge. Softened metal protects the tool by sticking or sliding on the surface. A flow zone takes over after the shear plane and is visible at the division of undeformed and deformed material. Thus the principal cutting action occurs at the shear plane, determined to a large extent by the cutting ratio between undeformed and deformed chip thickness.

3.3.4 Plasticity during turning

In machining, the amount of plastic deformation near the tool–workpiece interface is very high. The newly generated surface is clean and the imposed compressive stress is high. Regarding high temperature attained in the cutting zone over a large part of the interface, the workpiece material and the tool are interlocked and atomically bonded to such extent that normal sliding, as observed in boundary lubrication conditions, cannot occur. This condition is also termed 'seizure' in machining [113].

The normal sliding of workpiece material over the tool surface could not have occurred because of continuous contact in both hills and valleys of the tool surface. In most machining operations with steel or hardmetal tools, interlocking and atomic bonding preclude the possibility of sliding of workpiece material over a large part of the interface on cutting tools.

It is the reason for the use of addition of different solid machining aids to sintered material with the aim to suppress the seizure in PM machining and at least partly to convert the material flow over the cutting edge to sliding conditions.

The second partner in the cutting process is the tool. The tool is made from a material selected to be completely rigid under the stress and temperature conditions at the interface. Most cutting tools are metallic. High speed steel and hardmetal tools are used in more than 90% of all industrial cutting. With these, very strong metallic bonds at the tool–workpiece interface would be expected, especially when cutting times are relatively long, and the speed is not very low.

After cutting steel, iron and many other workpiece materials, the contact area on cemented carbide and steel tools is normally partly or completely concealed by adhering workpiece material (diffusion, high temperature). A layer of steel adheres strongly to both rake and clearance faces of the tool. Atomic bonding at the tool (hard non-metallic refractory materials)–workpiece interface is likely to be much weaker than with metallic tools.

With ceramic tools or tools with chemical vapour-deposited coatings discrete layers of workpiece material are not found usually. With these tool materials, however, the absence of adhering material on the tool does not usually indicate that seizure has given way to sliding at most of the interface [113].

Most of the energy needed in the metal turning process, which is more in detail analyzed, is expended at the shear plane; a concentrated shear occurs when the workpiece metal is forced against the cutting edge. There is a flow along the face of the tool–flow lines appear behind the deformed chip, after the shear plane, and the surface turns rough due to the varying strains in the metal [118].

The plastic behaviour of the metal through the shear plane is influential in the process since it affects the strain hardening of the chip and the cut surface. The metal is plastically deformed at a high temperature through 'hot working'. Structural change and work hardening are the main results. Work hardening increases the cutting force, reducing the shear angle, and produces a thicker flow over the rake face of the insert. The degree of deformation depends considerably upon the size of the rake angle of the tool. When the rake angle is small, the shearing force is high. The size of this angle and the area of the shear plane are thus influential to cutting performance. In practice, factors such as the rake angle and cutting data also affect the conditions of the shear.

Figure 3.27 shows characteristic plastic deformation of the formed chip and of the new machined surface of a wrought steel

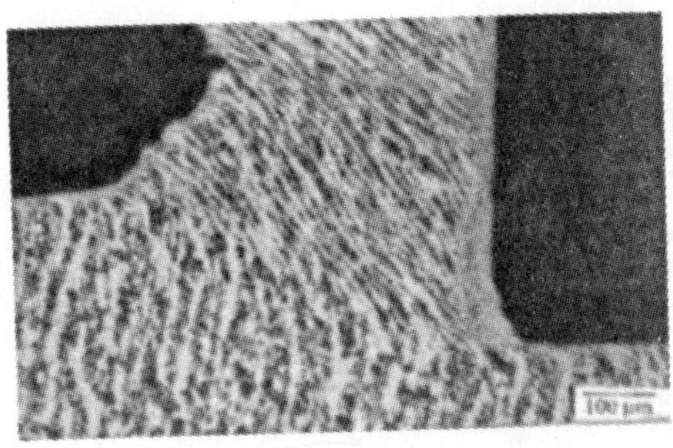

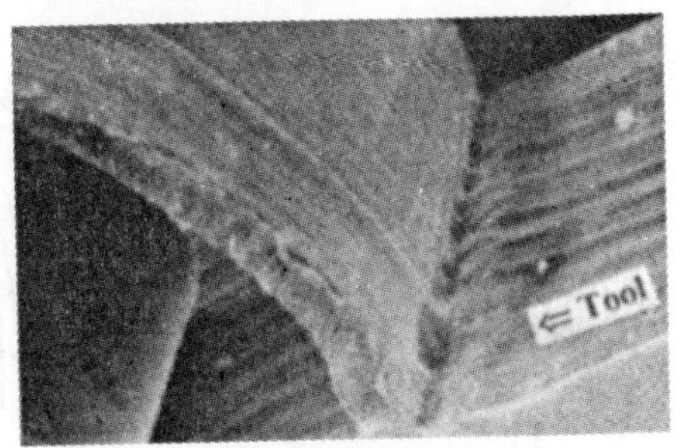

Fig.3.27 Characteristic plastic deformation of the chip formed in turning of wrought steel (C45). Optical micrograph. Nital etched (Courtesy of K. Vasilko).
Fig.3.28 (right) View of the formation of a chip in turning of a wrought steel workpiece (C45). SEM. (Courtesy of K. Vasilko).

and Fig.3.28 plastically the formation of a chip.

3.3.5 Heat factors and cutting process

The plastic deformation and friction between the workpiece and the tool generate heat and result in formation of thermal fields. These affect the cutting process and the cutting tool life. For the heat generation a higher amount of power is required. From this reason a more detailed view on the temperature state in the contact zone is required. Almost all energy used for plastic deformation of the workpiece and friction (up to 95%) at machining changes into heat. The heating of the cutting tool deteriorates its mechanical and cutting properties.

In the cutting zone there are three sources of heat:

- primary plastic deformation Q_1,
- secondary plastic deformation Q_2,
- friction between the back side of the workpiece and the tool Q_3.

These heat sources are shown in Fig.3.29. The heat of primary plastic deformation, which is the main source of heat at cutting, spreads into the cutting tool and the workpiece. The intensity of heat extension into the workpiece depends mainly on the cutting speed. The maximum heat is absorbed mainly by the chip, lesser by the workpiece and the tool and directly minimum by the envirnoment. With increasing cutting speed, considering the constant rate of heat conductivity in the material, the amount of heat the chip abducts increases and the amount of heat taken away by the workpiece decreases. The temperature of the machined surface of a workpiece is by this cooler at higher cutting speed.

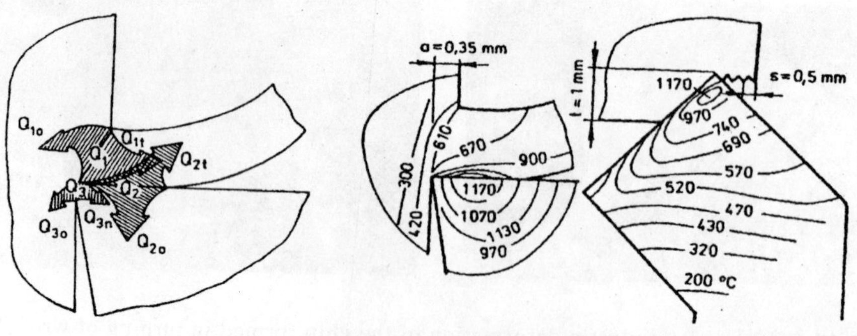

Fig.3.29 Sources of the heat from friction and deformation and their extension into the chip, tool and workpiece [112,123].
Fig.3.30 (right) Temperature field of the tool, chip and of workpiece in turning (wrought steel C45, v_c = 80 m/min, tool HM P20) [112,123].

The maximal effect on the formation of heat has the cutting speed, a lower one the thickness and the lowest the width of the machined layer beside the workpiece and tool material and geometry. Regarding three sources of heat, the temperature in the chip and in the tool is not uniformly distributed, as shown in Fig. 3.30.

It follows from it that the maximum temperature in turning steel is on the tool face in a short distance from the cutting edge. Figure 3.31 shows the temperature fields in the tool formed during machining of steel, titanium, and nickel under such conditions that the temperature in the contact zone between the chip and the tool was equal, namely 650°C.

The reason for this occurrence is apparently the difference in coefficient of friction between the machined and the tool material and the difference in plasticity of these materials under the cutting conditions used. This fact has a substantial effect on the plastic deformation in the cut zone.

The arrested zone in the contact of the chip and cutting face is the largest for nickel and the smallest for titanium. The high coefficient of friction between titanium and tool is the reason that the chip does hardly move on the tool, but leaves upright to the tool face in form of elements. In machining of steel the zone of maximum temperature is shifted from the cutting edge and, from the view of the face, is in the central zone of the contact plane. This results in much more homogeneous thermal loading of the tool, and therefore not even at high cutting speeds thermal failure of the

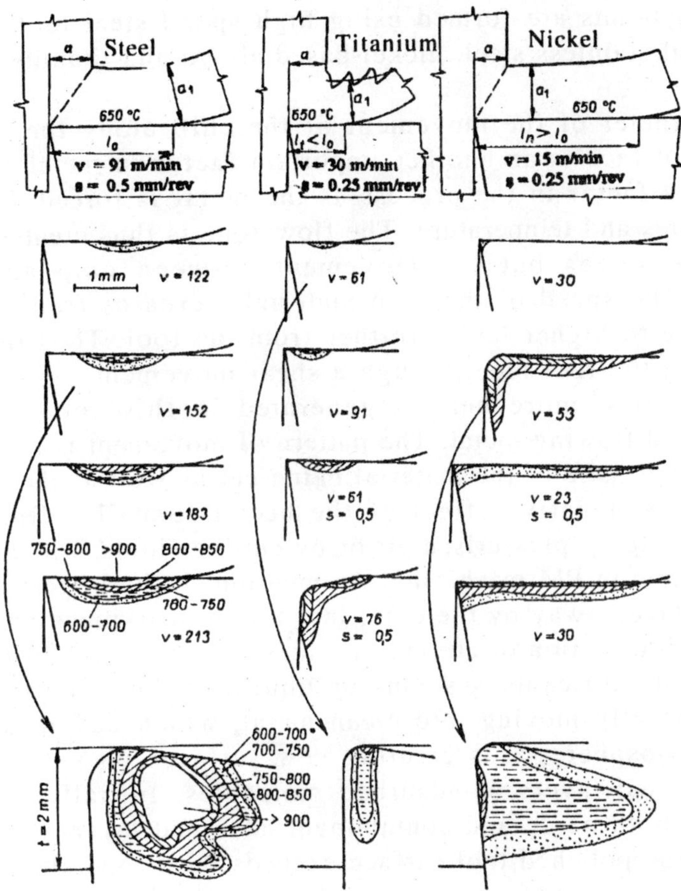

Fig.3.31 Temperature fields in the tool originating from machining of wrought steel, titanium, and nickel [123].

tool occurs. In machining of nickel a reverse situation compared with titanium occurs. The length of the contact between the face and the chip is large, and the zone of the high temperatures is spread along the total length of the contact, but does not reach a large depth. A singularity of this state is that the maximum temperature is reached directly at the cutting edge or in its vicinity (Figs.3.29, 3.30) which causes its plastic chamfering. The cutting tool thus loses its cutting properties. It is possible on this basis to assume that this effect should be considered to some extent for machining of structural PM nickel alloyed steels which is considered as more difficult compared to other PM materials without common contents of nickel. It is confirmed also by experience that very strong bonds are formed using high speed steel tools to cut conventional stainless steel, nickel-based alloys and titanium alloys [113].

The character of the movement of the chip along the contact with the tool face is an another important factor in metal cutting. Friction is a factor in the process as the metal is forced along at high pressures and temperature. The flow zone is thus created when the surface seizes but the movement between chip and tool continues. The speed of the chip material increases from zero at the interface to higher levels further from the tool. Thus the chip moves along the tool face through a shear movement.

Consequently, more heat is generated in this zone of both stationary and flowing metal. The pattern of movement is to a large extent characteristic of the material being cut as well as the cutting data use. The negative effect of the heat is usually eliminated through cooling by pressurised air or by cutting fluids, which is not fully realisable in PM machining. As mentioned, the largest part of the heat is taken away by the chip, less by the workpiece and tool.

Simultaneous action of cutting provides very good conditions for the access of lubricants, gaseous or liquid, to the interface. The tool is constantly moving into clean metal, which has no contact with the atmosphere. It is creating new surfaces on which there are no oxidised layers or adsorbed molecules. Initially the tool surface has both oxide and contaminant adsorbed layers.

Also, if the polished tool surface exceeds the temperature about 500°C during cutting, by reaction with atmospheric oxygen a layer of oxide can be formed on it. If, during the cutting operation, a jet of oxygen is injected towards the cutting edge below, the tool surface is oxidized up to the cutting edge and the oxide layer near the edge is very thick.

Unidirectional movement of the workpiece material tends to sweep such layers from the tool surface. The higher the cutting speed the less the possibility of any external fluid (gas) penetrating the interface, particularly near the cutting edge [113].

Removal of heat by conduction through the chip and through the body of the workpiece is likely to have relatively little effect on the temperature at the tool/workpiece interface, since both chip and workpiece are constantly moving away from the contact area allowing very little time for heat to be conducted from the source. The tool is the only stationary part of the system. It is the tool which is damaged by high temperatures and, therefore, in most cases, cooling is most effective through the tool. This is also the reason why the thermal conductivity of a tool material is a very important property, *e.g.* in high speed steels high thermal conductivity alleviating the need for extreme temper resistance.

3.3.6 Built-up edge

It is the condition of seizure which gives rise to one of the major types of chip formation. The built-up edge, which occurs mainly under industrial cutting conditions, can be formed with either a continuous or a discontinuous chip. The built-up edge is a dynamic structure, being constructed at successive layers greatly hardened under extreme strain conditions.

At certain conditions and materials, successive layers of the flow zone material will build up and harden on the tool face. The flow zone moves up and along with the top of the formed layer, and in this way, a built-up edge (BUE) is formed. Metal is pressure-welded continuously on the rake angle of the tool and eventually becomes unstable. It breaks off at a certain point in the process where upon the build-up of a new layer commences. BUE is a negative factor appearing in various forms and conditions in machining. It can usually be eliminated by altering the conditions of the machining process on which it thrives.

Certain temperature and cutting speed ranges promote the growth of BUE, and certain workpieces and tool materials are more prone than others. Soft metals such as Al and Cu are especially critical, as are low cutting speeds. High speeds soften the layer and replace it with a flow zone. The BUE can also take some of the cutting edge with it when it breaks off, thus blunting the tool. On the other hand, it may itself be hard enough to function as a cutting edge.

Some forms of BUE particularly affect the rake angle and also the chip thickness. Formation of BUE increases with larger rounding of the cutting edge (ER) but decreases with more positive rake angles, *i.e.* 'sharper' tools are less susceptible to the formation of BUE. The strength of BUE decreases with increasing temperature.

Adhesion of the workpiece on the tool contributes to the formation of BUE usually at high pressures on the frontal face of the cutting tool. This state is formed at an increase of cut depth, lowering of the rake angle and of the cutting speed. Some characteristic built-up edges formed at turning of wrought steel are shown in Fig.3.32. At low cutting speed where a built-up edge is formed, this body usually remains adhering to the chip and

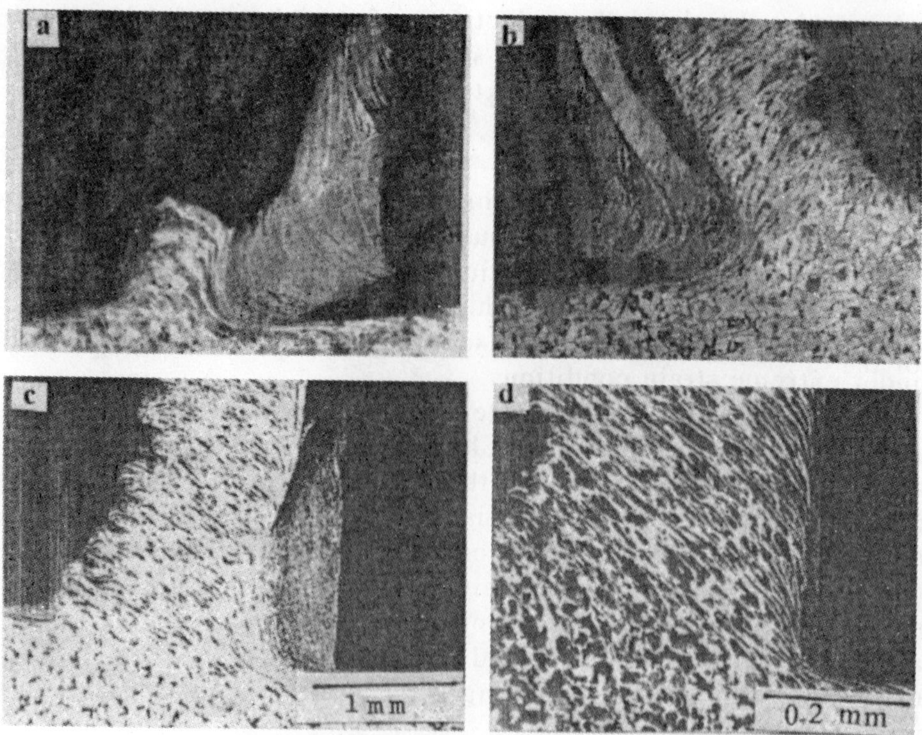

Fig. 3.32 Characteristic built-up sections through quick-stop after cutting steel. a – base form of a built-up edge, b – layered build-up [112], c – build-up after cutting 0.15%C steel at cutting speed 30 m/min, d – built-up edge after cutting 0.4% C wrought steel at cutting speed 180 m/min with CVD coated (Al_2O_3) tool [113].

workpiece material after a quick-stop, an separates almost cleanly from the tool.

3.3.7 Tool wear and tool life

All factors that affect the machining process of materials manifest themselves in final consequence in tool wear, *i.e.* in tool life, and by this in a marked measure in effectiveness of machining. Most of the known data concerning tool wear are relevant for turning or similar cutting operations, not for drilling which is mostly done in machinability testing of PM materials. Also the extent of existing machining data for PM parts is lower than for wrought materials, and therefore some forms of tool wear exist only to a minor extent. A general overview of the types of tool wear can contribute to a more detailed understanding of the relationship between the material machined, the tool material and geometry and machining conditions in PM machining.

Analysing tool wear and taking measures to ensure that there is optimal, balanced wear is important for the quality and performance, and thus the economy, of metal cutting operations. Wear affects how much and in what way the cutting edge deviates from the nominal dimension and the surface finish required.

A balanced, predictable development of wear over a long time is the normal goal. The way in which this development takes place determines the fluctuation of the machined dimensions within the tolerance area and thereby the frequency of positional adjustment of the cutting edge. There are five main wear mechanisms which dominate in metal cutting, as illustrated in Fig.3.33.

The tool concept including all functions in the cutting process and the practical realisation (*i.e.* chip breaker profile, stability and geometry of tool holder, insert style etc.) defines the wear process which can be divided into three main load factors: mechanical, thermal, chemical, and which manifest itself in various wear modes. In the following summary, nine special variants of tool wear are related to the above mechanisms [118]:

1. *Flank wear* takes place at the flank or clearance face of the cutting edge along the length of engagement. On this way it affects the position of the edge and the insert geometry. Its development can be positive in that it makes the edge sharper as machining progresses, however, after a certain amount of wear, increasing friction against the machined surface deteriorates the performance

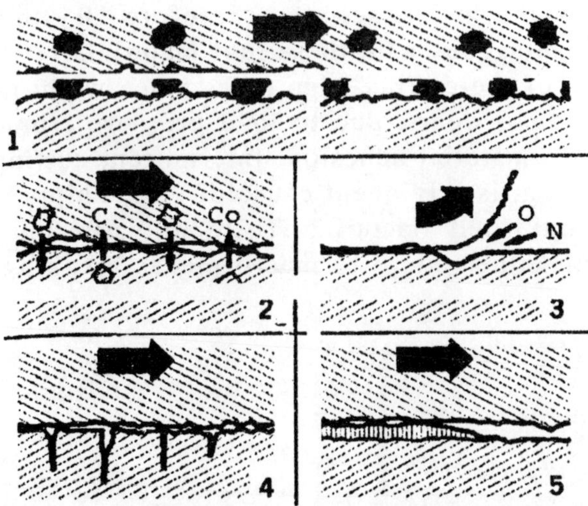

Fig.3.33 Schematic representation of five forms of tool wear during metal cutting [118]. 1 – abrasion, 2 – diffusion, 3 – oxidation, 4 – fatigue, 5 – adhesion.

of the edge. The main cause is the abrasive wear mechanism and the effect grows with higher cutting speeds. For cutting tests, flank wear is a frequently used criterion. A typical example of flank wear is shown in Fig.3.34.

2. *Crater wear* takes place on the chip face, where high temperatures and pressures prevail. Diffusion and abrasion wear mechanism cause tool material to be worn away and if this is allowed to develop excessively, a change in cutting geometry can occur, affecting chip formation and cutting forces, and weakening the cutting edge.

3. *Plastic deformation*, 'thermal softening', can take place as a result of a combination of high temperatures and high pressures on the cutting edge. High cutting speeds and feeds along with hard workpiece materials, or such that show high work hardening, leads to heat and high compression. ´Hot hardness` is a necessary property for the tool to withstand this effect. Once the deformation starts to take place, further deterioration follows as a result of edge blunting, higher loads, and even higher temperature, leading to an escalation of critical effects between the edge and workpiece.

4. *Thermal cracking* is the main type of fatigue wear due to intermittent effect. Esp. with low ductility tool materials, cracks can form on the cutting edge, leading to edge weakness and the risk of rapid breakdown due to failure. The application of cutting fluid is often a negative factor if temperature fluctuations are allowed to amplify during machining. Especially ceramics and cermets are susceptible to thermal cracking; therefore dry machining is frequently advantageous with these tool materials.

5. *Chipping* of the cutting edge takes place when the edge-line breaks, rather than wears, due to load pressure from the cutting action or due to material adhesion. Intermittent cutting, as *e.g.* in milling or sawing, is a frequent cause of edge chipping or uneven breaking, and ultimately fracturing. There are various degrees of this wear type with either micro to macro pieces breaking away from

Fig.3.34 Example of flank wear on a turning tool [124].

the cutting edge. In many cases the tool material type or grade is not suitable for the operation in question, indicating that more strength is required, esp. resistance to fatigue loading. Micro-chipping of a HSS tool at milling is shown in Fig.3.35.

6. *Built-up edge* formation is a wear type occurring largely as a function of temperature and cutting speed interaction. The workpiece plays an important role as does the tool material. Low carbon steels generally have a severe tendency to smear as have higher-carbon steels that have been excessively soft annealed. The surface finish is first to be affected negatively, followed by a change in cutting geometry and then edge breakdown. A typical feature is the emergence of scars in the machined surface that stop abruptly when the BUE breaks away.

7. *Notch wear* on the trailing edge is to a great extent an oxidation wear mechanism occurring where the cutting edge leaves the machined workpiece material in the feed direction. But abrasion and adhesion wear in a combined effect can contribute to the formation of one or several notches. There is a state of tension where the cutting edge leaves the workpiece material and the surface can be deformation-hardened leading to concentrated notch wear. Notch wear has a considerable effect on the surface finish of the workpiece as the notch leaves high peaks and burrs which rapidly exceed the permitted profile height in finishing. Excessively high cutting speeds for the tool material in question, combined with a large entering angle, are typical causes of notch wear. Notch wear may involve the effect of squeezing, also called side-flow. Notches are worn due to concentrated wear, spaced at a distance equal to the feed per revolution, Fig.3.36.

8. *Flaking* involves the coating being damaged at an early stage, usually because of demanding machining conditions and/or inferior coating. Poor adhesion of the coated layer on the grade, as well

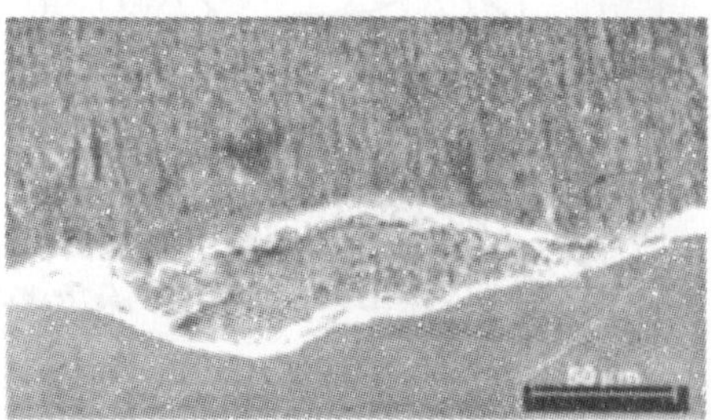

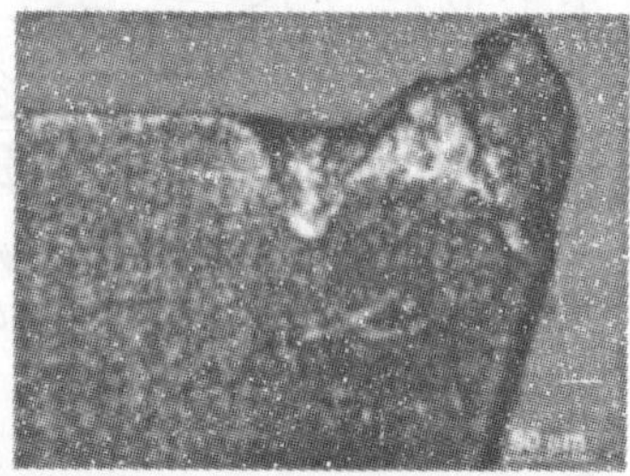

Fig.3.35 Microchipping of a HSS mill cutter [124].
Fig.3.36 (right) Notch wear of a turning tool [124].

as plastic deformation, lead to flaking when the workpiece material is smearing or if the cut is intermittent. If the exposed substrate is more susceptible to wear, this is also a critical factor. The tools with higher flaking resistance are required.

9. *The diffusion wear* in connection with the tool wear mechanism has been investigated to a minor extent. The diffusion wear mechanism has two reasons. The first one is the solution of the components of the cut tool in the workpiece (*e.g.* the solution of C and W) in the iron and the second one the solution of workpiece components in the cut material. Examples for hardmetal tool (WC) and Fe–C workpieces are shown in Fig.3.37. As shown in Fig.3.37 (left) the wear was caused through the surface formed newly by diffusion, which had a lower hardness compared with starting WC-surface hardness. In the second case, the diffusion of cutting material in the workpiece occurs. Diffusion occurs into the chip and into the formed new workpiece surface. It is typical example of the diffusion wear that results in cratering. The diffusion process depends on the material properties (relative solubility of the elements) and temperature in the cutting zone.

10. *The adhesion* (pressure welding) wear mechanism is characterised by the extraction of particles of the cutting material by the chip of the rake and of the free surface. It occurs at the relative movement of the tool and workpiece materials through the tangential stress of the particles which were extracted from both tool material and workpiece. The higher the strength of the workpiece the higher are the tangential stresses on the cutting edge and by this the more intense is the wear. Higher strength of the cutting material, the tool can be higher loaded and the wear is lower [110].

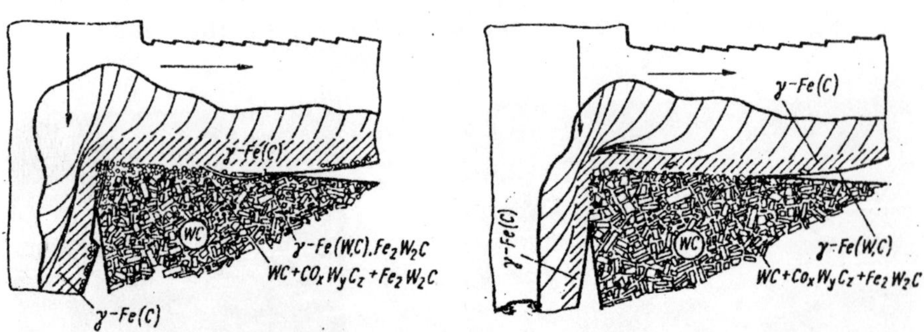

Fig.3.37 Schematic illustration of the diffusion process: left – at wear of the monocarbide hardmetal tools, right – of the cut tool material components in the workpiece [110].

Figure 3.38 shows the characteristic wear surfaces on a form tool, drill, end mill, and turning tool insert. These figures show how the wear process changes the geometry of these different types of cutting tools. Severe geometric changes that decrease the angle between the rake and clearance can weaken the tool so that the edge may suddenly fracture. More detailed characteristics of a drill wear which can be formed are shown in Fig.3.39.

The location and size of these wear surfaces play an important role in determining the useful life of the cutting tool. Localised stresses on cutting tool surfaces are a major contributing factor with regard to the location and size of wear surfaces. The magnitude of the state of stress in the cutting region also varies with time and creates a potential fatigue failure environment.

The ways to remedy the main characteristics of wear formed mainly in turning and similar cutting processes in wrought steels are listed in Tab.3.1.

It shows that the meaningful research and development efforts with regard to tool wear can contribute to the point that the life of a cutting tool can be basically predicted using base tool and work material properties regarding the cutting conditions on the basis of fundamental science and engineering principles.

The complex machining wear environment will have to be studied further and better understood by tribologists, PM material scientists, and manufacturing engineers. The process will become

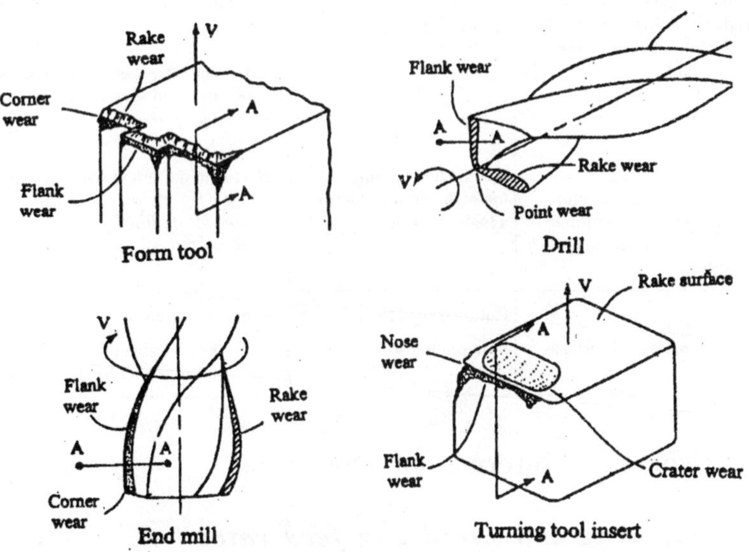

Fig.3.38 Wear surfaces on common tools due to the tool motion [125].

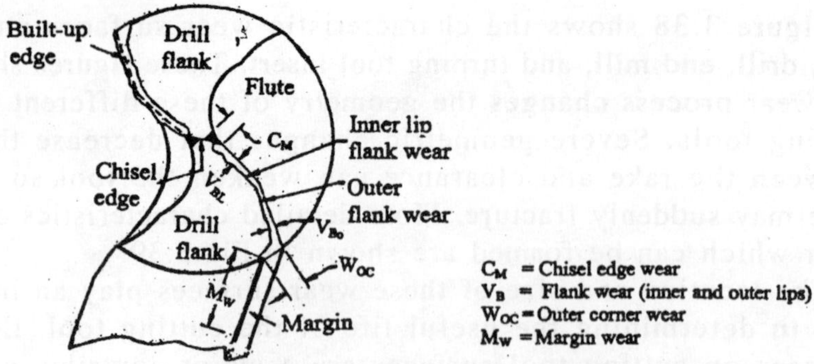

Fig.3.39 Drill point showing wear areas [126].

Tab.3.1 Some wear characteristics and the ways to remedy [127,128]

Tool wear	Possible cause	Remedy
Flank and notch wear		
a) Rapid flank wear causing poor surface texture or inconsistency in tolerance	a) Cutting speed too high or insufficient wear resistance of tool	Reduce cutting speed (but when machining heat resistant material with ceramics, increase cutting speed) Select a more wear resistant grade
b) Notch wear causing poor surface texture and risk of edge breaking	b) Oxidation, attrition	Select an aluminium oxide coated grade for steel machining For work-hardening materials, select a smaller entering angle or a more wear resistant grade
Crater wear Excessive crater wear causing a weakened edge. Cutting edge break through on the trailing edge causes poor surface texture	Diffusion wear due to high cutting temperature on the rake face	Select an aluminium oxide coated grade Select positive insert geometry First, reduce the speed to obtain a lower temperature and secondly, the feed
Plastic deformation Edge depression a) or flank impression b) leading to poor chip control and poor surface texture. Risk of excessive flank wear leading to insert breakage	Cutting temperature too high combined with a high pressure	Select a harder grade with better resistance to plastic deformation a) Reduce cutting speed b) Reduce feed
Built-up edges Rough, irregular machined surfaces	Cutting speed too low Material too soft	Increase cutting speed Choose heat treatment of the material such as to result in higher hardness (if possible)

the direct source of information about wear.

3.3.7.1 Tool wear and speed and feed rates

Cutting speed and cutting depth do not affect the stresses and the built-up edge in the workpiece to any great extent, but the feed rate does. Both small and large feed rates can give rise to material tension.

This is another reason for testing to find the optimum feed rate for an application. Edge sharpness and a positive geometry help to keep material stresses from building up during machining. A smaller entering angle gives rise to smaller build-up of material through a greater magnitude of the cutting forces.

Increased cutting speed (velocity) and temperature have a varying influence on different types of tool wear. Higher cutting speeds applied for increasing the machining productivity must be balanced against increased tool wear.

It should also be pointed out that tool wear plotted versus cutting time shows an initially fast wear stage, followed by an almost stationary wear stage, which precedes tool failure (final wear), Fig.3.40. Today, with modern cutting tools the first stage may be hardly discernible.

The wear mechanisms combine to attack the cutting edge in various ways depending on the tool material, cutting geometry, workpiece material properties, and cutting parameters. A basic analysis of the tool wear affected by many factors as mentioned also before, is an important strategy in optimising cutting performance and the required surface quality.

Regarding the wear mechanism, in many cases it is apparent that the main wear mechanism for continuous cutting of PM materials is abrasive flank wear. It is however necessary to assume that the tool wear is a more complicated process, which usually cannot be attributed to only one single wear mechanism.

For lowering the tool wear and for the optimisation of the cutting conditions and of the cutting materials it is therefore necessary to consider the wear mechanism of the cutting tool [130].

In general, tool suppliers provide rules for the selection of the best tool material, geometry, and machining parameters like speed and feed rate. For this the supplier must be aware in more detail

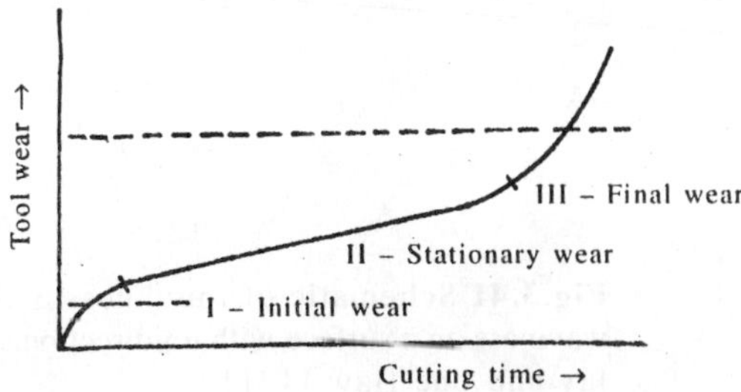

Fig.3.40 Tool wear–cutting time diagram [126,129].

of all properties of the PM workpiece materials being machined. The best solution would be joint machinability testing by the PM parts producer and the tool manufacturer, resulting in combined knowledge and experiences in PM materials and in machining.

3.3.7.2 Surface finish and surface integrity

Surface finish is concerned with only the geometric irregularities of surfaces of solid materials and the characteristics of the instrument for measuring roughness. Surface finish is defined in terms of roughness (irregularities in the surface texture, *e.g.* produced during machining), waviness (more widely spaced component of surface texture and may result from such factors as machine or work deflections, vibration, or chatter), lay (the direction of the predominant surface pattern), and flaws (unintentional, unexpected, and unwanted interruptions in the surface, *e.g.* cracks, nicks, scratches and ridges) as shown in Fig.3.41.

Surface roughness is closely tied to the accuracy or tolerance of a machined component. A close-tolerance dimension requires a very fine finish, and the finishing of a component to a very low roughness value may require multiple machining operations.

Virtually all material removal methods produce changed surface and subsurface conditions. The mechanically and metallurgically affected zones produced by material removal processes may also

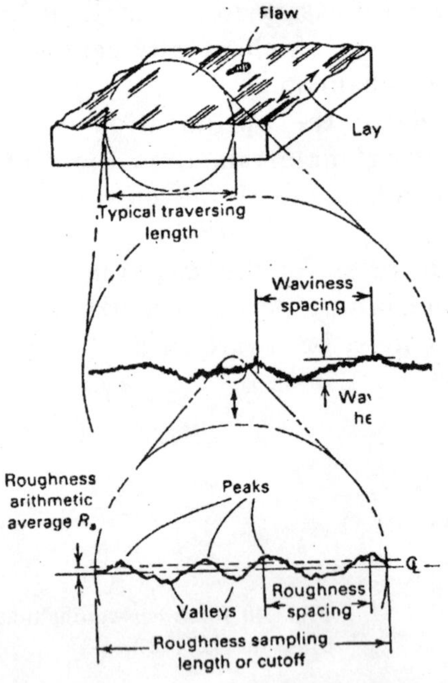

Fig.3.41 Schematic of roughness and waviness on a surface with unidirectional lay and one flaw [131].

extend into the surface to a considerable depth as a function of whether roughing or finishing conditions are used in the material removal process. The depth to which the changes reach is strongly affected by the respective parameters of the process.

Most often roughness is measured in terms of the arithmetic average, R_a, or the root mean square (rms) value, R_q. Roughness of a surface is measured in the plane perpendicular to the profile of the surface (ISO/DIS 4287/1). The typical shape of an effective cross (vertical) profile of the surface is shown in Fig.3.42.

In the case of porous sintered steels it should be kept in mind that the roughness of the pressed and sintered surface differs from the machined one. The latter has a rather symmetrical roughness profile while in the former the external surface is smooth and is interrupted by the pore channels, resulting in a roughness profile that is highly asymmetrical with regard to the hypothetical smooth surface. This has to be considered when comparing roughness values of sintered components and machined ones.

In common roughness measurements, the chisel has a standard radius of 1.27 mm on the edge of the chisel but it is wide so that it will not drop into the porosity of a PM surface and give a false reading by measuring the cavities as shown in Fig.3.43.

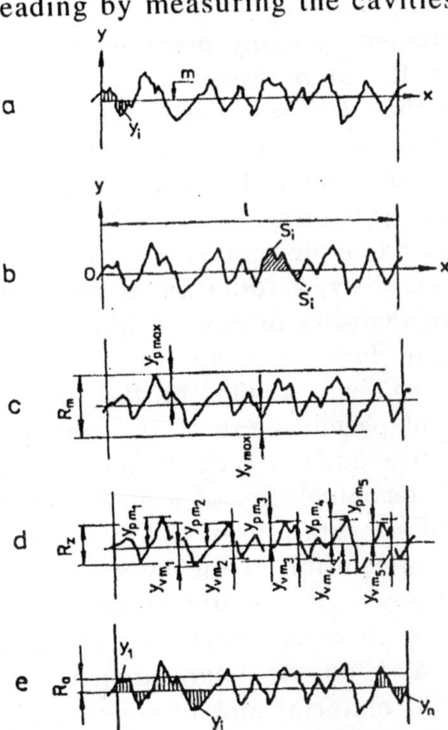

Fig.3.42 Recording of an effective profile of a machined plane (profile diagram) and parameters for its description. a – center part of profile, b – center arithmetic profile line, c – highest height of irregularities, d – height of irregularities, e – center arithmetic profile deviation [112,131].

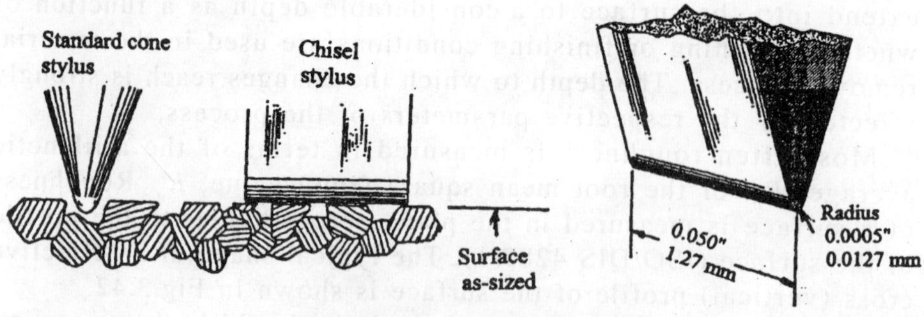

Fig.3.43 Left – Effect of chisel and cone styli on a standard as-sized PM surface, right – chisel stylus for surface measurement in PM [109].

3.4 MACHINABILITY TESTING

The machinability is not a property of a material, but a mode of behaviour of the material during cutting, and assessments of machinability should, therefore, specify the general conditions of cutting for which they have validity [117]. In that respect, the machining process can be regarded as a tribological one, and the relevant properties are therefore rather system-related than merely material-related.

The term machinability is explained by many manifestations of material during cutting as follows. It may be assumed that there is a specified quality of a material that can be clearly defined and measured by some criteria as an indication of the ease or difficulty with which this material can be machined. In fact there is no clearly-cut unambiguous meaning to this term.

Machinability is defined also as a complex physical property of a metal involving (1) true machinability, a function of not fully defined strength properties, (2) finishability or ease of obtaining a good finish, and (3) wear of the tool during cutting. These criteria are summarised in the statement: 'From any standpoint, the material with the best machinability is the one permitting the fastest removal of chips with satisfactory tool life and surface finish at low overhead cost' [102,132,133]. The machinability of a material can be defined also as 'the degree of disintegration of a chip' [108].

Machinability is in reality a manifestation of the interaction of workpiece and tool under effect of power. It is therefore influenced by both workpiece characteristics such as composition expressed in microstructure, mechanical, and physical properties, and machining parameters such as tool material and tool geometry, cutting speed, depth of cut, feed rate, and cutting fluids

[118,134,135]. The mentioned characteristics result in high complexity of the machining process in wrought steels, which is yet more complicated in machining of PM steels.

It follows that the practical evidence is that machinability cannot be determined from the knowledge of simple material and cutting tool parameters because of the combined effects of a large number of the factors mentioned before [130].

Problems arise because there is such a variety of machining operations, with different criteria of machinability which are not comparable. A material may have good machinability by one criterion, but poor machinability by another, or when a different type of operation is being carried out, or when conditions of cutting or the tool material are changed.

Machinability is the response of a metal (or more generally a material) to machining. It can be in general and also shortly described to machine well or have high machinability for any one or more of the following reasons:

- results in longer tool life under otherwise equal conditions when compared to other materials,
- produces well broken–up, short chips,
- provides good surface finish,
- gives uniform dimensional accuracy of successive parts,
- requires low power consumption in removing a given quantity of material,
- produces each part at the lowest overall cost.

Many definitions of machinability as the metal cutting process presented in general prove that the machinability is extremely complex, and especially in machining of powder metallurgy materials [106,136,137].

3.4.1 Measurement of machinability

Because machinability cannot be predicted solely from a knowledge of workpiece material and cutting tool properties but is in fact a system property, it is commonly determined empirically through machining tests which should however have as scientific character as possible.

For practice and also from the scientific point of view, machinability is a very important quality of materials that describes their various suitabilities for the changing of shape and/or properties by different machining procedures. There are two groups of criteria affecting the cutting process up to the finished part:

a) the criteria of cutting related to the workpiece:

- surface quality,
- manufacturing accuracy,
- functional integrity;

b) the criteria for cutting related to the tool:

- tool life (wear),
- tool defects (damage, thermal effects, etc.) [138].

Machinability is, therefore, a difficult term to define and can be measured by a number figure of merit which is proven by the following. The terms 'machinability index' and 'machinability rating' are used as qualitative and relative measures of the machinability of a steel under specified conditions. Several machinability indices oriented on special cutting process and conditions described in the literature are considered. Machinability judgements have been based on one or more of the following criteria:

- cutting tool life measured by the amount of material that can be removed by a standard cutting tool under standard cutting conditions before tool performance becomes unacceptable or tool wear reaches a specified amount,
- cutting speed measured by the maximum speed at which a standard tool under standard conditions can continue to provide satisfactory performance for a specified period,
- thrust and torque force measured by the power required to remove a unit volume (mass) of material under specified machining conditions,
- number of holes drilled before drill failure,
- comparisons with a standard steel based on experience in machine shops (often problematic),
- shape and quality of surface finish,
- feeds resulting from a constant thrust force,
- power,
- cutting interface temperature,
- chip shape,

all under acceptable overhead cost. However, even these entities remain partially undefined as measure of machinability and as the cutting tool life for their determination and, finally, they are not fully standardised [54,130,139,140–2].

Alternatively, the behaviour of the surface quality during machining of PM parts at the increased speeds and reduced chip loads can be used as an index of measuring the machinability.

Machinability ratings were also relied upon as aids for choosing machining conditions to be used on materials unfamiliar to production personnel [105,135,143].

The number of various machinability indices shows the already mentioned complexity also of PM material machining. Each of them has advantages and disadvantages. Probably, the number of holes drilled before drill failure in drilling test and the amount of material removed before tool failure in turning and similar processes under specified optimal conditions obtained are the simplest indices. Sometimes the test up to tool failure, especially in turning tests, were excluded because of the difficulty to be performed in a reasonably short time.

Using the power required to drill the various test materials as the indication of machinability drill life test is relatively inexpensive, the results are reproducible, and the shape of the test specimens is not a limiting factor. It must be joined with accomplishing index – number of holes drilled or amount of material removed.

Cutting forces and temperature measurements are technically interesting but are often also excluded because of the relative complexity of the required equipment especially in short time industrial tests.

Surface finish in PM components is very critical and it can play an important role especially in determining the effectiveness of a machining aid for the materials.

Cutting edge wear data as a single index, which is complicated to measure, provide therefore no sufficiently meaningful data for the least machinable materials. The reason is that both drill wear and turning tool wear are highly sensitive to the composition of the workpiece material properties [142,144,145].

Each machining index, if it has to be significant for PM material and by this way for the producer of a part, must be defined by cutting conditions in relation to the material.

In general, the amount of material machined before a tool fails is considered in some cases as the economically relevant criterion to evaluate the machinability also of PM parts [135]. Considering porous material, the amount of material machined expressed in mass and not in volume units can better define the process.

The productivity in machining is the most important factor and by this an index. This can be improved by increasing the time between process stops for changing tools which is the consequence of the tool life improvement, and by the reduction of scrap from defective parts or from parts used to adjust machine settings [130].

To the correct investigation of the machinability also of PM materials belongs also all process improvement methodologies which include some kind of process monitoring and corrective actions in case of anomalies in relation to the expected results. For example, an accepted practice for machining operations is statistical process control (SPC). In the case of final manufacturing operations, when the origin of the machinability problem is caused by the material behaviour, it happens quite often that the corrective action should have been taken in previous manufacturing phases such as powder mixing or sintering.

3.4.2 Machinability testing

The complexity of machining of PM materials is also the reason that up to present standardised machinability testing methods for all cutting processes for PM steels are lacking. It is not possible to expect the determination of a single dependent variable as an universally acceptable standard. Also, one type of machining test may produce 'acceptable' machinability while the same sample will fail in another type of machining test.

Except of the cutting conditions, which are common also for wrought materials, such as tool concept, *e.g.* chip breaker profile (usually not critical in machining of PM steels), stability and geometry of the tool holder, insert style, etc., the PM steels compared to wrought steels exist with almost infinite modifications of microstructures [146]. At the same time, the machinability test of a material is compared to the selection of the right tool and machining parameters, knowledge of the loads on the tool material, and the properties of the tool material together with an analysis of the wear mechanism. It follows from it that this is a complicated task especially for testing the machinability of PM materials.

It seems that that optimum solution will be the standardisation of the test system for each single cutting process, at least for drilling and turning. The results would be in most cases comparable. Also in this case there are many variables which cannot be realised under equal test conditions for some cutting processes. In as much as the drilling and turning are the most widely used cutting processes in the PM industry, the drilling test with all possible variables and the turning test are used to a greater extent and investigated more, and, therefore, they appear as first for the standardisation. The other cutting processes are not systematically investigated since their use in PM parts production is limited.

Some of the test criteria used are best suited to laboratory studies intended to elicit information about the effects of changes in microstructure, composition, or processing conditions on machinability [54].

3.4.2.1 Drilling test

Up to now the published data about machinability of PM steels are predominantly based on drilling tests. The reason for using a drill test is that it is fast and reproducible. The advantages of the drill life test is that it is relatively inexpensive, and the shape and size of the test specimen is not a limiting factor. It is also claimed that the drilling test is a much more sensitive test in detecting small differences in material microstructure than other cutting tests [134,147,148]. Surely each cutting method has own specific dynamic effects on the cutting of a material also with specific properties, and therefore it would be very tricky generally to transfer the results from one cutting test method to another.

The tests should be carried out on a rigid and stable drilling machine specially designed for drilling tests, often instrumented for these purposes, incl. clamping the workpiece. The rigidity of the drill must be considered.

Two aspects for drilling test can be applied. The first with the aim to determine the technically and economically optimum number of holes drilled with required diameter and depth and surface quality in a given part. The second one is to compare the drillability of different materials under given stable process conditions. Key parameters for the evaluated drilling operation are shown in Tab.3.2.

For the measurement of machinability in drilling, a number of indices have been proposed and used such as:

- number of holes drilled before drill failure occurs,
- time to drill a hole,
- drill life,

Tab.3.2 Key machining parameters for drilling operation [149,150]

Parameter	Metric		English	
	Unit	Formula	Unit	Formula
Drill diameter (d_d, D_d)	mm		in	
Feed rate (fm)	mm/min		in/min	
Depth of hole (l, L)	mm		in	
Drill speed (s, S)	rpm	$1000\ v_c/(\pi D_d)$	rpm	$3.82\ V_c/D_d$
Cutting speed (v_c, V_c)	m/min	$v_c = (\pi D_d s)/1000$	fpm	$V_c = 0.262\ SD_d$
Feed per rev. (f_r, F_r)	mm/rev.	$f_r = fm / S$	in/rev.	$f = fm / S$
Cutting time (t)	min	$t = l/fm$	min	$t = L/fm$
Rate of material removal (Q)	cm^3/min	$Q = \pi D_d^2 fm/4000$	in^3/min	$Q = \pi D_d^2 fm/4000$

- rate (amount) of material removal,
- productivity measurements.

The most frequently investigated cutting parameters are:

- drill (spindle) speed and drill diameter, the common parameter comparable with other results would be cutting (drilling) speed,
- feed rate, axial – thrust (drilling) force,
- torque,

which are measured as response functions in drilling tests. The other variables such as workpiece material, lubrication, tool material, and tool geometry must be held constant throughout testing of one type of material [151]. Under all test conditions, not only in drilling, material variables must be very substantially considered.

In some cases, a machinability index in drilling test of a material was calculated as the average of the numbers of holes obtained with five separate drills.

It seems that drill speed as one of the machining parameter expressed in *rpm* is an insufficient characteristic for the characterisation of the drilling conditions. The amount of removed material in term of cutting tool life under defined cutting speed should be the additional index [129,152]. The drilling force data are considered as a possible basis for a rating system.

Some indices like the 'time to drill a hole' are fast and inexpensive to develop, however, the variability of factors influencing the cutting process and of methods for measurement indicates that this may not be a very reliable means of quantifying machinability. It is possible to use this index for fast comparative testing, *e.g.* the drill geometry and cutting speed at constant feed on one material with determined properties or vice versa one drill type by drilling different materials as first step for more detail tests.

The exact measurement of the time to drill a hole is not simple if made manually. A similar unreliability problem can exist when measuring the surface finish only. Other indices based on 'forces exerted during machining' require more expensive instrumentation and a high initial set-up cost.

The 'drill life' tests have provided a good compromise between being fairly easy to conduct and having a reasonably good reproducibility and reliability. Further, this index provides valuable information useful in doing a productivity or cost analysis in relation to the number of holes drilled before drill failure. The preferred approach would be to use as a criterion the 'number of holes drilled' in which the 'drill life' is implicitly included (precondition

– defined workpiece material properties and cutting conditions).

Some of the mentioned methods are generally impractical, a method to suit the specific requirements of the machined part is normally used [126]. Average thrust and torque values have often be used to establish the relative drillability of different materials and to study the effects of various process variables on drill performance.

Basically, the test method should be as simple as possible considering also industrial practice aspects. It means to evaluate only one parameter for measurement. It is possible to suppose that such conditions are formed, *e.g.* at drilling with constant thrust (axial drilling) force, adapted for selected tested materials and drilling conditions. In this case the constant axial force is operated through a simple pulley-and-weight system to obtain the desired force. A schematic of such simple equipment for drilling test with constant thrust force, in this case of rectangular bars, is shown in Fig.3.44.

In this method, the drilling is by itself stopped (feed) when the drill wear (flank wear) reaches ~0.2 mm, *i.e.* drill fails. At this moment, the axial drilling force reaches the equilibrium with the radial friction force in the cutting zone. The 'number of holes' drilled before drill failure, or the length of the holes drilled (because in some materials it is not possible to drill minimum one through hole), is a machining index. Under these conditions other factors, such as drilling speed, drill diameter, and cooling can be considered as stable regarding the material variables which should be tested. In drilling test with constant thrust force the feed rate changes by itself in dependence on the machinability of the material. The feed rate decreases and the time for drilling one hole becames longer

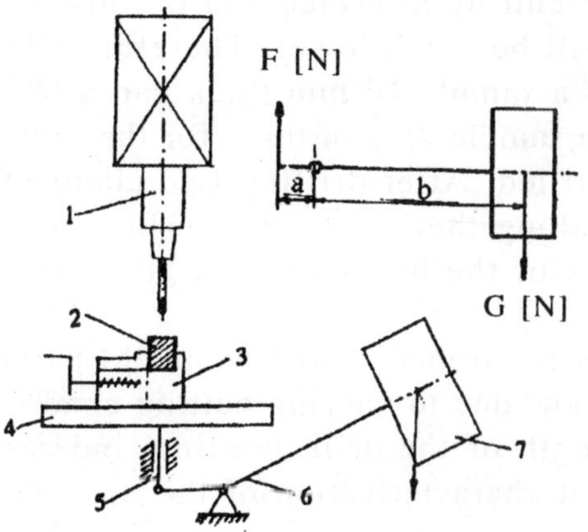

Fig.3.44 Scheme of equipment for drilling test of materials with constant thrust force [153]. 1 – drilling head, 2 – specimen, 3 – hand vice, 4 – turntable, 5– vertical leading of the turntable, 6 – leading arm, 7 – weight.

when drilling material with poorer machinability, which seems to be in this test method an additional measurable machinability index [153].

On the basis of a conception that the drill loaded with a constant thrust force can be compared to the indenter of a hardness tester a model for drilling test, shown in Fig.3.45, was proposed. This conception does not regard the fact that the indenter does not rotate and the penetration of a drill into the workpiece is the most complicated phase of drilling

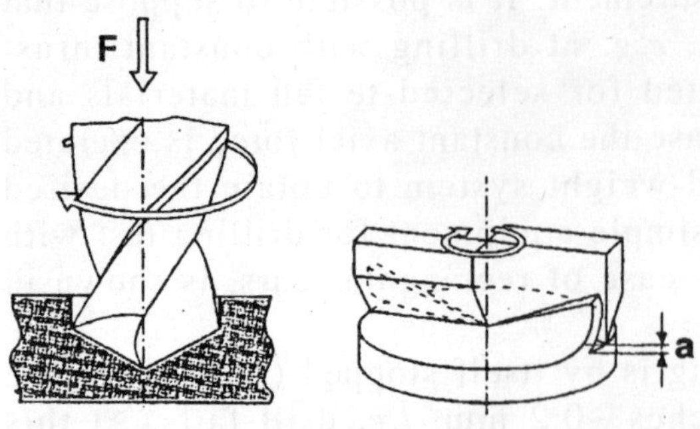

Fig.3.45 Model of a drill bit as indenter of hardness tester and of the material removed in each revolution [130].

Discs of various diameters up to 150 mm and up to 25 mm thickness are often used as samples for the drilling test. Compacting presses of high capacity are needed to produce these samples. In single cases, rectangular bars were used. The standard impact test bars (10 × 10 × 55 mm) could be used as standardised samples for drilling tests. In this case the drilling can be provided on the shear surface (in contrast to the press surface) corresponding to the face of a part being drilled in practice. Also the depth of the holes drilled in the parts in most cases does not exceed 10 mm. When drilling samples 10 mm thick compared to drilling samples 25 mm thick, the cutting interface and by this the temperature in the cutting zone will be much lower. Therefore, the results obtained in drilling test of a sample 10 mm thick can differ from those obtained in drilling of a sample 25 mm thick for the same total length or number of holes drilled. After drilling two clamped impact test bars as test samples along their contact interface it is possible to measure the roughness of the holes without additional cutting, Fig.3.46.

Comparison of the drilling tests performed by different PM parts producers or authors is very difficult due to varying cutting conditions (cutting speed, feed rate, length of the drill, cooling, indices etc.) and due to mostly insufficient characterisation of the material being tested.

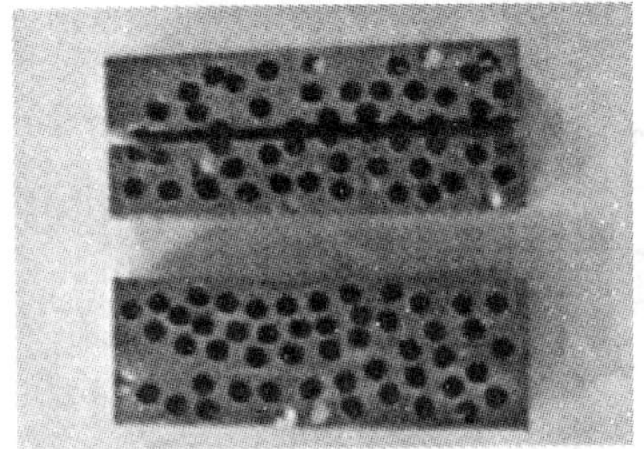

Fig.3.46 Two impact test bars as-clamped and as-separated after drilling test [153].

Drill speeds of 840 or 1000 rpm and a feed rate of 0.051 mm/ rev. are recommended for the drilling test of PM materials in the ASM Handbook [144,154,155]. In terms of lower machinability of PM steels in general lower cutting speeds are used. Drilling with lower cutting speed is more sensitive to the microstructure characteristics compared to drilling with higher cutting speed (higher drill speed and drill of larger diameter). In practice, the PM parts producers usually try to test a given part by drilling with the drill of the diameter for the hole required. In any case all the drills used in a drilling test series should be from one producer and from the same production charge.

The importance of the right selection of a machining index for evaluating the machinability of materials with widely varying compositions and by this with widely differing microstructure and mechanical properties must be considered more in PM machinability testing.

The designers and PM part producers can better refer to machinability expressed in terms of the number of holes drilled before drill failure. When testing materials of different density – porosity, the number of holes drilled could be supplanted by the mass (weight) of material removed.

3.4.2.2 Turning test

The turning operation consists of straight and face turning. Turning test for a number of PM materials is performed in a turning lathe. Specimens of cylindrical shape of various outer and inner diameters and lengths are used. To reach sufficient turning length more samples are coupled. Figure 3.47 shows the scheme of the arrangement for measurement of cutting forces in turning of sintered and/or of powder forged steel using coupled specimens.

The principle of a short time tool life test at face turning which could be used also in PM steel turning is shown in Fig.3.48. In this test the turning (constant workpiece revolutions) occurs from the

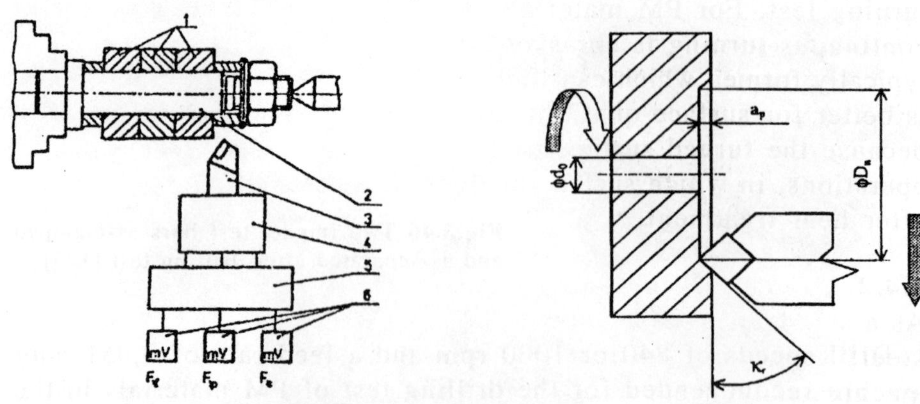

Fig.3.47 Scheme of arrangement for measurement of cutting forces in turning sintered and powder forged steels with coupled specimens [156]. 1 – workpiece, 2 – mandrel, 3 – tool, 4 – dynamometer, 5,5 – amplifier.
Fig.3.48 (right) Principle of the tool life short time testing at face turning [157–159].

surface of the center hole (r_0 radius) to the circumference of the ring shaped specimen (*e.g.* rollers) at constant feed and depth of cut. By this the cutting speed continuously increases. The test is finished at the cutting speed (v_c is maximum, D_0 diameter) at which the tool totally failed. It is possible to measure the cut time (T_{crit}) or the length or volume of the machined material $L_{crit,}$ $(D_0 - d_0)/2$). This short time turning test makes it possible to determine Taylor's relation between the cutting tool life T and the cutting speed v (often corroborated):

$$T \cdot v_{max} = c_T \text{ (constant) [160]}, \quad T = A \cdot v^k \text{ [161]},$$

where A and k are empirical constants, with k depending mainly on the tool material.

When testing specimens of lower diameter after finishing the first cut follows turning as at the beginning from the center hole (v_c minimum) up to tool failure. It is possible to test different tool material on one workpiece material or vice versa one tool type on workpieces with different properties and other related conditions, *i.e.* to test the machinability of different PM materials in turning with one or more tool types or under different cutting conditions.

For testing interrupted cut, rectangular specimens are used. Cutting speed and depth of cut in combination with cutting tool (material, geometry) and tested material are the main variables in turning test. For PM materials the main wear for the tool during continuous turning is abrasive flank wear; a depth of cut notch is typically formed which can limit the tool life [129]. The turning test is better for surface quality evaluation compared with drilling test because the turned surfaces are mostly exposed to other finishing operations, in which surface quality is also preferred, *e.g.* grinding after heat treatment.

3.4.2.3 Face milling test

As a standard procedure for the evaluation of the machinability of steels used for PM steels, in some cases the VOLVO standard machinability test is used. This standard procedure is basically a face milling operation modified in such a way that only one single cutting tooth is used on the milling head. Cutting geometry and other cutting conditions are subjected to tight specifications. The cutting tooth is made from high speed steel of type SIS 142756 mod. (similar to AISI-T5) with tight specifications of chemical compositions and heat treatment and with a hardness of 64–65 HRC. The position of the cutting tooth on the milling head, as well as the cutting geometry, are shown schematically in Fig.3.49.

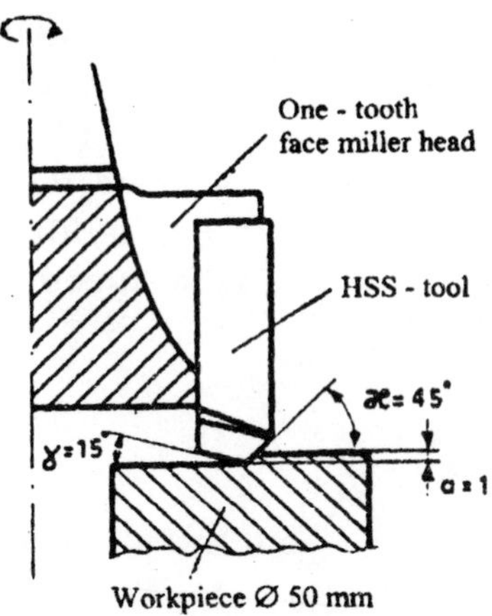

Fig.3.49 Set-up for VOLVO standard machinability test (face milling with one-tooth milling head) [118,162,163].

The following cutting data are applied: cutting depth: a = 1.0±0.05 mm, feed: s = 0.1±0.002 mm/tooth, cutting speed v_c = 20–200 m/min. This method was used mainly for testing of powder forged parts. The workpiece was normally a rod with a diameter of 50±0.1 mm which was clamped on the milling machine according to given specifications. A series of forged cylinders premachined to above specified diameter for testing PF materials, instead of one continuous rod usually can be used. Face milling was carried out under the stated conditions stepwise until the wear-land occurring on the flank of the cutting edge has reached a maximum width of $F = 0.7$ mm.

After each step, the cutting tool is taken off the machine, and the width of the wear-land is measured under a microscope. In this way, the wearing behaviour of the cutting edge and the wear-inducing character of the workpiece materials can be closely followed through the entire machining period.

Other machining methods in PM parts manufacturing are used to a substantially lesser extent.

4

Cutting Tools

A wide diversity of tools are in fact gathered under the words 'cutting tools', such as turning tools, drills, threading taps and dies, shaping, planing tools, scrappers, shaving tools and broaches, saws, milling cutters and hobs, counter bores and countersinks, reamers, files, grinding wheels. Summarised, the machining operation factors affecting the final part properties are:

– the part to be machined (workpiece – shape and material),
– the machine–tool, and the cutting tool [164].

Only an approach that includes the wide variety of workpieces which was described before also the different tool materials and geometries can bring in a broader view on complexity of machinability of PM materials. It was shown before that in some extent at increasing hardness of the PM material the increased wear of the cutting tool was noticed in general.

The productivity of machinery manufacturing industries has continually increased, in particular, from the dramatically increased cutting speeds chip removal rate enabled by the cutting tools. These performances are mainly linked to the study and developments of all types of cutting tools especially of hard and super hard materials.

There is a great variety of tools and materials in use in wrought steel machining which are applied in most cases with some modifications in geometry also in PM steel machining. The use of a specific cutting tool is depending on the workpiece material properties and the performance of the production units. General tendencies like dry machining, which is the special area in machining of PM materials, or hard machining which require beside other machining enhancement methods some new cutting tools to achieve the desired performance. It is a permanent interaction. The use of some tools for single processes in wrought steel machining is listed in Tab.4.1. Without further specifications, these data can, however, be accepted only as rough information.

Tab.4.1 Typical machining methods and cutting tools for some materials machined [127]

Tool	Machining process				
	Turning	Milling	Boring	Broaching	Reaming
HSS	●	●	●	●	●
HSS coated	O	O	●	●	●
HM	●	●	●	●	⊕
HM coated	●	●	●	●$_2$	⊕
Cermets	⊕	⊕	-	⊕$_3$	-
CBN	O$_1$	O$_1$	●$_1$	O$_1$	-

Remark: O only in exceptional cases, ● state of the art, ⊕ promising; index: 1 for hardened steel, 2 with post-grinding of the chip face, 3 use as guide-strip material

The industrial production also of PM parts always operates under a high cost pressure. In addition to the increasing requirements for the effectiveness and economic efficiency, environmental aspects of the technologies must also be considered. For machining, this means the increase of the output with minor formation of the chips and marginal use of cutting fluids. The cutting tools, especially the tool materials used, are of great importance in this sense. The systematic investigation of cutting tools used in machining of wrought steels including modified geometry for sintered steels, surely will be a contribution for successful machining especially of high strength and high hardness steels.

In order to enable a precise selection for a certain application, it is important to have knowledge of which failure mechanisms that constitute the greatest danger, especially what type of wear mechanism that is dominating. It can be assumed that hard workpiece materials cause abrasive wear and soft materials adhesive wear. Usually, the tool materials chosen are a compromise between the requested properties with respect to cutting conditions. Regarding specific properties of sintered steels, including the heterogeneity of microstructure, the corresponding tool material must be chosen for the required cutting process. For this reason, the microstructural composition of the machined workpiece is decisive for the behaviour and, by this for, the effectiveness of cutting tool materials used. At increasing strength of the material the reduction of the cutting tool life is noticed in general.

According to [165], the quality of a tool material is primarily based on compressive strength, wear resistance, red hardness and corrosion resistance in relation to the workpiece material.

It is necessary to note that data regarding all cutting tools and cutting tool materials are related mainly to conventional metal (ferrous, non-ferrous, etc.) workpiece materials. The wide range of cutting tool materials which are on the market and also in development with their specifics enable to select an appropriate cutting tool and cutting tool material for machining of all grades of PM steels under optimum cutting conditions. For machining of PM steels, the tool geometry should be modified compared to machining of wrought materials of similar hardness.

4.1 CUTTING TOOL MATERIALS

Any machining operation which involves the removal of metal by a cutting action requires that the material used for cutting tool will stand up to the rigours of the cutting action.

There are three basic problems to be overcome:

- the wear which takes place at the cutting edge (tribological characteristics),
- the heat generated by the energy required to form and to remove material (chips) from the workpiece,
- the mechanical and/or thermal shock involved in the cutting action.

In order to face the mentioned severity the tool must be made of a material having the following main properties:

- hardness (sufficient up to operating temperature) to combat the wearing action,
- hot strength to overcome the heat involved,
- sufficient toughness to withstand the interruption or vibration occurring during the machining process,
- chemical stability, especially towards the workpiece material [164].

The following materials from those generally used also for cutting in powder metallurgy are: high speed steels, hardmetals, cermets, ceramics, cubic boron nitride, diamond. Superhard materials on the base of diamond and of cubic boron nitride with their hardness and wear resistance outperform the other tool materials mentioned here [166].

The basic requirements for a cutting tool material are high hardness and high bending strength. Figure 4.1 allows comparison between the hardness and bending strength of the mentioned cutting materials (feed, depth of cut and shock resistance increase with bending strength; resistance to deformation and speed of cut increases with hardness).

Table 4.2 lists the characteristics of the cutting tool materials mostly used in PM machining. As seen, there is a compromise between performance and cost. Thus, the benefits from improved machinability need to outweigh the cost of the better tools to justify the use of the better performing, but more expensive tools.

The selection of cutting tools depends generally on the type of operation, the machining parameters, and the machinability of the workpiece. Typically, for PM materials, the three most widely used cutting tools are: high speed tool steels, TiN-coated steels, and hardmetal tools.

Hardmetals (cemented carbide) cover a wide range of hardness and toughness levels and overlap cermets and sialons at their harder end. In general, increasing hardness brings with it a reduction in

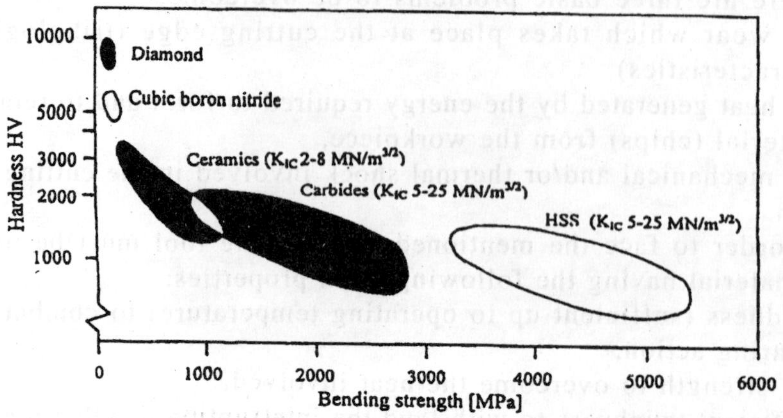

Fig.4.1 Hardness and bending strength of materials used for cutting tools [167].

toughness, however not necessarily of the bending strength. (For ultrafine grained hardmetals both hardness and bending strength increase but the fracture toughness decreases, Fig.4.1).

Generally the cutting materials have to exhibit high hot hardness and temperature resistance, wear resistance and high toughness. The use of a given cutting tool is related to the cutting forces formed in the cutting process of a workpiece regarding also the economy of the machining operation.

The selection of an optimum cutting tool is also affected by the cost of the cutting materials. It follows from the comparison of the cost of a cutting corner on the same type of indexable insert that where normal uncoated hardmetal is the cheapest – base 1, coated

Tab.4.2 Characteristics of cutting tools for PM steels [105,149]

TOOL MATERIAL	TOOL LIFE	SHOCK-RESISTANCE	RELATIVE COST
High Speed Steels (HSS)	Poor to Reasonable	Excellent	Low
TiN-coated HSS	Good	Excellent	Medium
Hardmetal/CBN	Good to Excellent	Poor to Reasonable	High

hardmetal 1.2, ceramic 1.7, sialon 3.7 and CBN and PCD are forty times as expensive [164].

4.1.1 High speed steels

The production of high speed and tool steels occurs by conventional ingot metallurgy and by powder metallurgy routes [168].

High speed steels have the lowest hardness and the highest toughness of all cutting materials in general use. Their hardness is brought about by a heat treatment process, so they are not naturally hard but can be brought into a 'fabrication state' (soft) and then, through suitable heat treatment, into the 'application state' (hard). In the latter state, the bulk of the matrix structure consists of highly tempered martensite within which carbides of several micrometers in diameter are embedded. The alloying elements – primarily W, Mo, and V – tend to combine with carbon to form in the matrix very strongly bonded carbides, in part during solidification – as primary carbides – but in part also during tempering of the quenched tool at relatively high temperatures at which nanometer-sized precipitates are formed, resulting in secondary hardening. HSS therefore obtain their high hardness not though martensite but through a precipitation hardening process. Since the secondary hardening occurs at temperatures of typically 500 to 550°C, the HSS tools can operate up to these temperatures without loss of hardness. If the temperature is raised further, the fine secondary carbide particles partially dissolve, alloying elements going into solution in the iron, and growth through Ostwald ripening occurs, resulting in overaging and an irreversible loss of hardness (this behaviour is typical for precipitation-strengthened materials such as, for example, Al–Cu and Al–Zn–Mg alloys). Therefore, if the temperature of the cutting edge rises above 600°C for some time, the high speed steel will soften and the edge will fail. For this reason, they are limited to comparatively lower cutting speeds in dependence on the workpiece material, *e.g* up to the order of 50 m/min for stainless steel and higher up to 100 m/min for carbon and low alloy steels. In the case of the machining of PM steels due to lower amount of removed material from a part this limit is not decisive.

The areas of machining methods for the application of a HSS tool regarding the principal required characteristics are illustrated in Fig.4.2.

In general, HSS steel tools are used for machining steels with the hardness up to ~200 HV 10 (Rm =~700 MPa). The major area of application for the use of high speed steel as the cutting material is drilling and tapping, substantially less turning. At least 80% of all drilling is done with high speed steel. The second important application area for high speed steel is milling [164]. In term of the cutting process the high speed steels are used mainly for machining operations at which the cutting forces vary in amplitude and frequency, *i.e.* typically in interrupted cut. It means in cases in which the high toughness and resistance to mechanical and shock heat loading is most important, at least more important than a high volume removal rate, and when the most dangerous failure mode is not wear but fracture.

4.1.1.1 Ingot metallurgy high speed steels

Traditionally, the grades of high speed steel available are divided into three groups. The grade using tungsten as its major alloying element is known as the 'T' (tungsten) series according to the AISI designation. The second grade contains molybdenum and considerably less tungsten is present. This is known as the 'M' (molybdenum) series of alloys. The third grade contains cobalt and can be either a T or an M series material.

The hardness of high speed steels after heat treatment is usually within the range 65 to 70 HRC. Their high toughness is a great advantage compared to other cutting materials and makes them

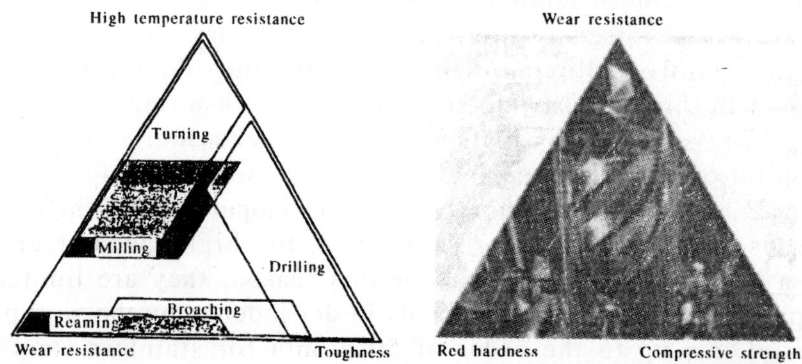

Fig.4.2 The areas of HSS tools for main machining methods [169]; (right) – Efficiency of machining tools (Böhler, Kapfenberg).

useful in a wide range of application.

Conventionally cast and wrought high speed steels are characterised by primary carbides irregular in shape and different in size. These are often large, angular and sharp-cornered and form chains and clusters in the microstructure. It is a consequence of the crushing of the eutectic carbide network, which is formed during solidification, during subsequent hot working of the ingot. The properties (esp. the bending strength) of a HSS semi-product are therefore anisotropic. They are higher in the rolling direction, *i.e.* parallel to the carbide stringers, than in perpendicular direction.

The composition of some high speed steel grades used also in machining of PM steels is listed Tab.4.3.

T1 high speed steel is restricted to comparatively low cutting speeds. M2 and M3 are evaluated as efficient and M36, M41 and M35 as high efficient high speed steels. M2 and M3 steels are popular for the turning, planing, shaping tool, for milling cutters. These alloys are popular for producing drills and they are also used predominantly for the production of taps. If hot strength is the main requirement such as in the machining of heat resistant alloys then M42 is used and/or M35 [170].

The harder grade M42 (HRC 67–69, approx. 1000 HV (Vickers diamond hardness) is used for tools with high abrasion and fatigue resistance of the cutting edge if this is the vital factor. Higher speeds will cause the temperature of the cutting edge to rise above the softening point. M42 is also regarded as the ideal substrate for coated inserts (especially TiN) made from conventional high speed steel, whereas M35 is currently the preferred material for powder

Tab.4.3 Chemical composition of mostly used ingot metallurgy high speed steels (0.45% Mn, 0.45% Si) [168,170]

Grade		Composition [mass %]					
AISI	DIN	C	Cr	Mo	V	W	Co
T1	S 18-0-1	0.70-0.80	3.80-4.60	Max 0.50	1.00-1.60	17.0-19.0	-
M2	S 6-5-2	0.90-1.0	3.80-4.60	4.50-5.50	1.50-2.20	5.50-7.00	-
M3 class 1	SC 6-5-3	1.15-1.25	3.75-4.50	4.75-6.50	2.25-2.75	5.00-8.75	-
M3 class 2	S 6-5-3	1.15-1.25	3.75-4.50	4.75-6.50	2.25-2.75	5.00-8.75	-
M36		0.90-1.00	3.80-4.60	4.50-5.50	1.70-2.40	5.00-6.50	7.30-8.70
M41	S 7-4-2-5	1.05-1.15.	3.80-4.60	3.30-4.30	1.50-2.20	6.20-7.70	4.50-5.50
M35	S 6-5-2-5	0.80-0.90	3.80-4.60	4.50-5.50	1.50-2.20	5.50-7.00	4.30-5.20
	S 3-3-2	0.95-1.05	3.80-4.60	2.20-3.20	2.00-2.70	2.40-3.40	-
M42	S 2-10-1-8	1.05-1.15	3.5-4.25	9.00-10.00	0.95-1.35	1.15-1.85	7.75-8.75
	S 10-4-3-10	1.15-1.30	3.80-4.60	3.50-4.30	3.00-3.70	9.50-11.0	10.0-11.5
T15		1.60	5.0	-	5.25	13.0	5.25

metallurgy high speed steel with a TiN coating.

The development of high speed steels alloyed with cobalt (HSCo – cobalt high speed steel) has led to grades that feature excellent high temperature strength properties without compromise to wear resistance and toughness. Cobalt increases the hot strength and hardness of the matrix [171]. Drills, milling cutters, reamers and taps manufactured from these grades meet the exacting demands of high productivity machining.

4.1.1.2 Powder metallurgy high speed steels

Compared to the classical ingot metallurgy method, powder metallurgy allows to get tools with improved machinability, better dimensional control during heat treatment and cutting performances (PM HSS). The primary advantage is however the much larger compositional flexibility that enables steel grades inaccessible by ingot metallurgy; if cast, these materials would be too brittle even for hot working, due to excessive carbide networks. The rapid cooling of small droplets ('micro-ingots') prevents significant segregation effects. The high speed steel powders are produced by gas (N_2, rarely Ar) or by water atomisation, resulting in spherical or irregular particle shapes, respectively.

The standard production route for PM high speed steels, originally introduced as the ASEA-STORA process is inert gas atomisation [172], canning in mild steel containers, and consolidation by hot isostatic pressing, followed by forging, extrusion, or rolling. The spherical gas atomised powders are particularly fitted for hot isostatic pressing, resulting in high bulk density of the powder in the filled can. HIP is commonly done at temperatures in the range of 1100°C and at pressures of 1000 to 2000 bar. Low oxygen content of the powders is crucial since any oxygen in the powder will be transferred to the as-HIPed material and may cause slag inclusions. Oxygen levels in the powders are usually <200 ppm, and handling is conveniently done under inert atmosphere. Furthermore, avoiding of slag inclusions through appropriate handling of the melt has improved the reliability of PM HSS grades [173].

Recently, spray atomised (Osprey) high speed steels have also become commercially available [174], although the production volumes are still comparatively small.

Water atomised powders, after an adequate annealing to improve the compactibility, are cold compacted in a die or cold isostatic pressed and then sintered, supersolidus sintering being commonly employed. This way allows getting near-net-shape tips with especially designed chip breakers, which optimise the cutting geometry of throwaway tips. Sintering is a crucial step [175,176], since full density has to be attained without too much carbide growth, and the typical sintering temperature intervals may be as narrow as 5–10°. Since the optimum temperature interval also depends on the carbon content, sintering is usually done in vacuum to avoid uncontrolled decarburization. The sintered tips are then subjected to annealing, austenitizing, quenching and tempering which lead to a hardness of 65–70 HRC. Table 4.4 gives the composition of some PM high speed steels.

The high speed steel produced by powder metallurgy methods are characterised by a uniform fine-grained structure with fine, isotropically and uniformly distributed carbides and mostly also larger volume of carbides, which features result in tools with high durability when correctly applied. The usually presented differences in carbide size and distribution between the conventional cast (ingot) and the powder metallurgy high speed steels is shown in Figs.4.3 and 4.4.

A more detailed view on size, shape and distribution of carbides in PM high speed steel is shown in Fig.4.5. The mean distance between the carbides in this steel was ~2 μm. The difference in size and distribution of carbides between the PM HSS (M2) prepared by HIP and extrusion without and with addition of niobium and of conventional HSS steel is shown in Fig.4.6.

The finest carbides were observed in the extruded PM HSS (1.9Nb) steel and as expected the coarsest in the conventional HSS.

In conventional HSS, a higher proportion of carbides are larger and angular shaped compared with those in PM HSS which are smaller, rounded and uniformly distributed. The coarse, angular

Tab.4.4 Composition of PM high speed steels produced via the HIP route

Grade		Composition [mass %]						Trade mark	Ref.
AISI	DIN	C	Cr	Mo	V	W	Co		
M3-2	S 6-5-3	1.2	4.0	5.0	3.0	6.4	-	ASP 23	10, 177
		1.3	4.2	5.0	3.1	6.4	8.5	ASP 30	
		2.3	4.0	7.0	6.5	6.5	10.5	ASP 60	
		1.5	4.0	7.0	4.0	-	-	Vanadis 4	
		2.0	4.0	2.0	5.0	14.0	11.0	*S290	178
		1.6	4.8	2.0	5.0	10.5	8.0	*S390	
		1.3	4.2	5.0	3.0	6.3	8.4	*S590	
M4		1.33	4.3	4.9	4.1	5.9	-	*S690	
M3-2	S 6-5-3	1.3	4.2	5.0	3.0	6.3	-	*S790	

Remark: *microclean

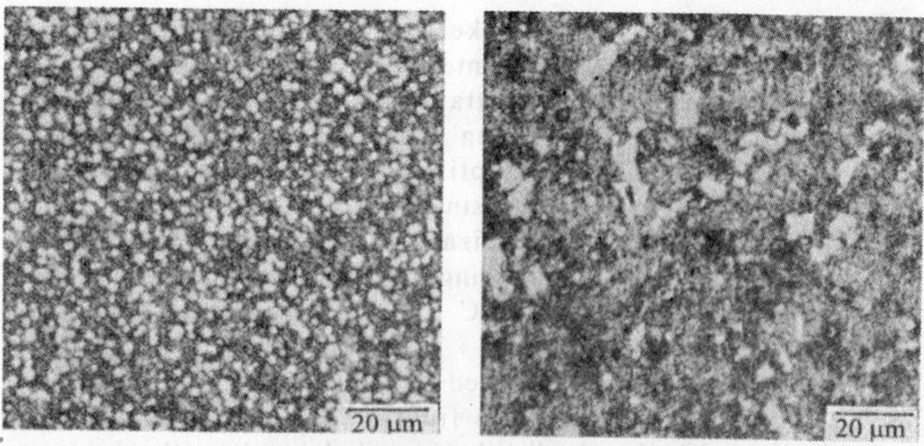

Fig.4.3 Microstructure of PM M390 microclean plastic mould steel in the as-hardened and tempered conditions [178].
Fig.4.4 (right) Microstructure of conventional (ingot) high speed steel (1.4528x105CrCoMo 18.2) in the as-hardened and tempered conditions [178].

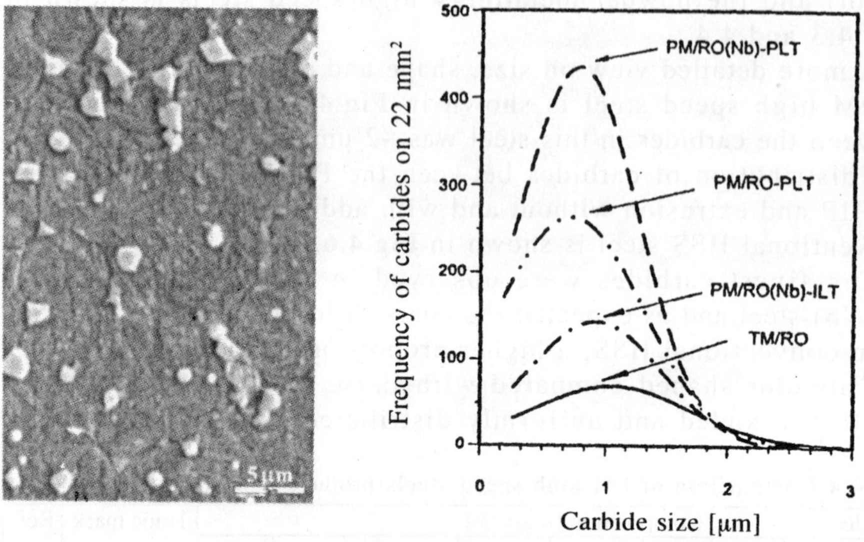

Fig.4.5 Microstructure of PM high speed steel M2 (HIP). SEM.
Fig.4.6 (right) Distribution curves and size for carbides in PM HSS (PM–PLT) (1200°C/20 min, 3 × tempering at 550°C/1 h) and in conventional HSS (TM), all base M2 steel [159,179]. (The carbide size distribution curve for PM steel should be revised – note of the authors of this book).

carbides offer advantages in metal cutting, especially of abrasive workpieces, but lower the bending and fatigue strength. Due to the bigger size of the carbides in an ingot HSS, it should have a better wear resistance than the steel produced via PM technology. As the powder technology gives the material a refined microstructure and by this an improved toughness, this can be utilised to increase the alloy content of the steel. With a maintained or improved toughness, the volume of carbides can be increased and/or another type of carbides with higher hardness can be obtained [180].

The composition of PM high speed steels can be the same as of ingot steels, as *e.g.* for M2/S 6-5-2, but some new grades were developed with the composition for special applications. Generally, the advantages of PM high speed steels are more pronounced for high alloy grades that are hardly or not at all accessible by ingot metallurgy.

Powder metallurgy high speed steels are much more easily machined than conventional grades because they tolerate higher sulphur content without any decrease of toughness and cutting performances. Their fitness to grinding is also much better because the carbides are finer and homogeneously distributed.

The bending strength of PM high speed steels is higher about 1.5 times compared with ingot high speed steel (*e.g.* 4000 MPa→3000 MPa at 65 HRC) [10]; a most welcome feature is the isotropic strength distribution.

The main application areas of PM HSS steels (*e.g.* 'microclean' grades) as heavy duty machining tools are: – pinion-type cutters of solid or composite design, – side-milling cutters, end-mills, – profile cutters, – hobs, – broaching tools of all kinds, – machine taps, – twist drills, – reamers [178].

In particular, sintered cobalt-containing high speed steel (also coded HSCo XP = ASP 30) produced using powder metallurgy (HIP) technology exhibits superior toughness and grindability. The use of XP steels is particularly advantageous when machining materials that are difficult to cut or when the material is extremely hard. Taps and milling cutters have particular advantage when made from XP grade steel [118].

Like all highly alloyed tool steels, PM steels are also very sensitive to notches, due to hardening and tempering to high hardnesss. For this reason, sharp corners and radii must be avoided when designing a tool to minimise stress concentrations in the material.

In spite of the mentioned advantages of PM high speed steels compared with conventional high speed steels, they did not find a broader area in machining of PM steel precision parts. The position of conventional HSS and PM HSS, respectively, in relation to their wear resistance and toughness at cutting is shown in Fig.4.7.

One critical feature of PM HSS is their significantly higher cost which may be 2–3 times that of ingot metallurgy steels (although here the higher content of the alloy elements must be considered). PM HSS are particularly successful if the cost of tool preparation and change markedly outweighs that of the starting material, as *e.g.* with complex-shaped tools. It is also an advantage of PM manufactured tool steels in general, and thus also high speed steels, can simplify the choice of material depending on the unique combination of wear resistance and toughness to the cutting process of a PM steel.

The choice between conventional and PM HSS tools lies in the decision between the improvement in cutting properties/tool life and the higher cost of PM steels. The most important characteristics of both materials is wear resistance under given cutting conditions. Generally, the wear resistance increases with increasing size, volume fraction and hardness of the primary carbides in the steel matrix due to larger size and more angular shape of the primary carbides in a conventionally manufactured high speed steel. It should have higher cutting efficiency than the same steel produced via the PM techniques [177].

The section of the PM HSS drill flank wear surface (0.2 mm) in drilling with constant drilling force of 333 N showing 'total' failure of the drill which stopped before finishing drilling the last hole is shown in Figs.4.8 and 4.9. The dropping out of the carbides on the cutting edge is visible and also the fracture of the cutting edge.

The comparable results in a short time test in face turning (Fig.3.48), (cutting edge durability – so-called frontal workability test according to ISO 3685/1977) with PM HSS inserts without and with cobalt addition and with conventional HSS inserts show the

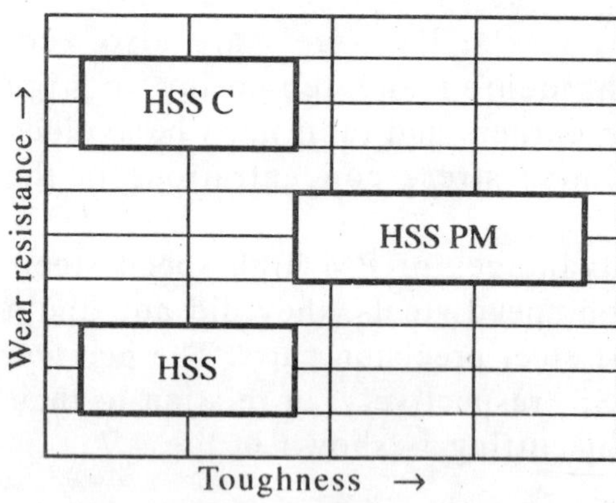

Fig.4.7 Wear resistance and toughness of conventional HSS (HSS), coated conventional HSS (HSS C) and PM HSS (HSS PM) [169].

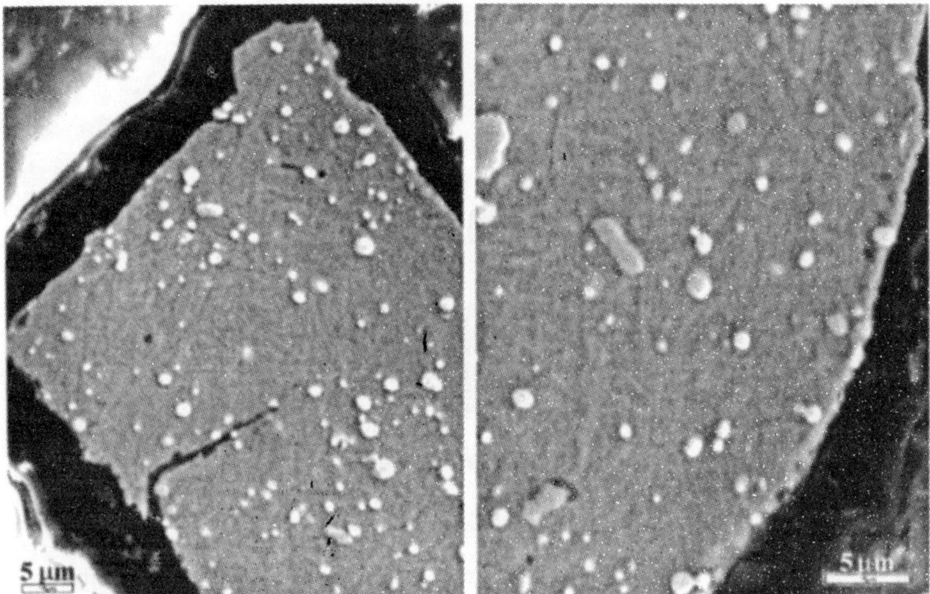

Fig.4.8 Section of the PM HSS drill (3 mm in diameter) flank wear. SEM
Fig.4.9 (right) Section of the PM HSS drill (3 mm in diameter) flank wear. As in Fig.4.8 – continuation along the failed cutting edge. SEM.

advantages of PM HSS for this type of machining, Fig.4.10. In this case, the 'critical wear diameter' is the largest diameter at which total cutting edge annihilation occurred within the cut from the inner hole diameter (beginning of the cut at constant spindle lathe revolutions). The addition of 1% Ti to PM HSS deteriorated its cutting properties.

As shown, in this tests of durability of the cutting edges of the inserts from all three types, PM HSS exceeded the value of D_{crit} for those of conventional HSS by about 50%. Co addition to HSS inserts did not affect the results compared to those attained without Co. With increasing starting cutting speed (increasing revolutions of lathe spindle), the wear of the inserts was increasing and the difference between PM HSS and ingot HSS bits was also increasing on the behalf of PM HSS bits.

The reinforcement of HSS matrix with refractory carbides such as TiC, VC, WC, NbC and TaC, which are the hardest is the way for improving the cutting performance of such HSS tools. These carbides are the most compatible with matrices like HSS as some of them (VC, WC) are primary carbides in this type of steel. So a kind of reactivity exists between the HSS as the matrix, and the

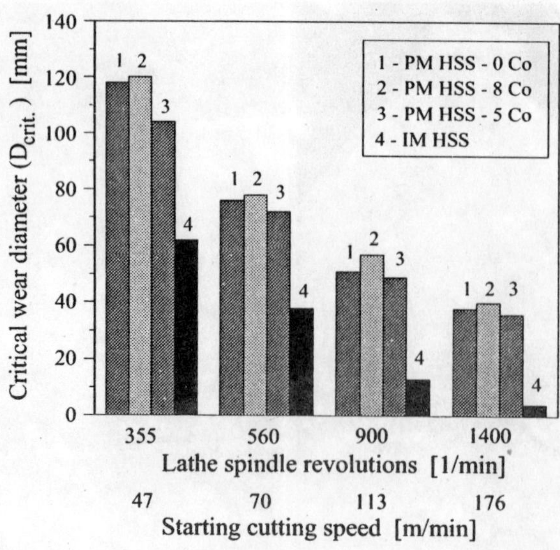

Fig.4.10 Critical wear diameter (D_{crit}) of the workpiece at which cutting edge annihilation occurred at short time face turning of wrought steel with PM HSS without and with Co- addition and conventional (IM) HSS (M2) inserts at different revolutions of the lathe spindle [158,159]. Workpiece: steel (C60E4, ISO 683/1-87; 0.45C, 0.25Si, 0.65Cr), 200 mm outer diameter, 40 mm inner (starting) diameter for cutting. *Test conditions: insert (12 × 12 × 6 mm) geometry (clearance angle α = 11°, rake angle = 0°, nose radius 0.4 mm); cutting conditions: depth of cut 1 mm, feed 0.2 mm/rev.; criterion – total loss of cutting ability – cutting edge annihilation at the cut of the workpiece on the critical wear diameter (D_{crit}).*

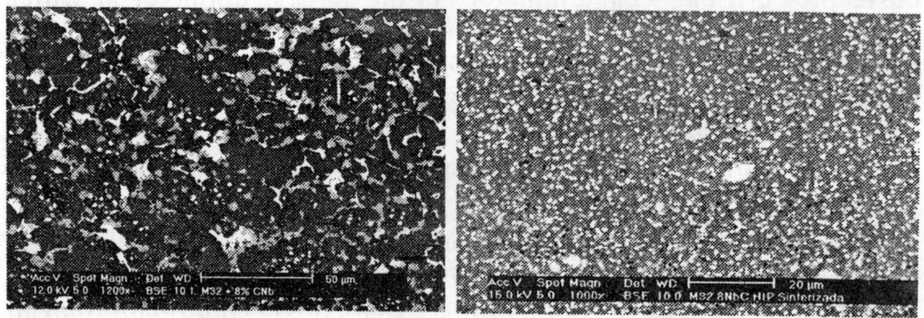

Fig.4.11 BSE images of M3/2 + 8 vol.% NbC, obtained by: left – conventional pressing and sintering at 1285°C; right – HIP + 1150°C/150 MPa [181].

reinforcement carbides. The improvement in wear resistance and mechanical features can be achieved with HIP technology (refined microstructure) as shown in Fig.4.11.

On the basis of powder metallurgy possibilities, it can be assumed that powder metallurgy technique for production of cutting materials will be able to prepare some grades especially designed for machining PM steels. In that respect, PM offers advantages since also small batches of special HSS grades can be manufactured under industrial conditions while in the case of ingot metallurgy at least one arc furnace batch has to be produced. This means that the powder route production of cutting tool materials will help more to find specific solutions for machining of powder metallurgy structural steel parts.

4.1.2 Hardmetals

Hardmetals, also known as cemented carbides, are, in the simplest variant, two-phase materials consisting of hexagonal WC as the hard and wear-resistant phase, contained as fine, angular grains, and a cobalt binder phase that cements the WC grains together. There is no international standard based on composition and mechanical properties for hardmetals. There is, however, an ISO standard for machining applications. The combination of the hard but brittle carbide and the soft but tough cobalt results in unique hardness-toughness combinations, and since their introduction in the 1920s hardmetals have become the most universal materials for metal cutting, rock drilling etc. [182–184].

Two main factors affect the cutting properties of a simple WC–Co hardmetal: the cobalt content and the grain size of the tungsten carbide. Depending on the application, the binder content may vary between 3 and 30 mass%, and the WC grain size may range from 0.3 to 30 μm. If increased red hardness and crater resistance is required, as *e.g.* for high speed turning, cubic carbides such as TiC, NbC and TaC are added to the basic WC. Due to their microstructure, hardmetals are naturally hard and do not lose their hardness irreversibly as do HSS; on the other hand, they are more difficult to machine and can only be produced by powder metallurgy techniques. Their properties depend strongly on the binder content, hardness ranging from 800 to 2400 HV 30 (83 to 95 HRA), bending strength from 600 to 3000 MPa, Young's modulus from 400 to 700 GPa, and thermal conductivity from 20 to 120 W/m·K (the latter indicating the considerably 'metallic' character of WC).

Hardmetals are manufactured by intense mixing of the starting carbide and binder powders – usually in ball mills or attritors – compacting in dies or isostatically, and liquid phase sintering in vacuum. In the final stage of the isothermal sintering process, gas pressure – usually about 70 bar argon – is introduced into the furnace to close the last remaining pores ('Sinter-HIP'). Drills are frequently manufactured by mixing the powders with a plasticizer, extrusion, shaping, debinding, and finally sintering.

Today, ISO standard R 513 ('Application of carbides for machining by chip removal') is used predominantly for standardisation; this standard is however based on applications rather than on compositions or properties. Therefore, the same class of hardmetals (*e.g.* P20) may include materials of considerably varying composition and mechanical properties.

This standard classifies applications by first taking the workpiece materials and nominating them into one of three main machining groups, designated P, M or K, (to which different colour codes have been allocated) and then subdividing each of these three groups into machining applications which are identified by a number prefixed by letter of main group [6,164].

P – colour code BLUE. This group is subdivided to P (01, 10, 20, 30, 40, 50). Highly alloyed workpiece materials which fall into this group are in terms of PM machining: steel including ferritic and martensitic and similar cast steel. The P applications need hardmetals containing TiC to combat the problems of cratering. The typical hardmetal for finishing operations will have a low Co content, *e.g.* 6% to 7%, a high TiC content of around 20% and TaNbC of the order of 10%.

M – colour code YELLOW. This group is subdivided to M (10, 20, 30, 40) which contains generally less TiC. Workpiece materials that fall into this group of hardmetals for machining are also in terms of PM machining: hard manganese steels, austenitic steels, heat resistant alloys, cast steels, non-ferrous alloys. The range of hardmetal alloys used for the M applications is much narrower, *i.e.* Co contents are from 6% to 9%. TiC from 4% to 8% and TaNbC from 5% to 9%.

K – colour code RED. This group is subdivided to K (10, 20, 30, 40). Workpiece materials which fall into this group can be partly: short chipping malleable cast iron, austenitic stainless steels and non-ferrous metals. The K30 and K40 applications require toughness and therefore need a hardmetal with a high Co content to withstand the shock. The very light finishing operations, K01, present no problems of toughness and so the hardest, most wear resistant, plain Co–WC grades are used, *i.e.* 5% Co and fine grain WC [164].

Within each main area as shown there are numbers indicating the varying demands of machining, from roughing to finishing, Fig.4.12. An overview of the application of individual grades of hardmetals, including those coated for machining of steels, is shown in Fig.4.13.

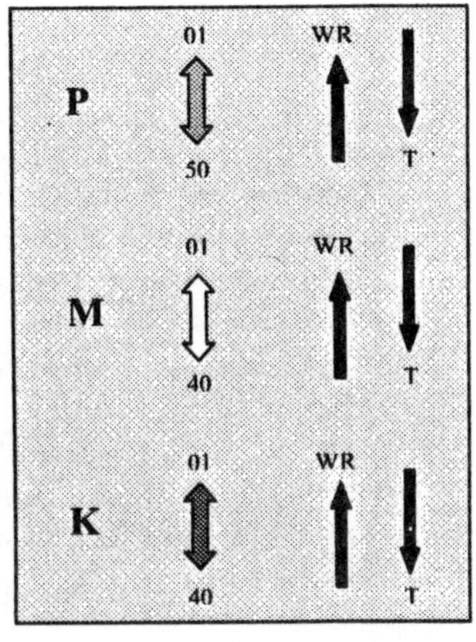

ISO Group	ISO Type	'Standard' HM	Multirange HM	Coated HM
P	P 01	F2; F1	T2	S 20 CN
	P 10	S1; S 1,1		
	P 15	—	S 20	
	P 20	S 2		
	P 25	—		
	P 30	S 3	S 25, S 30	S 20 CNA
	P 40	S 4	S 45, U 20	
	P 50	S 5		U 20 CN
M	M 10	U 1	S 30	
	M 15	—		
	M 20	U 2	U 20	
	M 30	U 3	S 45	U 20 CN
	M 40	—		
K	K 01	H 3		S 20 CNA
	K 05	H 2	H 05	
	K 10	H 1	H 10	
	K 20	G 1		U 20 CN
	K 30	G 1,1		H 10 CN
	K 40	G 2		

Fig.4.12 The groups of hardmetals (P, M, K) with numbers (01 to 40, 50) indicating the demands on wear resistance (WR) and toughness (T) [127].
Fig.4.13 (right) Overview of base hardmetals used in machining of steels [112]. P – materials forming long chip, M – materials forming short and long chip, K – materials forming short chip.

The hardmetal grades P40 and K10 are most often used in the manufacture of solid carbide drills. Micrograin K10 grades are today most common thanks to their good combination of hardness and toughness. In the hardest and most abrasive PM materials, carbide drills are essential to attain acceptable productivity and tool life. K10 micrograin grade cemented carbide typically consists of 10% cobalt and 90% tungsten carbide (WC).

Ultrafine cemented carbide with fine WC-crystals (~0.5 μm) exhibit a high bending strength (up to 4300 MPa), hardness (2000 HV 30) and compressive strength. The combination of these properties joined with combination with a hard layer, prevents diffusion and adhesive wear and, therefore, is suitable for turning of the heat treated steels at interrupted cut [164,165,167,185].

The main fields of application of hardmetal tools are shown in Fig.4.14. In the very broad field of application of this group of cutting materials the hardmetal tools for machining of metals are in the range of 0.5 to ~2 μm WC grain size and of 5 to 15% Co. Hardmetal reaming cutters, especially at the cutting speeds >30–40 m/min, exhibit significantly higher tool life compared with HSS cutters. The use of cutting fluids is necessary [151,186].

Hardmetal tools are available in a multitude of compositions and geometries, the major manufacturers offering specially optimised tools for the main machining operation. The information about the hardmetal cutting tools designed especially for machining high

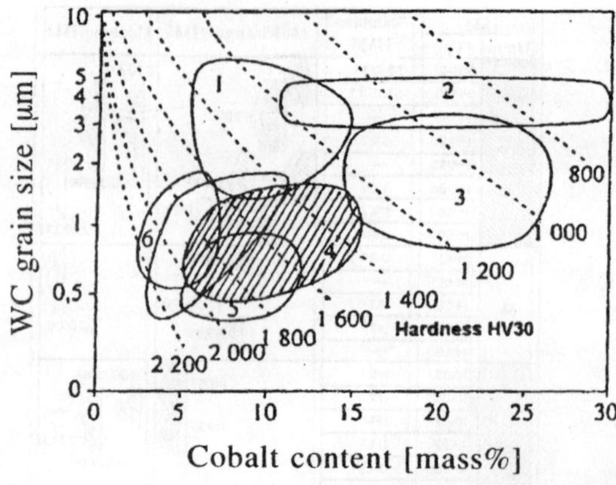

Fig.4.14 Fields of application of hardmetal tools according to the WC grain size and the cobalt content. 1 – roto-percussive drilling (mines and public works), 2 – hot rolling cylinders, 3 – cold forming tools, 4 – cutting tools for metals, 5 – cutting tools for composite materials, 6 – wood cutting tools [167].

strength-hardness PM steels are lacking. Sintered steels are too narrow an application range to make the development of special hardmetal grades really attractive for the hardmetal manufacturers. However, there are grades available that are recommended for cutting of PM steel parts. As an example, for turning of sintered steels, a leading hardmetal manufacturer recommends a P15 grade (HM) with about 8% cubic carbide (TiC–TaC) in the core but not in the rim, the inserts then being coated by medium temperature CVD, on top of the multilayer coating being Al_2O_3. Recommended cutting parameters are 90–100 m/min cutting speed and 0.25 mm/rev. feed, the cutting depth being adapted to the respective application [187].

It is recommended in general to machine PM high strength steels with hardmetal tools because HSS tools have in many cases low tool life [188] while, as mentioned above, machining with diamond tools is expensive and for carbon steel not suitable.

Some types of hardmetal inserts have pressed grooves into their rake faces with differing configurations of chip control, Fig.4.15.

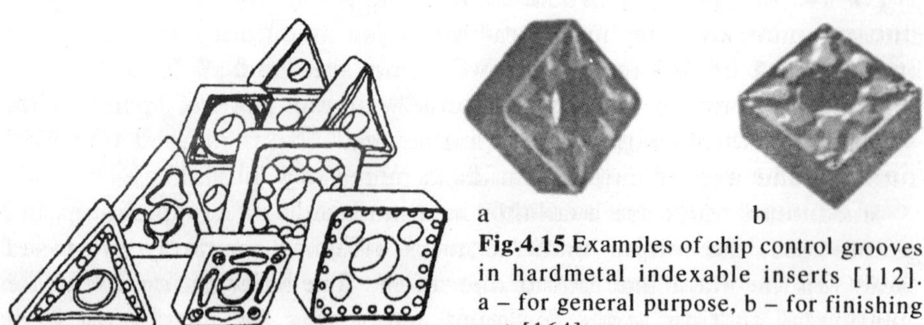

Fig.4.15 Examples of chip control grooves in hardmetal indexable inserts [112]. a – for general purpose, b – for finishing cut [164].

4.1.3 Stellite

Stellite is the trade name for a group of cobalt-based alloys which are naturally hard and do not require heat treatment to attain their cutting properties. One grade is now offered as Stellite Alloy 100. The composition and properties of this alloy are: 34% Cr, 19% W, 2% C, balance Co; hardness ca 950 HV, hot hardness 535 HV at 700°C compared to 175 HV for high speed steel, density 8.75 g/ cm^3.

Stellite turning tools are available as tool bits and as tipped tools with cutting profiles similar to those of the high speed steel but welded tools. The tool bits are solid stellite and the tipped tools are made from cast stellite tip which are brazed onto steel shanks in the same way that hardmetal tools have been originally manufactured.

Stellite tools are used to cut surfaces which are extremely difficult to machine with hardmetals and where the cutting edges of hardmetal would be viable to fracture. A typical example is machining of welds. The range of cutting speeds in which it will perform satisfactory is lower than that for hardmetals but a little higher than that for high speed steels (stellite has a narrow, specialised field of application).

The tipped tools are supplied with the cutting profiles and rakes already ground and ready for use on the machine [164].

4.1.4 Cermets

The term 'cermet' is used to designate a composite material comprising at least one ceramic phase and one metallic phase (WC-less). The term ceramics in this connection includes oxides, carbides, nitrides, borides as well as covalent compounds. These materials are manufactured by powder forming methods. Today, the term 'cermet' applies to hardmetals based primarily on cubic carbides such as titanium-base hard material phase (TiC, TiN, TiCN) with a Ni–Co (binder) phase [68]. The grain size of TiCN is usually in the range of 0.5 to 2 μm. Figure 4.16 illustrates the structure of a cermet. The cermets have the density in the region of 6.0 g/cm^3.

Each manufacturer has own compositions and many include further carbides such as Mo_2C, WC and TaC. The cermet tools or inserts have the hardness of ceramics around 1600 HV. Cermets can be manufactured as 'graded' materials: by varying the sintering atmosphere the C:N ratio in TiCN can be varied esp. in the surface regions, resulting in cutting performance similar to that of coated materials.

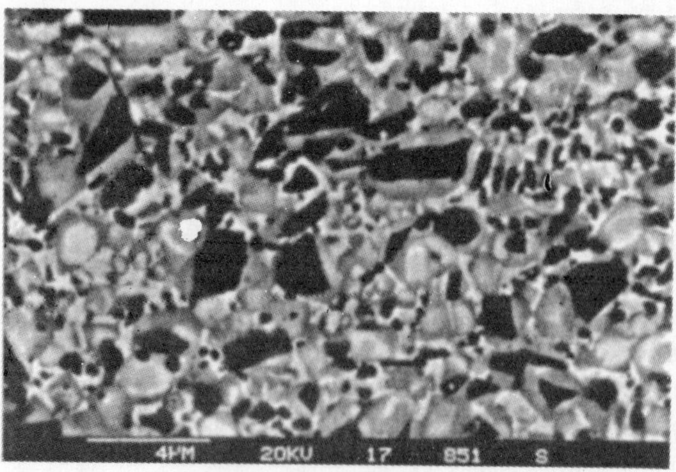

Fig.4.16 Microstructure of a TiMoCN–Ni – cermet. (Photo Courtesy W. Lengauer).

Cermets are capable of machining at speeds which exceed those applicable to hardmetals. They perform well with medium to light cutting in both turning and milling applications on steel workpieces. These cutting materials exhibit a high wear resistance and a low tendency for adhesion to the steel materials. Some cermets are used also for interrupted cut (high thermal shock resistance) and for milling, *e.g.* end-mills and for turning at high cutting speeds [164,165], although the resistance against thermal and mechanical fatigue is inferior to that of hardmetals. Compared to hardmetals they allow higher cutting speeds because of their higher hot hardness and plastic deformation resistance (at lower toughness), and they display a reduced edge built-up at lower cutting speed because of their chemical inertness. Cermets achieve high surface finishes and close size control in turning, grooving, threading, and milling applications. Expensive grinding operations are frequently replaced through machining with cermet cutting tools. Coolant is also applicable [167], but especially with interrupted cut dry machining is preferable to alleviate thermal shock loading.

If the individual phase hardness in workpiece material is less than 35 HRC then cermets can be used as they show a lack of affinity for the built-up edge to occur. Moreover, cermets can be utilised at the high speeds and light cuts demanded for PM near-net secondary machining operations, with suitable insert geometry. Most cermets are manufactured by cementing titanium carbonitride (TiCN) grains with a Ni–Co binder, although up to 11 alloying elements can be presented in the complete matrix. The TiCN principally provides the cermets hardness, as well as built-up edge resistance and, importantly, chemical stability.

The gap in composition between high speed steels and hardmetals is filled by composites consisting of a dispersion of TiC (or more scarely TiN) particles (50 to 75%) in a heat treatable steel matrix (chromium steel, martensitic Ni–Cr steel or maraging steel). The materials, marketed under the mark 'Ferro-TiC' (or 'Ferro-TiN'), are being used in many applications requiring any combination of wear, heat and corrosion resistance.

'Ferro-TiC' (Ferro-Titanit, Fe matrix + up to 30 mass% TiC) is rather softer as-annealed (35 to 50 HRC) so that blanks of this material can be easily machined to finished tools. Heat treatment allows then increasing the hardness to its usage level up to 73 HRC. Thus, worn tools may be annealed, reshaped and heat treated in order to be re-used mainly for wear resistant applications [33]. The softening of the matrix at annealing can cause the decrease in amount of the hard phases.

4.1.5 Ceramics

The group of ceramic tool materials includes sialons, aluminium oxide-based ceramics, and silicon nitride. They are very brittle for cutting of porous materials. In machining of powder based materials, ceramic tools were used perhaps only in machining of powder forged steels. The advantage of the ceramics is high hardness at high temperatures. Below 800°C, hardmetals have superior strength compared with ceramic

There are three groups of the cutting ceramics:

- oxide ceramics based on practically pure Al_2O_3,
- oxide ceramics with addition of ZrO_2,
- oxideless or nitride–carbide ceramics.

The alumina based ceramics are harder than hardmetals and, therefore, can operate at higher speeds without plastic deformation of the cutting edge occurring. All commercial alumina tools contain zirconia in concentrations ranging from 2 to 10% for interrupted cut or for machining steel. A further most common addition is TiC ('black ceramics') for machining hardened steel and case-hardened steel. Their higher hardness ranges make them more sensitive to shock. On hard materials they can produce surface finishes which are commonly obtainable only by grinding.

The sialon materials are based on silicon nitride (Si_3N_4+SiO_2+Al_2O_3+Y_2O_3). They perform well on cast iron at high cutting speeds but are not suitable for general steel machining. The main advantage of sialon is the increased chemical stability. Sialon tools are almost exclusively used for the machining of heat resistant alloys.

Silicon nitride (Si_3N_4) has a good toughness which permits a rough turning of grey cast iron with interrupted cuts and milling at high feed rates per tooth. It should not be used to machine steels. Hot pressing technique has to be used to achieve good strength. The hardness of this cutting material is ~1800 HV and its density is 3.2 g/cm^3. The grain size is in the range 2 to 3 μm. It has good wear resistance and cutting edge strength, it also has high resistance to thermal shock [164].

Zirconia (ZrO_2) toughened alumina (ZTA) is an example of dispersion toughened ceramic composites. The absence of any chipping over the cutting edge and the occurrence of closed crater contour have demonstrated the improved toughness and thermal shock resistance characteristics of ZTA ceramics [164,167,189].

The application of Al_2O_3+TiC known as black ceramics and ZTA (Zirconia Toughened Alumina) are classic examples of particulate reinforcement/dispersion strengthened materials. The improvement in the surface finish and a reduction in flank wear with cutting speed is achievable by using these cutting tools [116]. The use of ultrahard tools is necessary only in the case of heat treated or sinter hardened PM steels.

Ceramic cutting tools have been approved at machining of cast iron – turning and milling especially at extremely high speeds also in relation to the coated cemented carbides. Their use in PM production was not successful [165].

4.1.6 Cubic boron nitride

Cubic boron nitride (CBN) is employed for metal cutting as polycrystalline boron nitride (PCBN) and is used as a cutting material when hardmaterials become limited in the cutting speeds that can be employed and at very high workpiece hardness. The CBN material is prepared by pressure sintering the CBN powder with a very small amount of binder at high temperature (1500°C) and high pressure (5 to 6 GPa) when the hexagonal crystals are converted into a cubic structure which is extremely hard – of the order of 4000 HV (the hardest material with the exception of diamond). They are known as high CBN content and low (<65%) CBN content cutting materials respectively. The low CBN content cutting material has a greater resistance to wear than the high content material under light cutting conditions and yet it has a somewhat reduced hardness. This applies to hard workpieces, *e.g.*

in PM hardened steels. It does not perform well on soft steels and austenitic stainless steel.

CBN material has proved to be particularly adept at secondary machining, as have coated cermets and advanced multilayered cemented carbides, though to a lesser degree. CBN is thermostable up to 1200°C and does not suffer from the reactivity with iron as does diamond. Principally, these materials are employed because they resist abrasion, edge breakdown and built-up edge formation, whilst aiding in the production of good quality surface finishes.

Notably, CBN has shown superior performance when machining wrought materials such as alloy steels and certain cast irons with hardness of 45 HRC and above which, due to their high hardness and abrasion resistance, would be beyond the limit of most conventional tool materials. For PM applications, CBN can deal with matrix hardness of 25 HRC if the hardness of the microstructural components varies markedly. The hardness of the microstructure phases is a critical parameter because if it exceeds 50 HRC CBN can be utilised at any PM bulk hardness. Toughness is one factor that limits their production applications, specifically when interrupted cutting occurs, or if the porosity level is high. Any inserts must have strong edges, hence the use of edge preparations such as T-lands and large edge hones [190].

The use of CBN cutting materials is concentrated on cutting of hard materials. More detailed tribological investigation is needed for the explanation of the dependence between the microstructure of CBN-cutting material and the cutting process, including workpiece material. The selection of the CBN grade is important; tougher grades such as solid CBN can be used, but this may influence PM surface finish. The medium grain CBN is the most popular for PM secondary machining operations, *e.g.* finishing operation on sintered steel with a bainitic micro-structure [191].

By the use of CBN for machining of heat treated steel the surface quality and dimensional accuracy is obtainable equal to those achieved at conventional grinding and honing [164,167,190].

In summary, it should be said that if PM grain (phase) hardness goes beyond 50 HRC, CBN of medium grain size should be used for general secondary machining operations but, if the cost is a serious issue and particle hardness is lower than 50 HRC, a cermet tool material is preferable. Alternatively, if bulk hardness value is less than 28 HRC, an MTCVD cemented carbide with a coating process with TiCN can be employed.

As stated in [192], the reaction of CBN tool and MnS–MnO was observed and analysed by optical microscopy and line micro-analyses. The concentration of B decreased near the tool surface and Mn diffused into this zone of the tool until 80 μm inside. But S was scarely detected in this zone. Consequently, it was considered that CBN tool reacted with Mn or MnO, and the diffusion of B and Mn weakened the tool and caused a large wear in turning Fe–Cu–C–MnS steel. CBN tool reaction with FeO and a reacted zone of about 40 μm was also observed near the tool surface.

This physical–chemical behaviour of CBN material is important especially in cutting of sintered steels with addition of machining aids containing manganese, as, for example, MnS. Similarly sulphur addition to sintered carbon steel improve the machinability for most cutting tool materials except CBN tools. This need not be the case at a finishing operation.

Figure 4.17 shows some summarising results concerning the cutting speed, cutting tool material and life for structural steel and in Fig.4.18 for stainless steel.

According to these figures, CBN tool showed a very high cutting performance for sintered carbon steel at high cutting ranges. Sulphur addition to sintered carbon steel improved the machinability

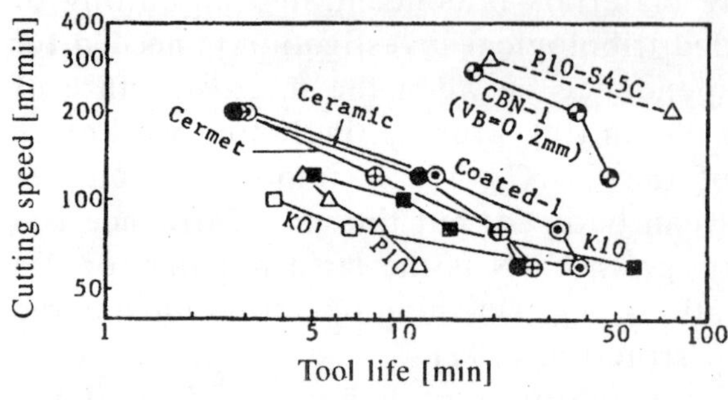

Fig.4.17 Tool life curves of various tested tools in machining of sintered Fe–2Cu–0.5C steel (density 6.5 g/cm^3) and wrought steel (S45C) [192]. *Turning conditions: depth of cut 0.5 mm, feed 0.1 mm/rev., criterion 0.3 mm average flank wear.*

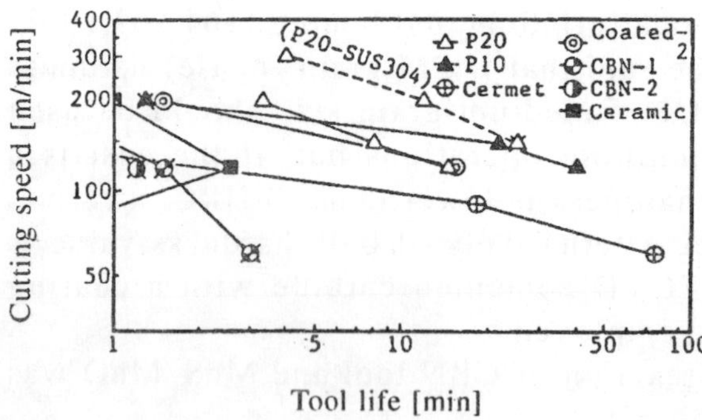

Fig.4.18 Tool life curves of various tools in machining of sintered and conventional stainless steel (SUS 304) [191]. *Criterion 0.2 mm average flank wear, depth of cut 0.5 mm, feed 0.2 mm/rev.*

for most tools except CBN. In machining of sintered stainless steel HM P10 tool showed longer life than other tools, and especially at low cutting speeds below 150 m/min.

4.1.7 Polycrystalline diamond (PCD)

The hardness of polycrystalline diamond approaches that of natural diamond but is not equal to it. PCD is the hardest material we know and has superior abrasion resistance to any other cutting material.

Polycrystalline diamond is used to 'tip' cutting inserts. They are brazed at temperatures which should not exceed 800°C at which the stability of diamond starts to decrease and with synthetic diamonds, thermal damage may occur through catalyst inclusions. It is able to cut at very high speeds on abrasiv nonferrous workpiece materials, such as hypereutectic Al–Si cast alloys. It should be not used on steels and cast iron and by this not on sintered steels. Diamond reacts easily at high temperatures with carbide-forming and carbon-dissolving metals, such as iron and sintered steels [164,193].

4.2 TOOL COATINGS AND HARDENING PROCESSES

4.2.1 Tool coatings

The standard tool materials used for machining operations such as high speed steels and hardmetals offer excellent hardness–toughness combination but the hardness and the chemical resistance are not quite up to the level of ceramic cutting tools which on the other hand are rather brittle. A very successful method of improving the cutting performance of tough tool materials is coating with hard ceramic layers. These coatings – mostly carbides, nitrides, and oxides – offer improved wear resistance and also inhibit interdiffusion between tool and workpiece. Thus the positive cutting properties of ceramics, as, for example, high hardness and chemical inertness, and those of the base materials, high toughness and mechanical and thermal fatigue resistance, can be combined.

The coatings are considerably harder than normal hardmetal and so the cutting edge has more abrasion resistance, and therefore they do not rub away as the chip is flowing over the rake face. The materials which are used as coatings all have an extremely low solubility in iron at the temperatures which arise during machining and so no cratering occurs as a result [164].

It is however essential that the coatings adhere well to the base material and are not removed also by severe cutting conditions, *e.g.* interrupted cutting. This problem, esp. chipping of the cutting edge, initially plagued the coated hardmetal inserts. Furthermore it must be kept in mind that the high cutting performance is retained only until the coating is worn away; afterwards the behaviour of the base material is to be expected. Therefore, regrinding of coated tools is not a practical method; this does not play a major role for hardmetals anyhow but may be a limitation for the use of HSS tools.

The selection of the coating material as well as of the coating technique depends on the machining operation and the base tool material. In the latter case, esp. the maximum temperature applicable during coating without adversely affecting the base material properties is a critical parameter. This holds particularly for high speed steels which, to ensure geometrical precision, have to be coated in the as-heat-treated state. Quench-and-temper heat treatment after coating would result in distortion; furthermore the hard but brittle coatings would tend to peel off during quenching. Therefore, the maximum coating temperature for high speed steels is ~550°C, *i.e.* it corresponds to the tempering temperature for these steels. For hardmetals, which do not irreversibly soften during heating, coating at, for example, 1000°C does not cause problems.

Common coating processes are either physical vapour deposition (PVD) or chemical vapour deposition (CVD). PVD processes, mostly based on sputtering techniques, can be applied for a larger range of coating compositions and operate at lower substrate temperatures, *i.e.* the base material to be coated is not thermally overloaded, and the residual stresses are mostly compressive which improves adherence, but since the processes are oriented, 'shadow' effects may occur that result in irregular coating thickness esp. for geometrically complex tools. CVD processes result in more regular coating since there are no 'shadow' effects and enable also multilayer coatings but generally need higher temperatures, and the residual stresses tend to be on the tensile side, especially at high deposition temperatures. Therefore, considerable effort has been directed towards development of lower-temperature CVD processes, in an attempt to combine the advantages of both PVD and CVD processes. Thin layers, which are necessary for very sharp tools (when only thin chips are top be removed), are preferably deposited by PVD processes [194].

Today, the main processes are [165]:

- standard – high temperature – CVD (HTCVD) with a working temperature of 900–1000°C,
- medium temperature CVD (MTCVD) operating at substrate temperatures of 700–900°C,
- plasma-assisted CVD (PACVD) operating typically at 550°C

(*i.e.* it is also applicable for high speed steels),
- PVD techniques working typically at 500°C and below.

Both for CVD and PVD, industrial-scale equipment is commercially available as state-of-the art, PVD equipment being more sophisticated and therefore more expensive. Also the deposition rates of PVD are markedly lower, resulting in lower throughput for a given coating thickness. Therefore, PVD is used mostly for thin (3–5 μm) and special coatings on temperature-sensitive tools while CVD is applied for standard coatings with average thickness (5–12 μm) on mass-produced hardmetal tools.

The most common coating materials for hardmetals are TiC, TiN, TiCN and Al_2O_3 deposited by CVD. TiC and Al_2O_3 are very hard (>3000 HV). The well-known gold-coloured TiN is slightly less hard (about 2600 HV) [164] but offers a lower friction coefficient against metallic workpiece materials [127]. In practice, single-layer coatings are used only rarely on hardmetals but multilayer CVD types prevail since for a given coating thickness many thin layers result in better toughness than a few thicker ones. Typically, a thin TiC layer (*e.g.* 0.5 μm) is deposited directly on the hardmetal substrate (TiC adheres better to hardmetal than TiN), and then the atmosphere composition is changed to result in a transition layer of TiCN and finally TiN. In many cases, Al_2O_3 is then deposited on top, or even alternating layers of Al_2O_3 and TiN. Thicker, multilayer types dominate for non-interrupted cuts while for interrupted cuts such as milling, thinner coatings with less numerous layers – *e.g.* TiC–TiCN–TiN – are preferred. A very important item is the preparation of the substrate for optimum adhesion of the coating; today, hardmetals with graded structure, *e.g.* Co-enriched near-surface zones, are available for coatings.

For high speed steel tools, *e.g.* drills and taps, thin single-layer TiN coatings are common, deposited mainly by PVD processes. TiN shows much better adhesion on high speed steels than on hardmetals. In addition to TiN, also TiAlN (futura nano top) coatings, with improved oxidation stability important for, for example, dry machining, are applied today by PVD processes, as are other hard phases, *e.g.* such consisting of CrN, making use of the ability of PVD to deposit a large variety of hard phases.

Diamond coatings deposited by low-pressure CVD methods, from CH_4–H_2 gas mixtures, are applied on tools for machining of non-ferrous metals such as Al–Si alloys, but for machining of iron base materials the low stability of diamond in contact with iron and carbon is detrimental. C–BN CVD layers would be very desirable but are not yet state of the art.

Both CVD and PVD coatings positively affect the cutting performance of tools. The hard ceramic surface layers, with fairly low thermal conductivity, enable higher cutting speeds without thermal overloading of the base material, and the generally lower friction coefficient against metals lowers the generation of heat at the contact faces. Furthermore, the high hardness reduces abrasion, *i.e.* mainly the flank wear. Interdiffusion between tool surface and workpiece, esp. the chip, is inhibited due to the high thermodynamic stability of the coatings, resulting in reduced adhesive, esp. crater, wear. For machining of porous PM materials, the reduction of microfatigue effects on the cutting edges by the coatings is a major item [45].

In machining practice, coated tools enable significantly higher cutting speeds than similar uncoated ones [195]. For coated hardmetals, a performance increase by up to a factor of 9 has been reported [164]. Furthermore, coating broadens the application range for tools. Therefore, of the indexable hardmetal inserts 80% are coated at present. In any case, the tool and the coating can be specially combined to result in optimum performance for any given machining operation.

For powder metallurgy steels, data about the machining behaviour of coated tools as compared to uncoated ones are scarce. However it can be expected that also here the advantages of coatings can be brought to bear, both in the case when hard, abrasive sintered steels have to be machined and also in the case of softer workpiece materials when adhesive wear prevails; in the latter case the lower coefficient of friction should be a major advantage [194].

Figure 4.19 shows examples of coatings on IM and PM high speed steel and on hardmetal.

Another possibility for machinability enhancement is coating cutting tools with a nickel–phosphorus solid lubricant (chapter 6.3.1).

The proper selection of the cutting tool, including the coating, is a very important and complicated task. As shown, at present time there are many coating systems on HSS, hardmetal, and cermet-base materials. Therefore, it may also be more complicated for users to choose the optimum cutting material for the solution of its machining task or to use the chosen cutting material optimally especially for machining of PM steels.

Different cutting tools have different machining characteristics and properties. For instance, Fig.4.20 illustrates the improvement in machinability of a PM alloy when using a TiN-coated high speed

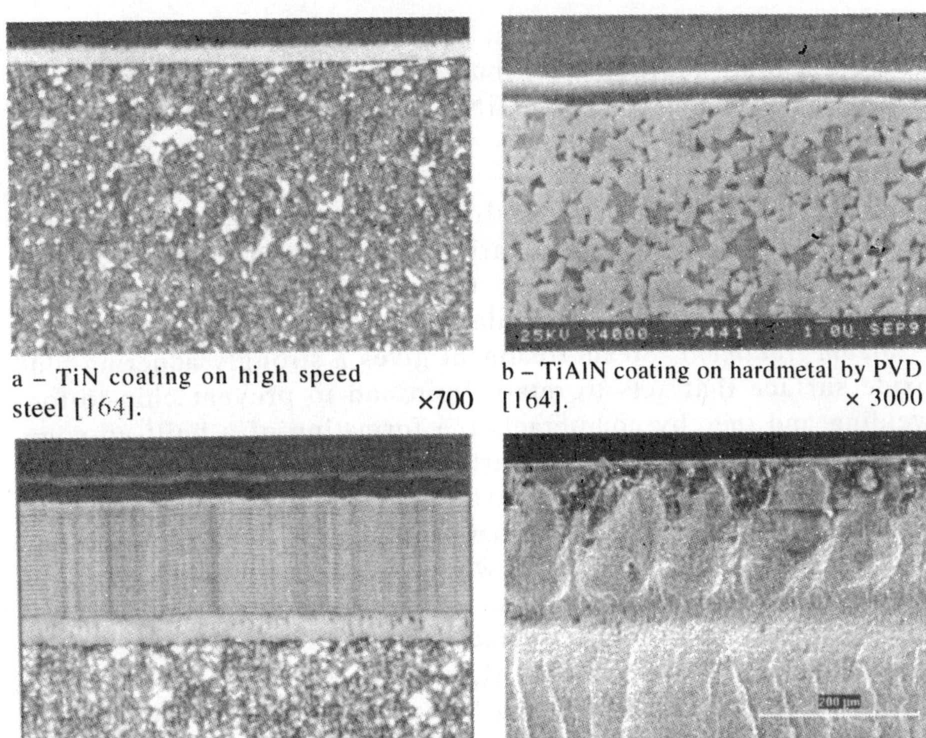

a – TiN coating on high speed steel [164]. ×700

b – TiAlN coating on hardmetal by PVD [164]. × 3000

c – 51-layer coating structure of Valenite's new SV421 cutting insert [196].

d – Bending fracture surface of TiN coated PM HSS (M2) (HIPed) [197].

Fig.4.19 Illustrations of various coatings on high speed steel and hardmetal substrate.

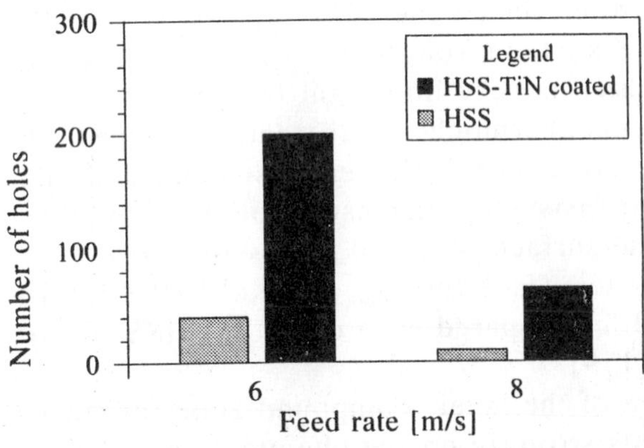

Fig. 4.20 Impact of tool selection on machinability of PM alloy (FC-0208). Density 7.2 g/cm^3 [acc. to Ref.105]. *Drilling test: HSS 14 mm drill-bit without and with TiN coating.*

steel drill. (The feed rates seem to be high for drilling this steel).

At both feed rates and cutting speeds, the TiN-coated drill had a significantly higher life than the uncoated drill. At a faster feed rate and cutting speed, the number of holes drilled by both types of drill was lower ~70% lower. Lower feed rate and drilling speed were favourable for improvement in drilling with coated HSS drills. The 0.5% MnS addition contributes more also to the decrease of relative cost of drilling with TiN coated drills compared to HSS drills by about 50% [105].

4.2.2 Heat and surface hardening treatment processes

Tools may be treated with a variety of surface treatments beside coating. The following surface treatments of cutting tools have been used to good effect, particularly in machining of PM materials.

Steam treatment. Steam treatment gives a strongly adhering blue oxide surface that acts to retain fluid and to prevent chip-to-tool welding and thereby counteracts the formation of a built-up edge. This is an advantage particularly for machining of softer, less abrasive PM materials (ASC100.29, ASC100.29-2% Cu and Distaloy AE). Steam treatment can be applied to any bright tool but its most useful applications are with HSS drills and taps.

Nitriding. Nitriding is a process used to increase the hardness and wear resistance of the surface of a HSS tool by introducing nitrogen atoms into the material, causing a N-enriched diffusion zone to form at and below the surface. It is particularly suitable for taps used in abrasive PM materials. It is applied to twist drills when is it desirable to increase the strength and wear resistance of the cylindrical lands [118].

Plasma nitriding. Plasma (ion) nitriding works at substantially lower temperature in comparison with the coating techniques and thus eliminates the negative effect on the HSS substrate. Also the outer dimensions and the surface roughness of a tool are not affected by the process, which is very important for accurate tools. Plasma nitriding is the proper thermochemical treatment especially for high performance application with high wear resistance and high dimensional accuracy. Wear resistance increase through an implantation of nitrogen ions in the surface of a tool is possible. However, in the frontal machinability test (face turning, Fig.3.48) HSS inserts did not attain higher tool life compared with other PM HSS inserts tested shown in Fig.4.10 [179].

The thickness and nature of the layer (compound zone) produced by plasma nitriding depends strongly on the chemical composition of the material. The materials alloyed with Ti, Nb, V, W, Mo, Cr and Al are suitable for plasma nitriding because their nitrides are disperse and hard.

Laser surface treatment of a high speed steel (investigated T15 steel) enables to attain a strong secondary hardening and the temperature of the secondary peak of laser-surface-melted steel (LSM) was higher than after conventional heat treatment. Improved properties of the steel by refining of the microstructure, decreasing the proportion and average size of carbide particles, increasing the supersaturation of austenite and martensite, and eliminating inclusions were attained [198].

4.2.3 Residual stresses

The residual stresses in cutting tools, mainly in high speed steel tools, which originate during production, heat treatment and at grinding, have a large effect on tool life. These stresses can be reduced up to eliminated by, for example, vibrations with the frequency corresponding to the resonance frequency of the vibration of the tool in a special vibrator [199].

The residual stresses in a cutting tool can be eliminated also by the vibrations which are developed by the effect of a magnetic field. The tool is exposed to the effect of the alternate magnetic waves in an electromagnetic coil saturated by the alternate current of the network frequency (50 Hz). Under their effect the tool ascends in to the vibrations and by this the residual stresses in the tool are taken away. Using this method, the life of hardmetal and high speed steel tools was increased by 20–220%, depending on the workpiece material. It was recommended to expose the tool before the first use and after every grinding [200].

To approve this method, the part shown in Fig.4.21 was milled with shank millers exposed to the effect of a magnetic field. Table 4.5 gives the number of parts machined by the shank millers exposed to the effect of the magnetic field with an amplitude of A = 15736 A/m and B = 23603 A/m. 150 components were regularly machined by milling with the millers not affected by the magnetic field. A higher number of parts by ~35% was machined with the shank millers affected by the magnetic field applied at a lower amplitude of 15 kA/m. The form of magnetic waves,

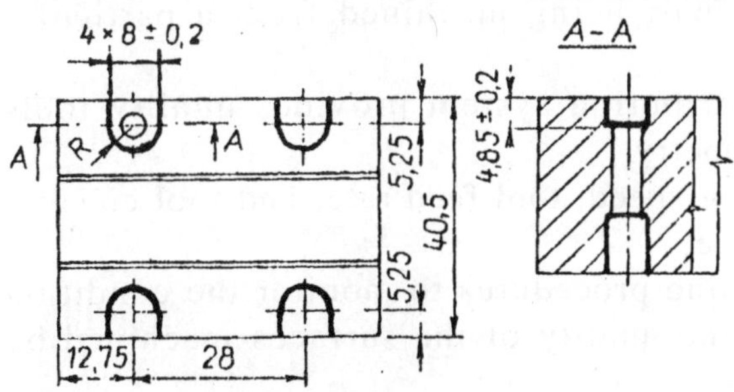

Fig.4.21 Component machined with the shank millers 8 × 20 mm affected by a magnetic field. Workpiece material SAE 52100 (bearing steel) [201]. *Cutting condition: speed 17 m/ min, dry machining.*

Tab.4.5 Number of parts machined with HSS 4.8 mm in diameter shank millers for each state affected by magnetic field at alternating voltage 220 V/50 Hz [201]

Amplitude [A/m]	Shank miller					Average
	1	2	3	4	5	
	Number of components machined					
15736	210	218	170	189	223	202
23603	157	102	54	204	196	143

frequency and time of exposure must be adapted to the weight, form and the size of the tool which should be exposed to the effect of magnetic induction.

4.3 CUTTING TOOL WEAR AND TOOL LIFE

In evaluation of the effect of some factors, as, *e.g.* microstructure heterogeneity and proportion of particular microstructure constituents on the behaviour in cutting of sintered materials, the base material characteristics of the workpiece which cannot be exactly determined for all occasions, must be considered. At all evaluations of the life of a cutting tool the temperature in the cutting zone must be considered. The temperature increases with increasing porosity due to decrease in thermal conductivity of the porous material and due to increasing cutting speed and by this increases the abrasive tool wear [202,203].

Cutting tools wear because loads on the wear surfaces are high and the cutting chips and workpiece that apply these loads are moving rapidly over the wear surfaces. The cutting action and friction at the contact surfaces increase the temperature of the tool material, which further accelerates the physical and chemical processes associated with tool wear. In order to remove the unwanted material as chips, these forces and motions are necessary; therefore, cutting tool wear is a production management problem for manufacturing industries. To successfully manage machining processes, it is required to:

- select the proper machine tools and cutting tools to produce the geometric features in a part being machined from a particular material, *i.e.*:
- ensure that the tool distribution system provides quality tools having the required geometry,
- specify the correct cutting speed, tool feed rate, and tool engagements with the workpiece,
- establish on-line or off-line procedures to monitor the condition of the cutting tool and the quality of the surfaces machined by the tool,

- have maintenance procedures that ensure consistent machine tool operation,
- take into account the cost of machine tool operation and tool use, permitting a clear idea of the economic objective for the machining system in relation to the workpiece material [125].

In the following, some examples for selection of cutting tools are given, Fig.4.22. No systematic collection is available in the literature, especially for PM machining. Uncoated and cotated high speed steel, hardmetal, cermet and other cutting tools designated for machining of wrought soft plain iron up to hardened steels (55–62 HRC) and stainless steels including various non-ferrous materials can be also used or effectively adapted for machining of PM steels.

4.3.1 Effect of cutting tool type in PM turning

The effect of cutting tool material on the cutting tool life as observed in turning of sintered Fe–1.5Cu–(0.5,0.7)C steels based on reduced and atomised iron powder (density ~6.6 g/cm^3) and stainless steel (SUS304, 6.5 g/cm^3) particularises prevailing data regarding the turning of wrought steels. The turning was performed with several kinds of hardmetal tools and with tools like cermets, CBN, TiC and TiC–TiCN–Al_2O_3 coated hardmetals under different conditions (cutting speed, feed and tool geometry) and the results were evaluated in relation to cutting force, cutting temperature, tool wear, tool life, surface finish and surface quality. Summarising the results of the tests, the following conclusions were obtained [190,203]:

- the cutting temperature of a sintered porous material is much higher as compared to wrought steel due to its lower thermal conductivity affected also by alloying,
- CBN tools showed an excellent cutting performance for sintered carbon steel at high cutting speed range. The different carbon contents in sintered steels hardly affected the life of CBN tools. The tool life of CBN with ceramic bonding was more than ten times longer than those of the other tools at high cutting speed range (cutting speed varied from 60 to 300 m/min),
- tool lives of carbide K10 and K01 became very long at low cutting speed. In this case, a fairly large and stable built-up edge was formed at 60 min/min and prevented tool wear. The surface quality of wrought steel turned with a tool with and without a built-up edge formed is shown in Figs.4.23 and 4.24,
- the quality of finished surface of sintered carbon steel was im-

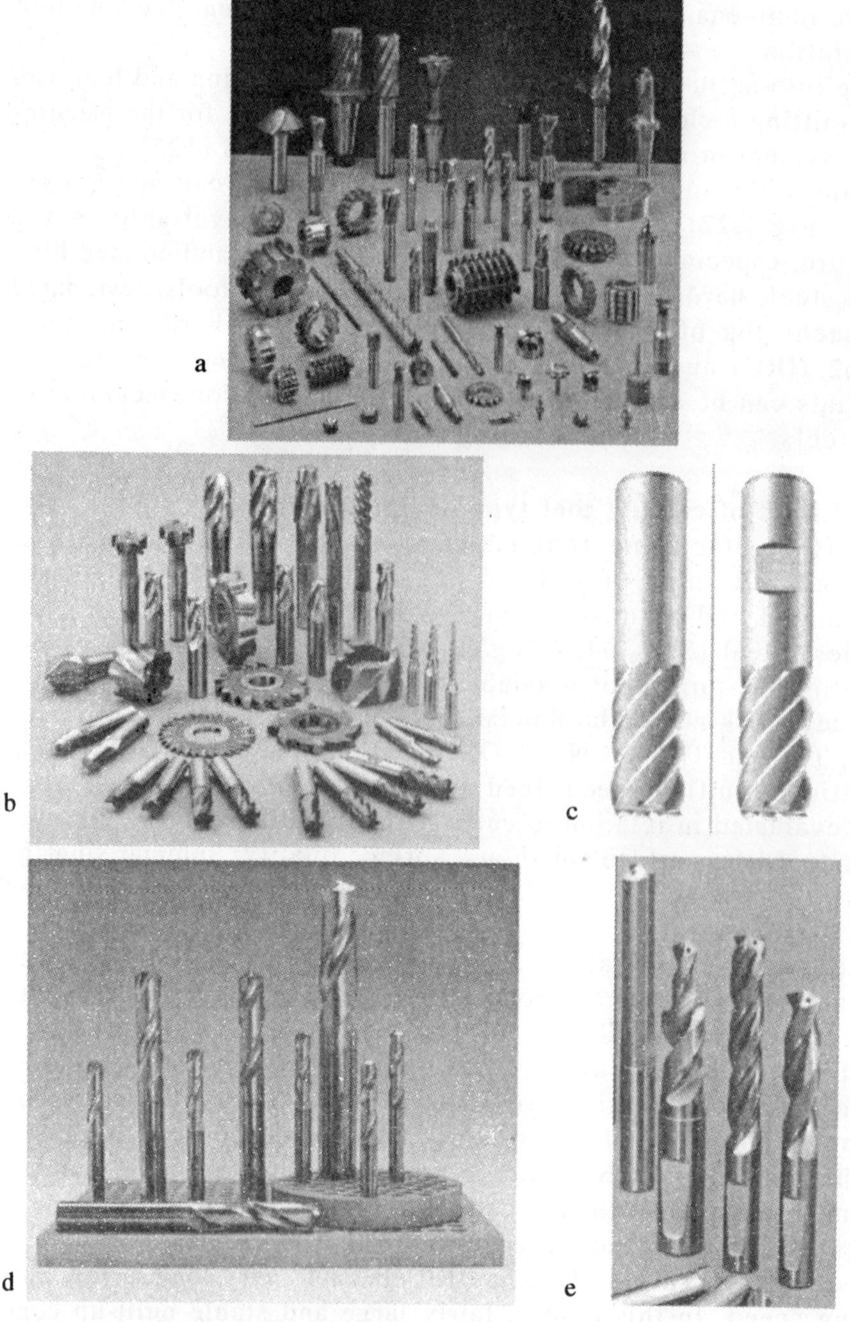

Fig.4.22 Illustrative examples of some cutting tools. a – various HSS (M35, M42, T15, ASP) coated and uncoated tools, b – PM HSS-(8.5, 10.5) and HSS-E tools, c – solid cermet end mills, d – HM drills with the multiphase core MTB 90, e – fine grain HM drills with oil holes.

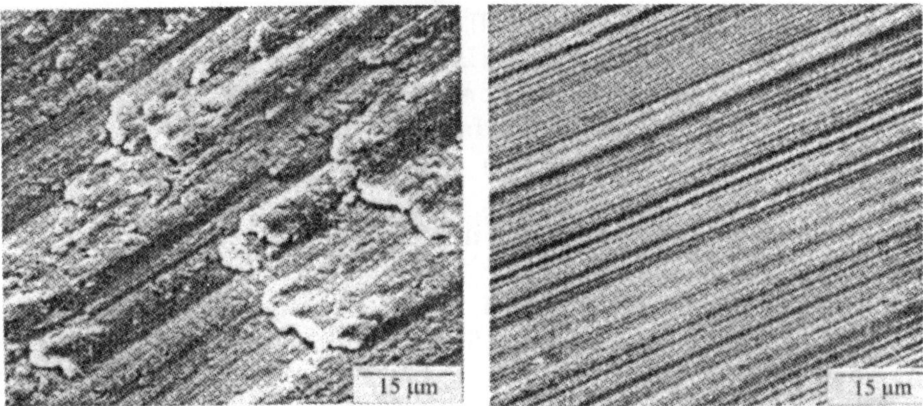

Fig.4.23 Surface of a turned wrought steel covered with the particles of the built-up [112].
Fig.4.24 (right) Machined surface of a material as in Fig.4.23 formed in a built-upless zone [112].

proved when high cutting speed and large rake angle were adapted, and the small pores remain on the finished surface without crushing. Surface finish is affected also by frittering (breakout) caused by plan approach angle and tool nose, occurring normally at the exit point from the cut,

- in the case of sintered stainless steel, work hardening was much less compared with that of conventional stainless steel due to the porosity and the life of carbide tools increased under the condition of small feed and low cutting speed,
- a work hardened layer was found in each workpiece material and can be related to the 'deformation cutting theory' explaining the effect of porosity on machinability of PM steels.

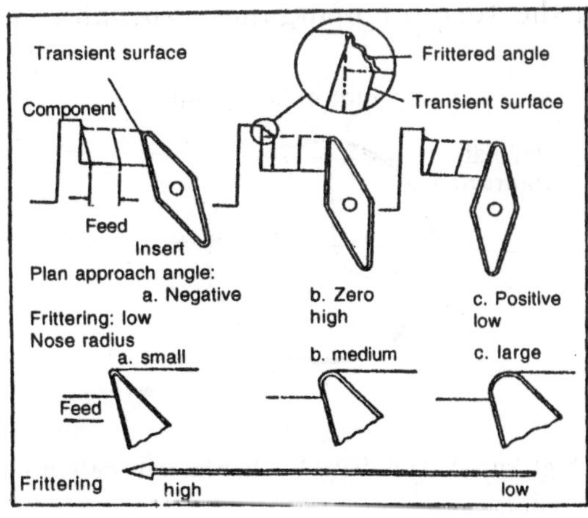

Fig.4.25 Effect of plan approach angle and tool nose size on the incidence of 'corner frittering' when turning a PM part [190].

As shown in Fig.4.25, a small insert nose radius will exacerbate frittering, as a result of the axial force being magnified along a shorter cutting edge length. Increasing the feed rate will increase the pressure on the transient surface leading to higher frittering and poorer surface finish.

4.3.2 Effect of drill geometry in PM drilling

4.3.2.1 Effect of cutting edge

The efficiency of drilling PM materials was increased using a standard but sculptured edge (SE) coated carbide drill shown in Fig.4.26.

Drilling tests with an 8 mm drill (the difference lies in thickness of the web and in clearance of the drill) were performed on the following material grades: Fe–0.8C (F–0008), Fe–2Cu–0.8C (FC-0208) and Fe–2Ni–0.5Mo–0.5C (FL-4605) also with sulphur addition (density 6.3 g/cm^3, hardness 29–50 HRB). The following results were obtained:

- using the SE drill system, PM materials could be machined at significantly higher speeds and feeds than are usually considered acceptable,
- drilling results with a carbide drill for the mentioned structural steels were very similar up to the number of holes drilled in steel AISI 1118. The major difference was in machine load regarding also the workpiece material,
- when the high speed steel drill was used, each material had different average number of holes per drill. The highly alloyed workpiece material resulted in the lowest tool life; the iron–carbon material had the longest tool life. The number of holes was significantly less than for the carbide drill,
- when sulphur was added in the form of manganese sulphide or

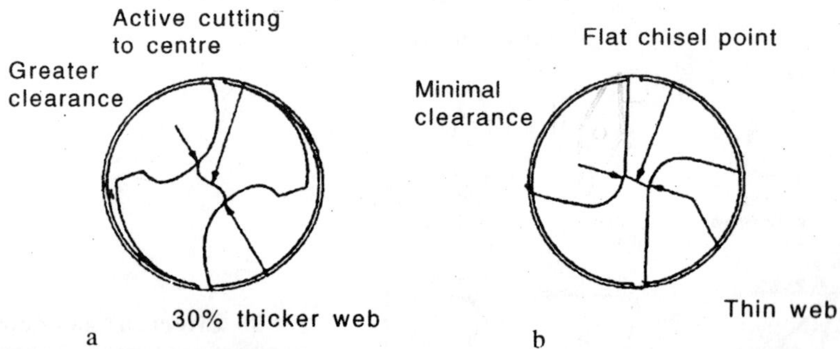

Fig.4.26 Drill point configuration. a – high speed steel drill, b – hardmetal sculptured edge drill [204].

was prealloyed, high speed steel tool life was extended. Within the number of holes drilled with the carbide tool, it appeared that the sulphur addition has little effect. The amount of this life extension varied with the material selection. Under the given test conditions, there seems to be a trend for prealloyed sulphur to show greater improvement in tool life than admixed manganese sulphide,

- the cost per hole drilled decreased in drilling using the hardmetal drill sculptured edge compared with the high speed steel drill.

The shape of the chip in drilling, mostly in machining long chipping materials, can be changed by the splitted cutting edges of a drill as shown in Fig.4.27.

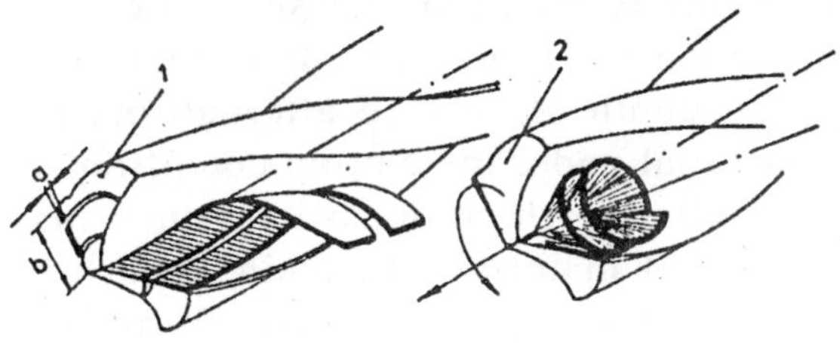

Fig.4.27 Effect of a split cutting edge on the chip shape in drilling. 1 – drill with the splitted cutting edges, 2 –drill with the continuos cutting edges [112].

4.3.2.2 Effect of drill point angle

For testing the effect of drill point angle the HSS 9 mm drills with 29° helix angle, but with 119° and 135° drill point angles were used. The drilling tests were performed with materials as Fe–C, Distaloy SA and AB and PNC45 without and with resin impregnation, all with 0.5% graphite addition (density 7.0 or 6.2 g/cm^3). The following results can be concluded [202]:

- the highest cutting forces (torque and thrust) were measured for Distaloy SA (sponge iron powder base),
- resin impregnated materials, especially Distaloy AB (atomised iron powder base) reduced cutting forces compared with non impregnated ones, more with 135° drill point angle,
- using a drill with 135° drill point angle was more effective in drilling of Fe–C material compared with alloyed materials,
- the effect of drill point angle on cutting forces and by this on drill life is very closely linked to the degree of alloying.

4.4 CUTTING TOOLS, WORKPIECE MATERIAL AND SURFACE INTEGRITY

Dimensional accuracy and smoothness of machined surface (surface integrity) depend on many final factors taking part in machining of a PM material including the cutting tools used.

For highly critical parts, it is mandatory to make individual, specific evaluations. Process parameters that provide surface integrity should be applied selectively to critical parts or to critical areas of given parts to help minimise cost increases.

The base assumption for the highest dimensional accuracy and surface quality of a machined part are minimal machine deformations, thus high stiffness, minimal tool wear and minimal cutting forces. The last can be to reached by reducing the chip thickness to minimum [165].

The surface integrity guidelines should be primarily intended for application to metal removal processes used for final surface generation rather than roughing cuts. It is important, however, to know the type and depth of surface alternations produced during machining of a material under the cutting conditions used, especially for sintered materials for which densified zones may be formed.

In machining of PM materials the main remark is orientated on surface quality of holes drilled because an additional finishing operation to improve the surface finish is not acceptable due to cost reasons.

In general, emphasis is placed on tool wear, but the effects of machining on component integrity can be much more important and obvious means of assessing the geometric, topographical, and metallurgical characteristics of the machined component and of the machining operation as a whole. The surface quality can be linked to other forces generated while cutting and to the vibration that attends machining [27].

4.4.1 Effect of drill geometry on surface integrity

Effect of drill speed and of drill point angle at constant helix angle on surface integrity was tested in Fe–C, Distaloy SA and AB and PNC45 materials without and with resin impregnation. The tests with HSS 9 mm drill with 119° and 135° point drill bit angle at drill speed of 152, 273, and 394 rev./min and at feed rate of 0.09, 0.11, and 0.20 mm/s were performed. Summarising the following results were obtained [205]:

- best roundness quality for Distaloy AB-0.5C steel was achieved with drill bit 119° at medium drill speed and at the highest feed rate. Conversely, low feed rate and low speed combined to result in poor surface finish,
- in Distaloy SA, poorer surface quality was received with 135°

drill bit causing some enlargement in diameter. No apparent effects in surface quality and roundness was detected with regard to drill speeds and feed rates,
- in Fe–C (ASC100.29) material the best results were obtained in the intermediate speed range with the 135° drill bit,
- better surface finish in phosphorus steels was obtained with the 119° drill bit. Low drill speed and feed rates gave poorer diametral accuracy. Drilling of P-containing steels resulted in poor surface finish with a relatively high hole diametral variation,
- resin impregnation improved surface quality at intermediate drill speeds and feed rates without any significant effect of the drill geometry.

The results show that the drilling conditions including drill geometry must be adapted to the composition and real properties of the workpiece material. The explanation of this is based on a relatively larger scatter of the results obtained in general in drilling.

4.4.2 Effect of metallurgy factors on surface finish

Machining ferrite – a very soft material having extremely low resistance to plastic flow – gives rise to a large lateral flow and built-up edge formation resulting in very poor surface finish. One way of avoiding these conditions is to increase the matrix flow resistance, *i.e.* increase hardness (carbon, alloying). Consequently, increasing matrix flow resistance will result in a measurable control over surface finish.

The surface finish of iron-base powder metallurgical parts can be significantly improved to a critical level by carbon and nickel. Increasing carbon and nickel level beyond the critical level results in a deterioration of surface quality, *e.g.* at 0.7% C and 4% Ni. The amount of carbon and nickel content required are interdependent [140].

In Fe–(0.3–0.7)C–(0–4)Ni steels at 0% nickel increasing carbon content resulted in a degradation of surface quality. However, increasing nickel from 0 to 4% resulted in improvement of surface finish with increasing carbon content. The surface finish is not influenced by carbon and nickel individually but also by a strong interaction present as iron carbide and nickel in solid solution in the iron.

The results for the Fe–Ni–Mo powder steels were indifferent in that they were neither good nor bad in most surface integrity tests. However, the surface profiles of Fe–Ni–Mo compacts were characterised by an irregular profile with a high hardening effect and an incorrect cutting action by the drill. Flank wear on the drill was lowest with the Fe–Ni–Mo compacts due, it is thought, to a stable built-up the edge on the drill and not to the high bulk hardness. This is the reverse of the situation for plain iron, where the wear rate was severe and the drill point exhibited built-up edge with blue temper to the flank [206].

Surface integrity is also affected by the presence of sulphur in the material. The most susceptible powder to hardness increase during drilling was the sulphurized variety, the least sensitive being the plain iron. With respect to both roundness and surface texture parameters, the sulphurized powder showed superior drilled hole characteristics. Significantly the most regular surface profile occurred with the sulphurized compacts.

Effect of porosity on surface finish is very broad. The porosity in the compact can destabilise the drill, encouraging vibration, which is detrimental to the drilling process and by this to the surface finish. Increase in drill wear, which is compounded by the lack of cutting fluids (in most cases), causes several unwanted side effects. Thus, drilling forces can increase, which in turn tend to cause the drill to wander axially off centre in a helical manner as the drill generates a hole.

In general, the surface quality of parts depends greatly on the cutting speed. Surface quality (surface roughness) can be improved by increasing the cutting speed and reducing feed. This applies especially to soft materials as the consequence of the built-up formation [207]. Machining under the conditions of the formation of a larger built-up edge which breaks off periodically, results in poor surface finish, Fig.4.23.

To ensuring the best conditions for enhancing machinability, optimising hole geometry, surface finish and reducing residual stresses caused by work hardening may be through the control of drill rigidity and reduce the effects of vibration and of drill point geometry (point modification, helix angle variation, short series twist drills or multicoated and carbide drills) and metallurgy, together with the recommended sintering behaviour [107,116,140,152].

5

Factors Influencing the Machinability of PM Steels

The investigation of machinability in wrought steel is aimed at, and very deeply investigated mainly on, proper cutting process parameters, including tool geometry and material and adequate machine. A special attention is there oriented on automation of the cutting processes for optimum productivity. The machinability of PM steels is in general regarded to be poor compared to wrought steels. This comparison does however not explain the state sufficiently from technical and material points of view. Except for iron–carbon materials (and for the only recently introduced Cr–Mo steels), PM steels generally do not use the same alloying as wrought structural steels. They are alloyed predominantly with copper, nickel, phosphorus, and molybdenum and are also alloyed in much higher amounts compared to wrought steels. If the differences in diffusion solubility of the alloying elements will be taken into account, then comparison in machinability of both base material systems is not adequate even if the porosity is excluded. The machinability of PM steels corresponds to their physical and material characteristics which are special and unique and with them the materials are subjected to machining.

The machining of PM steels is more complicated since these materials are affected by more factors than wrought steels, which however are not sufficiently defined in relation to the machinability. Therefore, it is not possible directly to use machining conditions for wrought steel of comparable chemical composition or mechanical properties for machining of PM steels. Between the many factors affecting the properties of a PM part, very often it is not possible to choose those having the decisive effect on machinability. It is often stated that the machining of PM parts in prevailing number of cases results in excessive tool wear and by this in excessive cost and poor surface finish. The definitive demonstrations for a such statement are missing. The machining processes of sintered compacts offer unique problems that in many cases cannot be defined in advance, and only a limited number of them could be investigated [139].

From this reason also the possibility for improving the machinability of PM parts lies mainly in the principles of powder metallurgy, *e.g.* in PM specified material properties and special additions regarding the machinability. Then the complex machining conditions, defined by the cutting tools, by their geometry and material, and the basic machining conditions and machines should be adapted to the characteristic material properties. All machinability re^ults of sintered materials are new contributions to the knowledge about the interaction between the tool and the workpiece, perhaps with modified properties and adapted machining conditions.

For a detailed analysis the base factors and characteristics influencing the machinability of PM material in a cutting action are shown in Fig.5.1. As shown the machinability of a sintered material depends upon the same factors mentioned for machining of wrought and cast materials but very differently affected by the special properties of PM materials.

As shown in the case of cutting a metallic material using a tool with regard to complete effectiveness in machining of a part, three

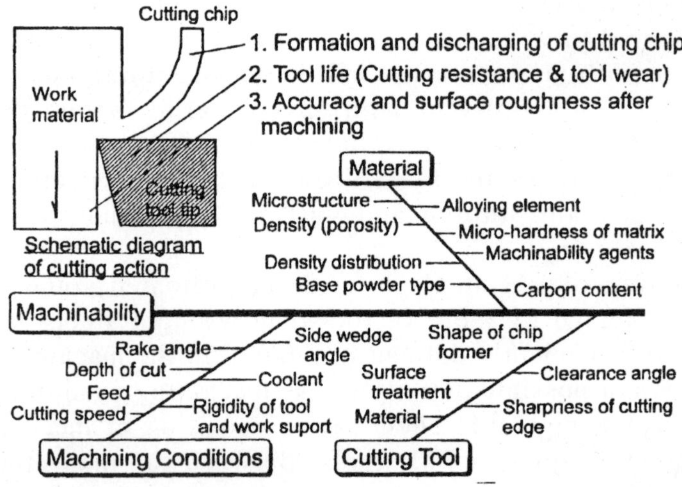

Fig.5.1 Fundamental problems in machining and factors affecting the machinability of PM materials [208] (Ishikawa's diagram for machining of PM steels).

problems occur. The 1st problem is the formation of the cutting chips, the 2nd is the problem of the tool life, and the 3rd the problem of accuracy and surface roughness of the product. By improving all these problems in parallel, improvements in the quality and efficiency can be achieved.

For optimising machining processes it is necessary to identify the influencing parameters. The machinability is mainly influenced, independently of wrought or powder metallurgy materials, by the three interacting systems – workpiece, tool, and cutting conditions:

Workpiece. The material and the geometry of the workpiece mainly affect the machining properties. With PM materials, more factors affecting machining are involved than with cast and wrought ones. How and why the properties of PM materials differ from those of wrought or cast materials was partly explained before and will be further described in following.

Tool. The system 'tool' consists of the material (and coating, if present) and of the tool geometry. The substrates and coatings have to be adjusted to each other and to the geometry. The geometry is often dictated by the machining quality necessary for the planned application.

Cutting conditions and machine system. The machine system is closely linked to the cutting conditions. It can restrict the choice of tools and cutting conditions. For example, only efficient machines with sufficient stiffness and fast spindle speeds enable the use of high performance tool materials, since these materials are often very sensitive to vibrations and only work cost efficiently at high productivity rates.

Cutting parameters, feed, speed, and depth of cut, and the machining conditions have to be chosen with respect to the demands of the application. Both cost efficiency and product quality have to be well balanced for an optimum machining result [209].

To the secondary factors influencing the machinability of PM materials belong also the parameters such as tool geometry, cutting tool material, and machining parameters, their influence roughly corresponding to those in the machining of cast and wrought metallic materials. On the other hand they depend on the PM material itself and are effectively adapted for machining PM steels [108].

In terms of the PM workpiece properties, there are two groups of factors influencing the machinability of PM steels. The first one is formed by the base starting materials and the processing routes used which are characteristic for manufacturing the PM parts, and each of them has some specific effect on machinability. This group of factors is that which is responsible for the machinability of PM steels in positive or negative sense. The second group is formed by special additives to the material that act as machining aids (occasionally also called 'machining agents') for enhancement of machinability, which is an advantage for powder metallurgy since these additives can also be introduced as admixed powders and not only through the melt as is the case with ingot metallurgy.

By this the machinability of PM parts is in the first group of factors influenced by chemical composition, microstructure, mechanical and physical properties of sintered materials as in wrought steels but with other effects causing poorer machinability of PM materials. All machining parameters have also their impact on the machinability. In connection with the other mentioned factors, the material properties cause instability in the cutting process, and, finally, increase in tool wear.

On this the primary major factors affecting the machinability of sintered materials are based which are responsible for their properties, and are considered in evaluating of also the machinability, and can be divided in two more detailed groups are:

- manufacturing and processing technique including compaction, sintering conditions, and sintering atmosphere,
- material characteristics including starting iron powders and alloying elements, total alloy chemistry, resulting microstructure, mechanical properties, and porosity (shape, size, distribution of pores and of non-metallic inclusions).

Of all these many factors, alloy chemistry and microstructure have been characterised as the most important factors that affect machinability [180,210]. Considering the variations in all material and processing conditions common in powder metallurgy the complexity of machinability of PM steels can be seen. Each result obtained in machining PM steels is a manifestation of the effort to make this problem clearer and if possible to standardise the general machining tests conditions [188].

Addition of alloying elements usually results in good, up to very good mechanical properties but worse technological ones, in this case, machinability. Therefore, also for the analysis of the machining processes, it is necessary to stress that the development of new materials for PM parts production, which will need additional machining, would be followed also by the development of machining tools and corresponding processes including the search for new cutting tool materials. It means that the result of all machining processes applied as the finishing production step (excluding some heat treatment processes) has a decisive role for the later application behaviour of the component, of a sintered part in shape, dimensional tolerances and in finish [104].

Due to various and complex interrelations between these factors and their influence on the behaviour of the material in cutting, a clear statement on the influence of single parameters on the machinability on the basis of the present knowledge can hardly be made [108]. From this point of view it is necessary to analyse in greater detail the effect of the single factors on the machinability of PM parts as a base for possible more concentrated ways for improving machinability.

Opposite to the poor machinability of PM materials an advantage in machining PM parts is the substantially smaller amount of material that has to be removed in form of chips by particular cutting process from a part compared with production of the parts from wrought and cast materials by machining. The cutting machines in PM are more simple and the additional machining process makes possible to produce effectively the PM parts as shown on daily experience. It is caused mainly by shorter total time due to smaller amount of material abated by single machining operation.

The number of variables present when determine the machinability of sintered ferrous parts is so great that it is not possible to declare only one property which is deciding for machinability and to focus on it the reliability. From this point of view it would be important to evaluate the machinability of PM materials by many factors together, which is possible to measure [107,139] but it would be very complicated and often not effectively realisable. The simpler methods for testing the machinability for main machining operation in PM machining used, *e.g.* in drilling and turning including base workpiece material properties will be more effective.

Efficient final machining carried out only by the producer of sintered parts, effectively with specialists in machining, based on the general knowledge of the mentioned interaction, may be one of the methods of producing even more complicated parts by PM procedure more efficiently from the viewpoint of material and cost. The producer knows all aspects concerning the material, processing and final microstructure, mechanical and other properties of a PM part. In many cases machinability may be a dominating factor in material selection. Thus, there is considerable interest and activity in measurement and improvement of machinability in PM steels [10,107,130,137,211]. Generally, machining by the parts producer yields better results than outsourced machining unless the machining shop has sufficient experience with PM products.

5.1 EFFECT OF PROCESSING CHARACTERISTICS ON MACHINABILITY

The effect of all technological and physical–metallurgical processing characteristics taking part in the production of PM materials is specifically involved in final properties of a being machined part. It means that they must be considered for machinability testing of PM parts. Many factors exist which must be regarded to varying extent.

For general factors are considered some PM processing and materials characteristics which can have a general validity in PM machining. The effect of the processing factors on machinability can be evaluated only through the final properties of the sintered parts.

5.1.1 Compaction

The effect of compaction pressure and technique on machinability of sintered steels can be expressed indirectly through the resulting density/porosity, which is regarded in general as the main feature of a PM part for deteriorating the machinability. Particularly the compaction of iron based powders towards high density and high green strength is a precondition for green machining, as *e.g.* in the case of warm compaction.

5.1.2 Sintering

In terms of machinability, the effect of sintering the compacts at low (up to <1000°C), medium, or high temperature for a corresponding time affects to a great extent their machinability through the physical, microstructural, and mechanical properties attained through sintering. In particular the alloy element distribution, which is strongly affected by the sintering parameters, can have a pronounced impact on the machinability as will be shown below.

The variations in sintering conditions that unavoidably occur in industrial practice may cause significant differences in machinability of sintered materials due to changes in mechanical and physical properties [108]. The variations in machinability may be caused by, *e.g.* small differences in sintering conditions (temperature, time, composition of the sintering atmosphere, cooling rate) and by different compounds added for improved machinability which can be one of the undefined factors [152]. Also sintering at different manufacturers under nominally equal conditions may result in significant differences in part properties and by this in machinability.

In particular the cooling rate of a sintered part in connection with sintering conditions has a significant effect on machinability due to change in hardness (especially for carbon-containing steels) and in possible oxidation of open pore surfaces in the compact (see chapter 2.4). Machinability will also be affected if the parts are either over- or undersintered.

5.1.3 Sintering atmosphere

The type of sintering atmosphere used for sintering a material must be considered in relation to the starting oxygen content of powder admixture and the final properties of the material, ultimately also to the as-sintered oxygen content. From this point of view, the effect of sintering atmosphere on machinability can be in some cases very marked.

The oxygen content in the material is one of the characteristic of a sintered material in relation to the machinability as a result of sintering atmosphere. Any variation in the oxygen, nitrogen, and carbon contents – which may occur also during the sintering process through reactions between the atmosphere and the material components, influences the chemical and mechanical properties and by this also the machinability of the parts depending on base alloying and processing. Some common effects on properties of a sintered material result from the base composition of an atmosphere which are [142,212]:

- materials sintered in endothermic gas atmosphere or in another carburizing atmosphere (higher carbon content) have poorer machinability than those sintered in hydrogen–nitrogen or pure hydrogen atmosphere. Similar effects are found with nitriding atmospheres as, for example, nonpurified dissociated ammonia for low alloy steels, but also plain N_2 or N_2–H_2 for stainless steels. It is finally the effect of the carbon content (and/or nitrogen) in the atmosphere that affects the properties of the material workpiece,
- decarburization and carburization of surface layers of a sintered part as an effect of atmosphere can cause major problems to various degrees in machining.

As regards decarburization, it is necessary to note that hydrogen, if not polluted by humidity, is not directly decarburizing agent. In principle, hydrogen reacts with carbon to form methane

($2H_2 + C = CH_4$). This reaction is however rather irrelevant in practice, compared to the effect of humidity. In case of hydrogen not sufficiently dry (or if air can penetrate to the heating zone), a significant decrease of the carbon level will be originated by H_2O, following the reaction ($C + H_2O = CO + H_2$) which becomes more effective at higher temperatures since the thermodynamic stability of H_2O decreases with temperature while that of CO increases [213]. Therefore, avoiding decarburization is more difficult at higher sintering temperatures. This is frequently done by addition of carburizing compounds (propane, methanol) to the atmosphere (typically N_2–H_2). However here it must be considered that the carbon potential that is just optimum at the sintering temperature is strongly carburizing at lower temperatures, and surface carburization therefore may occur especially during slow cooling, with resulting impact on machinability. Today, unintentional carburization of PM parts is more a problem than decarburization [213,214].

Another, unavoidable carbon loss is caused by carbothermic reduction of the oxides present in the starting powder, *i.e.* by reaction of carbon with the oxides, in part also hydroxides in the base metal powder system [31]. As described in chapter 2.4.1, these carbothermic reduction reactions occur in three different temperature 'windows', which are different for plain carbon steels and Cu, Ni, and/or Mo alloyed ones compared to Cr–Mo alloyed steels [46,215]. Typical oxygen levels are in the range of 0.05–0.10% for water atomised powders, and the resulting carbon loss amounts to about 0.06%. For sponge iron powders, the oxygen content and thus the carbon loss are higher – about 0.20% and 0.15%, respectively, and the latter is markedly affected by the sintering temperature, higher temperatures resulting in more complete deoxidation but also in higher carbon loss [63]. Generally, however, this 'intrinsic decarburization' is well predictable and homogeneous over the entire cross section.

Similarly, cracked ammonia, which was the most convenient sintering atmosphere in the past and in some cases also in the present time, is a reducing atmosphere to sinter many metallic powder materials. The dissociation reaction gives a pure, low-moisture content atmosphere with the residual ammonia typically below 250 ppm which for most steels, especially carbon containing ones, is acceptable with regard to its nitriding effect. As long as the moisture content is low the atmosphere is nearly neutral with respect to carbon [31,216].

The steady state during sintering differs from a true equilibrium here in that only a partial, or local equilibrium established at the metal–vapour interface. In systems with continuously flowing atmosphere the steady state condition is most frequently achieved [31]. The situation inside the parts is totally different, indeed, in these regions equilibria can be established only between local carbon and local oxygen or residual water vapour.

When sintering in an endothermic atmosphere and in gaseous hydrocarbons the situation is definitely more complicated as far as the carbon equilibrium is concerned. Under this respect, carbon monoxide can be carburizing or decarburizing, depending on temperature and the required final carbon content or graphite addition. Of course, CO_2 and H_2O are decarburizing, while methane is carburizing [31].

Carbon decrement in sintered parts can also be linked to alloying elements. For instance Si and Ni increase the chemical activity (a_c) of carbon, whereas other alloy additions, like Cu, Mn and Cr, decrease the carbon activity. An increase of a_c means that the equilibrium carbon content in % is lower compared with plain iron. The presence of Ni and Si results in a decrease of the carbon content of saturated austenite (and pearlite as well). For this reason, the addition of more than 0.7% carbon to nickel containing PM steels should be avoided, to prevent any possible formation of brittle carbides at the grain boundaries, and of course also to avoid the formation of higher amounts of retained austenite.

With different design of sintering furnaces and in many cases when the atmosphere flow is kept too high, the large differences in parts properties can occur. In such cases the microstructure, for instance of a copper alloyed material with carbon, can range from ferrite–pearlite mixes to austenite–bainite. Diffusion alloyed powders containing nickel and molybdenum allow to get hard phases even at relatively low cooling rates.

When variations of this nature occur, it is virtually impossible to establish a standard machining procedure and still obtain good tool life, surface finish, and a constant production rate [152].

Also, differences in properties between the surface areas and the core of a sintered part can occur at high density parts (insufficient contact of the atmosphere with the core – closed pores) [217], large volume part, size and geometry (see Tab.2.5). It should happen that high density and big compacts could be badly sintered at the core [31,218], due to markedly slower heating rate in the cores and thus drastically reduced isothermal sintering time (if the nominal sintering temperature is reached in the core at all). For this reason, the ideal equilibrium stages should be considered almost completely valid only of for the 'skin' of the parts. This state can occur also when sintering metal powder systems which require a 'pure' atmosphere for continuous taking away the gas reaction products mainly during heating period, which is not secured, for example, when sintering under a getter. Examples are Fe–Cr–Mo steels which require much lower CO equilibrium partial pressures than, for example, carbon steels for carbothermic reduction of oxides (typically 10^{-3} bar compacted to 1 bar), and CO diffusion from the pore channels thus controls the progress of deoxidation [63]. The correct sintering of chromium alloyed steels is possible using N_2–H_2 mixes or pure hydrogen.

The data about the final oxygen content in a sintered part alloyed with elements with low affinity for oxygen (Ni, Cu, Mo) in relation to the type of sintering atmosphere are lacking. It is not to be expected that an oxide film is formed on open pore surfaces in a sintered iron part alloyed with mentioned elements, at least not with the thickness that can deteriorate the machinability.

The oxide films on pore surfaces that can be formed at cooling of, for example, Cr-alloyed steels in insufficiently pure atmosphere may lead to increased abrasive wear of cutting tools. The practical and defined experiences about this effect are lacking because these materials have not been, or have been in a very limited range, tested for machinability. It is very complicated to determine the effect of the oxide surface layer (oxygen content) formed during cooling on the pore surfaces on the machinability of alloyed steels. This is a special case of the formation of impurities in powder metallurgy parts during sintering itself, the formation of oxides as reaction between sintering atmosphere and some alloying elements during cooling mainly with the elements with high oxygen affinity. In reality, it is not necessary to expect any effect of these oxides on the machinability of the parts.

The oxygen content in parts alloyed with chromium depends on sintering temperature and carbon content as listed in Tab.5.1.

As shown, the final oxygen content in a Cr-containing steel was greatly affected by sintering temperature and carbon content. The lowest oxygen contents (0.07 and 0.03%) decreasing with increasing carbon content at 1200°C do not indicate additional reoxidation during cooling; it is the reduction stage during sintering that determines the final oxygen content, *i.e.* primarily the maximum temperature attained [27]. The cause of high oxygen content after

Tab.5.1 Oxygen content in as-sintered Fe–3Cr–0.5Mo–XC steel (base prealloyed Astaloy CrM powder) as a function of sintering temperature and initial carbon content [37]. Sintering in N_2 atmosphere (purity of sintering atmosphere not defined)

1120 °C, 30 min			1200 °C, 30 min		
Starting carbon content [mass %]					
0.20	0.35	0.50	0.20	0.35	0.50
Oxygen content [mass %]					
0.52	0.28	0.58	0.12	0.07	0.03

sintering at 1120°C independently of carbon addition should be analysed.

For 3% Cr–0.5% Mo–C steels it has been shown that the removal of surface oxides takes place at about 1000°C, and this carbothermic reaction enables formation of stable metallic bridges [219]. However, a second reduction maximum is found at about 1250°C that was attributed to the removal of internal oxygen. Metallographic studies confirmed that the starting Cr–Mo alloyed powder contains fine oxide inclusions within the particles [64,66], Fig.2.31. These inclusions are still present after sintering at standard 1120°C but disappear when sintering at 1250°C or higher. It can therefore be assumed that the machinability of Cr–Mo alloyed sintered steels will depend on the sintering temperature; however, so far no studies have been made in that respect.

Some short, preliminary conclusions can be made in terms of sintering atmosphere *vs.* machinability:

- decarburization and carburization – the best surface finish was obtained with nitrogen, endothermic gas or $80N_2$–$20H_2$ atmospheres; the surface finish is also improved with a higher (0.8% C) content,
- tooling forces in turning are lowest for $95N_2$–$5H_2$ atmospheres compared to, for example, hydrogen–nitrogen ones,
- sintering of materials alloyed with carbide and nitride forming elements (Mo, Cr, V) in a nitrogen-based atmosphere can cause deterioration of the machinability of the parts (although there is very little experience in that field),
- decreasing the hydrogen content in the sintering atmosphere retains more sulphur in the surface layer if sulphur in some form is present in the material.

The adaptation of the sintering atmosphere to optimise the machinability of the PM parts of some composition is dependent on the type of machinability testing desired. Since the method for machinability evaluation is often determined by the part producer, the best powder composition and sintering atmosphere should be selected to give the optimal results for each particular machining operation.

5.2 EFFECT OF MATERIALS CHARACTERISTICS ON MACHINABILITY

One of the most important topics in machining is the possibility for the characterization of the workpiece material properties in relation to machinability. The microstructure as result of chemical composition and processing, surface characteristics, and presence of inclusions or pores all can influence the machining characteristics of the workpiece to different degrees [105]. The porosity, which is an indispensable feature of virtually all pressed and sintered materials, is regarded as the primary factor affecting the machinability of PM steels.

5.2.1 Effect of porosity on machinability

The porosity of sintered materials is regarded as the first and often as the dominant factor deteriorating their machinability. It is the reason for more detailed analysis of the effect of pores, of porosity in cutting process. In general, operations like machining are strongly influenced by the inherent porosity of conventional powder metallurgy parts. Porosity affects not only the physical and mechanical properties of the material but also strongly influences its machinability. Therefore, the differences in machinability between wrought and PM steels are often explained by a very simple view, *e.g.* by the presence of inherent porosity with its effect on other physical and mechanical properties [104;137, 142,148,152].

The poor or poorer machinability of PM parts compared to fully dense ones is in general explained on the basis of the interrupted-cutting theory, the hard-inclusion theory, and thermal-conductivity theory. These notions on the effect of porosity on machinability of PM steels are difficult to isolate, to evaluate independently from other factors taking part in machining, and then to determine its real effect on the cutting process. In spite of it, a more detailed view on the possible effect of each of these factors on the mechanism of cutting process of a PM material will make possible to characterise more effectively also the effect of porosity with its characteristics on machinability considering in parallel the properties of the machined material.

Porosity with 'adverse' effects on mechanical and other

properties of a sintered material manifests itself also by other effects which are expressed in poor machinability and by this in high wear of the cutting tool. Porosity of PM materials as one distinct characteristic is, therefore, regarded in general for the most important and base factor causing the deterioration of the machinability of sintered materials, *i.e.* as a factor significantly changing directly the cutting process and indirectly the thermal conductivity [210]. Much of the current research and development activities on machinability, therefore, focus on the influence of porosity of a sintered part [148,220]. The detailed comparative analysis of all factors relevant besides the porosity can show another view on the problem. However, detailed work related to porosity and machinability of the material appears in general to be lacking. The relatively simple determination of porosity (density) of a material is an advantage for making possible analysis and relation to the machinability.

Porosity as the most deteriorating material property in machining of PM materials significantly influences cutting forces, tool wear, cutting temperature, and chip formation in dfifferent ways. This reality is systematically discussed and confirmed by research and practical results. In general, it is mentioned that higher relative density of a porous material improves machinability. It is assumed that as full density, *i.e.* zero porosity, of a material is approached, machinability may be close to that of wrought materials of corresponding properties. Confirmation of such statements regarding different composition is, however, also lacking. These factors should be considered in evaluating the machinability of PM parts. The effect of porosity on machinability of a material should be evaluated in parallel with a detailed microstructural characterization.

In general, it is necessary to note to porosity-machinability relationships that density (porosity) variations across and along the machined part simply cause the loss of cutting efficiency regardless of the mechanism of failure of tool material by the cutting process [45]. Also the dimensional accuracy of the machined part is reportedly reduced by porosity.

At present, already two theories can be presented explaining the mechanism only for the direct effect of porosity on the cutting process of a sintered part. The first one is based on the notion that the porosity causes an interrupted cut in a porous body and by this deteriorates the machinability as in similar macrointerrupted cutting of wrought (fully dense) material – *interrupted cutting theory*. The second one is based on experimental analysis of machined surfaces, showing the plastic deformation of the subsurface area of a newly formed surface, eliminating the pores in this layer – termed *deformation cutting theory* by the authors of this book.

5.2.1.1 Interrupted cutting theory

It has been assumed that the detrimental effect of porosity leads to a constantly interrupted cut as the tool edge breaks out of the metallic phase into pores, immediately re-entering the metallic phase, and this intermittent loading–unloading results in (mechanical and thermal) fatigue should also to produce undesirble chatter and vibrations which accelerates the tool wear. This leads to poor surface finish, which should be linked to other aspects for the cutting [148,210,220]. The conception of interrupted cutting action using a tool as in turning is illustrated in Fig.5.2. The data presented demonstrate different views on the effet of porosity on machining when it occurs surely as interrupted.

It shows a sequence of a discontinuous contact between the tool and the workpiece. According to this idea the tool cuts the material passing one pore after another – interrupted cut. This effect is regarded most significant since the tool edge breaks out of the workpiece into pores, being unloaded, and on re-entering the metal is rapidly loaded once more. This action of successive small impacts on the cutting edge should cause more rapid tool failure than continuous cutting operations – impact fatigue mechanism.

This should produce also undesirable chatter and vibrations and introduce a destabilised regime to the cutting edge, leading to a shorter tool life [148,159]. Porosity also introduces microscopic shock and impact loading to the cutting edge as it passes over pores

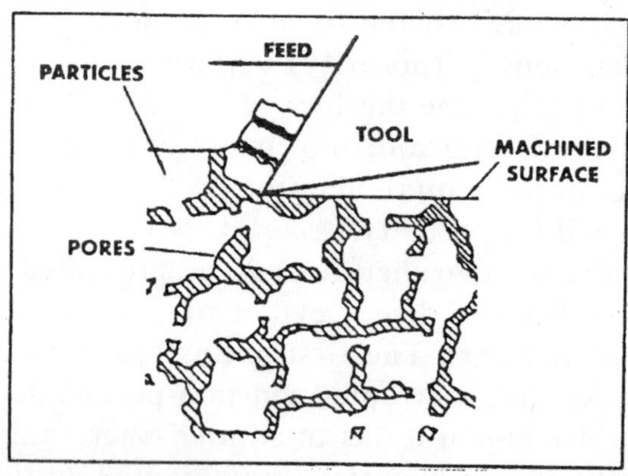

Fig.5.2 Schematic for machining of porous PM material [211].

and re-enters the material along the tool/chip interface leading to a shorter tool life [108].

As the sharp cutting tool edge encounters a pore in the microstructure, a small segment of it is relieved elastically, and is impact loaded again when it contacts material on the other side of the pore. This results in microfatigue of the cutting edge (mechanical and heat effects), which leads rapidly to chipping and dulling of the tool edge [107,139]. It is assumed in some works that due to larger surface area of pores in a sintered material compared with wrought steel this fact could probably increase the potential for physical and chemical reactions between the tool and workpiece that may accelerate wear [136,137]. This cannot occur because the contact surface of the tool with the chip under cutting used is always equal [129].

By this conception the cutting process includes 'microblows' at the cutting edge in turning porous iron. The tool cutting edge should be chipped in amounts approximately equal to the 1 μm pore size of the material (which would mean that the cutting edges progressively chip at 1 μm width). Chipping was attributed to microcrumbling, occurring as the tool passed through voids within the porous part. Abrasive tool wear was claimed to be due to the non-continuous contact between the machined surface and the tool material at the flank face, which allowed abrasive particles coming only from the cutting tool to enter this wear junction [142,219].

However, when discussing the interrupted cut it is necessary to note that the area of engagement of the cutting edge in PM machining is substantially larger than the average pore size. In general, the pores tend to be 1–10 μm, only in extreme cases up to 100 μm, which is at least 20 times smaller than the tool contact area in the cutting zone [142]. In all cases the contact of the tool with the material is much larger than the size of particular pores. Therefore, interrupted cutting is minimum very limited. Experimental confirmation of this theory would be not simple, and perhaps, therefore, is lacking. In any case, the illustration in Fig.5.2 is misleading if the three-dimensional character of the machining geometry is neglected; it can be assumed that, according to the rules of stereology, in a sintered steel with 15% porosity at any given time 85% of the total cutting edge are engaging metal, and only 15% are moving within pores (although the location of the 15% changes extremely fast).

Thus, it shows that the interrupted cutting model does not adequately explain the dynamic characteristics of the system tool/

workpiece. It is also the proof that the cutting process cannot be characterized only as interrupted, otherwise the chips would be extremely small, in the order of the powder particle size.

5.2.1.2 Deformation cutting theory

The consequence of all machining processes is the formation of a special surface deformed and work hardened layer on the workpiece as known in machining of wrought steels. This is combined with the occurrence of the high level of compressive stresses after cutting with the tools with geometrically defined cutting edges and of tensile stresses, *e.g.* after grinding.

The properties of this layer can be markedly changed compared with the starting physical–chemical properties of the workpiece. It has been shown that during machining of wrought steels heavy deformation and resulting work hardening occurs, as shown in Fig.5.3.

The machined rim layer is formed by the mechanical–thermal effect of the cutting edge on the workpiece material. This can be for wrought steel by the following structure presented:

- absorption layer, oxide layer ($t^* \approx 1 \ldots 30$ nm),
- straining (flowing) layer, strongly hardened layer ($t^* \approx 1 \ldots 5$ µm),
- work hardened surface layer ($t_v \approx 10 \ldots 500$ µm), area with far-distant retained machining stresses [110].

Fig.5.3 Cross-section of the subsurface of a turned workpiece (wrought steel Ck 45, Rm = 600–900 MPa, ~180 HB) showing the nature of the deformation layer [110]. *Turning conditions: HS123 tool, cutting speed 120 m/min, feed 0.1 mm/rev., depth of cut 1 mm, flank wear of the cutting edge 0.45 mm.*

These observations are based on microhardness measurements and on X-ray analysis of the stresses formed. It was also shown that a high plastic deformation of the machined random layer causes a significant increase in dislocation density. It is known that in the unalloyed steels the transformation layers can be formed with the hardness corresponding to martensite. During cutting of the ferrite–pearlite steels the ductile ferrite grains contribute more to the total deformation of the layer. The cementite lamellae closely to the surface are more frequently destroyed.

A markedly higher microhardness closely to the machined surface was determined as the proof of the deformation and work hardening caused by turning as shown in Fig.5.4. Formation of an adiabatic-shear, nanocrystalline white layer was observed on as-machined surfaces of the workpieces.

The consequence of all machining processes is a significant increase of the stress state in the workpiece which affects all following processes including the surface finish. There is no reason why this characteristic of the machined surface of the wrought steel workpiece could not also be applied for sintered steels. The range of a such state in PM workpiece properties can be partly affected by porosity if the heterogeneity of the microstructure is not taken into account.

Also after grinding an annealed Fe–C steel the microhardness of the surface 600 HV and below the surface (0.06 mm) 200 HV was measured [112].

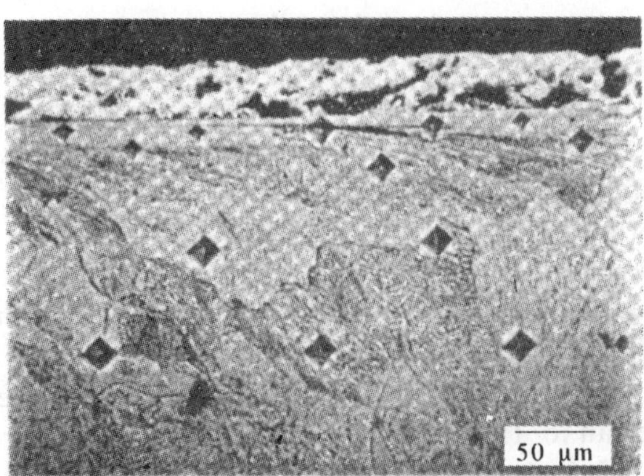

Fig.5.4 Deformed and work hardened (change in microhardness indentations) surface layer in turning of wrought Ck 60a steel (Rm = 700–1050 MPa, ~210 HB) [110]. *Turning conditions: cutting speed 102 m/min, feed 0.2 mm/rev., depth of cut 2 mm.*

This knowledge concerning the deformation of the wrought steel surface with its complicated character in dependence on cutting conditions can be accepted as a basis for the investigation of the PM cutting processes regarding mainly the differences in workpiece material properties. It means that these aspects can also be valid for sintered steels regarding the effect of porosity.

Considering the mentioned knowledge it is possible with a greater certitude to say that on the basis of some experimental observations that the 'deformation cutting theory' explains more exactly the effect of porosity on machinability of the sintered parts. The examination of machined material has shown the presence of a layer beneath the freshly machined surface that contains little or no porosity. The action of the cutting tool has compressed the material sufficiently to remove the remnant porosity. The plastic deformation of the subsurface area can be seen optically on cross-section of the being machined surface and determined by the measurement of microhardness, although the depth of the visible porosity reduction exceeds the depth of the measurably work hardened layer.

When cutting a sintered porous part, a larger area of material is deformed than in wrought steel, at the expense of porosity. It is assumed that to perform larger deformation in a porous material, a higher force is needed to transport the material and to fill the pores, which expresses itself negatively also in an increase of temperature in the contact zone of the cutting pair (tool and workpiece). In contrast some results attained mainly in drilling tests showed that the material with higher porosity was machined with lower cutting forces. It means that lower forces are needed for the transport of material during deformation in a porous material. These estimations are presented without taking into account real cutting conditions and material properties.

The deformed material exhibits higher hardness, and the cutting of this layer needs higher force compared to those needed for cutting of undeformed material. It follows from it that with increasing porosity the width of the deformed layer increases but the microhardness of the deformed layer can be lower than at lower porosity because the force was used firstly for the transport (= pushing) of the material into the pores at lower work hardening. With decreasing porosity, the deformation state of the material in the cutting zone should approach the state for the wrought material, in which no transport of the material proceeds.

Deformation of the machined surface of Fe–1.75Ni–1.5Cu–

0.5Mo (Distaloy SA, AB) material at turning was examined [139]. The underlying structure of the material contained little or no porosity, thus revealing surface deformation caused by the machining operation. The deformed layer was hardened to a certain width. The action of the cutting tool compressed the material sufficiently to remove the porosity.

It means that the cutting edge never came in contact with one single pore nor suffered intermittent cutting because the cutting edge itself always contacts pre-densified, pore-free, material. The micro-hardness of the pearlite in the deformed layer of the material which showed most consistency was 371 HV, and near the surface to 241 HV at at depth of 100 μm for the dry machining conditions. The apparent hardness of tested materials was in the range 162 to 223 HV10. Also the microhardness increase for the swarf was measured. The Distaloy SA appeared to be easier to machine than the AB variety which indicates a possible effect of the base iron powder grade (sponge or atomised). Contrary to expectations, materials compacted at 500 MPa were easier to machine than those compacted at 700 MPa. This could confirm that at the lower density the cutting force acted more in material pushing into the pores and less in work hardening.

The presence of pore deformation zones confirmed that the tool cutting edge actually cut fully dense materials, which resulted in a gas tight machined smooth surface. The more porous materials gave in this case longer tool lives and lower cutting forces, so the ease of deformation of substructure may be of consequence. It means that the machining conditions which reduce resultant deformation, *i.e.* in cases when material is simply shifted into pores, tend to increase the insert life, in this case in material with higher porosity [139]. These results contradict the notion about poorer machinability of materials with higher porosity if the microstructure is not taken into account.

The deformation cutting theory explaining the effect of porosity on machinability was proved also by measurement of densification (water displacement method) of chips formed in turning (cutting speed 100–160 m/min, depth of cut 0.6 mm) of Fe–0.8C material with 1 to 8% porosity compared with wrought steel. The densification of the chips increased with higher porosity of the part. It follows from it that the chips independently on initial porosity should have the density approaching to the full density of the machined material. Cutting conditions must be regarded.

The change in density of the chips can also be affected by the temperature increase in the cutting zone. The temperature increase reduced the mechanical strength of the material in the zone, which became softer. It diminished the resistance of the material to deformation, and of the friction between the tool cutting edge and the chip and by this to the increase in densification. This feature was specially proved at cutting speed over 110–140 m/min [221].

The deformation of subsurface area also at drilling was observed and work hardening determined by microhardness measurement. Figure 5.5 shows a cross-section of the surface of a hole drilled in ferrite–pearlite (Fe–C) steel. A relatively larger deformation was observed in ductile ferrite grains. Figure 5.6 shows the deformed surface layer of a hole drilled in Distaloy SA-C steel. Visually a smaller extent of the deformation was observed in alloyed material what need not be in agreement with the work hardening of this layer. Figure 5.7 shows the bottom of the last unfinished 'blind' hole in drilling test of a Fe–C steel when the drilling (feed) was stopped due to the drill failure. The conical shape of this hole indicates a marked failure also of the dril land. The deformation of the machined surface is initiated yet at the beginning of the cut in drilling, Fig.5.8. This is an abrupt of a deformed layer at the entering of the drill into the workpiece.

A large deformed area is characteristic for this zone showing that the drilling occurs with a large plastic deformation of the material. As shown, the newly generated surfaces of the holes were work-hardened and densified below the cutting edge and along

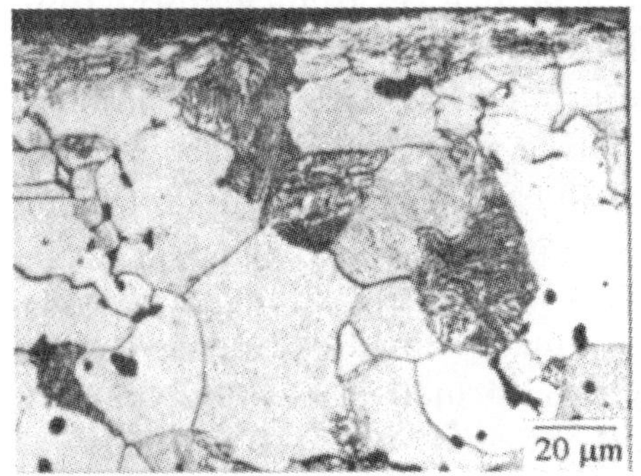

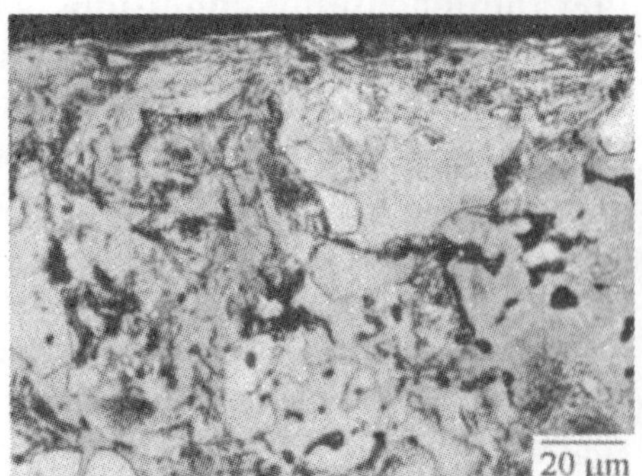

Fig.5.5 Section of a deformed surface layer of a hole drilled in Fe–0.29C_c steel based on ASC100.29 iron powder. Density 7.04 g/cm^3, Rm = 220 MPa, 74 HV 10; v_c = 8 m/min. Optical micrograph. Nital etched.

Fig.5.6 (right) Section of a deformed surface layer of a hole drilled in Distaloy SA–0.29C_c. Density 6.90 g/cm^3, Rm = 572 MPa, 187 HV 10; v_c = 8 m/min. Optical micrograph. Nital etched.

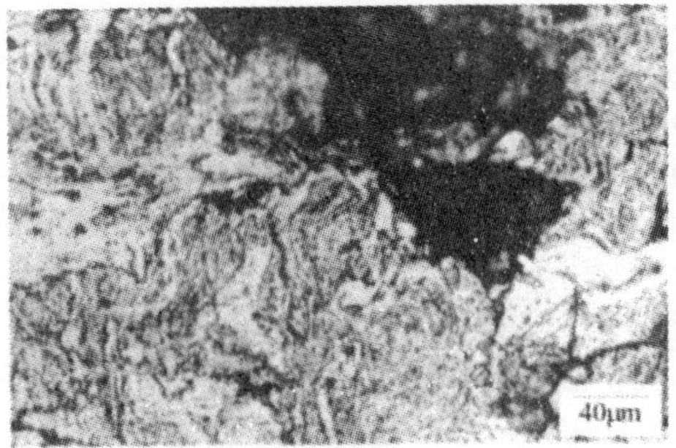

Fig.5.7 Microstructure of an extremely deformed layer in the bottom of an unfinished hole drilled at drill failure in sintered Fe–0.32C_c steel. Density 6.99 g/cm^3, Rm 229 MPa, 80 HV 10. Optical micrograph. Nital etched.
Fig.5.8 (right) Detachment of a layer at the beginning of drilling – at entry of the drill into the material as in Fig.5.7 with higher feed. Optical micrograph. Nital etched.

the surface of the holes. The width of densification area increased with porosity and drill wear.

The deformation and work hardening of the freshly drilled surface was confirmed by microhardness measurement as listed for some iron based materials in Tab.5.2.

The increase in microhardness of the deformed hole surface layer is in agreement with the deformation cutting theory of the effect of pores on the machinability. The increase in microhardness is large which proves that this layer has 'new' metallurgical properties. It can be assumed that in machining, in the following rotation the cut occurs in deformed material with new properties, not in undeformed which would be the case following the interrupted cuttine theory. It might therefore be concluded that if deformation occurs in addition to densification, sintered steels should be machined with large feed to cut behind the deformed

Tab.5.2 Microhardness (HV 0.01 or HV 0.025) of the core of the samples and of the deformed surface layer in the holes drilled. Density 6.82-7.08 g/cm^3; alloy: 1-3 sponge iron powder (SC 100.26), 5 and 6 atomised iron powder (ASC 100.29). XC = graphite addition [mass %]. MnX – machining aid. *Drilling test: HSS 3 mm drill, drill speed 850 rpm, constant thrust force 333 N*

Alloy No.	Composition	Microhardness		Alloy No.	Composition	Microhardness	
		Core	Deformed layer			Core	Deformed layer
1	Fe-0.3C	101-292	269-719	5	Fe-0.3C-0.5MnX	104-363	312-457
2	Fe-0.3C-0.5MnX	118-332	403-714	6	Fe-0.7C-0.5MnX	110-333	186-495
3	Fe-0.7C-0.5MnX	103-393	342-626	7	Dist. SA-0.3C	119-312	172-484
4	Fe-0.3C	105-387	287-706	8	Dist. SE-0.5MnX	206-807	576-893

layer of the previous pass (as it is done when machining austenitic stainless steels). If, however, deformation occurs, there should be no effect of the feed since the cut would be made into deformed and work hardened material anyhow. Figure 5.9 shows the deformed surface layer with the microhardness indentations of a drilled hole.

According to the deformation cutting theory, as the density of the material increases the thrust force and torque increase linearly. According to it, it is also presumed that at low density the cutting forces are low because excessive work hardening induces cracks in the material and makes it brittle. The initial increase in cutting force at low density was due to the work hardening of the material caused by its densification. As the density increases, the material becomes stronger and the extent of the work hardened layer is reduced. Less cracks being generated, the forces required to cut the material increase due to the intrinsic resistance of the material. At high density, densification under the tool is minimised and the hardness of the densified layer becomes independent of the density of the part. The applied stress forced the closure of pores without welding the surface brought into contact with the tool [222].

In relation to this simplified explanation of the deformation cutting theory, it must be taken into account that the extent of the deformed zone will be greatly influenced and modified by the chemical composition of the material and by the final heterogeneity of the microstructure, *e.g.* when soft and different hard phases form the microstructure in different proportions.

Subsurface deformation could be minimised in machining operations by using very low cutting speed/feed regime for the material. If deformation is avoided then there is no significant elastic recovery of the material and less tendency, *e.g.* for the drill bit or tap to jam and break. Deformation of the PM specimens during turning was impossible to avoid, where the cutting speeds are ten

Fig.5.9 Section of a deformed surface of a drilled hole in Fe–0.56C_c alloy; density 7.08 g/cm^3, 105 HV 10. Optical micrograph. Nital etched. *Drilling conditions: HSS 5 mm drill, 850 rpm,* v_c *= 13 m/min, feed 0.11 mm/rev.*

times those used in drilling [139].

In respect to the interrupted cutting theory, it is assumed that interconnected porosity provides a path for cutting fluids to escape from the cutting area when used. This reduces their ability to cool and lubricate the cutting edge. It may reduce their ability to wash chips from the cutting area [130]. Considering deformation cutting theory, this effect is minimised because the working surface is densified and by this the connection of the cutting fluid with the open pores is interrupted as in full dense materials.

The presented results confirm that the deformation model does not adequately explain the reduced machinability of PM materials caused by porosity independently of the particular properties of a material being machined, but it is much nearer to the reality compared to the interrupted cutting theory.

These results and observations only show that all factors must be taken tnto account together for the decision about the positive or negative contribution of each to the machinability of PM materials [142]. It is possible from this point of view to assume that for the deformation of a material with higher porosity a lower force is needed compared to lower porosity material. Therefore, it is possible to say that the detailed analysis of the effect of porosity on cutting process of PM material is lacking. This is also due to the many different methods used for testing the machinability of PM steels.

Experimental results which present also microstructure analysis of the machined surface of a material support the deformation theory. It will be necessary to analyse and determine the forces causing the deformation of a sintered material with various physical and mechanical properties at specific cutting conditions in relation to porosity.

The presence of a deformed layer enables drawing some conclusions. The elimination of the porosity ahead of the cutting tool means that the cutting edge never comes into contact with the porosity nor suffers intermittent cutting. Subsurface deformation also results in a machined surface that it will be gas tight and free of pores. Observation of the wear pattern on tool inserts shows that there is some elastic springback of the material when the insert nose has passed. This can lead to early failure with some insert/ material combinations. The elastic recovery of the material can result in the radial reaction force exceeding the cutting force, which causes a deterioration in machined surface quality and leads to rapid tool failure [139].

Conclusively, the following can be stated:

- **According to the interrupted cutting theory the machinability of sintered steels should be the better the higher the density is.**
- **According to the deformation cutting theory, the cutting edge always cuts into 'fully' densified material, and the machinability depends on the properties of this densified area. If the densification results in a larger width of the material pushed into the pores combined with lower work hardening, as in highly porous materials, machinability is improved.**
- **Thus, both theories yield contradicting predictions for the machinability–porosity relationships; most experiments rather support the deformation cutting theory.**
- It is generally accepted that in porous materials deformation occurs during machining as in wrought steels. The extent of this deformation beside the cutting conditions may be more affected by the porosity and PM material properties. The task is however to determine how and to which degree the deformation in a porous material and the resulting work hardening affects the machining process.
- Because the interrupted cutting model does not adequately explain phenomena exhibited in PM machining, these phenomena regarding the deformation cutting theory including work hardening are better discussed in terms of following offsetting parameters: the influence of porosity on thermal conductivity, cutting temperature, bulk strength of the workpiece, cutting forces, chip formation, and generated surface finish. The tool material with its geometry and cutting conditions are unavoidable factors taking part in the cutting process

These factors (not individually) should be considered in evaluating the machinability of porous PM parts.

5.2.2 Effect of porosity on thermal conductivity

The thermal conductivity of PM steels is generally lower than in wrought steels, due to pores. The porosity of PM materials influences also the specific heat, taken per volume. At full density, the thermal properties of PM steels converge to those of corresponding wrought compositions.

The thermal conductivity of PM materials has been related to the conductivity of fully dense materials through the equation [142,223]:

$$K_{B} = K\ (1-B\cdot p),$$

where K is the thermal conductivity of porous material, K_B is the thermal conductivity of a fully dense material, p is the volume pore fraction, and B is a constant depending on the shape and distribution of pores.

This equation predicts that the thermal conductivity has a simple linear relation with porosity although non-linear relationships have been obtained. A decrease in thermal conductivity with increasing porosity leads to lower heat conduction within the machined workpiece.

The pores contribute towards a reduced thermal conductivity of the compacts, resulting in a slower dissipation of heat from the chip/tool interface [104,210,224]. The deteriorating effect of thermal conductivity with increasing porosity of a material is, therefore, regarded for one of the reasons for low or poor machinability of porous sintered material. In lower density materials, having lower diffusivity coefficients, higher amounts of heat remain in the cutting zone when they are machined [225].

Generally, however, the effect of porosity on the thermal conductivity should not be overestimated; it is most pronounced in the case of low sintering temperatures, as shown by thermal diffusivity measurements [43] and by measurement of the electrical conductivity [226] to which the thermal conductivity is linked by the Wiedemann–Franz law. The latter investigations have indicated that at least for well sintered materials the effects of the matrix (alloying, heat treatment state etc.) on the thermal conductivity can easily outweigh those of the porosity.

The reduction in thermal conductivity, locally rapidly increasing the cutting edge temperature, may result in clattering and edge deformation, and accelerate tool wear. Local hardening of the workpiece may make subsequent finishing cuts more difficult [130,136,137].

As a result, wear can be expected to increase with porosity. It is also possible for a temperature increase to have an appreciable effect on oxidation and other chemical reactions on the generated surfaces including those with lubrication films when used. This may in turn further affect tool wear by a temperature increase, depending on reactions at the tool/workpiece and tool/chip interfaces (see Figs.3.29 and 3.31).

Thermal conductivity is linked to the properties of the base material, and by this also the content of alloying elements further lowers it. For this reason, the lower thermal conductivity of a sintered material, as a result of the porosity, results in higher temperatures in the contact zone and can intensify many different tool wear phenomena. Therefore it is important in machining PM parts as one of the causes for a large tool wear and its early failure. In any case also the effect of alloy elements, which may outweigh that of the porosity, on the thermal conductivity has to be considered.

The temperature in the cutting zone is one of the most important factors affecting also the formation of the chips. Figure 5.10 shows that with increasing porosity of the being machined workpiece at turning the temperature in the cutting zone increases by sôme times as thermal conductivity decreases (although the level of temperature increase seems to be very high considering the low porosity levels). The temperature in the cutting zone increases also with increasing cutting speed. The base result of the temperature increase is the deformation of the material and the friction of the chip on the flank of the tool.

By increasing porosity and at the same time decreasing specific heat and thermal conductivity of a material, temperature-dependent wear processes, *e.g.* diffusion and oxidation, at the cutting tool are enhanced which leads to a reduction of the edge life and a rapid tool failure [105].

For drilling, changes in temperature with density for new and worn drills at various depths affect in different degree. Drill temperatures were significantly higher at the higher cutting speeds and also increased with hole depth. In addition, there was a greater difference in temperature between the new and worn drills at greater hole depth [104].

5.2.3 Effect of porosity on tool wear and cutting force

Practical experiences in general confirmed the deteriorating effect

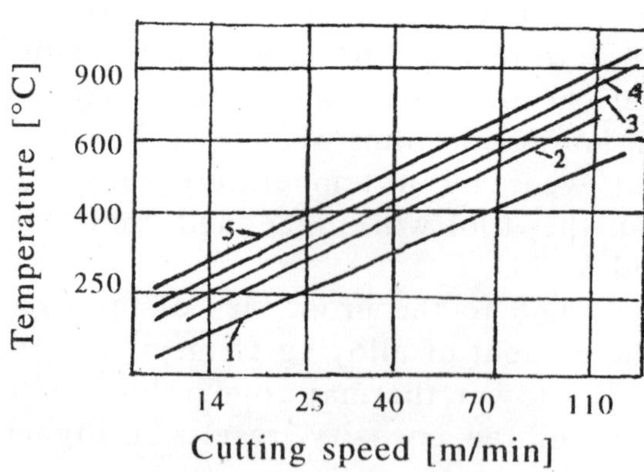

Fig.5.10 Dependence of temperature in the cutting zone on cutting speed and porosity at turning of powder forged Fe–0.8C steel and of corresponding wrought steel [227]. 1 – wrought carbon steel, sintered steel – porosity: 2 – 1, 3 – 3, 4 – 5, 5 – 8 [%].

of porosity on tool wear which should be evaluated with the thermal state in the cutting zone and with the deformation of the cut zone as combined effect of porosity. An increase in flank wear with porosity was observed for both turning and drilling of sintered and powder forged iron at relative densities in the range of 92–99% [228].

All consequences of porosity shorten markedly the tool life. However, systematic tool wear measurements involving PM workpiece materials are seldom found in the literature [126]. It is clear that wear data and tool life relationships in wrought materials cannot be expected to be valid for porous materials because of the porous nature and mostly of the different composition.

It is necessary to note that the final effect of porosity on machinability of a porous material may be evaluated together with the cutting method and conditions. For example, it was observed that while grinding 70 to 80% dense iron material that the grinding ratio of removed workpiece/tool volume increased with increased porosity [142].

An increase in flank wear with porosity was observed in turning of sintered iron at relative densities in the range of 92–99% [229]. However, no correlation between flank wear and relative density in the range of 82–90% was reported for both the turning and drilling of sintered iron [230].

A marked effect of porosity on flank wear and on number of holes drilled (drill life) for 304 stainless steels was determined, Figs.5.11 and 5.12.

The worst results were obtained at the lowest relative density (70%). At a relative density of 90%, lower flank wear increased

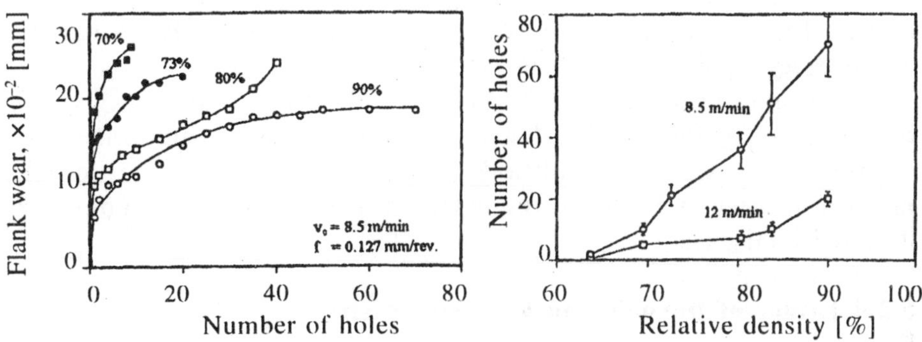

Fig.5.11 Flank wear growth as function of relative density for 304L stainless steels [126,142]. *Drilling speed 8.5 m/min, feed 0.127 mm/rev.*

Fig.5.12 Effect of relative density on number of holes drilled for 304L stainless steels [108,126,142]. *Drilling speed 8.5 and 12 m/min, feed 0.127 mm/rev.*

as a function of the holes drilled. Increase in drilling speed caused a marked decrease in drill life (low number of holes drilled). In these cases it could be possible to presume the interrupted cut.

At 64% relative density of 304L stainless steel the drill failed while the first hole. Lower density materials failed primarily because of flank wear while margin wear predominated in drilling the higher porosity materials. The drill life was affected by the cutting speed. At 12 m/min cutting speed the tool life was reduced by 50 to 90% compared with that at the 8.5 m/min speed [108,142].

On the other side, a non-linear relationship between the thrust force and the density was found when drilling sintered 304L stainless steel, Fig.5.13. The highest torque was needed for drilling the material with relative density ~70%. The exact tool wear measurements involving PM workpiece materials are seldom found in the literature.

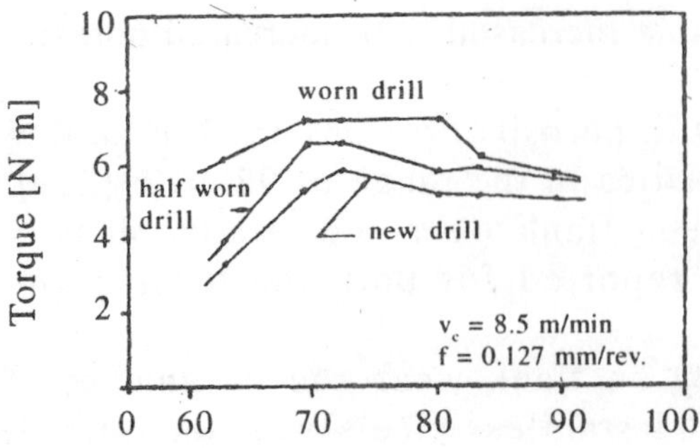

Fig.5.13 Effect of relative porosity on drill torque at three tool life fractions for PM 304L stainless steel workpieces [142].

In another case, the effect of porosity on machinability was attributed to the fact that an increase in porosity results in a decrease of the average shear stress in the cutting zone. Thus, cutting forces should decrease and thereby machinability increase as porosity increases.

In terms of significant differences in alloying of stainless steels, on one side, including low density and of common structural steels, on the other side, the mutual comparison and transfer of the machining results between them will not be possible without very detailed analysis.

5.2.4 Effect of porosity on surface finish

Porosity has a strong influence on the surface finish of machined parts. The surface finish of a porous material can be expected to have characteristics that are different from those of a wrought material, when cutting conditions for both material are identical.

This is because porosity as mentioned causes variations in tool wear mechanism, in material work hardening, and in densification at the surface, and in higher temperatures at the cutting surface when machining cannot be performed in the presence of a cutting fluid [129,136].

On the other side in certain powder metallurgical composition, surface finish can be degraded causing a smeared surface-through bulk transportation of material leading to an impairment of porosity, which could be important, *e.g.* for self-lubricating bearing applications [138].

In any case, it must be taken into consideration that the matrix of PM steels is hardly ever really identical to wrought steel, even in the case of nominally identical composition (at least the Si and Mn contents being different), and therefore also comparison of the machinability hardly depicts the porosity effect exclusively but the differences observed are at least in part caused by the different matrix materials involved. Statements that phenomena observed with PM steels, as *e.g.* lower surface quality, are caused by the porosity have thus to be regarded with some care, at least unless all relevant material data are given and mainly cutting tool and cutting conditions. Basically the surface finish is increased at each porosity level when the cutting speed increases and depth of cut decreases.

5.2.5 Effect of porosity on chip formation

Porosity affects chip formation and chip continuity. Chips tend to densify, depending on cutting parameters. In general, chips produced in PM machining are discontinuous, especially at intermediate and low densities [129]. In cutting PM material to an edge as occurs during turning, the tool deforms some of the workpiece material which then separates as a chip. Large stresses built-up as the layer which is to become the chip approaches the cutting edge. Elastic and then plastic deformation of the metal occurs as the cutting forces reach the yield strength of the material [157].

In general, the higher the material density, the longer the chip. The precise characterization of the length of the chips in dependence on porosity is lacking. The material properties must be considered. The formation of continuous chips in machining high density materials cannot be confirmed in all cases without knowledge of additional factors.

Chips produced during machining of low density (porosity >35%) 304L PM stainless steel are very small – less than 0.2 mm in their longest dimension.

In powder forged parts approaching the theoretical density (~99%) the chips are also short and, therefore, non continuous which is an advantage for turning in automatic lines. Usually the surface zone is turned that contains the so-called residual (surface) porosity of the part not exceeding usually ~0.2–0.3 mm in depth in which area the pores can be oxidised at heating to the forging temperature. This contributes to the formation of short chips.

The chips formed in machining porous materials are short in spite it that the cut occurs in a deformed layer because the surfaces of the compressed pores are not cold welded. The formation of short chips in cutting process of PM materials caused by porosity can be regarded as a generally positive effect for handling, with the exception of very fine chips formed at drilling of some materials.

5.2.6 Effect of composition on machinability

For assessing the effect of chemical composition on machinability, it is necessary to take into account also the effect of the base iron powder grade. As mentioned above each iron powder grade, in dependence on production and preparation method and on raw materials used, is characterised by its own specific physical, structural and technological properties. The base characteristics of iron powder grades are common and they do not have a direct relation to the machinability. In reality the possible relationship between the properties of the starting iron powder grade and the machinability of a sintered part cannot be simple, due to the multitude of parameters involved. Also the results of some tests in turning and drilling related to the effect of iron powder grade are not fully comparable due to various test conditions [139].

Perhaps only through the relationship of the iron powder grade to the final sintered microstructure and by this to the mechanical properties, the effect of iron powder grade on machinability could be partly defined. Inasmuch as the base properties of each iron powder grade were expressed in as-sintered properties of a part, by this they also affect machinability.

In general, the basic iron powder grade may have a strong effect on wear of the tool flank made of cemented carbides during turning. It was stated that sponge iron powder causes lower wear of the tool in turning than water atomised powder in spite of higher content of non reducible inclusions coming from the ore (possible amount and size), which can be detrimental to machinability, if sufficiently hard [10,162]. It will be necessary to evaluate the reason for different machinability of sintered parts in dependence on base iron powder grade as for mechanical properties, which is not sufficiently explained. The prealloyed powders form a special group of base powders. The machinability of sintered steels based on these powders was tested in a small range, *e.g.* for Fe–2Ni–0.5Mo–C steel.

The chemical composition as a characteristic of the sintered material can be accepted for machinability only as basic information. The real effect of the composition on the physical and mechanical properties and, consequently, on machinability is expressed primarily in the microstructure formed during sintering. The knowledge regarding the known or assumed effect of single factors taking part in formation of the microstructure with constituents of special physical–metallurgical properties of a material can contribute to formation of a material with adequate properties required, but yet not determined for acceptable machining.

Here, it must be kept in mind that compared to wrought 'plain' iron (without additional refining), the starting iron powder grades contain less, *e.g.* manganese <0.3% (for atomised grades even <0.15%) and virtually no Si. An 'unalloyed' wrought steel should contain up to 0.8% Mn, 0.5% Si, 0.06% S and 0.08% P and small amounts of Al, Ti and Cu. This shows a main difference in unalloyed PM and wrought steel.

Except for ferrite–pearlite materials, the characterisation of the microstructure of alloyed systems by any common technique, by some simple figure or directly through a physical property in relation to the machinability of a material is very difficult. The machinability of materials with a microstructure formed by some defined microstructure constituent, as *e.g.* plain pearlite, has not been investigated.

Knowledge of the composition of a material and the various phases occurring mainly in formation of a characteristic microstructure can help to [149]:

- select the optimum material from a variety of choices,
- define their properties as well as application and operation restrictions,
- to be active in the development of new material compositions which requires also the development of new technologies and machining processes.

Table 5.3 shows the influence of some particular chemical elements on the machinability of wrought steels. Figure 5.14 shows

Tab.5.3 Influence of some chemical elements on the effécts in wrought steel and on cutting ability with special emphasis on cutting forces, chip shape, and tool wear [138]

Chemical element	Effect in steel	Machinability criterion			Cutting ability
		Cutting forces	Chips	Wear	
Carbon	Pearlite/cementite	o		o	o
Copper	Increases toughness and strength at higher temperatures			o	o
Nickel	The formation of fine grained structure			o	o
Molybdenum	Carbides forming, fine grained structure				o
Manganese	(Slight carbide forming), MnO, MnS				o>1 %, • <1 %
Silicon	SiO_2	o	•	o	o
Chromium	Carbides forming,, fine grained microstructure				o
Sulphur	MnS, FeS, and other sulphides	•	•		•
Phosphorus	Cold segregations	•	•	o	•
Lead	Ocurrence of insoluble metallic inclusions (Te, Bi, Sb)	•	•	•	•
Nitrogen	Nitrides forming		o	o	o

Remark: effect - • favourable, o detrimental

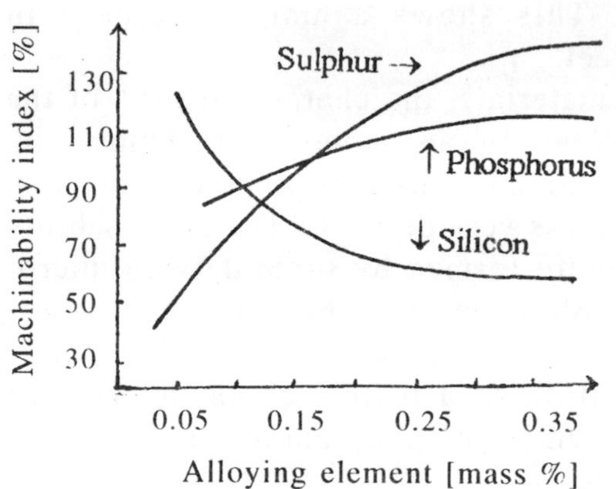

Fig.5.14 Influence of sulphur, phosporus and silicon content on the machinability index of wrought free cutting steels [138].

separately the effect of sulphur, phosphorus and silicon on the machinability index of free machining steels.

The data in Tab. 5.3 and in Fig.5.14 make it possible to consider the extent to which they are comparable with those attained in the machining of PM steels alloyed with the mentioned elements. The combined effect of complexly alloyed wrought steels on machinability can markedly differ from those alloyed with particular elements.

It is known that complex PM alloy steels, containing Cu, Ni, Mo and C, have lower machinability than those alloyed by particular elements. This also applies to steels prepared from diffusion alloyed powders, and to high-nickel steels in general [10].

In the diffusion-alloyed materials, machinability is adversely affected especially by the nonuniform distribution of nickel with the resulting presence of microstructural constituents with widely varying hardness, typically from ferrite, to martensite (see below). In combination with higher carbon contents, the cutting force increased markedly for all compositions. Insufficient characterization of the microstructure or some other property of these steels may cause a number of problems in production if machining is needed.

Generally, all alloying elements added for improving strength, hardness and other related properties of the steel tend to deteriorate its machinability.

In the following, some informations are given about carbon, copper, and manganese which are considered also to be machinability-enhancing.

5.2.6.1 Effect of carbon on machinability

In advance is necessary to note that in term of machinability, sintered plain iron parts or such with carbon below 0.1% are usually poorly machinable because of low hardness and high ductility and tendency to tear easily in all machining operations [152]. Formation of built-up edges occurs easily esp. in the case of low cutting speeds [207]. Carbon-free material would therefore be the best material in terms of abrasive tool wear, however, excessive galling occurs in drilling, tapping, etc., and the tool must be changed for these conditions, not for abrasive wear.

Carbon, which is the primary and also cheapest alloying element in steels, is considered to be a former of specific microstructure constituents, and by this has a dominant effect on the machinability of carbon steels, chiefly because it governs strength, hardness, and ductility. Increasing the carbon content of steel increases its strength and the unit power consumption for cutting.

Wrought low-carbon steels in the 0.15 to 0.30% C range are usually machined satisfactorily in the as-rolled, as-forged, annealed, or normalized conditions with a ferritic–pearlitic structure. The medium-carbon grades, containing up to about 0.5% C, machine best if an annealing treatment that produces a mixed structure of lamellar pearlite and spheroidite is performed. It is a result of a suitable ferrite/pearlite ratio for achievement the highest machinability of a carbon steel. If the structure is not partially normalised, the strength and hardness may be too high for optimum machinability. In wrought steels with carbon level higher than 0.55%, a completely spheroidized structure is preferred. Hardened and tempered structures are generally not desirable for machining.

Carbon steels nearly always have better machinability than alloy steels of comparable carbon content and hardness. Steels hardened and tempered to hardness levels greater than 300 HB are an exception to this observation; under such conditions, alloy steels have superior machinability, which is usually attributed to, first, the higher te ıpering temperature required to temper an alloy steel to a specified hardness level and, secondly, nonuniformity of the tempered microstructure due to limited hardenability in carbon steels.

Regarding the very low solubility of carbon in α-iron, carbon forms cementite (Fe_3C) in the form of lamellae in pearlite. With increased amounts of carbon in steel the content of cementite increases as well. Cementite possesses a microhardness of approximately up to 1150 HV. Cementite lamellae spacing affects all mechanical properties and by this also the machinability of the material; the finer pearlite plate spacing, the harder the material and the shorter the tool life. Pearlite is a harder microstructure constituent than ferrite and generally causes higher (abrasive) tool wear. Higher carbon levels produce much finer, almost irresolvable pearlite.

On the other hand, a built-up edge is less common when machining pearlite than when machining ferrite. Hard constituents, such as massive carbides or oxides, can be very abrasive to the cutting tool; such particles generally accelerate tool wear [54]. Ferrite can be readily cut and causes little tool wear, but it also contributes to the formation of a built-up edge on the tool and a relatively poor surface finish on the workpiece, although here also the cutting parameters are of major importance. Spheroidised structures can behave similarly, and thus excessive spheroidising is not recommended for machining [231] but large quantities of massive carbide particles can cause significant wear on the tool.

In sintered steels, carbon has a similar effect on machinability as in wrought steels with the exception of some singularities mainly in alloyed steels. Carbon should be present in all steel parts, with the exception of cases in which the strengthening and hardening effect of carbon is undesirable (or where carbide formation is unwelcome as in stainless steels, for example). The effect of carbon must be considered in cases in which sintering causes a reduction of its content; this is to be kept in mind when compositions are given because for sintered steels the nominal (*i.e.* starting) carbon content is usually given which may be higher than the as-sintered (combined) one. Apart from the presence of pores, the microstructure of sintered iron–carbon steels is virtually identical with that of corresponding wrought plain carbon steels. Characteristic microstructures of sintered Fe–(0.3, 0.7) graphite steels are shown in Figs. 5.15 and 5.16.

A lamellar pearlite structure was found to be the best for machining at the defined carbon contents, *e.g.* high tool life at low to medium carbon level sintered steels [142]. For drilling different types of carbon steels, an increase in carbon content resulted in an increase in the cutting forces. The effects of varying combined carbon levels on tool wear *vs.* machining time are shown in Fig.5.17. Tool wear increases rapidly with the combined carbon content in the range (0.0–0.85)% C. This means that the tool life depends to a great deal on the content of carbon and alloying elements due to higher hardness of cementite and by this on the microstructure constituents and on their mutual proportion.

A carbon content of up to 0.25% in steel can favourably affect the machinability as it generates the desired brittleness by cementite in the cutting zone. Sharp cementite lamellae in the pearlite can conceivably accelerate abrasive tool wear of cutting surfaces,

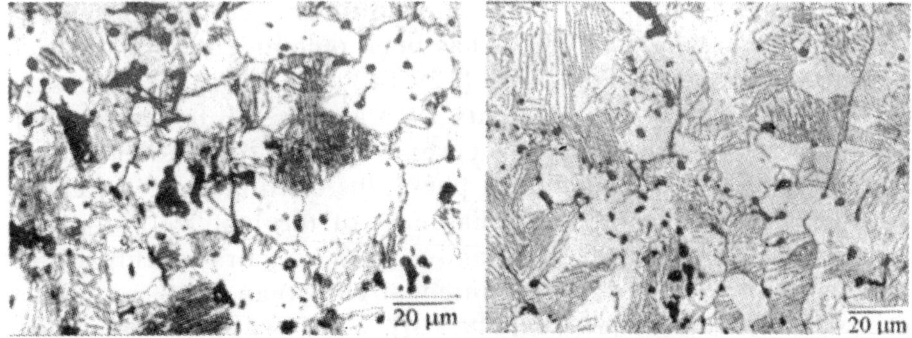

Fig.5.15 Microstructure of sintered Fe–0.3% graphite (0.28% C_c) steel. Iron powder SC100.26, sintering 30 min at 1120°C, cracked ammonia, density 6.97 g/cm³. Optical micrograph. Nital etched.

Fig.5.16 (right) Microstructure of sintered Fe–0.7% graphite (0.59% C_c) steel. Iron powder ASC100.29, sintering 30 min at 1120°C, cracked ammonia, density 7.02 g/cm³. Optical micrograph. Nital etched.

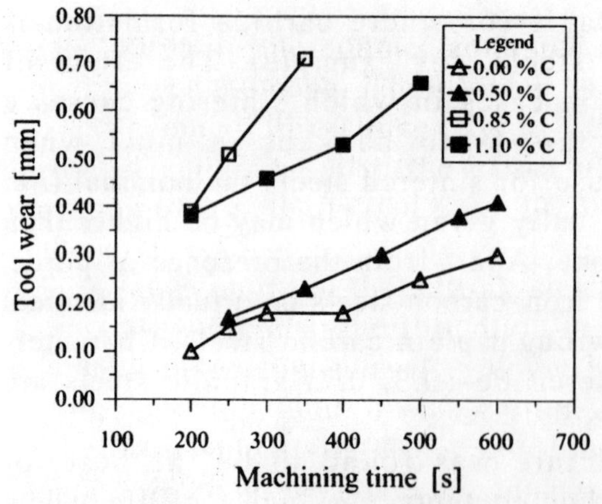

Fig.5.17 Dependence of nose tool wear land of sintered iron on carbon content and on machining time in turning [152].

preventing at the same time the formation of sulphide and oxide–sulphide layers in the crater of the rake face. This can be achieved if the cementite lamellae are broken into smaller, globoidal shapes by means of heat treatment. In the soft ferrite base, they can be pushed aside more easily by the tool. A change in standardised steel chemistry from minimum to maximum, *e.g.* from 0.2 to 0.6% C, changes markedly the ferrite/pearlite ratio and by this the machinability [136,232]. As mentioned, the spacing between pearlite lamellae as a function of the cooling rate and of alloying also affects the machinability (lower spacing, poorer machinability) of the material because of the differences in hardness.

A combined carbon content of 0.2–0.6% is recommended as optimum for good machinability of sintered iron–carbon steel. In some cases, considering the cutting conditions used, and providing a better chip breaking effect, it has been recommended to restrict the carbon content to 0.1–0.3% [129,142]. The drilling forces at these intermediate carbon contents were about 20% lower than for plain iron. However, when the carbon content reached the level where pearlite was a predominant phase, the drilling force increased again because of the increased hardness of the material.

In general, a carbon content above 0.6% impairs machinability and the steels with a carbon content higher than 0.9% require a spheroidized structure [10,104,149,220].

It must, however, be kept in mind that the effect of carbon on the machinability of PM steels is not universal. It depends on the base composition iron–carbon or iron–carbon-alloy elements systems, on the real microstructure constituents formed, and on their proportion. In alloy steels, ferrite is hardened by the solid solution of alloying elements and cementite in pearlite $(Fe,Me)_3C$ (Me:Cr,

Mo, Mn,.., but not Cu, Ni) which is harder with finer spacing of lamellae beside other constituents. By increasing the content of carbon or by adding alloy elements which form hard compounds, the machinability can deteriorate greatly even though the other mechanical properties are not drastically changed. Of course sinter hardening effects induced by alloy elements, such as formation of bainite or martensite, also have a profound effect on the machinability; however, these changes are commonly discernible *e.g.* from the hardness and strength.

As stated above (see chapter 5.1.3), one of the major problems is undesirable carburization or decarburization to various degrees – especially non-uniformity (surface areas of a part are more decarburized than inner areas). When variations of this nature occur, it is virtually impossible to establish a standard machining procedure and still obtain good tool life, surface finish, and a constant production rate [152].

The carbon content also affects surface roughness, although its effect can be greatly modified by the nature of the cutting operation or by the cutting conditions. In wrought steels, low values of surface roughness resulting from machining can be most easily achieved with carbon steels containing approximately 0.25–0.35% which agrees well with the optimal carbon content for machinability of PM steels [129,142].

The surface finish in dependence on the carbon content was poor (R_a of 4.5–9.2 μm), especially for plain iron and Fe–0.85% C. The surface finish for Fe–0.5% C and Fe–1.1% C steels was in the range 4.5–6.9 μm [152]. The highest combined carbon content gives the best surface finish. However, it is all quite poor. The feed rate does not influence the finish to any extent. When the finish is poor at the low feed rates, it generally remains constant as the feed rate increases, even to the extent that the finish at the highest feed rate is better than the theoretically possible as shown in Fig.5.18.

The surface finish (H) depends upon the feed (s), and the corner radius (r). It can be theroretically calculated by the geometrical ratio [143]:

$$H_{theor} = s^2/8r.$$

In Europe, R_a is the most common unit for the characterization of the surface finish, and corresponds to approximately $H/4$. As shown in Fig. 5.18, the results presented for porous materials do not correspond to the presented theoretical ratio for the surface

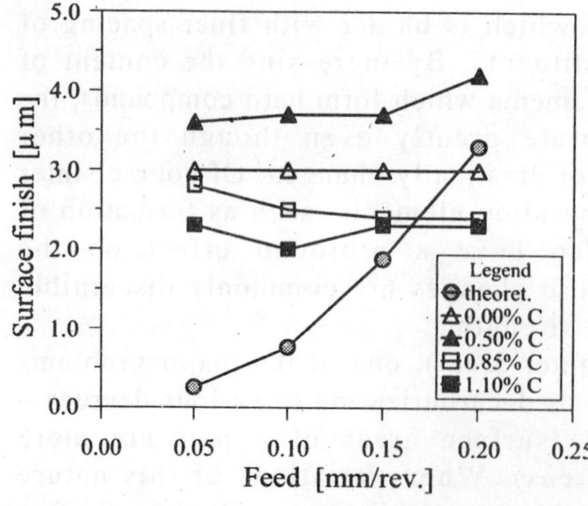

Fig.5.18 Surface finish *vs.* feed rate in turning for new carbide cutting edges in dependence on combined carbon content. Sponge iron powder (Ancor MH100), density 5.8 or 6.2 g/cm³ [107].

finish valid for wrought full density material.

On the basis of the technical and practical experiences it can be recommended to keep the combined carbon content in sintered steels at a minimum of 0.2–0.3% when carbon is a permitted element. In all cases, the combined carbon content, characterized by the microstructure constituents formed, should be a parameter in the sintered material and not the amount of graphite added [157]. As mentioned, the optimum machinability is exhibited by the plain iron–(0.3–0.6)% C steels. When pearlite is the predominant phase, the drilling force would increase again because of the increased hardness of the material.

The effect of each alloying element in combination with carbon on machinability is specific, as a consequence of the microstructure character formed, and in most cases it is very difficult to determine since it also depends on, for example, the cooling characteristics of the sintering furnace used, the loading of the sintering boxes and the surface-to-volume ratio of the component [32].

5.2.6.2 Effect of copper on machinability

Copper as the most important metallic alloying element in sintered steels results in a strengthening effect but in contrast to other alloying elements also improves machinability to some extent. Therefore, copper in sintered steels is accepted also as a machinability enhancer. Metal deformation is improved, and it is frequently assumed that the thermal conductivity of the matrix is increased by it. This is attributed to the fact that copper is a much better conductor of heat than (not determined by adequate measurements), for example, the hardness of copper containing carbon-free steels is higher than of Ni-containing steels.

The effect of nickel on tensile strength and hardness in iron is lower than that of copper, because of the considerably lower diffusivity of nickel in iron and also the fact that nickel martensite is only slightly harder than ferrite [10]. Since up to 8% Cu is dissolved in iron, data concerning the thermal conductivity of the copper containing steel would be a contribution for better understanding the machinability of these steels. In general, machinability of Fe–Cu steels is lower compared with plain carbon steels due to higher strength and hardness, but is higher compared with those for steels of the equal strength and hardness alloyed with other elements [77,157].

For the real improvement of machinability, generally copper addition is required to provide free, not dissolved, copper in the sintered part. The addition of copper for increase of machinability as mentioned before is a common method. It means that copper mainly in combination with carbon is regarded to offer beneficial effects, beside increase in thermal conductivity, controlling also both pore size and compact shrinkage, which, in turn, is a simple aid to machinability [107,152].

Also, an addition of 2 to 3% Cu separately without further machining aid reduced the drilling thrust force by 13%, while additions of copper up to 1% effectively reduced drill wear [129]. There is, however, no information on how copper affected the machinability of a Fe–Cu–C steel. Therefore, the amount of the copper addition required depends on both the cooling rate and combined carbon content.

5.2.6.3 Effect of manganese

Manganese as a strengthening element in steel in low additions is also used as an effective machining agent in the presence of an adequate sulphur content in PM steels. As a machining agent, the manganese content must be sufficient to ensure that all sulphur present will be combined to MnS and FeMnS (for wrought steel a mass percentage ratio Mn/S of about 15 is established). Of course, the Mn consumed by formation of MnS will not be available for modifying the steel matrix.

In general, the addition of 0.5% manganese (without any sulphur addition) to the iron containing less than 0.6% graphite (carbon) can markedly improve machinability by reducing the cutting forces and tool wear in drilling (reduction of drilling force by ~50%) and turning [142] but increases it for mixes containing 0.9% C. When a high sulphur content is accompanied by an increase in manganese content, a better surface finish is obtainable, which results in an improvement in dimensional accuracy. The beneficial effect of large, globular sulphides on the machinability of steels was confirmed [233].

The existence of non-metallic inclusions due to the presence of manganese in mixed Fe–Mn–C steels in term of machinability has not been proved. The non-metallic inclusions could be present when industrial ferromanganese was used as a manganese carrier. In terms of low manganese content, the possible effect of the amount of non-metallic inclusions on the properties and machinability of an alloy is not known but cannot be negligible.

5.2.7 Effect of microstructure on machinability

It may be assumed that the machinability of the PM steels could be described better by the microstructure, which is the product of all PM processing conditions, than by some mechanical property. The characterization of the microstructures is very difficult and in most cases not realisable in practice, and is some cases also for research in relation to machining not only of PM steels.

The importance of microstructure and composition to control the machinability has been long established. The PM process lends itself to an easy manipulation of both composition and, through the processing conditions, of the microstructure of the sintered material. The microstructure of PM materials can be modified and controlled predominantly via the sintering process. As mentioned, sintering influences the pore structure, surface chemistry, and the distribution of elements and phases within the microstructure of the material.

With its microstructure the workpiece material is in contact with the cutting tool. It follows from it that the microstructure is responsible for the machinability if cutting conditions are held constant in the first step. All single factors and processing conditions mentioned before are expressed in the character of the microstructure. From this reason, when speaking about the effect of single variables, their effect on the formation of the microstructure as a product of many material and processing factors must be regarded. It is well known, that these factors cannot be simply characterised for machinability. The microstructure is a dermonstration of the complex effect of base iron powder properties, processing, composition and, in many cases, of very complicated microstructures formed during sintering.

The microstructural heterogeneity in sintered materials prepared from admixed or diffusion bonded powders represents the most complicated state for the characterization in relation to the machinability of a material; this type of microstructures is not found in wrought steels, and therefore no machining guidelines can be derived from classical metalworking. Furthermore, the microstructure of the workpiece material also changes during the machining process as shown in the deformed layer during cutting, expressed also in an increase in microhardness [105].

The heterogeneity of the microstructure of alloyed PM steels results in possible parallel presence of widely different microstructural constituents, from soft ferrite up to hard martensite, in various proportions in dependence on the base composition, alloying technique, and sintering conditions. Between these two microstructural constituents are pearlite, upper and lower bainite increasing toughness properties, and sorbite. For the practical characterization of the property and for the identification of a microstructure constituent, microhardness testing and microanalysis are the simplest tools. Data for the proportion of particular constituents in the microstructure of all materials, especially of higher alloyed steels are missing; however, these proportions depend very much on the respective manufacturing, esp. sintering, conditions, and published data would therefore be of limited use.

Ferrite in plain iron and partly in Fe–C alloys differs from the ferrite hardened by solid solution of an alloying element soluble in iron. Ferrite as the mirostructure constituent with lowest hardness is strongly deformed in surface layer in machining. The microhardness of ferrite increases with alloying in dependence hardening effect of the alloy element (see Fig.2.10). The microhardness of pearlite formed by ferrite and cementite lamellae is further increased in an alloyed matrix. Similarly, the microhardness of other microstructural constituents increases in dependence on type and amount of alloying element including its solubility in iron and mutual reaction with other elements. Martensite is usually characterized by high microhardness (~800–1000 HV) but can be also softer, *e.g.* nickel martensite. Retained austenite can also be present in the microstructure. It follows from it that the microstructure of a PM alloyed steel can be characterized by the microhardness values which may range from ~100 to 1000 HV. It shows the impact of microstructure differences upon machinability and the reasons why is it not well known or understood. The main disadvantage of the heterogeneous microstructure is, therefore, the uncontrolled machinability of such material.

In terms of machinability of the material with heterogeneous microstructure, the effect of the mentioned microstructure constituents on machinability or on some factor acting in machining are not known. In machining, during one rotation the cutting tool must pass the microstructure constituents with the given extremely varying microhardness. It is, however, not possible to directly assume that the microhardness of a microstructure constituent is a sufficient characteristic for the machinability, for example, it is not known why a medium carbon,(~0.3–0.5)% steel has optimum machinability. There was suppressed the negative effect of soft ferrite (tearing effect) by a not known positive effect of pearlite. The cause may be a compromise effect of ductile ferrite and harder pearlite portions and different friction properties in the cutting process between them. A similar case is also the positive effect of copper on the machinability of Fe–Cu–(C) steels.

The physical properties of single microstructural constituents which really affect the machinability are not characterized mainly in terms of machinability investigated. The increase of the temperature in workpiece and tool is a consequence of friction of the workpiece–tool friction pair under load used. With the aim to determine more in detail the sliding characteristics of a microstructure by a special device ('grater') the sliding properties of single phases were investigated [234]. It means that for the friction in the cutting zone at the applied load and form, also the workpiece microstructure with its characteristic friction properties is responsible. The lowest friction coefficient is exhibited by pearlite followed by bainite. Martensite exhibited the highest friction coefficient. Perhaps this research way could be effective for the formation of special microstructure contribution to the more detailed analysis of the process proceeding in the cutting zone. It should be a microstructure exhibiting lowest friction and by this way lower temperature in the cutting contact zone and by this lower wear of the tool at the cut.

The disadvantage in the microstructure of PM steels can be also an advantage because the microstructure can be modified by processing and alloying not only in term of functional properties but also in term of machinability, not regarding the addition of machining agents which is a separate area of interest. This possibility of PM so far has not been an object of focused research, perhaps due to lacking of the base characteristics of phases in term of machinability. It means that not only material composition but mainly the knowledge of the specific properties of the microstructure constituents formed is, therefore, a most important factor in PM machining [129,152]. This will be also a contribution in orientated standardization of materials and machining conditions for PM steels.

As a very simple statement it could be said that the PM process lends itself to easily modify the microstructure of the material more favourable for machining by changing the processing conditions. At the given porosity, the microstructure, including the pore structure, can be changed by changing the alloying technique, alloy elements and their amount, or by adding some liquid-phase forming additive, excluding the existence of hard eutectics (of course, always regarding the required dimensional tolerances and mechanical properties). Furthermore, the type and amount of phases in the microstructure can be controlled and modified by adjusting the sintering conditions, including the atmosphere, and the cooling rate and/or with the use of subsequent heat-treatment, as shown in Fig.5.19.

In this case, a slow cooling rate of 9°C/min resulted in improved machinability of this material with lamellar pearlite mixed with a divorced eutectoid carbide microstructure.

The possibility of changing the chemical composition is in reality limited due to the fact that selection of material for the component is mainly based on the mechanical and toughness properties and that the limits given by material standards have to be regarded. The more correct conclusion is that PM steels are different and it is possible to assume that their microstructure, except the machining aids, can be optimised frequently by modification of the machining conditions and the cutting tools. Thus, the knowledge required to

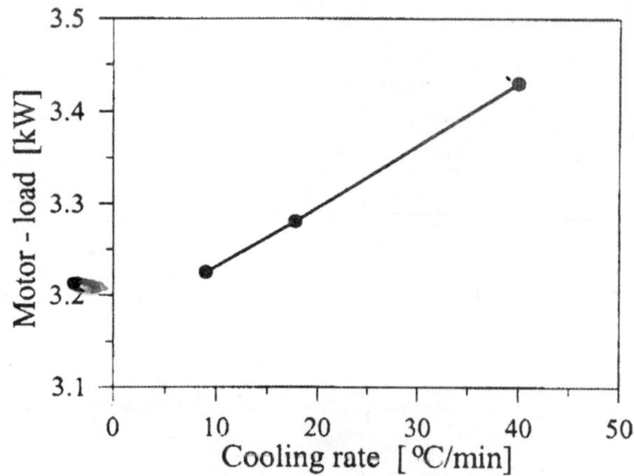

Fig.5.19 Effect of cooling rate during sintering on the power required for machining (motor load) of FC-0208 alloy with 0.5% MnS addition. Density 7.2 g/cm^3 [105]. *Drilling test: TiN coated hardmetal 14 mm drill, drilling speed 33 m/min, feed 350 mm/min.*

machine PM steels has to be acquired and efficiently disseminated widely. Materials that are easier to deform yield better tool lives than stronger and harder materials, and materials with low carbon content and/or high copper content produced, in general, the 'best' results.

Finally, as regards composition, it is necessary to note that an unavoidable part of the microstructure are also the pores as microstructure constituents with zero elastic modulus (Young's modulus). In connection with porosity, it has been stated that the deformed layer generated during the cutting possesses a new microstructure with higher hardness which is a product of the starting microstructure and the cutting forces. In this connection it is necessary to analyse the particular microstructure constituent properties and their effect on the deformation processes at cut because they affect by this the newly formed microstructure.

The usual controls of the sintering process, as resulting hardness, combined carbon content, or dimensional change, are correlated to machinability but as is known from the practice they are not satisfactory characteristics and will not reveal all causes that will generate problems later at the machining operation [130].

A comprehensive summary of the many different microstructures found in sintered, heat treated and powder forged steels including some after machining is shown in Refs.235–238. However, new materials with new microstructures are continuosly developed.

6

Measures to Improve the Machinability of PM Steels

Measures to improve the machinability of PM materials can be derived from the influencing parameters. At present, three methods are very common in the PM industry to achieve this goal which are [209]:

material properties	–	free-machining additives (machining aids),
	–	pore filling (infiltration, impregnation),
	–	micro-cleanliness improvement (effect not sufficiently defined),
production processes	–	microstructure modification (alloying, heat treatment, annealing, normalizing),
	–	presintering,
	–	green machining,
machining operations	–	cutting conditions,
	–	tool (material and geometry, coating).

From all these methods, the main focus in investigations has been oriente l towards the effect of machining aids on machinability.

The first and second groups of methods are proper powder metallurgy methods regarding its possibility and greatest chances for improving the machinability of PM steels. They affect the workpiece properties by many routes which adapt the microstructure and properties of the workpiece for improved machining. Today, the use of some intermediate state of the parts for machining, *e.g.* presintering or green machining, also belongs to the powder metallurgy methods.

The first group of measures is to improve the machining properties of the workpiece by admixing machinability additives to the powders, by filling and sealing the porosity or by improving the micro-cleanliness of the powders. From these the use of machining aids, some of them especially developed for this application, seems to be the simplest one and, therefore, is in the foreground of attention of the powder metallurgy community. Common methods to seal the porosity such as copper infiltration or resin impregnation belong to the PM methods by controlling the microstructure and composition of the part with the aim to improve the machining however they must be finally cost effective. A marked improvement of micro-cleanliness of powders cannot be expected because purity is a requirement for all applications independently of machining, and today's PM ferrous powder grades have attained a very high state of purity, which is simply a precondition for compacting, sintering and, finally, machining.

Changing the production process for a given product is a promising method. The modification of the microstructure by alloying and sintering conditions with regard to machinability has not been sufficiently investigated although, as describe above, the microstructure and the corresponding properties are decisive for machining.

Therefore, improving the machinability by controlling the composition and microstructure of the parts during sintering is an attractive option in PM, although of course the main requirements towards dimensions and mechanical properties have to be met. On the other side it is, for example, well established to modify the microstructure by a heat treatment before machining or to add a presintering step followed by machining and to sinter afterwards. Both production routes have the disadvantage of an additional production step but may ease machining very much. Green machining is a new technology, which works without additional steps [209].

The third area of the methods for improvement of the machinability of the parts is the optimisation of the machining process itself, which can be either an optimisation of the cutting conditions or an optimisation of the tool system or even both, always in relation to the properties of the workpiece. If it should be a contribution for improving the machinability of PM steels, it must be realised in a very close relation to the first two methods, not as appendix.

In reality, it is not possible therefore to detach one of the other method. They are investigated and performed as a common problem and task in serial steps, from the simplest one to the more complicated.

6.1 MACHINING AIDS IN POWDER METALLURGY

In order to improve the machinability of all sintered materials, and especially of high strength steels, adding machining aids is a widely used method for machinability enhancement of all kinds of PM steels independently of porosity.

The special additives which are reported as lubricants in cutting process, are either mixed-in during the mixing stage or incorporated into the melt at the production of atomised powders. The first method is useful and technologically simple to add to the base powder mix by blending, prior to compacting and sintering of parts, resulting – if properly done – in the uniform distribution of controlled-size particles of a machining aid in the material. The role of these additives differs from the effect of the lubricants used in die compaction of powders.

Compounds referred to as lubricant are surface-active and flow agents can also adversely affect the effective uniform rapid die filling. As machining aids, they are most effective in die filling and at the earliest stages of the compaction stroke where particle rearrangement occurs [239].

Another advantage of mixed powders in general is greater flexibility since new machining aids compositions are easily incorporated into existing processing procedures, there is no need to change the powder supplier, and in general mixed powders are easier to compact. The production of powder prealloyed with some machining aid, mainly with sulphur with corresponding manganese is one of the methods used. This results in increased machinability and tool life. Even with lower density PM parts, the improvement in machinability can be dramatic. In this case, it is essential that the machinability-enhancing additives minimise the effects of porosity and non-metallic inclusions. This advantage of the PM routes is used to maximum range for machinability enhancement, in order to eliminate the disadvantages of PM, *i.e.* the previously mentioned machinability problems.

The machining aid, if it should be effective, must in general fulfil two main requirements:

– mechanical and physical properties and dimensions of the sintered component are retained at the same values as for material without a machining-enhancing additive,

– improved machinability [104,107,130,162,240].

Today, an acceptable machinability additive must have further desirable characteristics for fulfilling the previous requirements as follows:

– non-toxic, no adverse environmental effects,
– inexpensive, economical for use in PM parts production,
– thermodynamically stable in contact with all the common mix compositions and sintered steels, up to high temperatures or, if reactive, should form a machinability-enhancing compound (at least during cooling),
– no detrimental effect on the muffle, belts and internal walls of the sintering furnace and atmosphere in the furnace, with no burn-off,
– should minimise the detrimental effects of porosity and microstructure heterogeneity and the often mentioned non-metallic inclusions on machinability and by this way to contribute to a significant improvement of machinability.

These main requirements on the role of a machining aid in cutting process should be the criterion for an option of an element or compound with the physical–chemical properties fulfilling these conditions [104,240,241].

In general, not all machinability aids are ideal following the listed criteria because they have in most cases a minor or also higher adverse effect on dimensional characteristics and compromise sintered properties. In reality they are mostly non-metallic inclusions in the material with chemical–physical properties differing from those of the matrix, with a specific role in the cutting process. From this reason practically it is necessary to take into account also a deteriorating effect of all machining aids on the base material characteristics. The tolerable limit of it is in all cases individual for each part, with the basic requirement to avoid pronouncedly negative effects on the essential service properties of the part.

The additives that have been investigated, and to some degree used in PM part production, are such as S, MnS, MoS_2 (sulphides in general), sulphates, and others represented by Ca-containing compounds, Bi, Se, Te, Pb, hBN, and some special compounds developed only for this purpose. A special group is formed by glasses and resins for impregnation to fill and seal the porosity for significant improvement in the machinability of PM parts, and in general compounds containing group VIIB (F, Cl, Mn, Br, J) elements [118,135,157].

The most important elements that can affect the type, shape, and size of the characteristic inclusions and thus also the machinability of steel, reducing tool wear and built-up edge formation during cutting, are: sulphur, phosphorus, tellurium, boron, selenium, etc. Other alloying elements, such as carbon and manganese, are usually not taken into consideration in connection with non-metallic inclusions, since they affect predominantly the strength properties of steel and therefore impair the machinability [149]. Only copper and, to some extent, manganese and phosphorus are used in PM on a small scale not only as alloying elements but also for machining enhancement. Boron, except in BN compound, is used as a sintering activator through liquid phase formation. This results in the formation of hard eutectic phases and, consequently, in a decrease of machinability. The boron liquid-phase sintered Cr and Mo steels, having a microstructure without a eutectic, can be the object of machining [76].

The large number of machining aids used and investigated proves that the machining of PM steels in general is a more complicated process than it is recognised [137]. For this reason, the data on the chemical–physical properties of machining aids which are important for machining of PM steels, are described separately.

6.1.1 Role of machining aids in machining process

The role of machining aids in the machining process of a sintered part must be known from the beginning because the ideal free-machining aid performs several functions in the cutting process.

The first is to promote microcracking and fracture of the chip/workpiece interface ahead of the cutting tip and by this to affect wear mechanism as shown in simplified form in Fig.6.1. It means to promote chip fracture (chip breaker) and prevent welding of the hot chips to form continuous swarf since the removal of continuous swarf from a machine is usually very complicated, therefore all free-machining wrougt steel grades are optimized for formation of short chips. In sintered steels, porosity, in spite of its negative effect on machinability in general, contributes to the formation of a short chip [242].

A second function is to prevent built-up edge formation. A built-up edge can form in an area where local cutting forces and temperatures promote welding. The built-up edge changes tool geometry and cutting conditions. It is often unstable and, as it forms and then decays, *i.e.* usually breaks off, changes surface finish and dimensional tolerances.

A third function is to act as complex lubricant and a barrier to

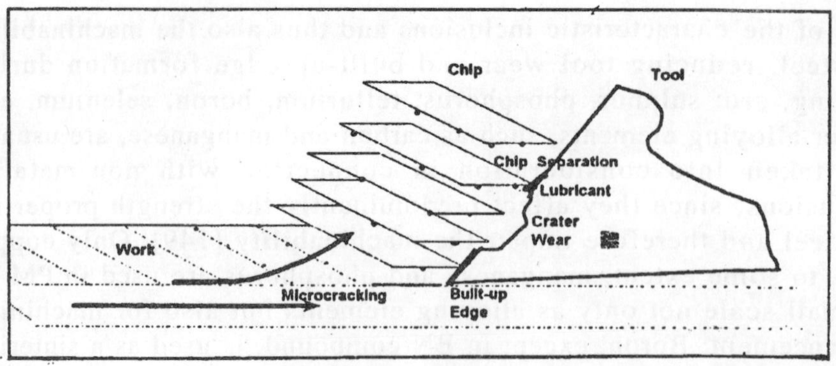

Fig.6.1 A simple illustration of the effect of a machining aid in the cut [136,137,242].

diffusion in the region of the tool face behind the cutting edge where 'crater wear' normally occurs. In this region, the chip moves rapidly across the tool face. The heavily deformed chip can rapidly abrade material from the tool, especially if it contains hard second phase particles such as cementite or alloy carbides. Thus, it is beneficial for the free-machining aid which may spread to form a temporary coating over the tool surface. This coating may act as a barrier to thermally promoted diffusion of mobile species between tool and chip, which can drastically change tool properties locally and is the main mechanism for crater formation.

In general, from technical point of view, the machining aids have to decrease the tool/chip friction and by this way cause a decrease in cutting forces and the temperature in the cutting zone and also a result in improved surface finish and generally in improved machinability of a material [243]. It should be noted that the properties of the machined layer of the material change (work hardening), as explained in chapter 5. This change in the material properties is a very important factor which is frequently neglected when studying the friction state of a cutting pair. In most cases, only the starting material properties are considered.

From this point of view, it would be very important to investigate the friction properties of all constituents taking part in the machining process and their mutual reaction, *i.e.* the workpiece material with defined mechanical and microstructure properties, including machining aids and tool material up to seizure as in sliding process (friction pair).

High temperature in the cutting zone becomes an important consideration because some additives such as, for example, polymer materials with low melting temperatures used to close porosity can improve surface lubrication and by this decrease the wear in the cutting zone. In some cases, the reactions between an additive at high temperature can change the microstructure of the machined material and could lead to the production of parts having undesirable microstructures [129,142].

The beneficial effect obtained by the addition of a machining aid can be explained by the role of MnS, which was most frequently investigated. Manganese sulphide 'inclusions' act as stress raisers in the machining shear zone to initiate cracks that subsequently lead to fracture of the chip. These inclusions are also known to deposit a layer on the wear surface of the cutting tool. It means that in machining operations, MnS acts as a lubricant, Fig.6.2, although it is not a solid lubricant as, for example, MoS_2 though it minimises the tool/chip friction, reducing tool wear [147,148].

At lower or mild machining conditions the protective layer is not formed, and the inclusions (MnS), remain separate even after cutting. At increasing machining conditions the MnS from the chip or workpiece material is softened and extruded and starts to coat the tool surface during the machining process of powder metallurgy parts, thus having a lubricating effect at the tool (rake face)–chip surface. The reduced tool–chip friction results in lower cutting temperatures, cutting forces, and reduced tool wear. All mechanisms jointly contribute to the improved machinability. As no major built-up layer of manganese sulphide is actually observed, it can be hypothesised that the coating formed is continuously being removed and re-built during the machining process. By this the chip breaking is facilitated under the formation of short friable chips. The short chips are characteristic for machining of PM steels as a

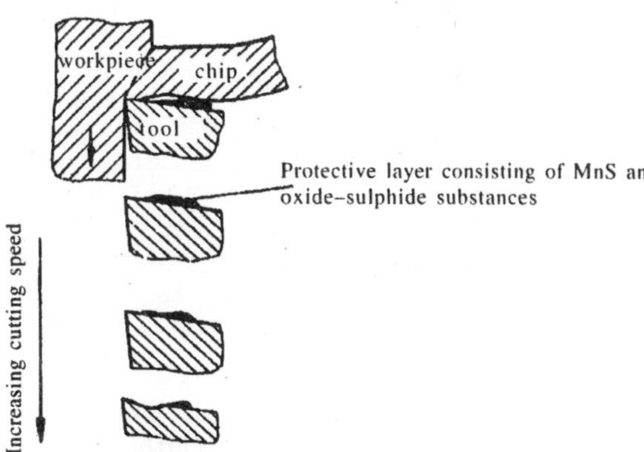

Fig.6.2 Conditions in the tool–workpiece–chip cross section showing protective layer consisting of MnS and/or other oxide–sulphide substances at increasing cutting speed in turning of wrought steel [138].

consequence of the role of the machining aid as chip breaker (the same reason as in free-machining of wrought steels) and of the porosity.

The lubrication effect of a machining aid is dependent strictly also on cutting speed and feed as shown schematically in Fig.6.3. It illustrates the functional dependence of machining conditions and the development of the protective layer from non-metallic substances (also the machining aids) on the rake face of a cemented carbide tool. The machining conditions between the lines 1 and 2 are such that they enable, due to pressures on the rake face, the softening and retaining of the protective layer. In the areas under line 1 the machining conditions do not enable the softening of non-metallic inclusions and thus do not enable the formation of the protective layer – so this is the area of intensive wear. The area of machining conditions above the lines assures however extreme softening and owing to high pressures also the extrusion of non-metallic substances from the crater wear area on the rake face. Thus, it is the area between lines 1 and 2 that enables the development of an efficient protective layer of non-metallic substances.

Special attention with regard to the role of a machining aid should be paid to the porosity; also here significant differences can occur compared to fully dense wrought steels. As an example, admixed MnS particles, and other thermostable aids are also located in the sintering interparticle contacts and mainly in the pores in dependence on the density of the sintered parts. The smaller size

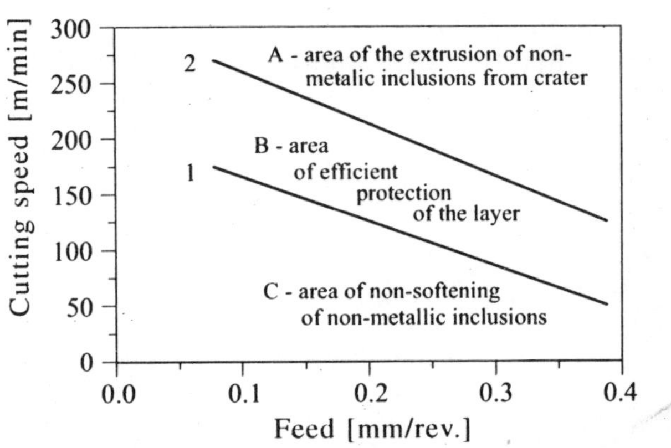

Fig.6.3 Areas of the effect of machining conditions in wrought steels (cutting speed – feed relation) on the development of the protective layer in cutting (turning) [138].

of the MnS particles allows a large number of the particles to remain within the pores even at high densities, not only at the common 6.8–7.0 g/cm^3. The presumption that the majority of the MnS particles will be accommodated within pore sites and has a minimum impact on parts properties also in parts with a density of about 7.5 g/cm^3, should be analysed in more detail due to its relevance to mechanical, in particular fatigue, properties.

Since the particles are soft and deformable they conform to the pore geometry, and they can undergo extensive plastic deformation at temperatures encountered in typical machining operations. The MnS particles readily deform in the primary deformation zone ahead of the tool–workpiece interface, resulting in a reduced shear-flow stress in this region. The pure iron particles have a beneficial effect insofar as they anchor the MnS particle in the sintered microstructure, thus avoiding pullout when the machine tool passes the particle [105,225,244]. It is apparent that for the completion of this role ('anchorage') the aid should be dispersed uniformly throughout the workpiece, preferably as fine particles. The effect of an anchorage of the machining aids on the powder particles is usually lower on harder ones.

The uniform distribution of fine machining aids in the compact is desirable but in practice hardly obtainable, and it is virtually impossible in the case of admixed machining aids since the cores of the matrix powder particles of course remain free of any machining aid. Furthermore, it was demonstrated in many cases that, for example, MnS forms clusters and agglomerates in the compact as is also possible in the case of some other fine additions in a mix with the base powder.

If these agglomerates are incorporated in the interparticle necks formed, they may act as non-metallic inclusions in the load-bearing cross section, which change the character of the interparticle neck fracture and by this the mechanical and physical properties of the part, mostly adversely.

The effect of some machining aid on machinability of a sintered material should be in all cases evaluated considering also its effect on relevant physical and mechanical properties since by this way can be decided about the effectiveness of the aid.

Global information stating that the machining aid has no or only a minor detrimental effect on certain properties of the material is not sufficient as a basis for an adequate decision. It is necessary to note that there is insufficient knowledge about the effect of some machining aids on the service properties of a sintered part in spite of a sufficient machinability.

In general, the additives used as machining aids contribute to lower cutting forces, longer tool life, and better surface finish what is the reason for their use. Some alloying elements reducing friction at the tool/chip interface, tool wear and built-up edge formation during cutting are acceptable. There is also a tendency for chips to break more readily [129].

6.1.1.1 Effect of machining aids on compressibility

Machining aids in a powder mixture can firstly affect the compressibility. This effect is generally not affected by the composition of powder mixtures, and is considered in general to be minor to negligible. Most of the additives can contribute to the compressibility of the iron powder. The density of the machining aids must be considered for assessing the green density of a powder mix which is usually changed more by admixing further powders. The addition of frequently used MnS has in general only a minimum effect on green density of the alloy compositions.

Usually iron powder compositions required a marginally higher compacting pressure to achieve the same absolute green density when MnS was added to the iron powders due to the effect of the lower density of MnS [25]. It must, however, be considered that this lower absolute density does not necessarily mean higher porosity since the main effect is due to lower theoretical density of the mix when adding low density powders, as stated above. In Fig.6.4 the theoretical densities of Fe–MnS and Fe–PbS powder mixes as given as a function of the MnS/PbS content as well as some experimentally determined green density values; as can be seen the green density at first slightly increases with the MnS content, and the porosity markedly decreases; only at higher MnS contents the porosity does not decrease any more. The same behaviour can be found with admixed PbS; in this latter case higher percentages, up to 10%, have been investigated, and a consistent increase of the porosity has been found, indicating that the adverse effect of admixed fine powder outweighs any lubricating one.

6.1.1.2 Machining aids – base characteristics

The features of some agents that improve (deteriorate) machinability according to Ref. 208 are listed in Tab.6.1. The additives were classified as metal-matrices, sulphide-base materials, oxide-base materials, fluoride-base materials, etc. For the metal-base-materials (Pb, Bi, Se, Te, etc.) they are soft metals but include many harmful

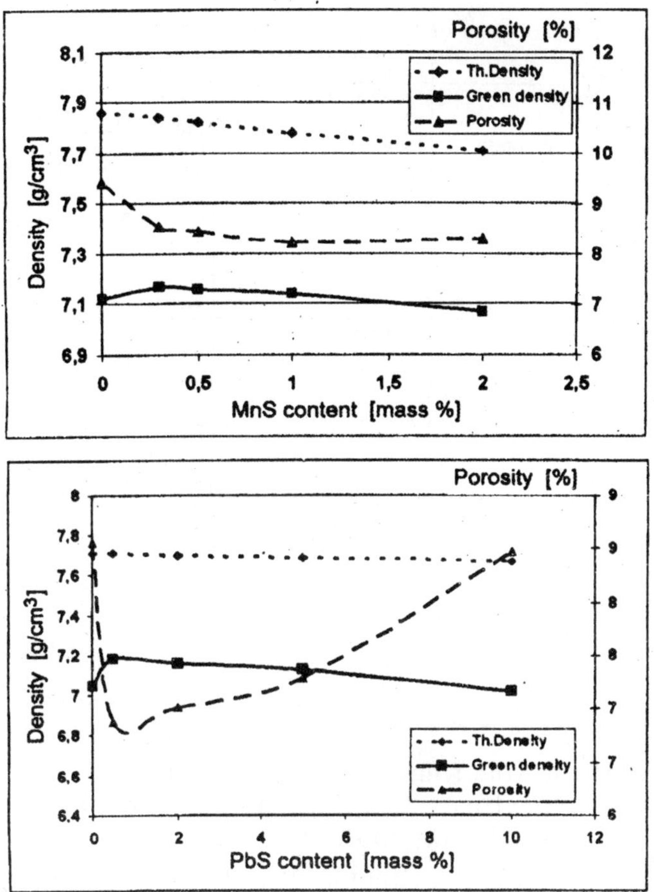

Fig.6.4 Theoretical and experimentally determined green density/porosity: a – Fe–MnS, b – Fe–PbS powder mixes. Atomised iron powder, 0.5% EBS as lubricant, compacted at 600 MPa [205].

substances. For sulphide-base materials, there are MnS, FeS, CuS, MoS_2, etc. Many sulphides are harmful substances, and also in a sintered atmosphere, particularly by the reaction with hydrogen at high temperature, the generation of H_2S, etc. is of great concern. In the oxide-base materials, there are many stable ones, as compounds, when they exist together with other elements in materials during sintering in an atmosphere used, they are difficult to decompose. It is also expected that these harmful substances would be scarcely scattered at sintering. However, because SiO_2 and MgO are very hard, the abrasion of tools become extremely severe if they are present in a material. CaO powder is strongly hygroscopic and since swelling of the green compacts thereof and

Tab.6.1 Effects and problems of the base machining agents groups [208]

Material	Machinability agent	Effect	Problem
Metal	Pb, Bi, Se, Te	Lubrication Chip breaker	Harmful material Sublimation during sintering
*Metal sulphide	MnS, FeS, CuS, MoS_2	Lubrication Chip breaker	Harmful material, Sublimation during sintering Inhibit sinterability (ex.)$2MnS=2Mn+S_{2(g)}$ $2H_2+S_2=2H_2S_{(g)}\uparrow$ $2CO_{(g)}+S_2=2COS_{(g)}\uparrow$ $C_{(s)}+S_{29g0}+CS_{2(g))}\uparrow$
Oxide	CaO, SiO_2, MgO Enstatite ($MgSiO_3$)	Chip breaker Tool protection	High safety SiO_2, MgO: High tool wear CaO: High hygroscopicity (ex.)$MgSiO_3+MgO+SiO_2$
Fluorite	CaF_2	Chip breaker	Resolving during sintering (ex.)$CaF_2+½O_2= CaO+F_{2(g)}$ $H_2+F_2=2HF_{(g)}\uparrow$
Others	BN Graphite precipitation or dispersion	Lubrication Tool protection	Resolving during sintering (ex.)$BN+H_2O=B_2O_3+NH_{3(g)}\uparrow$

Remark: Enstatite ($MgSiO_3$): Magnesium metasilicate–base oxide. Densities (g/cm^3): Te = 6.24, Se = 4.4, Bi = 9.8, Pb = 11.4, Mn = 7.3, Cu = 8.9. *According to the crystal system, only MoS_2 is a solid lubricant. The meaning of (ex.) is not disclosed in [208].

a hindrance of the sintering reaction occur, the CaO powder is not suitable for practical use. In the oxide-base materials, there is a material called 'enstatite' $MgSiO_3$. This material is one of the mineral materials having a slipping property – talc. During sintering although CaF_2 is very stable, with ΔG (1250°C)= –480 kJ/g-atom F. In particular, the thermodynamic stability of these agents and their starting characteristics, mainly in the oxide-base materials, must be taken into account.

According to [240], manganese sulphate $MnSO_4 + 4H_2 = MnS + 4H_2O$ can be also an attractive machinability aid when admixed into powder mix and by this into the parts. A disadvantage is that the addition of $MnSO_4$ to the furnace feed tended to upset the reduction reaction of the iron oxides.

In general, the effect of a particular machining aid on the machinability of a sintered material is affected not only by its mechanical properties but also by the respective machining operation and the cutting conditions applied.

6.1.2 Basic metal sulphide characteristics and effects

6.1.2.1 Effect of sulphur

Sulphur forms a base component in all sulphide machining aids but in some technically interesting cases is added also separately to the iron powder to form corresponding sulphides in the compact.

The economic incentive for addition of sulphur, *i.e.* to achieve higher rates of metal removal and longer tool life, has led to the development of free-cutting wrought steels. The sulphur is in this case added to the molten steel. There is significant improvement in machinability when resulphurised carbon steel is substituted for a plain carbon steel of approximately the same carbon content. In carbon steels, the sulphur content is usually restricted to a maximum of 0.05%. In the manufacture of resulphurised steels, sulphur is deliberately added to achieve the desired sulphur level. The most common level range of sulphur content in resulphurised steels is 0.08% to 0.13%, but some grades permit a sulphur content as high as 0.35%. The improved machinability is the reason for using sulphur as a machining aid in PM part production.

In contrary to the other machining aids, admixed to the base powder in form of special fine powders, a positive effect of sulphur addition on the density and mechanical properties of sintered iron is known. Sulphur (~0.1%) appears to promote sintering of Fe–S systems and by this way increases the density and produces very round pores. A (low) portion of the sulphur dissolves and diffuses into the iron matrix, and a portion of sulphur reacts with iron to form iron sulphides or, above the Fe–S eutectic temperature, the liquid phase, and with manganese sulphides. Higher addition of sulphur up to ~0.5% affects the fraction of closed pores and by this way the tensile strength of sintered iron increase by ~25% [242,245,246,]. Since the solubility of sulphur in iron is very low, and the Fe–S eutectic temperature at 988°C, with 31.6% S, Fig.6.5, is lower than the common sintering temperatures, a eutectic melt forms during sintering even at rather low sulphur levels, the solubility of S in austenite being as low as about 0.01% at the eutectic temperature, Fig.6.6.

To enhance machining, sulphur is added:

- in elemental form in small quantities in iron powder mixes forming iron and manganese sulphides in sintered alloys, and
- in the form of prealloyed powders containing a higher level of sulphur and manganese [149].

With regard to machinability enhancement, the sintered materials or powders are in this case characterized by the sulphur content. Sulphur does not affect machinability directly in the elemental form added to the powder mix or into the melt but as MnS and/or $Fe_{1-x}S$.

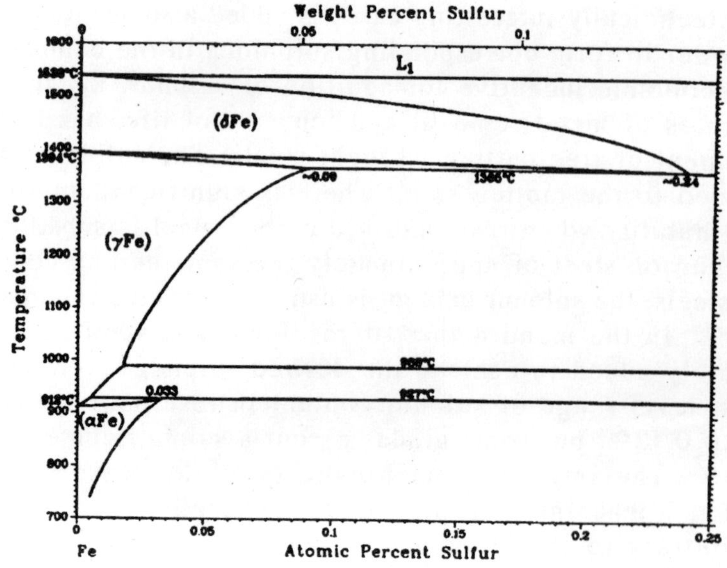

Fig. 6.5 Fe–S binary phase diagram [247].

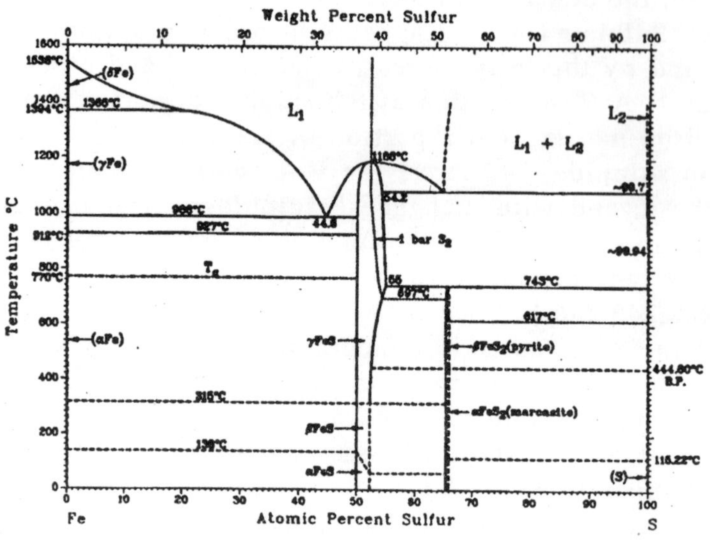

Fig.6.6 Fe-rich section [247].

In these steels, the sulphur content must be intentionally high and the manganese content must be high enough to ensure that all the sulphur is finally present in the form of manganese sulphide (MnS) 'inclusions' (in ingot metallurgy Mn/S ratio ~15/1 to prevent the formation of FeS). This ratio of Mn/S is not the case in machining of PM steels. When MnS inclusions are extremely deformed in the shear plane and in the flow zone adjacent to the tool surface, they contribute to higher cutting speeds, longer tool life, good surface finish of machined parts, lower tool forces, and produce chips which can be more readily handled. These steels perform more consistently than the non-free-cutting steels in a wide variety of operations [248].

Machinability can be improved by adding sulphur to both sponge and atomised iron powder based alloys. The presence of sulphur inclusions causes the material matrix to deform more easily, facilitating crack propagation and reducing cutting forces and by this way results in an improvement in tool life and in surface finish. This is the same affect on the built-up edge on the rake face as mentioned for MnS [129,142].

The effect of sulphur is the same as in general of all machining aids. For Cu–C containing steels with 75 to 90% density based on sponge iron powder, the addition of sulphur in the range 0.25 to 0.6% reduced cutting forces by as much as 25% in turning. A reduction of 12 to 20% in thrust force and improved surface finish were observed during drilling 90% dense sintered iron. It should be investigated how such low sulphur content is present in the microstructure of iron and some alloy systems, probably as an eutectic network.

Sulphur addition to the iron powder as a machining aid should be related to the carbon content. While close to optimum improvement in machinability can be obtained in 0.5% combined carbon containing parts through an addition of 0.5% sulphur, for plain iron parts and for such containing 0.85% combined carbon a sulphur addition of 1.0% is required [107]. The higher sulphur content permitted machining with higher speeds and heavier feeds resulting in improvements in machinability by up to 25% (based on drilling tests) (formation of manganese sulphide inclusions) compared with conventional sintered steels .

In general, significant reductions in cutting forces were obtained when 0.25–(0.5–0.6)% S was added to iron powder containing very low manganese levels of 0.05–0.06% [248]. In this case, sulphur should have been present as an eutectic network of $Fe_{1-x}S$.

It is known that a considerable loss of sulphur may occur during sintering [243]. Since hydrogen in the sintering atmosphere desulphurises the surface of the part through the formation of hydrogen sulphide mainly during heating, only the sulphur content at the moment of actual machining, *i.e.* the content after sintering, should be taken into consideration. The difference in sulphur content in the cross-section of the part, in particular between core and the zones near the surface, is a disadvantage of this form of sulphur addition. Some environmental problems may also exist with the exhausting atmosphere. This is why this form of sulphur addition to the powder admixtures is not commonly used.

Sulphur-prealloyed iron and some alloy powders eliminate the disadvantage of adding sulphur to the admixture in the elemental form. In this case, the loss of sulphur at sintering is significantly lower.

6.1.2.2 Thermodynamic stability of sulphides in ferrous matrices

Metal sulphides form a group of compounds used especially as machining aids in powder metallurgy. Thermodynamic stability is in general one of the properties that determine the area of the application in machinability process. This stability is relevant for the machining process by itself but also for sintering, since the sulphides are usually admixed to the base powder(s). Table 6.2 gives the basic data for several sulphides and other machining aids; the greatly varying thermodynamical stability of the sulphides is clearly discernible, especially when comparing the Gibbs free energy of formation per g-atom sulphur.

As a further example for the practical importance of thermal stability the mass loss of some sulphides processed for 30 min in hydrogen (dew point –35°C) and argon (dew point –55°C) in the temperature range of 200 to 1100°C is shown in Tab.6.3. Depending on the base element, a considerable mass loss of up to 60% is reported with increasing temperature. These high temperatures can be reached in microvolumes in the cutting zone. During sintering in hydrogen, in argon, or in vacuum the sulphides can fully or partly dissociate – and even evaporate, taking into account the high

Tab.6.2 Properties of various sulphides and other potentially used machining aids

Material	Mol. [mass]	Density [g/cm^3]	ΔG [kcal/mole]	ΔG [kcal/g-atom S]	Fp [°C]	Max. solub. in Fe [mass %]
Cu_2S	159.15	5.6	–19.0	-19.0	1100	Cu: 8
MnS	87.00	3.99	–42.7	-42.7	1610	Mn: 100
MoS_2	160.06	4.8	–56.2	-28.1	1185	Mo: 35
PbS	135.26	7.5	–24.0	-24.0	1114	Pb: very low
Sb_2S_3	339.68	4.12	–41.8	-13.9	550	Sb: 10
ZnS	97.44	4.1	–46.04	-46.04	1020	Zn: 42
C (graphite)	12.01	2.26	-	-	3730	2.1
Pb	207.19	11.34	-	-	327	very low
BN	24.82	2.34	–60.8	-	>3000	B: very low
CaF_2	78.08	3.18	–278 (?)	-	1423	Ca: Very low

Tab. 6.3 Mass loss in hydrogen and argon and temperature stability of some sulphides in vacuum in contact with iron and its oxides. T – Temperature of the reported formation of iron sulphate in the contact zone sulphide in vacuum – iron film [249]

Sulphide of	Mass loss [%] at temperature [°C] in hydrogen and argon												T [°C]
	Hydrogen						Argon						
	200	400	600	800	1000	1100	200	400	600	800	1000	1100	
Bismuth	5.5	18	10.3	-	-	-	0.15	0.39	8.9	10.0	12.0	-	-
Cadmium	0.7	4.8	1.2	20	53.5	-	0.4	0.7	2.5	3.0	8.2	15.0	520
Antimony	3.2	8.2	20	76	-	-	3.2	6.0	7.2	26.0	-	-	310
Cobalt	3.5	15.4	3.3	50	60.4	-	0.2	10.4	12.2	47.0	55.0	-	340
Copper	4.9	6.8	7.8	8.3	9.3	-	3.5	3.7	3.9	4.4	4.6	30.0	450
Molybdenum	0.4	1.0	1.5	2	3.7	10	0.4	1.0	1.2	1.5	2.6	6.0	-

vapour pressure of sulphur, and its role in the machining by this vanishes. The metal set free by dissociation of the sulphides can then be dissolved, contributing to alloying of the matrix. More detailed knowledge concerning the formation and properties of these films is necessary; it could contribute to the better characterisation of the chemical and physical properties of a machining aid.

Most of the sulphides are thermodynamically unstable during sintering and react with the iron matrix, forming liquid phase as predictable from the phase diagrams given in, for example, Ref. 250. The sulphide added decomposes, and the metallic component is dissolved in the base matrix while sulphur forms a Fe–S eutectic which during cooling solidifies, forming the sulphide $Fe_{1-x}S$. This behaviour has been observed in particular with MoS_2 and also with other sulphides such as Cu_2S, Sb_2S_3 and ZnS [251,252]. MnS, however, is stable up to high sintering temperatures.

In all sulphides, during sintering the sulphur content is reduced to different degrees in dependence on sintering atmosphere, although in general the manganese sulphide is regarded as thermostable. Some authors also claim that MoS_2 is thermostable but this clearly disagrees with the ternary system [250] and with experimental results [207]. Furthermore, the data listed in Tab.6.4 also show that, with the exception of MnS, the mentioned machining agents (S, MoS_2) are not completely stable during the sintering process but there is always some loss of sulphur.

The data show that the manganese sulphide is stable in an endothermic sintering atmosphere. A significant portion of elemental sulphur was lost during sintering. Given the sensitivity of physical properties to the sulphur content, it appears that considerable care should be exercised in furnace loading, atmosphere and scheduling to avoid the possibility of contaminating a furnace or subsequent

Tab.6.4 Sulphur recovery for sintered Fe–0.8C (F–0008) steel based on sponge iron powder; density 6.1 g/cm^3, 1120°C, 30 min, endothermic atmosphere [201]

Addition [mass %]	S		MnS		MoS_2	
	Added	Measured	Added	Measured	Added	Measured
	Sulphur recovery [mass %]					
0	0	0.01	0	0.01	0	0.01
0.25	0.25	0.17	0.09	0.11	0.10	0.04
0.50	0.50	0.41	0.18	0.18	0.20	0.12
0.75	0.75	0.60	0.28	0.26	0.30	0.18

premix composition by sulphur or its compounds [242].

According to [204], the addition of sulphur to iron powder in an amount of up to 0.4% caused an increase of the tensile strength of Fe–S compacts (porosity 15%) from 162 MPa (no S) to 203 MPa.

For MnS containing parts, a drop in sulphur content has been observed if the carbon content was above 0.6% C, *i.e.* in connection with graphite [253]. These results suggest that when the amount of added graphite exceeded 0.6%, the thermodynamic equlibrium between the sintering atmosphere and the powder mix constituents is affected, and a small proportion of the MnS is decomposed to Mn and S. Manganese will be dissolved in the iron matrix, or oxidised in vapour form with oxygen from the sintering atmosphere to MnO or to other manganese oxides when also considering the sublimation of manganese due to its high vapour pressure during sintering [5]. The addition of 0.5% MnS to iron–graphite mixes affects the carbon pick-up of the ferrous powder. It was observed that, for the same amount of added graphite, the combined carbon contents of base mixed materials containing MnS were lower than those of MnS–free mixes [253]. The same effect is associated also with the MoS_2 aid. A more detailed analysis of the thermodynamic of these systems should be performed bearing in mind also the carbon potential of the sintering atmosphere.

6.1.3 Manganese sulphide characteristics

Manganese sulphide (MnS) is the most well known and widely used compound for the machinability enhancement in PM materials. There are three methods for bringing manganese sulphide into the sintered part:

- the mentioned addition of sulphur to the iron powder containing manganese,
- production of prealloyed atomised powders containing manganese

with higher sulphur content,

- addition of manganese sulphide in form of fine powder to the iron powder admixtures of all compositions.

The third one is, from the technical viewpoint, the simplest and most widely used method. As usual, each of these methods has advantages and disadvantages. As mentioned before, the thermal stability of all machining aids and between them also of manganese sulphide plays an important role in the cutting process.

The manganese sulphide as a chemical compound was and is the topic of permanent investigation. Manganese sulphide added to the iron or alloy powder in form of fine powder has been found to be a very effective machinability enhancing compound. It fulfils the main requirements asked for a machining aid as mentioned before. Basically it has no detrimental effect on the muffle, belts, and internal walls of the sintering furnace, nor does it have detrimental effects on the atmosphere in the furnace. The addition of manganese sulphide results in improved machinability with very minor or no loss at all in mechanical and other properties of parts in general, with no or minimum burn off of manganese sulphide regarding the sulphur content reduction (as shown, for example, in Tab.6.4). It has a minimal effect on dimensional changes in the parts. It is not necessary to change the tooling. The MnS particles are inert chemically in the microstructure and have a cubic NaCl-type crystal structure [135,147,241,244,248]. With MnS in contact with iron, there is in principle a mixed sulphide (Fe,Mn)S in equilibrium with the iron matrix that becomes the more stable the lower its iron content is. The very low Mn content in the matrix keeps the MnS stable during sintering even at high sintering temperatures, sintering up to 1300°C in H_2 having been performed without any noticeable effect on the admixed MnS [251].

The thermodynamic stability of MnS depends on the Fe content in mixed sulphide (Fe, Mn)S. Its becomes more stable the lower the iron content is, as clearly visible from the solidus and liquidus temperatures, Fig.6.7. The ratio Fe:Mn is a function of the Mn content in the matrix, but due to the different affinities of Fe and Mn towards S, a fairly low Mn content in the matrix already results in Mn dominating in the sulphide, Fig.6.8; at 0.15% Mn in the iron matrix – as common for PM iron powder grades – the Fe:Mn ratio in the sulphide is about 1:99, and the solidus temperature of this sulphide is >1500°C compared to <1200°C for $(Fe_{0.5}Mn_{0.5})S$. Thus, it can be concluded that the very low Mn content in the matrix keeps the MnS stable during sintering even at high temperature. It

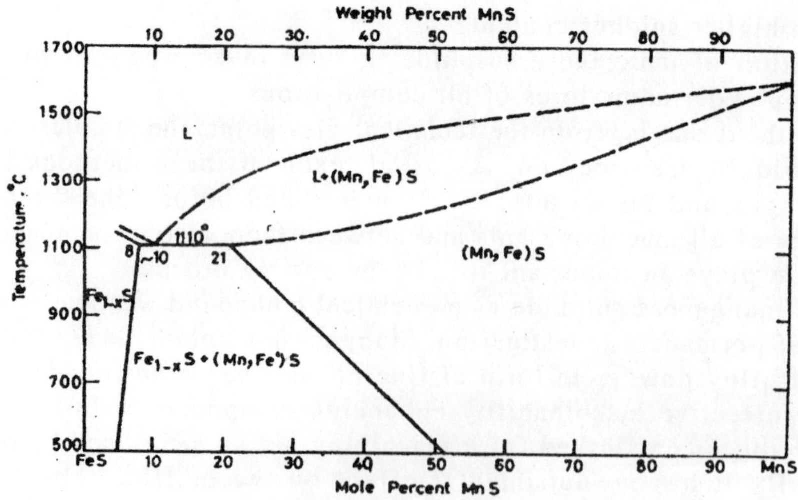

Fig.6.7 FeS–MnS pseudobinary section [250].

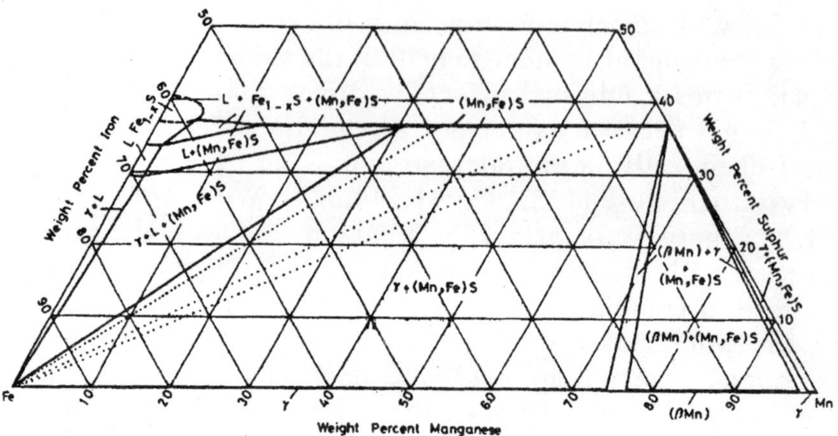

Fig.6.8 Ternary system Fe–Mn–S; isothermal section at 1100°C [250].

can be stated that MnS possesses the chemical–physical properties needed for an effective machining aid.

In spite of the presented relatively high enhancing effect of MnS on machinability of PM parts, the requirements on machining of high strength materials with a hardness above 200 HV have created the need for an additive which enhances the machinability even more than Mn [107,130,135,154,248]. It is a very severe task for powder metallurgy. It is should be expected that some effective machining aid added to powder mixtures for the materials with high strength and hardness will be developed. These improved machining aids will be based on a broader concentrated investigation of related materials in combination with appropriate cutting conditions.

The beneficial effect obtained by the addition of manganese sulphide can be explained by the fact that MnS inclusions act as stress raisers in the machining shear zone to initiate cracks that subsequently leads to fracture of the chip. MnS inclusions are also known to deposit a layer of MnS on the wear surface of the cutting tool. These layers may act to reduce tool chip friction which results in lower cutting temperatures, cutting forces and reduced tool wear, or act as diffusion barriers to reduce tool wear. All three mechanisms contribute to the improved machinability [148].

At present, there are two main grades of manganese sulphide available in practice, *i.e.* low purity and high purity manganese sulphide. Their basic chemical and physical properties, showing the main differences, are given in Tab.6.5. The low purity manganese sulphide is in the literature usually termed often as MnS (Elkem – ERATEM, Höganäs) or only standard MnS (low purity) and high purity MnS as, *e.g.* MnS+ (Pyron). The basic difference between these two grades of manganese sulphides is the iron and oxygen content, the 'high purity' grade, despite its designation, containing significantly more Fe than the standard MnS grade.

6.1.3.1 Low purity manganese sulphide

Low purity manganese sulphide (standard) contains some impurities (~3.5%) including calcium, magnesium, iron, silicon, carbon, nitrogen, and aluminium [148,225,255]. The impurities are present as complex oxide inclusions, which are hard and detrimental to machinability. It means that the portion of MnS in the product is ~80% and ~11% are manganese oxides. The melting point of this MnS is 1610°C. The typical average particle size should be around 5–6 μm and the specific surface area ~1.50 m^2/g. In reality the

Tab.6.5 Typical chemical and physical properties of MnS powders (MnS – low purity, standard; MnS – high purity) [254]

Manganese sulphide	Chemical composition [mass %]					Particle size [μm]		Colour
	Mn	S	O	Fe	MnS/ (Fe,Mn)S	Median:D_{50}	Median:D_{95}	
MnS (standard)	63	35	1-2 max	-	~63	5	12	Dark green
MnS (high-purity)	59-64	34	1 max	4-6	99	6	20	Bright green

larger particle size decreases also the quantity of MnS particles available for improving machinability [244].

MnS is very stable in dry air, even at elevated temperatures. The main disadvantage of MnS standard type is that it does deteriorate in the presence of moisture at all temperatures – it is hygroscopic. It should be noted that MnS powder does not directly react with the oxygen in air. It tends to absorb moisture in the air and reacts with the absorbed moisture, resulting in higher oxygen content up to 12%. Under such conditions MnS agglomerates and therefore cannot be used i . machining enhancement. This effect is enhanced by the small particle size in this state and large specific surface area of the additions normally used in PM. The precautions must be taken to ensure that it is not exposed to humid environments for any extended period of time. MnS containing drums should be kept closed to prevent any possible deterioration [147,148,241,254].

When stored in an appropriate fashion, for example in sealed gas-tight containers, both MnS and the mixes of iron powder with them show no pick-up of oxygen content. When either the MnS powder or the iron powder mix are left in direct contact with a highly humid atmosphere, there can be significant increase in the oxygen content of the powder, and MnS powder can agglomerate to massive and rather hard lumps. It is recommended that MnS powder should be stored as shipped–in a plastic bag, with a desiccant inside, and the bag tied. This experience follows from the observation that the MnS powder stored in open beakers in a glove box under high humidity conditions (80–90%) showed a 9% increase in oxygen content in just 12 days. Usually at ambient temperature and humidity of 20–60% the oxygen content of MnS powder is about 1%. Simply drying MnS powder that has been previously allowed to absorb moisture does not restore the powder to its original state [110,241,256], especially as agglomerates of fine particles remain. The formation of MnS agglomerates which are hard to destroy in air of common humidity (~70%) is shown in Fig.6.9.

The manganese sulphide grades produced initially for its use for machinability enhancement in single charges contained ~63% Mn, 29–37% S and ~3% oxygen. It follows from it ~80% MnS and ~11% manganese oxides with the specific surface area 1.54 m^2/g [225].

6.1.3.2 High purity manganese sulphide

High purity MnS is similar to low purity grade when added to typical PM premixes, it has very little effect on the green and sintered properties (tensile strength properties, impact toughness,

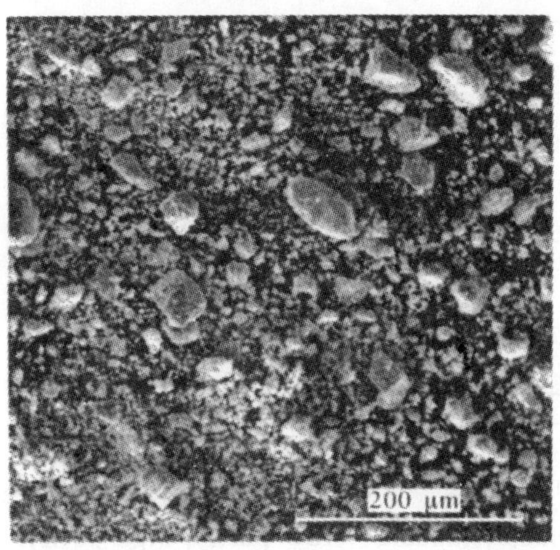

Fig.6.9 Manganese sulphide agglomerates formed by long-term storage in air. SEM.

fatigue strength, and fracture thoughness) of different materials prepared under varying processing conditions. It is also stable under both conventional and high temperature sintering conditions and does not react with the material matrix, and it is much more resistant to degradation in moisture which is its main advantage compared to standard MnS.

High purity MnS contains 4–6% iron in the manganese sulphide matrix and is therefore less hygroscopic, more resistant to oxidation, and is more stable than the previously mentioned manganese sulphide and contains no secondary phases. Oxidation resistance of the MnS+ results in less particle agglomeration and lumping, as the consequence of its resistance against absorption of moisture. It is produced by a process that involves the thermochemical reaction of manganese and sulphur in the presence of iron with subsequent controlled solidification [241,244,254].

6.1.3.3 Surface coated manganese sulphide

Since the main problem of MnS is its tendency to form agglomerates in humid environment, deactivation of the surfaces by thin organic coatings is also a feasible way to enhance its stability. Such measures are well known in the case of, for example, electrolytic copper powders; for MnS used as a machining aid, a suitable coating technique has been developed recently [257]. This coating affords a persistent protection against the environment. In Fig.6.10 the mass gain of various MnS grades is plotted as a function of the storage time in air of defined humidity. It is clearly visible that the standard grade starts to pick up oxygen immediately

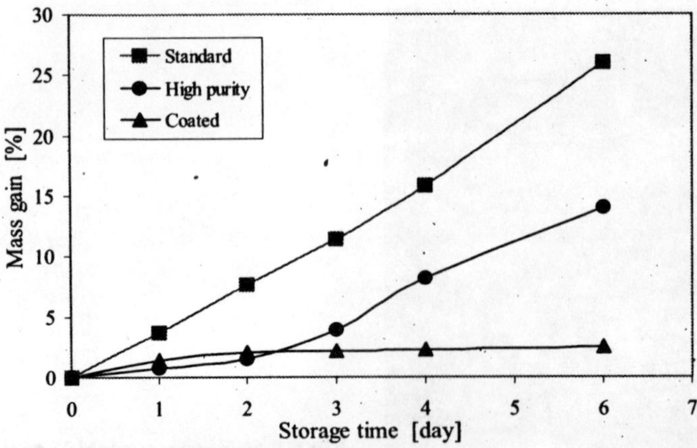

Fig. 6.10 Mass gain of various MnS grades (standard – low purity, high purity, surface coated), as a function of storage time in air of 50% humidity [257].

and at a considerable rate. The 'high purity', Fe-containing grade at first behaves rather inert but after some incubation time reacts at virtually the same rate as the standard grade. The surface coated MnS, in contrast, remains largely unaffected even after extended periods, indicating that the protective organic layer in fact passivates the fine MnS powder against the effects of humid environment.

The effect of this surface protection stands out still more clearly if the microstructures of sintered steels are compared that have been prepared using different MnS grades. Figure 6.11 depicts a sintered FeCuC steel that has been prepared with addition of three kinds of MnS (see Fig.6.10). The very large MnS agglomerates stand out clearly when the standard grade is used, Fig.6.11a. For the steel shown in Fig.6.11b commercial high purity, Fe-containing MnS was used. The effect of the lower reactivity is evident, significantly smaller MnS inclusions being observed; also here, however, the fine original particles have been somewhat coarsened. If the coated MnS powder has been used, Fig.6.11c, the particles remain predominantly fine, hardly any MnS inclusions >50 μm being observed.

6.1.4 Sulphides of other elements

6.1.4.1 Molybdenum disulphide

Molybdenum disulphide, found naturally as the mineral molybdenite, is a typical solid lubricant with a layered crystallographic structure similar to that of graphite. MoS_2 is also not wetted by water and is very stable against acids. The use of MoS_2 as lubricant causes

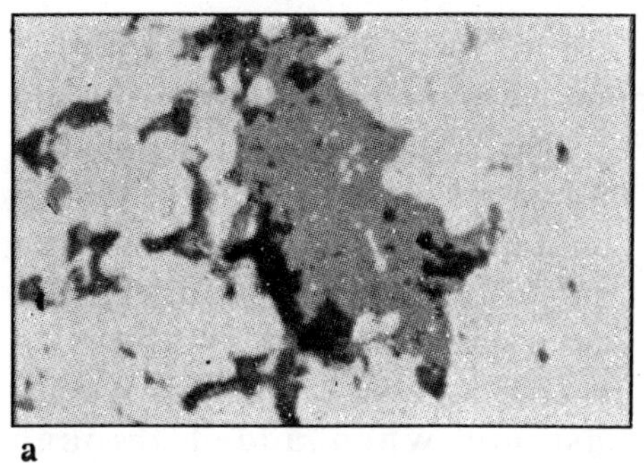

a

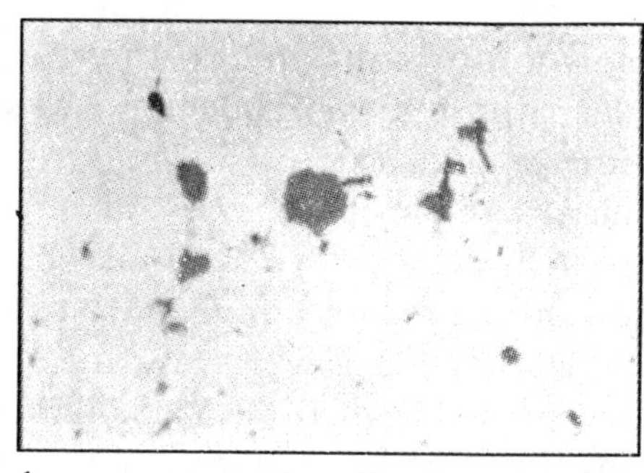

b

c

Fig.6.11 Metallographic sections of FeCuC–0.8% MnS materials with different MnS grades. Atomised iron powder, compacted 600 MPa, sintered 1 h at 1120°C in H_2. a – standard MnS, b – high purity MnS, c – surface coated MnS. Optical micrographs, unetched, ×400.

the decrease of coefficient of friction, of wear of the friction pair, of the temperature in the contact zone and results in increased seizure resistance. The purity of MoS_2 is important. When the purity is 99.4%, the coefficient of friction is 0.05, and at the purity of 98.4% it is as high as 0.2. The purity of MoS_2 as lubricant used for machinability enhancement should be minimum 99.5% and the particle size not below 1 μm because at lower sizes the particles lose the lubrication properties. Minimum 5% MoS_2 in a material eliminates the danger of cold welding (seizure) of a friction pair.

As stated before, MoS_2 is not stable thermodynamically in a ferrous matrix. It was reported that by the processing of iron–MoS_2 mixture at 1000 to 1120°C for 2 h in dissociated ammonia a protecting layer of metallic molybdenum was formed on the surface of each iron particle due to dissociation of MoS_2 [258]. This underlines that MoS_2 was not thermostable.

In the system Fe–10% MoS_2 (a content not used for machinability enhancement), molybdenum disulphide was preserved in the temperature range of 1000 to 1100°C in hydrogen for 60–90 s with its lubrication properties retained. Soaking at 1150°C for 60 s caused partial interaction of MoS_2 with the iron matrix. Prolonged sintering for 90–120 s caused full decomposition of the solid lubricant under the formation of Mo_2S_3 which can also decompose to molybdenum and sulphur. Sulphur reacts with iron and forms FeS (strictly speaking, $Fe_{1-x}S$). The proportion of the mentioned sulphides, evaluated as inclusions in the material taking part in machining, depends on the amount of MoS_2 in the mix and on the sintering conditions (time, temperature, atmosphere) [259].

Due to its lower stability at higher (sintering) temperatures in the presence of iron, MoS_2 is not a particularly suitable additive for machinability enhancement, at least not when added through admixing it to the starting powder (with MoS_2 in a cutting fluid the effect is surely different). At standard sintering conditions, it rapidly decomposes, Mo being dissolved in the matrix and the sulphur forming after cooling a sulphide network. Nevertheless, this sulphide network also positively affects the machinability and MoS_2 is investigated as a machining aid. In Ref.207 it was shown that for turning of sintered iron at low cutting speeds, addition of 2% MoS_2 resulted in a decrease of the surface roughness R_a from 7.7 to 1.5 μm, although at the time of machining there was definitely no more MoS_2 present but simply iron sulphide. Its thermal stability must therefore be known for the application under high temperature, *i.e.* at sintering temperature, Tab.6.3.

This confirms that to exert a positive effect on machinability, machining aids not necessarily must be solid lubricants in the sense that they have a layered crystallographic structure; also other sulphides seem to be effective, as clearly shown by MnS with its cubic NaCl structure that not at all resembles the layered structure of the classical solid lubricants.

In addition to machinability improvement, molybdenum disulphide as one of the best solid lubricants is used predominantly for the bearings working without a direct fluid lubricant or under the conditions of limited lubrication, *e.g.* in fact as a lubricant similar to oil or grease. The morphology of MoS_2 particles is shown in Fig.6.12.

The lubrication effect of MoS_2 is retained also after sintering if not totally dissolved, as shown on samples tested in dry friction, Fig. 6.13. The test of a dry friction pair can partly simulate the friction in the cutting zone apart of the temperature. The powder forged samples from Astaloy A powder with 0.6% C_c for the tests were prepared with addition of 0.5, 1 and 2% MoS_2 and for comparison from a HIPed high speed steel (M2). As shown, in structural steel the wear decreased significantly at 1% MoS_2 addition compared to 0.5% which can be regarded as insufficient (smaller width of

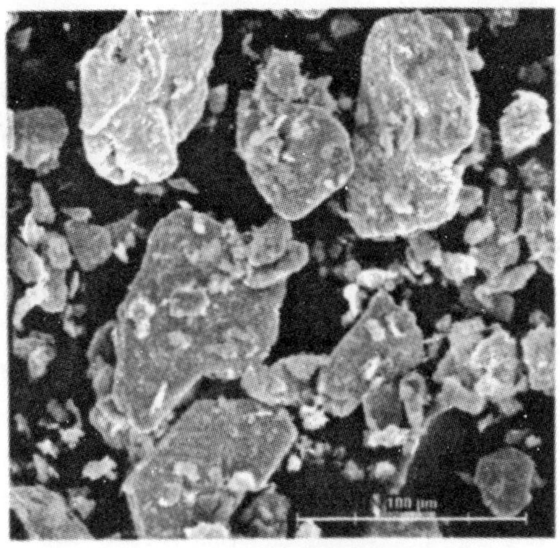

Fig.6.12 Morphology of MoS_2 powder. SEM.

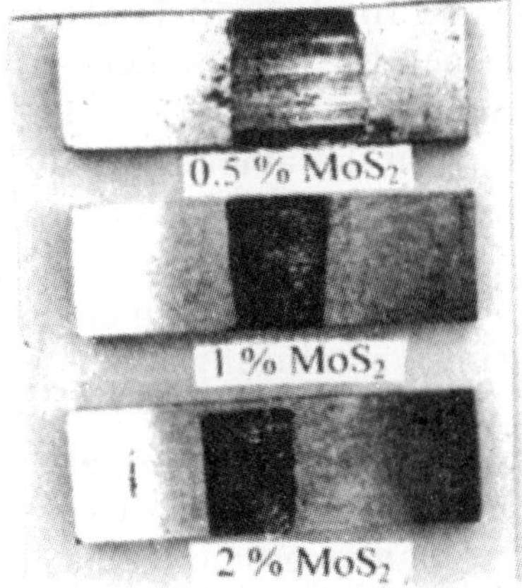

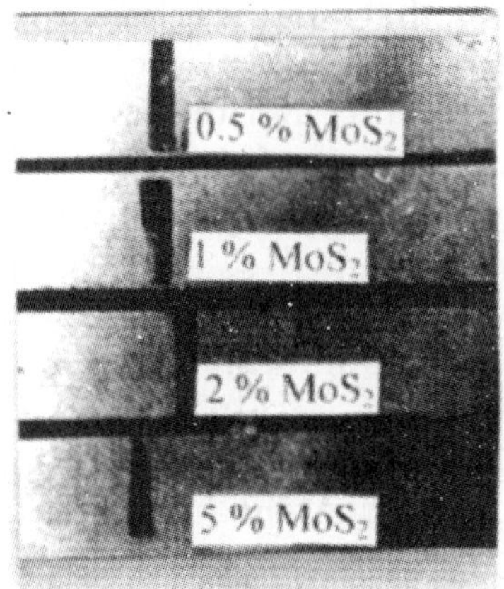

Fig.6.13 View on the traces after dry wear test of: a – powder forged Fe–2Ni–0.5Mo–0.6C steel (Astaloy A powder); 1120°C, 30 min, density 7.82 g/cm^3, 180 HB; b – HIPed HSS (M2) steel (64 HRC) [260]. Dry friction test (disc–block method): SAE 52100 steel (59 HRC) (disc)–impact test bar) (block), load 100 N, sliding distance 1000 m.

friction traces). None of MoS_2 addition used in HSS steel samples affected the dry wear (equal width of friction traces), *i.e.* the lubrication effect of the molybdenum disulphide was also observed in the full density material but with a difference in dependence on hardness of the workpiece affected by previous processing routes and possible decomposition of MoS_2 during sintering.

6.1.4.2 Other sulphides

In addition to MnS and MoS_2, other sulphides are also used in practice, *e.g.* in friction materials, such as resin-bonded brake pads. For the use as machining aids, several of them have been investigated [207,254], including Cu_2S, Sb_2S_3, and ZnS as well as PbS which will be discussed below, together with Pb.

In the case of other sulphides, their thermodynamic stability during the sintering process was generally as unsatisfactory as that of MoS_2. In all these cases, decomposition occurred with formation of sulphide eutectic melt and, after cooling, the typical network of $Fe_{1-x}S$. The metallic constituent was dissolved in the matrix since the solubility in iron was rather high, at least at sintering temperature.

This effect is particularly visible in the case of Cu_2S since in Fe–2% Cu_2S compacts the Cu distribution after sintering was virtually identical to that of sintered Fe–Cu, as shown in Fig.6.14. This indicates that for these machining aids the final result is always a microstructure of iron (containing the metallic component of the machining aid) and the eutectic grain boundary network of $Fe_{1-x}S$, the area fraction of the network depending on the amount of sulphur added through the machining aid, minus the amount lost through evaporation and H_2S formation.

Table 6.6 gives the dimensional and mechanical properties of plain iron containing 2 mass% of various machining aids; when comparing the results it must be kept in mind that these 2 mass% result in different effective sulphur contents for each material since the sulphur content in the different sulphides naturally varies. It stands out clearly that the decomposing sulphides promote expansion

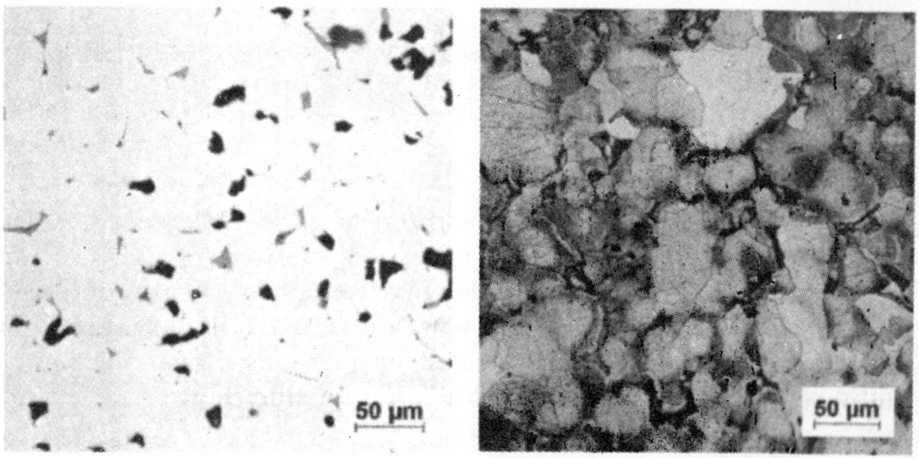

Fig.6.14 Sections of Fe–2% Cu_2S. Atomised iron powder, compacted at 600 MPa, sintered 1 h at 1120°C in H_2; left – unetched, right – Nital etched [207,251].

Tab.6.6 Dimensional and mechanical properties and density of Fe–2 mass% machining aid [189,210]. Atomised iron powder, compacted at 600 MPa, sintered 1 h at 1120°C in H_2, ρ_g – green density, ρ_s – sintered density

Machining aid	S nom. [mass %]	ρ_g [g/cm^3]	ρ_s [g/cm^3]	Δl/l [%]	KC [J]
MnS	0.74	7.07	7.07	-0.17	11.6
MoS_2	0.80	7.03	7.06	+0.11	18.0
Cu_2S	0.40	7.09	7.02	+0.20	>31
Sb_2S_3	0.57	7.01	6.77	+1.25	9.2
ZnS	0.66	6.97	6.89	+0.08	25.3
PbS	0.27	7.10	7.12	-0.04	>31
- (Reference)	-	7.12	7.25	-0.24	>31

during sintering; the structural integrity, indicated by the impact energy, however, does not suffer too much except in the case of Sb_2S_3; here probably the low density also plays a major role. Compared to the decomposing sulphides, MnS affects the impact energy rather adversely, which shows that solidified sulphide eutectic apparently results in better bonding to the ferrous matrix than do admixed MnS particles.

Calcium sulphide. Calcium sulphide is one of the most stable sulphides which is used also in ingot metallurgy as an aid to free machining. Perhaps due to its high sensitivity to humidity it was not applied in PM machining. In a PM steel for testing its effect on machinability calcium sulphide was prepared by reaction sintering. Calcium carbonate (an abundant), and molybdenum disulphide (a sulphur donor) were added to the powder mixtures (Fe–Cu-graphite) pressed to valve guide compacts and sintered at 1000°C in an H_2–N_2 atmosphere. Calcium sulphide was formed by the following chemical reactions [261]:

$$CaCO_3 \rightarrow CaO + CO_2 \uparrow$$
$$MoS_2 \rightarrow Mo_2S_3 + [S]$$
$$CaO + [S] \rightarrow CaS + [O]$$
$$CaO + 3/2\ S \rightarrow CaS + 1/2\ SO_2^*$$
$$CaO + MoS_2 + C \rightarrow CaS + Mo_2C + CO \uparrow.$$

(*determined by the authors of this book)

It was determined that a mixture of molybdenum and calcium sulphides formed in the fine pearlitic microstructure.

6.1.5 Characteristics of bismuth, selenium, tellurium and lead as machining aids

Bismuth, selenium and tellurium as machining aids are often used in ingot metallurgy due to their advantageous physical–chemical properties. It is reasonable to use these elements also in powder metallurgy parts. Their addition improve machinability but they are not available in standard grades of steel. These additions are expensive. Selenium and tellurium are often used as a replacement for sulphur and/or in combination with lead. Se and Te are in the same group of the periodic system of elements as is sulphur and are therefore similar to it, as are also the respective compounds such as, for example, MnS and MnSe. Typical percentages of either element would be ~0.02 to 0.25%. Both elements exert beneficial effects by promoting the retention of globular-shape sulphide-type inclusions (selenides, tellurides). Both selenium and tellurium, present in small quantities in steel, strongly segregate to grain boundaries and severely embrittle them [157] unless bonded as stable selenides and tellurides; their behaviour in steels is also rather similar to that of sulphur.

Normally, sufficient manganese in the steel is required to produce the soft compounds MnSe, and/or MnTe whose presence is necessary for chip nucleation. It is suggested that a combination of both effects – soft precipitate formation and elemental segregation – is responsible for the enhanced machinability observed in wrought as well as in PM steels (although the latter effect has to be regarded with some scepticism regarding its impact on the mechanical properties) [157].

6.1.5.1 Effect of bismuth

In wrought steels, bismuth in an amount of 0.02–0.2%, present in the elemental form, increases the machinability of steels without impairing their mechanical and toughness properties. These steels can be gas-carburized and nitrocarburized. The low melting point of bismuth (271°C) (the temperature which can occur in cutting area) and the melt therefore formed locally decrease the cutting forces and contribute to the breakage of the chips. By this the processes in the cutting zone are affected. Bismuth 'spreads' on the chip/tool contact as a consequence of the difference in the coefficient of thermal expansion between bismuth and the iron base matrix. Adhesive wear and also the abrasive effect of the material on the tool decrease. The pronounced improvement caused by (elemental) bismuth is attributed to the spheroidization of sulphide inclusions and to liquid-metal embrittlement of grain boundaries by liquid bismuth, during cutting.

Bismuth also avoids the difficulties associated with, for example, lead as a machining aid because is not toxic, but its higher price than lead must be adequately accounted for. Bismuth has a little deteriorating effect on density and mechanical properties of PM materials. The addition of bismuth is accepted in general as an effective machinability enhancer for materials based on iron powders.

For improving the machinability of PM steels bismuth is used in larger amounts than in wrought steels. The results regarding the addition of 1–2% Bi on turning of powder forged Fe–C steels based on sponge and atomised iron powders are given in [262]. Bismuth melts at the sintering temperatures and segregates along some silicate inclusions or MnS particles (if present), or they exist as particular inclusions [262]. The use of Bi made it possible to decrease the sintering temperature to 1000°C which also lowered the loss of Bi, and the forging temperature of the parts from 1100–1150°C to 1000–1050°C. In turning the decrease in flank wear by 2–6 times (decrease in abrasive and adhesion wear), and an increase in cutting speed by 20–75% compared with the machining of materials without Bi addition were achieved (80–100 m/min). Higher additions of Bi were not more effective. Similarly, the drilling thrust force decreased by 25–50% and the tool wear by 50–80% in working with 90% dense carbon steel [142,244]. The addition of bismuth improved the surface finish. The bismuth addition is more effective in the machining of difficult-to-machine materials, especially those with ~0.8% carbon.

6.1.5.2 Effect of tellurium

As stated in a number of studies [233], tellurium additions in amounts of 0.05–0.25% enhance the machinability of PM steels without any marked effect on mechanical properties. The use of elemental mixtures is impractical due to the excessive tellurium loss (up to 50%) during sintering, as a consequence of the high vapour pressure (on the level of that of zinc) (boiling point 994°C).

According to the phase diagram, Te is soluble in Fe up to several percent, and thus can be expected to be present in solid solution (unless the Mn content is also high). It is not clear how dissolved Te acts and whether there is a pronounced tendency to grain boundary segregation. As stated in [263], tellurium like sulphur, lead and selenium present in small quantities in steel, strongly segregates to grain boundaries and severely embrittles them. Normally, sufficient manganese is required to produce the soft compounds – MnS, MnSe, MnTe, etc. whose presence is necessary for chip nucleation. In copper-containing steels, Cu_2Te plays the same role.

Tellurium makes visible the MnS in a ball form in a material in optical metallography. X-ray analysis confirmed the existence of tellurides and sulphides in the steel and MnS contained up to 2.5% Te (solid solution in MnS) at 0.26% of Te addition to the powder mixture, and of sulphur in MnTe about 3% [58]. An increase in the impact strength of tellurium-containing steels was also recorded. If the Te content was higher than its solubility in MnS, tellurium segregated as of tellurides MnTe. Therefore, as stated above, sufficient amount of manganese in steel is needed, as is the case with sulphur. Together with manganese, tellurium forms complex Mn(S,Te) compounds which favourably affect cutting at high speed [263]. Tellurium should therefore be added to PM steels as the MnTe compound. In this case, tellurium is retained in the form of precipitates with a significant improvement in machinability.

A copper addition of 0.5% or more was found necessary to prevent the loss of tellurium from the PM steels during sintering in an endothermic atmosphere without inducing cracking also during standard forging tests [240,243]. From this point of view, the Fe–Cu–Te compound as a machining aid was regarded for a new PM alloy in form of Fe–6Cu–1Te, Fe–0.5Cu–0.1Te and in the form of an atomised Cu–5Te master alloy [222]. In copper-containing steels, Cu_2Te probably plays the same role in machining, although there is no detailed information on the stability of this telluride in a steel matrix during sintering. An addition of tellurium to the steel increased the machinability by 10–15 times with the decrease in mechanical properties (U.T.S., T.R.S.) less than 10%. The quality of the machined surfaces was increased. Small additions of tellurium in form of Cu_2Te to sponge iron powder as well to atomised steel powder grades gave a substantial improvement in machinability. In presence of copper an addition of 0.1% Te was sufficient to obtain optimum results, to reduce the drilling time by 60% [240].

Important improvements in machinability were recorded after adding tellurium in steels with the carbon content in the range 0.1–0.6%. However, a significant drop in sintered strength was observed, and very small changes in the tellurium content caused drastic fluctuations in other properties [243,253]. Furthermore, the forged specimens showed appreciable improvement in machinability when compared with samples without tellurium. Tellurium is also used to alloy effectively the steel for case hardening with subsequent tempering.

6.1.5.3 Effect of selenium

Selenium is virtually insoluble in ferrite but in austenite the solubility is higher, about 1% maximum. In the case of selenium a sufficient manganese amount is also required to produce the soft compound MnSe, whose presence is necessary for chip nucleation. With manganese and sulphur, selenium forms Mn(Se,S) which favourably affect machinability. In addition, it is possible to obtain better surface quality and more favourable shapes of the chips. With addition of selenium of up 1.5%, thrust and torque decreased in drilling down to 30% or even more [129,262]. In austenitic steels, selenium is often used as a substitute for sulphur [149].

When comparing sulphur, selenium and tellurium as machining aids, tellurium was found to be the most effective additive while sulphur was the least effective since it resulted in some reduction in mechanical properties [118]. The toxicity of Te and Se is eliminated when added in the form of a stable compound [142].

6.1.5.4 Effect of lead

Lead is added to the molten carbon steel during teeming of ingots or, sometimes, to the ladle. Because lead is insoluble, or nearly so, in molten steel, a fine dispersion of lead particles develops as the steel solidifies. The lead is usually found near or around the sulphide inclusions. Most of the resulphurised grades can be produced with an addition of 0.15 to 0.35% Pb.

Handling of leaded powder blends will require stringent safety precautions due to lead's toxicity. Additional safety probl ms may result from lead loss during sintering through evaporation (high vapour pressure; boiling point 1755°C). Bismuth as a compound avoids these difficulties (although there are also evaporation losses (boiling point 1559°C).

Lead as an additive is also used for enhancing the machinability of sintered ferrous materials as well as of brass, being insoluble in iron and brass. With PbS, the sulphides are more rounded. It may be supposed that PbS is decomposed during sintering, at the standard sintering temperatures the equilibrium phases being austenite, liquid Fe–S eutectic, and liquid lead [258], but PbS may to some extent be regenerated during cooling (especially slow cooling) since Pb is insoluble in iron [251]. In any case, also with PbS the typical sulphide eutectic is observed, at least at higher

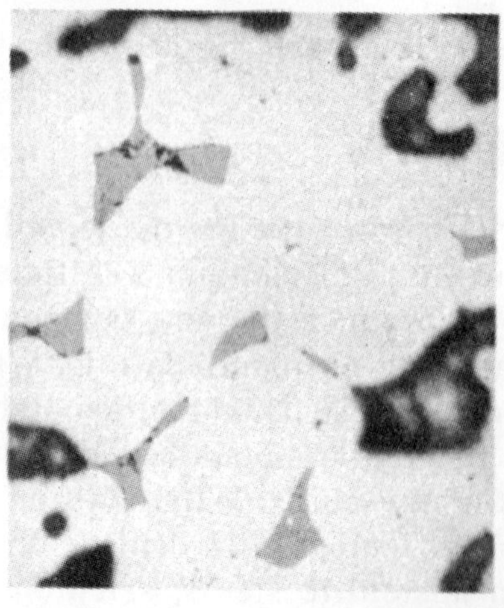

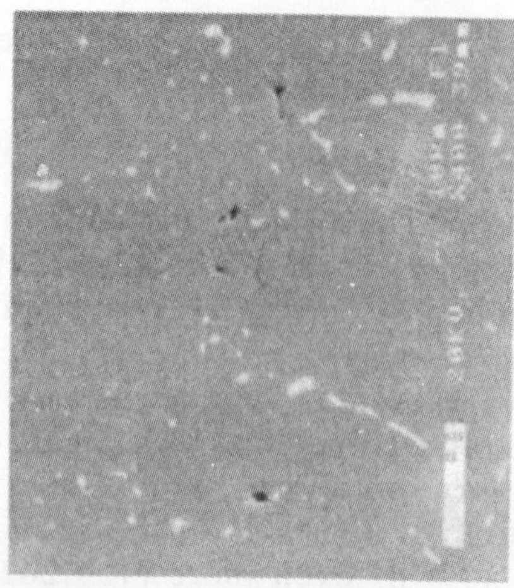

Fig.6.15 Section of sintered iron with admixed Pb as a machining aid. Atomised iron powder, compacted at 600 MPa, sintered for 1 h at 1120°C in H_2 [207]; left – Fe–0.8C–10PbS, optical micrograph, ×750, nital etched; right – Fe–2Pb, ×350, SEM.

contents while Pb is contained rather as rounded, droplet-shaped inclusions, Fig.6.15.

The machinability of PM carbon steels is improved by adding up to 2% Pb; additions of 0.5 to 2% Pb reduced drilling thrust by 25 to 50%. However, even such small amounts of lead can lower mechanical properties due to the presence of inclusions at the initial particle surface [129,142,243]. Lead improves the machinability of both straight iron and combined carbon-containing materials. Suitable alternatives are copper or sulphur.

In [221], turning tests at low cutting speeds were also performed with Fe and Fe–C containing various amounts of Pb and PbS, respectively. It was shown that both additives resulted in lower surface roughness compared to plain iron or Fe–0.8% C; however the surface finish was not quite satisfactory, at least not at the rather low speeds used here to more clearly reveal the effect of machining aids. The detailed analysis showed clearly the inferior finish in turning [207].

Finally, when Pb can be used, an increased nose radius on the cutting edge and/or a lower feed rate will considerably improve surface finish. Both Pb and Bi had a significant beneficial effect on drilling forces for densities above 6.25 g/cm³. It was necessary to add an optimum addition of either element to obtain an effective reduction in cutting force of a material with defined composition and properties [240].

6.1.6 Boron nitride, MnX and resin impregnation as machining enhancers

6.1.6.1 Boron nitride

Hexagonal BN is thermodynamically very stable; it decomposes at T>3000°C, although it has been shown that in the presence of metals, BN can decompose in vacuum at much lower temperatures, forming a boride eutectic [264]. BN added as fine powder (<1 μm) tended to form agglomerates but did not decompose at the standard sintering temperature of 1120°C; however, there was decomposition and formation of liquid phase if the sintering temperature was raised above that of the Fe–B eutectic (about 1177°C) and sintering was done in a nitrogen-free atmosphere, *e.g.* in pure hydrogen. Possible decomposition of BN is listed in Tab.6.1. If the atmosphere contained also nitrogen, decomposition of BN was suppressed. In carbon-containing steels, BN apparently inhibits the dissolution of carbon in the iron matrix; the microstructure of a steel with, for example, 0.8% C resembles that of a steel with a markedly lower carbon content, as shown, for example, in Figs.6.16 and 6.17. If the amount of BN was reduced, or if BN was added in the form of coarse particles (50–180 μm), this effect was less significant [149]. Regular carbon dissolution was also observed in the case of BN decomposition (sintering at 1280°C in H_2), at least in the case of low to moderate BN contents while at 2% BN also under these conditions retarded carbon dissolution was discernible. Usually, fine fractions of BN are used as a machining aid, typically <1 μm.

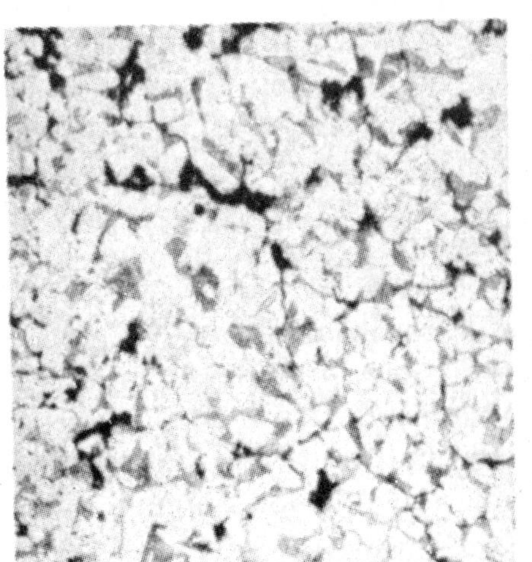
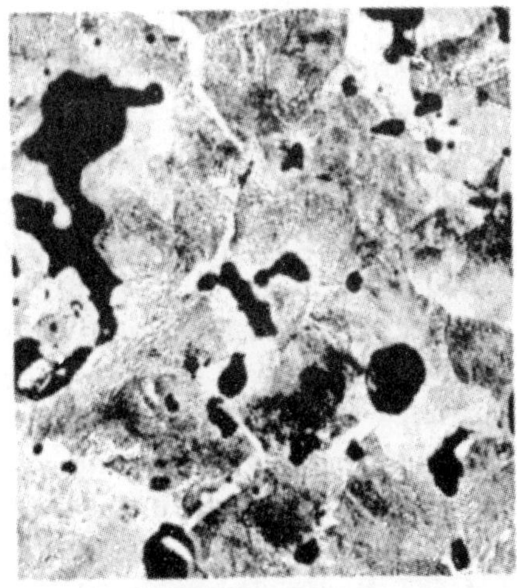

Fig.6.16 Sections of sintered steel Fe–0.8% C steel containing fine 0.5% BN. Atomised iron powder, compacted at 600 MPa. left – sintered at 1120°C, right – sintered at 1250°C [207]. Optical micrographs. Nital etched, ×200.

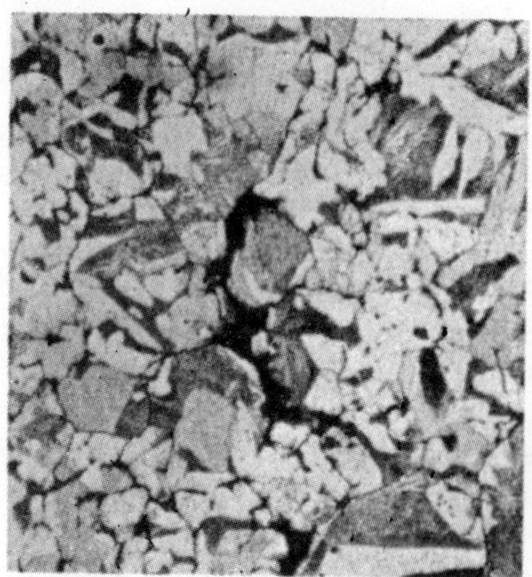
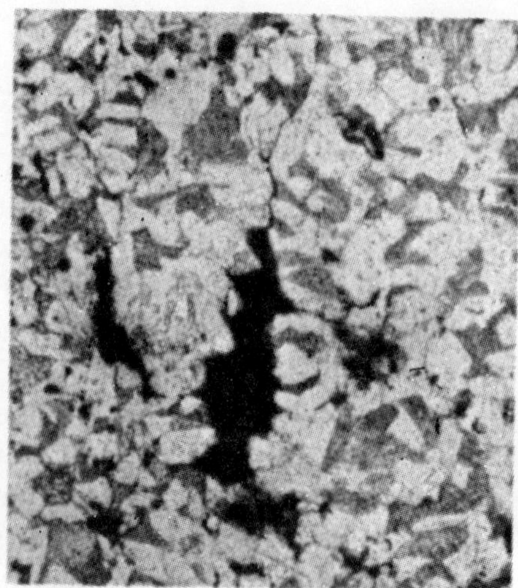

Fig.6.17 Sections of sintered steel Fe–0.8% C containing 2% fine BN. Atomised iron powder, compacted at 600 MPa. left – sintered at 1120°C, right – sintered at 1250°C [207]. Optical micrographs. Nital etched. ×200.

6.1.6.2 MnX

The additive known under the trade name MnX (developed by Höganäs AB) is a machining aid without disclosed chemical composition of the compound. At present, it can be purchased only in powder premixes. MnX is usually added in amounts of 0.2 to 0.5% to powder admixtures for machining enhancement, and its effect is often compared to the corresponding effect of MnS. In [209] it was stated that the chemistry of MnX is actually the compound calcium fluoride, CaF_2. Possible decomposition of CaF_2 is listed in Tab.6.1. On the other hand, in [264] CaF_2 was found to be stable in iron up to a sintering temperature of ~1280°C, which agrees well with the thermodynamic data (the Gibbs free energy of formation is about –280 kcal/mole at room temperature and –230 kcal/mole at 1250°C which is almost the same as for CaO). Furthermore, Ca is virtually insoluble in iron which also acts against decomposition. The presence of calcium was analysed in some particles in sintered Distaloy type and other Fe–0.5% MnX systems as shown in Figs.6.18 and 6.19 which could indicate that MnX is in fact CaF_2. The presence of fluorine was not determined.

6.1.6.3 Resin impregnation and infiltration

Resin impregnation is a process that eliminates or reduces porosity of castings and PM parts by saturating interconnected pores with

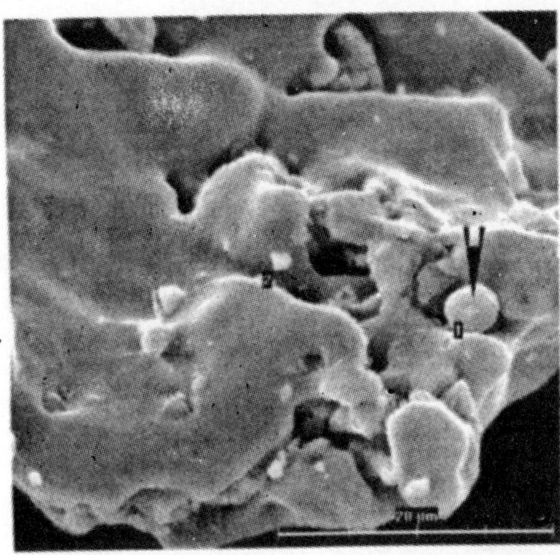

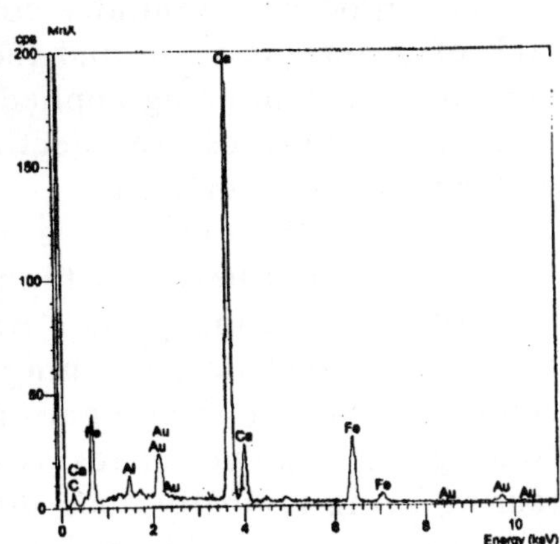

Fig.6.18 A spherical particle containing Ca in sintered Distaloy AE–0.5% MnX material (arrow), SEM.
Fig.6.19 (right) Spectrum of the particle designated in Fig.6.18. LINK microanalyser.

liquid resins. Low-viscosity resins are capable of good penetration reducing, therefore, also the porosity of PM parts (pore size <125 μm) – to produce leak-tight parts. Generally, resin impregnation is done to seal-off the porosity for different reasons, *e.g.* for the use in hydraulic systems; however, an added benefit of the process is to improve machinability of the part which is extremely desirable [105]. However, it is expected [265] that impregnation should also result in:

- improvement of corrosion resistance, the parts are impermable to gases and fluids,
- preparation of porous surface for subsequent surface treatments or coatings, *e.g.* electroplating,
- improvement of machinability of parts with the aim as presumed to eliminate interrupted cutting, which was considered as the main cause for the deterioration of the machinability of sintered iron-base materials.

In most cases, resin impregnation is regarded as an effective method for improving machinability. It should be mentioned there that to confirm this statement it will be necessary to investigate a larger number steels of different alloying and properties. Sintered density as the main characteristic of a sintered material is not a significant factor affecting the machinability of resin impregnated materials in the range 6.7–7.2 g/cm^3.

Impregnation methods. The common methods of resin impregnation are: wet vacuum, wet-vacuum pressure (the wet vacuum/

pressure process essentially consists of submerging the part in a tank of anaerobic resin and then applying vacuum), dry vacuum pressure (vacuum being applied before contact of the porous parts with the resin), pressure injection (forcing the resin into the pores without prior evacuation).

Anaerobic impregnation is generally recommended for volume manufacturing where the highest quality control standards are required for sealing parts. 'Anaerobic' means without oxygen, opposite of aerobics, meaning in the presence of oxygen. Four common types of impregnation materials are: – sodium silicate (waterglass) – polyester resins – low-viscosity heat curable resins (curing being typically done in hot water at, for example 95°C) – low-viscosity anaerobic resins (curing being afforded by catalytic metallic surfaces in the absence of oxygen). Of these, anaerobic resins are the most widely used for PM parts [151].

Performance. Plastic impregnation generally has little or no effect on tensile strength and ductility, since resin sealant in impregnated do not contribute to mechanical behaviour. In general it is stated that the resin impregnation provides a substantial (30 to 60%) reduction in cutting forces in all cases. Lower drilling, in generally cutting forces and longer tool life (decrease in flank wear) and better surface finish was obtained with impregnated parts. Impregnation also reduces chip thickness and adhesion, improves finishes, helps active consistent finish dimensions, and improves dimensional control of parts. Another possible effect of impregnation is a reduction in cutting temperatures due to absorption of heat by the plastic.

From the micrographs it seems that in resin impregnated materials, deformation during tapping was significantly less than in non-impregnated variants. This indicates that the deformation and work hardening during machining was linked to the closing of pores. In porous systems, where the pores cannot be closed, densification by the tool edge seems to have resulted in more pronounced work hardening than in impregnated systems where the pores cannot be closed. In the latter case, there should also be a positive effect on cutting forces, lower forces being necessary to cut the less work hardened material.

The precise reason for this result is not sufficiently cleared. It is supposed that natural lubricity present in thermoset plastic resins used in most PM impregnation processes results in improved machinability. It is possible to accept another explanation of the effect resin impregnation of the parts on their machinability which is probably more accurate. It was proved before that during cutting of a porous material the hardening and densification of a newly formed surface occurs. By this the material is pressed, the pores are closed. A practically full density material is formed behind the contact zone. In the case of a resin impregnated part, impregnation prevents material hardening and densification during the cutting process [141]. Therefore, the cutting process occurs in the material with starting properties which can be partly defined under relatively improved conditions compared with unimpregnated material cutting.

It is conceivable that the polymer becomes liquid at high temperatures near the cutting edge and forms an adhering film between tool and workpiece that exerts a lubricating effect.

PM components which require impregnation of pores or those which need to be coated after machining cannot be machined before in the presence of a cutting fluid [135].

Oil impregnation. Under the reduced atmospheric pressure, the pores of the parts are evacuated from air, and the parts are then dropped into a bath of hot or cold oil. Oil impregnation is often used for sintered components that are subjected to wear, like cams and gears. The oil impregnation can contribute to the machining improvement in some cases, but after it the oil usually must be taken off in term of further processing or loading of the parts. The addition of MoS_2 to oil can more increase improvement in machinability of the parts under previous conditions.

6.2 PRODUCTION PROCESSES

6.2.1 Modification of microstructure

Microstructure modification for machining enhancement is part of the PM production processes. Wrought steels are frequently heat treated by producers to ensure optimum machinability. Soft annealing or normalizing treatments are used to produce a relatively coarse pearlitic or spheroidal microstructure that possesses good machinability. Higher carbon steels require longer heat treatment cycles intended to produce coarse carbides dispersed in ferrite. In contrast, low carbon steels may be partially hardened to produce a microstructure that is less ductile and adhesive than fully annealed steels [137].

The importance of the microstructure and composition of PM materials for controlling machinability was recognised a long time ago. Regarding the fact that microstructures of PM steels are in such way different, also at identical composition, that single data from one alloy are not applicable for another one with narrow composition. For this reason, the problem of microstructure modification up to now in general has been discussed only in some cases and only very shortly [104]. The main aim of an oriented microstructure modification of a PM steel is to eliminate or minimise the fraction of hardly machinable microstructure constituents, such as martensite, and in a heterogeneous microstructure to lower the high concentration peaks of some alloying elements.

In this connection, it is interesting to consider the behaviour of chromium as a relatively new alloying element in PM (Cr-prealloyed powders), and of nickel as one of the oldest one. Chromium in PM steels, like in wrought steels, activates an increase in the stabilisation of supercooled austenite. This results in lowering of the critical quenching rate in the final phase. Nickel adversely affects the stability of (supercooled) austenite at decomposition [82].

The previous results showed that most/or all of research works aimed at microstructure modification were carried out to modify the microstructure by adding various solid machining aids. It is not a real modification because the base microstructure of a sintered or powder forged material with its specifics exposed to machining and causing its deteriorating was not modified by this. The aids are situated mostly separately in the pores and their function is – lubrication of the cutting surface in the cutting zone – chip/tool not affecting base microstructure character. It is a special form of microstructure modification. Only in a minor extent some of them react with the base alloy matrix under sintering conditions and by this way contribute to the usually unwanted microstructure modification and increase (or decrease) to varying extent the mechanical properties of the alloy.

Soft annealing and other forms of heat treatment of PM materials with the aim to improve machinability are done only in a limited range for porous materials but mainly for powder forged steels which are characterised by high hardness after fast cooling from forging temperature. Therefore, they must be soft annealed or spheroidized. A speroidized microstructure of a powder forged steel is shown in Fig.6.20.

Annealing can change the failure mode of a tool towards adhesive wear. The drill appears then to 'stick' to the workpiece when cutting the annealed test pieces, rather than overheating or breaking.

The present limited results show that the base heat treatments employed did not change the microstructure and machinability of

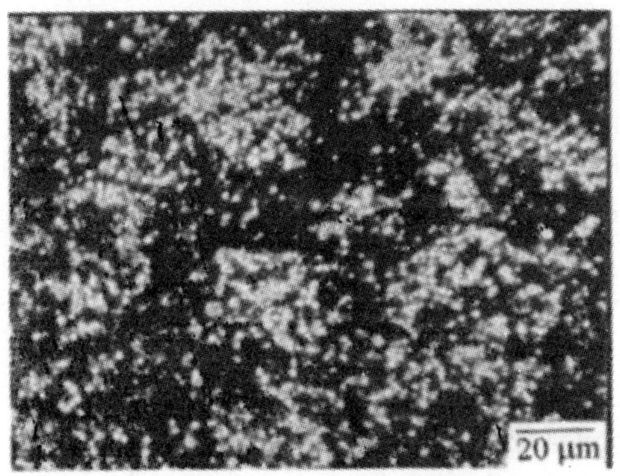

Fig.6.20 Section of powder forged (Fe–1Cr–0.7Mn–0.2Mo–0.8C) steel. Spheroidisation annealing 4 h at 750°C in N_2. Optical micrograph, dark field. Murakami etched.

PM porous steels sufficiently. The results did not indicate that the extra step is justified. Finer lamellae spacing in pearlite can deteriorate machining. Further study is necessary to define the optimum microstructure and heat treatment for machinability for particular alloys except Fe–C materials in which the role of pearlite is clear. It is perhaps the most important task for PM research considering machinability of PM steels with a complex view on machinability and mechanical and other properties. General statements about special treatment conditions on machinabiity of PM steels are not sufficient. An extra treatment process is justified only where parts require extensive machining prior to heat treatment. Furthermore, it must be granted that for components that are not subjected to further heat treatment, the treatment to optimize machinability does not compromise the application related, *e.g.* mechanical properties.

Additional heat treatment process for machinability enhancement must be effective. If due to some heat treatment in molybdenum–nickel steel, *e.g.* drill life increased from 3 (without heat treatment) to 15 holes (with heat treatment), this does not appear to be practically significant. However, if a similar relative increase, *e.g.* from 30 to 150 holes would be obtained under simple heat treatment conditions, *i.e.* at the equal ratio 1:5, it may be of practical benefit [137].

As mentioned, the main problem in the machinability of PM materials is the heterogeneity of the microstructure formed by alloying and processing beside the porosity with its own specific aspects. The consequence of it is also the different behaviour in machining of admixed and prealloyed powder base alloys of the same chemical composition. From this point of view also the – up to this time relatively few – published isothermal and/or continuous cooling transformation diagrams for PM steels (TTT eventually

CCT-diagrams) are not sufficient to design really efficient annealing, sinter hardening and heat treatment cycles (most important microstructure areas). They describe an average microstructure, consisting of many phases, not the transformation course of single phases forming the microstructure of the alloy. These single phases take part in the machining process also when they are deformed and form a new microstructure which can be characterized by microhardness measurement only. The possible starting heterogeneity of the microstructure which can differ due many reasons also at the same composition will be reflected in the results. The relatively simplest way for modifying the microstructure is controlled cooling from the sintering temperature.

6.2.1.1 Controlled cooling

Modification of the microstructure to improve machinability can occur by changing post-sintering cooling rates considering the specialty of the material.

The cooling rates which can be employed are those which represent the practical limits using conventional water jacketed and convection cooled sintering furnaces. The post-sintering cooling rate can have a measurable effect on the microstructure, machinability and mechanical properties of the material which must be always considered. The affecting the machinability of a material is possible only through the adequate change of the microstructure, which should be analysed and specified. In dependence on composition and mainly on the carbon content the proportion of the microstructural phases (ferrite up to martensite) can be significanlty changed by the cooling rate.

Usually, reduced cooling rates, achievable by using a temperature controlled water jacket cooling zone, can be used to substantially improve machinability, but with decrease of strength properties (suitable proportion of ferrite and of coarse lamellar pearlite, divorced carbides). An increase cooling rate can be used to improve tensile strength properties reflecting the presence of bainite up to martensite in the microstructure. These material characteristics must be considered when evaluating machinability [266].

The modification of the microstructure can occur also with tempering and simple annealing treatments which produced less improvement in machinability than tool improvements. The annealing treatment at 870°C changed the failure mode of the HSS drill. The drill appeared to 'stick' to the workpiece when cutting the annealed workpieces, rather than overheat [137]. Tempering had a lesser effect on the microstructure and properties of the Fe–0.85Mo (Astaloy 85Mo) + 2% nickel steel than anticipated.

The microhardness values suggest that a form of precipitation hardening may have occured in the pearlite and bainite areas of the microstructure. This may account for the increase in yield strength and loss of ductility with increasing tempering temperature. In contrast, the nickel-rich areas show a slight reduction in microhardness on tempering.

It is possible to expect that increasing 'tempering' temperature to 650–700°C, subcritical annealing or controlled transformation annealing after sintering may improve machinability further [137].

The published results can be accepted only as informative since the conditions for cooling rate changes differ between the sintering furnaces, in relation to the workpiece composition and final properties and also in relation to their shape and mass. These illustrate the lack of knowledge of the fundamental behaviour of PM steels at heat treatment in relation to machining. There is very probably a tendency to choose other methods for machinability improvement instead of heat treatment processes as separate processing steps. Heat treatment is however an attractive measure if it can be included in the sintering process.

6.2.2 Green machining

Green machining is one of the production processes applied in the machining of PM parts. Green machining offers one of the alternative routes to minimise the machining problems with PM materials. Green machining is virtually a must for those PM components that are sinter hardened since machining after sintering is impractical in this case.

Successful green machining requires sufficient green strength of the compact, usually higher than that of standard compacts even in the case of high compacting pressures (up to 800 MPa). Die wall lubrication technology is a possibility here as is the use of special lubricants which both are however still under development. Today, parts subjected to green machining are prepared mainly by warm compaction technology. Even if the higher green strength obtained with the warm compaction technology facilitates successful green machining, there are still some limitations left due to stability problems. The limitations restrict the handling of the parts and require a careful adaptation of the cutting parameters and the tool geometry to the material properties [267]. (It can be expected that these limitations will hold also for compacts prepared by the above mentioned new techniques). The posibilities and limits for green machining of PM parts will be best shown on the basis of the results attained with appropriate materials presented in chapter 7.

6.3 EFFECT OF MACHINING OPERATIONS

As stated above, for wrought steels enormous efforts are directed towards optimizing the machining process with regard to selection of tools and machining parameters. For sintered steels, the volume of machined parts is lower and the diversity of the materials is larger, therefore this optimisation process has been performed only to a very limited degree, and still less has been published, most of the practical experience of the parts manufacturers being kept confidential [118].

The effect of machining parameters and cutting conditions including tools on the machinability of PM steels cannot be generalised due to a considerably larger number of factors affecting the workpiece material properties compared with wrought steels. The acceptable results are those attained under some defined machining parameters of a workpiece with specific characteristics which could be applied also for a group of materials. Unfortunately, detailed information about the effect of cutting speed, feed, cutting fluids, etc., is still lacking. Therefore, further results for machining of main groups of PM steels are desirable in order to contribute to better knowledge of the effect of machining parameters on the effective machinability of PM steels as will be discussed in the following chapters.

6.3.1 Tool coating

Another possibility for the machinability enhancement besides the very enlarged PVD and CVD coating methods which were presented in chapter 4, is coating the cutting tools with a solid lubricant. The most effective antifriction coating of the tools for turning was a nickel–phosphorus solid lubricant deposited on a cemented carbide insert by electroplating. Cutting tests were performed on a heat treated wrought high manganese steel. A large decrease (halving) in the wear of the tool was recorded at all tested cutting speeds up to 60 m/min (depth of cut 0.65 mm) [268].

It can be assumed that this method, which does not exert any effect on the workpiece properties, could be an effective contribution in the nearest future to the machinability of PM steels, especially for those with poor machinability and by this way also for high-alloy diffusion alloyed steels and Mn and Cr steels.

7

Machining of Sintered Steels – State of the Art

As mentioned, the machinability of PM steels is improved by exerting a focused effect on the workpiece properties, through addition of different free machining additives closing pores by impregnation and infiltration, through the modification of the microstructure by alloying and by adequate heat treatment processes, combined with tool material and cutting conditions. The addition of a free-machining aid and testing of its effect on the machinability of a PM material in relation to the given cutting conditions seems to be the simplest method but must always be regarded together with the effect on the dimensional and mechanical properties of the workpiece due to mutual interaction.

Possible combinations of these routes provide, on one side, many possibilities for actual improvement in the machinability of PM steels. On the other side, the use of a relatively large number of machinability indices in connection with different machining methods enlarges and, consequently, complicates the view on the problem. In such a case, it is very often impossible to determine the effect of one individual factor on the machinability of a material. For this reason, many results attained in machinability improvement are valid only for the tested state, which is often insufficiently defined, in particular regarding the workpiece properties.

In terms of characterisation of material properties in relation to machinability, it is possible to define two groups of PM steels. However, these two groups cannot be compared in the area of the application in production. The first group includes iron–carbon steels, which can be characterised quite easily on the basis of hardness and the proportions of ferrite and pearlite. The second group is formed by alloyed steels. It is generally known that the alloyed materials are more complicated for characterising the workpiece properties in relation to the cutting process. These depend on starting powders, alloying and the alloying method, sintering conditions and enhancement methods, including various machining aids used. Therefore, in the case of alloyed materials it is essential to obtain more information on the effect of methods for improving machining, which result from all technological steps in material processing, in final mechanical properties, dimensional stability, microstructure in relation to the different cutting processes as third factor coming into account in machining PM steels. In addition, all three interacting areas, studied with the aim to improve the machinability of PM parts, have a stronger impact on alloyed than on iron–carbon steels. They all act together in the cutting process on the tool life and the quality of the machined surface under all aspects of effectiveness.

The data concerning the basic properties and machinability of PM steels ranging from simple iron–carbon to alloyed high-strength and high-hardness as-sintered and as-heat treated steels without and with the use of some machining aid or some special cutting method can contribute to judgement on the actual technical and economical effectiveness of the methods used and can help in further improvement of machinability. The results presented in this chapter on the effect of a particular machining enhancer mainly on machinability and, in actual cases, on mechanical properties show the advantages and disadvantages and the limits of individual methods for improvement of machinability in relation to the relevant alloy with its required properties.

7.1 PLAIN IRON AND IRON–CARBON STEELS

Plain iron is one of the most difficult materials to machine because of its softness and tendency to tear easily in all machining operations and the tendency to cold welding, *i.e.* to form built-up edges in machining. One method of treatment for improved machinability is to repress or otherwise densify the area to be machined. In repressing or densifying the part, a degree of cold-working occurs, thus obtaining better machinability. This coining increases density and cold-works the area to be drilled and/or tapped so that tearing of the threads, common in low density compacts, is reduced [113].

Iron–carbon steels seem to be relatively simple materials in term of effect of a machining aid on properties and on machinability. The properties of Fe–C steels are simpler for the characterisation especially in microstructure also in relation to mechanical properties compared to alloyed steels, and, therefore, the results obtained with them can be used also for reference information about the effects of some machining aid or of other method to improve the machinability of sintered steels in general.

7.1.1 Effect of manganese and carbon

Both carbon and manganese have a special position in machining of sintered steels as was stated before. Manganese (elemental) in small additions is regarded to be a machining enhancer. Their single or combined effect on machining of a material can form a basis also in relation to the added machining aid as will be shown. Firstly, the relation between the drilling torque and the density of sintered iron with an addition of 0.5% manganese for the 1st and 50th hole drilled is shown in Fig.7.1.

Raising the density from 6.4 to 7.0 g/cm³ increased the initial torque significantly and this trend persisted through the drilling of 50 holes. The increase in torque up to 6.7 g/cm³ was slower. The increase in torque with decreasing porosity of the material is opposite to base views on the effect of porosity on machinability of a sintered material. On the other hand a small decrease in drill wear with increasing density from 0.06 to 0.05 mm only as manifestation of manganese addition was observed. From these results it can be pointed out that force measurements alone cannot always be used to predict tool wear in machining of PM materials.

The combined effect of 0–0.9% carbon content (added as graphite) and of 0.5% manganese on thrust force in drilling for Fe–

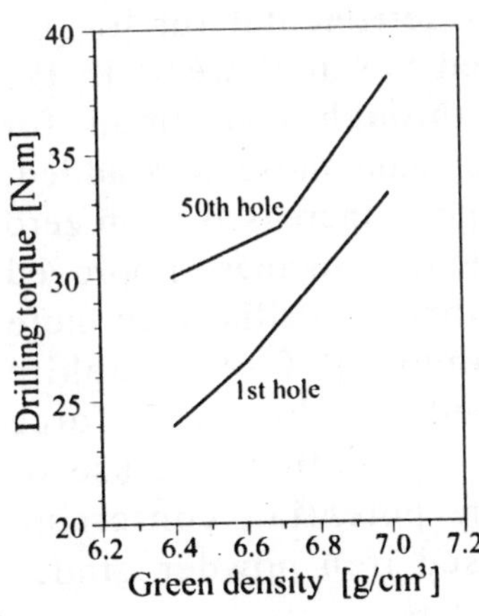

Fig.7.1 Effect of density of Fe–0.5Mn steel on drilling torque [144,155]. *Drilling test: HSS 4.3 mm drill, drill speed 1000 rpm, v_c = 13.5 m/min, feed 0.05 mm/rev., mineral oil lubrication.*

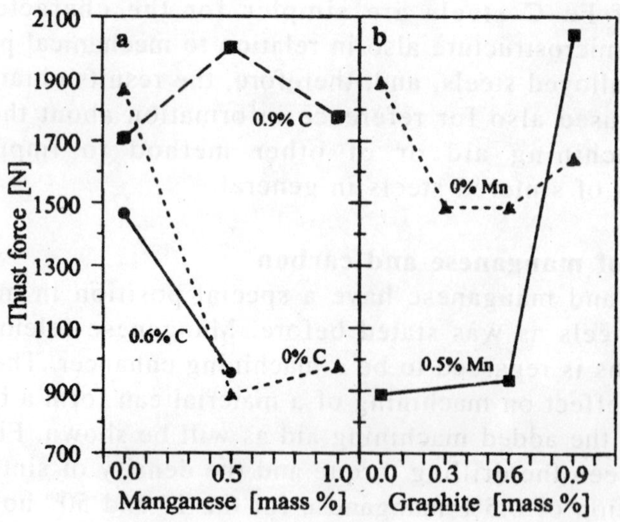

Fig.7.2 Effect of: a – nominal carbon content (given as added graphite) in dependence on manganese (electrolytic) content and b – of manganese addition in dependence on nominal carbon content (graphite addition) on thrust force in drilling of Fe-Mn-C sintered material (density 6.87 g/cm³) [144,155,230]. *Drilling test: HSS 8 mm drill, drill speed 1000 rpm,* v_r *= 21.1 m/min, feed 0.05 mm/rev., mineral oil lubrication.*

Mn-C compositions is shown in Fig.7.2. A manganese addition caused a decrease in thrust force for plain iron and in some extent at all levels of carbon content. The beneficial effect of 0.5% manganese on the drilling force in iron-manganese-carbon compositions with carbon content up to 0.6% was determined.

When the carbon content was raised to 0.9% (nominal, added graphite), a sharp increase in drilling force was recorded. It shows how the carbon level affects the thrust force and the combination of 0.9% C (added) and how 0.5% Mn was detrimental for it.

The addition of manganese also affected the drill wear. In the absence of manganese, drill wear passed through a minimum for 0.6% graphite added. However when 0.5% manganese was added, the drill wear increased continuously as graphite increased from zero to 0.9%. The actual drilling tests indicated that manganese improved the tool life for graphite levels <0.6%, but that the drills wore more rapidly at the 0.9% graphite level (optimum at 0.6% graphite addition with 0.5% manganese as observed also by thrust force measurements) [144,155]. For certain mixes studied, the use of manganese appeared to offer considerable potential, conferring advantages similar to those of resulphurised iron powder grades [230].

However, no analysis has been made of the way by which manganese in small amounts in the presence of very low (natural) sulphur contents without sulphur addition affects the microstructure and properties of Fe–Mn–C steel and by this effectively machining. The reason for this behaviour of manganese in machining PM steels should be more in respect to the microstructure formed in the workpiece. The machinability of low carbon steel is considered to be the best when it contains ~0.20% C, 0.90 to 1.00% Mn with 0.15 to 0.20% S. This sulphur content can be accepted directly as a machining agent.

7.1.2 Effect of S, MnS and MoS_2

Sulphur, manganese sulphide, and molybdenum disulphide are most widely used as machining aids. As mentioned above, combined testing and evaluation of the effect of some machining aid on mechanical properties and machinability of a material presents more informations about the interactions between material properties and cutting tool and the cutting conditions.

7.1.2.1 Effect of S, MnS and MoS_2 on mechanical properties

The effect of elemental sulphur, manganese sulphide and molybdenum disulphide on the transverse rupture strength and hardness of Fe–0.6C (F–0008) steel, based on two iron powder grades which were also tested for machinability, is listed in Tab. 7.1.

In this case, the difference in the density and mechanical properties of tested materials must be regarded. All three aids

Tab.7.1 Effect of machining aids (S, MnS, MoS_2) on transverse rupture strength (TRS) and on hardness (HRB) of F-0008 steel based on atomised (Ancorsteel 1000) and sponge (Ancor MH100) iron powder. Sintering 30 min at 1120°C, endothermic atmosphere (acc.to Ref.242)

Aid addition [mass %]	Iron powder (density)											
	Atomized (6.8 g/cm³)						Sponge (6.1 g/cm³)					
	Machining aid											
	S		MnS		MoS_2		S		MnS		MoS_2	
	TRS [MPa]	HRB	TRS [MPa]	HRB	TRS [MPa]	HRB	TRS [MPa]	HRB	TRS [MPa]	HRB	TRS [MPa]	HRB
0.0	541	54	537	53	552	51	340	24	352	23	348	22
0.25	525	52	536	53	447	52	441	36	365	24	397	35
0.50	592	53	592	53	526	58	439	35	370	27	475	44
0.75	564	53	564	53	537	58	430	35	354	27	470	44
1.00	592	56	592	55	606	68	435	40	363	26	496	46

increased TRS in atomised iron powder system by 10–12% and by 28 and 43% in sponge iron powder system (lower density) with sulphur and molybdenum disulphide, respectively. The effect of manganese disulphide caused the highest increase of both properties tested. The results also show that sulphur and molybdenum disulphides were not thermostable. The manganese sulphide was stable under endothermic atmosphere sintering. It appears that a significant proportion of the molybdenum disulphide reacted with the ferrous matrix and decomposed, and molybdenum diffused into the matrix, as described in chapter 6.

The sulphur formed a sulphide network (as in the case of elemental S added) that remained at the grain boundaries which however is not too detrimental with respect to mechanical strength and can favourably affect the machinability. It might be supposed that the liquid sulphide eutectic increases the load bearing cross section in the sintered material.

Hardness values depend on the amount and kind of machining aid but also reflect the effect of base iron powder grade but in this case of different density. Both sulphur and manganese sulphide increased the hardness of the material slightly. Increasing molybdenum disulphide additions caused a marked increase in the hardness of the F-0008 material based on atomised iron powder markedly and still more with sponge iron powder base compared to the hardness of the parts without MoS_2 addition. Sulphur and molybdenum disulphide had generally similar effects with sponge iron powder, in this case with lower density (which is not surprising since both additives result in formation of a sulphide network). Manganese sulphide was relatively neutral to the hardness.

For explaining the higher strengthening effect of these additives on sintered iron of lower density a more effective reaction of the sintering atmosphere with the material components with higher open porosity can be assumed, *e.g.* a larger reaction surface increasing surface activity because the surface of interconnected pores at 6.9 g/cm^3 density of a sintered iron compact is ~300 × 10^3 mm^2/mm^3 compared to ~160 mm^2/mm^3 at 7.2 g/cm^3 density, respectively [269]. At the lower density of the sponge iron-based materials a more beneficial effect of sintering activation through the liquid phase can therefore be assumed. This effect is generated by S and MoS_2 but not by MnS.

Molybdenum disulphide and manganese sulphide had opposing effects upon dimensional change. Increasing molybdenum disulphide content increased growth of the dimensions, which might indicate introduction of Mo into the sintering interparticle contacts as in the case of transient liquid phase sintering of Fe–Mo–C, which also results in expansion [40].

Similar effects on transverse rupture strength were determined with iron powder (Atomet 28)–graphite mixes (6.7 g/cm^3) using a 0.5% MnS addition. This addition at various combined carbon content between 0.3 to 0.9% caused a decrease in TRS of less than 5%. Similarly, with various combinations of MnS + MoS_2 in a total amount not exceeding 0.5%, with increasing MnS content the transverse rupture strength decreased by ~10% and hardness by ~8% [242,243,253].

It follows from this that the effect of the machining aids tested on the mechanical properties was not uniform. It is also affected by the iron powder grade and porosity of the compacts. The various effects of machining aids on mechanical properties could be explained by the differences in base properties of iron powder grades used and processing conditions.

The effect of machining aids on the dimensional stability of the parts plays an important role in practice. In this case, contrary to manganese sulphide, sulphur and molybdenum disulphide caused linear swelling of the dimensions with increasing amount of carbon from 0.2 to 0.8% C_c for F–0008 atomised and to 0.5% C_c for F–0008 sponge iron powder. In general, all machining aids result in less shrinkage of a part by a constant value independent on carbon content and compacting pressure. The MnS-containing compositions showed slightly less shrinkage than the base Fe–C-mix (Atomet 28 iron powder, 0.5% MnS) for all carbon contents (0.3 to 0.9%) by ~0.05%. In some cases less shrinkage occurred at higher carbon content, *e.g.* from ~0.3% lin. at 0.3% C to 0.1% lin. at 0.95% C [243,253].

7.1.2.2 Effect of S, MnS, and MoS_2 on machinability

The effect of the mentioned machining aids on the machinability of Fe–0.6C parts with mechanical properties given in Tab.7.1 is listed in Tab.7.2.

The machinability of parts based on atomised iron powder (with higher density) was generally higher, also without machining aid, in spite of higher hardness compared to the sponge iron powder base parts of higher porosity.

It is presumed that sponge iron powder particles are softer than water atomised particles which can explain why sponge iron powder based materials exhibit a greater tendency to stick to a cutting tool

Tab.7.2 Effect of 0.5% machining aid (S, MnS, MoS_2) addition (corresponds to 0.5%, 0.19%, 0.20% sulphur content), and of iron powder grade on hardness (HRB) and on machinability (number of holes drilled – N.h.) of Fe–0.5C parts; density: 6.1 g/cm³ (sponge, Ancor MH100), 6.8 g/cm³ (atomised, Ancorsteel 1000) [242]. *Drilling test: HSS 3.5 mm drill, drill speed 3000 rpm, v_c = 33 m/min, feed 0.08 mm/min, depth of hole 12.7 mm*

Iron powder	Machining aid							
	None		S		MnS		MoS_2	
	N.h.	HRB	N.h.	HRB	N.h.	HRB	N.h.	HRB
Sponge	19	24	158	35	33	27	285	44
Atomised	83	54	418	53	890	53	718	58

[142]. The difference in (micro) hardness between atomised and sponge iron powder grades, which could be reflected in the machinability of as-sintered iron, was not documented (see Tab.2.3). The physical–metallurgical explanation for this effect is lacking and is a task for further investigation; of course it is rather difficult to compare the machinability of specimens with such widely differing sintered density; the effects of density and that of the starting powder grade cannot be easily separated. Machinability testing of PM steel specimens with identical density but produced from atomised and pure sponge iron powders, respectively, would be an interesting task!

In this case, manganese sulphide caused the smallest increase in drill life in sponge iron powder based steel. For steels from atomised powder, however, manganese sulphide and molybdenum disulphide produced a greater improvement in drill life than sulphur. In testing, a combination of both additives with 0.3% MoS_2 and 0.2% MnS resulted in somewhat better machinability compared to MnS or MoS_2 only, 1000 holes to failure, than the individual additives [242]. The difference in density between atomised and sponge iron powder based parts must be considered but it is not sufficient for the explanation the differences in machinability of these materials (890 *vs.* 33 number of holes drilled both with MnS addition).

Starting differences in physical, chemical and technological properties between sponge and atomised iron powder grades were also visible in the cases of some machining additions in their different effect on machinability which cannot be explained only by different density of the tested part (6.1 and 6.8 g/cm³). The as-sintered properties of parts made from both types of iron powders must be taken into account not only the starting properties of the powders, *e.g.* different hardness of iron particles if in reality exists in an extent with a possible effect on machining. Generally, the differences in machinability resulting from different iron powder grades – atomised vs. sponge – are described in several publications but not explained. Here, more detailed investigations are necessary.

Higher amounts of additives are required to improve the machinability of stronger, harder matrices. The loss of the additive during sintering has to be considered. A very small loss of sulphur during sintering was observed as the manganese content was controlled. When the manganese content was too low, the sulphur formed iron sulphides which are more easily reduced by the hydrogen sintering atmosphere to form H_2S and by this the expected effect of sulphur on machinability was lowered which phenomenon is particularly unwelcome since it emerges primarily near the surface, *i.e.* where machining is carried out most frequently.

Under identical machining conditions, an addition of 0.5% MnS (high purity) to the sponge iron powder base steel Fe–0.7C_c (F-0008, iron powder Pyron D-63; density 6.3 g/cm^3) increased its machinability from 17 to 100 holes drilled, but the increase was smaller when MoS_2 was added. The 0.5% MnS addition showed a greater improvement in machinability for the slower speed and feed rates [135]. An explanation for this result is not presented but it would also be very important in terms of productivity in machining.

The effect of carbon content in the range of 0.3 to 0.9% in combination with 0.5% MnS addition on the drilling thrust force in Fe–C parts is shown in Fig.7.3.

An initial decrease of thrust force took place until the carbon

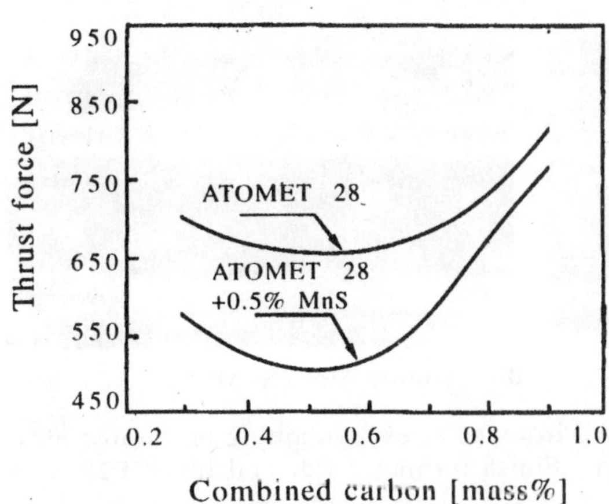

Fig.7.3 Effect of combined carbon content and of 0.5% MnS (high purity) addition on thrust force in Fe–(0.3–0.9)C parts (iron powder ATOMET 28); density 6.7 g/cm^3, sintering 30 min at 1120°C, rich endothermic atmosphere [243,253]. *Drilling test: HSS 6.4 mm drill, drill speed 800 rpm, v_c = 18 m/min, feed 0.05 mm/rev., rectangular bars, depth of hole 8 mm.*

content reached about 0.5%. This effect is a consequence of the proportion of pearlite in the microstructure, favouring chip breakage and by this enhancing machinability. At higher carbon content when pearlite was the predominant phase, the thrust force increased proportionally to combined carbon content, because of increasing pearlite fraction and by this higher hardness of the material.

Manganese sulphide addition improved the machinability by drilling of Fe–C parts for the whole examined carbon content range. The optimum effect was attained again at 0.5% C, where with 0.5% MnS addition a reduction in thrust force by 25% was reached compared to the MnS–free reference material.

When taking the surface finish of the materials as a criterion for machinability (see also 7.1.6), the effect of machining aids can also be revealed. In Ref. 205 finish turning of sintered bars was used as a machinability test, low cutting speeds being applied to result in more severe cutting conditions. As can be seen from Fig.7.4, addition of MnS resulted in markedly better surface finish compared to the reference material although a minimum MnS content of 0.5% seems to be necessary. Addition of only 0.3% MnS resulted in

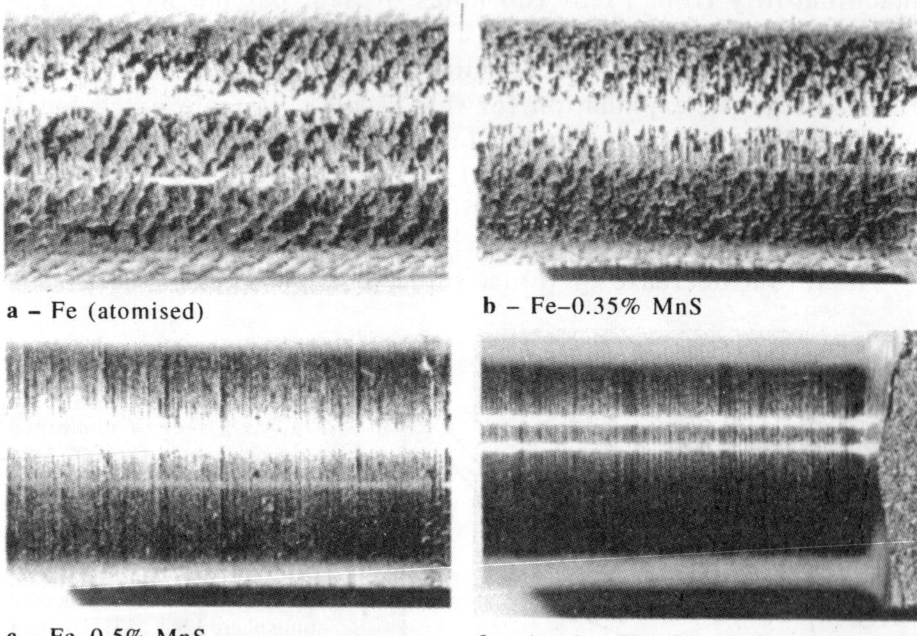

a – Fe (atomised)

b – Fe–0.35% MnS

c – Fe–0.5% MnS

d – Astaloy Mo–2% MoS_2

Fig.7.4 Machined surfaces of sintered iron and steels containing machining aids [207] (Courtesy of MIBA Vorchdorf). Finish turning, hardmetal insert P20, v_c = 15 m/min, 0.15 mm/rev.

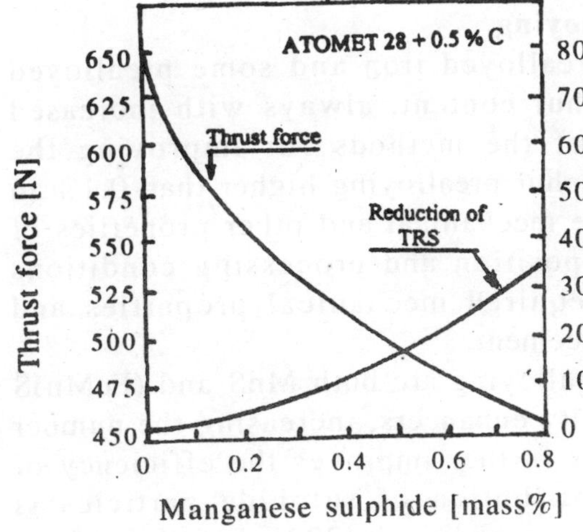

Fig.7.5 Effect of added MnS on thrust force in drilling and on transverse rupture strength (TRS) of Fe–0.5C material (iron powder ATOMET 28) [243,253].

almost the same poor surface finish as with the reference material without any MnS. Increasing the MnS content above 0.5% resulted in lower surface roughness; however the improvement is comparatively small.

Addition of MoS_2 resulted in surprisingly good surface finish as visible from Fig.7.4d although metallographic investigations had confirmed that MoS_2 had completely decomposed and only $Fe_{1-x}S$ had remained, which fact once more confirms that machining aids must not necessarily be 'solid lubricants' in the strict sense. However, it has to be kept in mind that in the MoS_2 containing material the sulphur content added is markedly higher than with MnS, although 2% MoS_2 still give better finish than 2% MnS. Also here the considerable effect of the starting powder is evident: the water atomised Fe–1.5% Mo steel powder resulted in significantly better surface finish than the atomised iron powder grade.

A decrease in thrust force in Fe–C material with MnS addition can be joined with a decrease in transverse rupture strength values, Fig.7.5. The thrust force for Fe–0.5%C parts decreased with increasing MnS addition, however at the expense of an associated decrease of TRS [135].

As shown in such cases a compromise between machinability by drilling and transverse rupture strength for Fe–C parts can be reached, *e.g.* at 0.5% MnS addition and 0.5% combined carbon content.

7.1.3 Effect of sulphur prealloying

The production of sulphur prealloyed iron and some prealloyed powders with increased sulphur content, always with increased manganese content, is one of the methods for improving the machinability of the parts. Sulphur prealloying higher than 0.1% in iron powder must not decrease mechanical and other properties of the parts. The chemical composition and processing conditions should be adapted for the required mechanical properties and maximum machinability enhancement.

The products of sulphur prealloying are both MnS and (FeMn)S which are efficient machinability enhancers; increasing the number and dispersion of the particles further improves the efficiency of sulphides. The improved distribution of sulphide particles is desirable under severe machining conditions [222]. It follows from it that there is no difference in effect of MnS on machinability of PM steels whether it occurs in interior or on the surface of powders. As far as the authors know, no micrographs of steels from sulphur prealloyed powders are available that reveal optically visible MnS particles.

An example of the powders prealloyed with manganese sulphide, it means with higher sulphur content, are prealloyed powders such as MP 37, and MP 36S, which are commonly used in free-machining steels. Transverse rupture strength and hardness values, base chemical composition and machinability rated in number of holes drilled for both powder steels are listed in Tab.7.3. The purpose of a higher Mn/S ratio in MP 37 powder is to ensure that all sulphur

Tab.7.3 Chemical composition, transverse rupture strength (TRS), hardness (HRB) and machinability in drilling (number of holes drilled and average time to drill 25 holes) of sulphur prealloyed commercial iron powder grades MP 37 and MP 36S (Domfer) without and with copper addition. Density 6.6 g/cm^3, sintering 20 min at 1120°C, nitrogen-based atmosphere [248]. *Drilling test: HSS 6.35 mm drill, drill speed 2300 rpm, v_c = 46 m/min, depth of hole 25 mm, constant drill thrust force of 684 N*

Iron powder grade	TRS [MPa]	HRB	Composition [mass %]				Number of holes	Av. time for 25 holes [s]
			C	S	Mn	O		
MP 37	645	60	0.68	0.24	0.94	0.35	82	8.1
MP 36S	381	55	0.70	0.38	0.38	0.15	*200	6.2
MP 37 + 2 % Cu	932	80	0.70	0.21	0.94	0.35	32	9.9
MP 36S + 2 % Cu	773	79	0.72	0.38	0.38	0.15	40	9.3

Remark: *drilling finished at 200 holes drilled without drill failure

present is in the form of MnS inclusions.

The transverse rupture strength of MP 37 material (higher Mn content) was considerably improved over that of MP 36S bars for all mixes. This may be due to the strengthening effect of higher manganese content, possibly by bonding the sulphur to a stable compound. On the other hand, the MP 36S alloy has a lower manganese content and exhibited better machinability compared to MP 37. It was proposed that the lower mechanical strength of the interparticle bonding in MP 36 S compacts promoted easy removal of material, thereby improving machining by reducing machining forces [29]. In this case it is necessary to define the requirements also for the mechanical properties or only for the improvement of machinability of a material.

Higher transverse rupture strength and hardness of tested materials with higher Mn content gave lower number of holes drilled.

Both MP 37 and MP 36S with 2% Cu addition exhibited an increase in transverse rupture strength and hardness and a marked decrease in machinability in comparison to those without copper addition. For both Cu-free and Cu-alloyed materials it seems that sulphur bonded to Mn results in higher strength but lower machinability than sulphur present at least in part as $Fe_{1-x}S$, probably to some extent as sulphide eutectic.

A large investigation of machinability in drilling was made with sulphur prealloyed iron powder 400 MS (atomised iron powder, A.O. Smith-Inland; 0.16% S, 0.21% Mn, 0.02% Pb) with 0.45 and 0.8% C compared to plain iron powder (0.02% S) in dependence on density of as low as 5.5 to 7.0 g/cm^3. The thrust force was increasing with increasing density from 5.5 to 7.0 g/cm^3 by ~100% in both – in sulphur prealloyed and without sulphur prealloying materials. The results contradict the interrupted cut theory about the effect of porosity on machinability of sintered steels. Increased carbon content from 0.45% to 0.8% caused an increase in thrust force by ~40% at the density of 5.5 g/cm^3 and by ~25% at the density of 7.0 g/cm^3 of the material. Sulphur prealloying resulted in lower thrust force by ~25 to 40%, more in lower density parts, compared to plain iron powder. Reduction in cutting force in drilling at v_c = 23 m/min due to sulphur addition was less significant at the higher carbon content [141].

7.1.4 Effect of drilling conditions for steels prepared from different iron powder grades

The following data show that the machinability is in a great extent affected not only by the base properties of sintered Fe–C material and by manganese sulphide addition but also by the machining parameters, in this case in drilling.

The effect of machining conditions in drilling of iron powder grade, density, and of 0.5% MnS addition on the number of holes drilled before drill failure is listed in Tab.7.4.

The addition of 0.5% MnS caused a significant increase in machinability. The results show that the machinability of these steels was affected in a great extent by the drilling conditions – drill speed and feed rate. Poor machinability of the materials without MnS addition is also evident. Minimal difference were observed in the machinability of D-63 samples with a density of 6.3 and 6.8 g/cm³.

The very high number of holes drilled in atomised base iron powder steel of the density 7.2 g/cm³ with 0.5% MnS addition has to be investigated in more detail. Surely it cannot be explained, *e.g.* by lower porosity compared to sponge iron powder steel with the density of 6.8 g/cm³. A detailed analysis of this state of the cutting pair in the cutting zone would be a contribution for better knowledge of machinability of PM steels. The high carbon content of 0.8% for these materials has to be stressed.

Tab.7.4 Number of holes drilled for Fe–0.8C steel + 0.5% MnS addition in dependence on iron powder grade, density, and drilling parameters. D-63 – sponge iron powder, A1000B – atomised iron powder. Hardness ~57 HRB for D-63 base powder, and ~64 HRB for A1000B iron powder at the density of 6.9 g/cm³ [135]. *Drilling test: HSS 3.5 mm drill diameter, depth of hole ~11 mm, samples: rectangular bars and rings*

Alloy No.	Iron powder	Density [g/cm³]	Cutting speed [rpm/m/min]	Feed rate [mm/rev.]	MnS [mass %]	
					0	0.5
					Number of holes	
1	Sponge	6.3	1000/11	0.36	250	725
2			3000/33	0.24	50	467
3			3000/33	0.36	42	325
4		6.8	3000/33	0.36	57	440
5			4000/44	0.47	26	107
6	Atomised	7.2	4000/44	0.47	25	2514
7		6.9	4000/44	0.47	35	*

Remark: *no value given

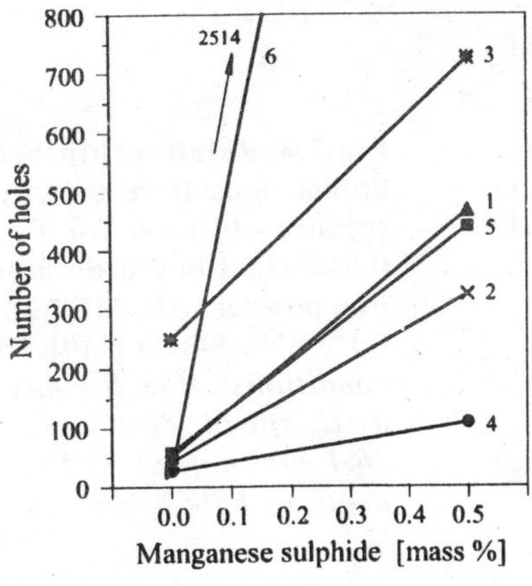

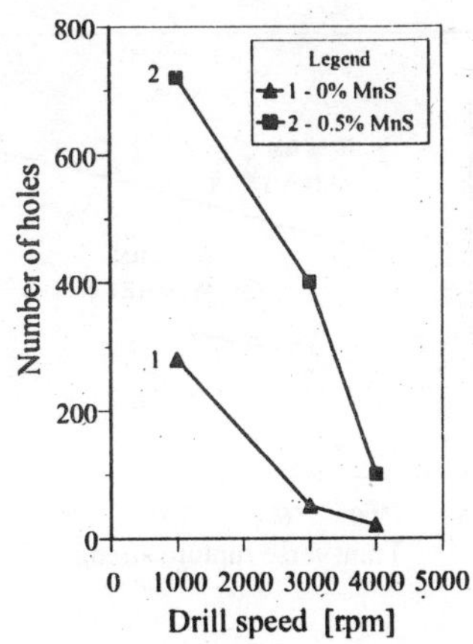

Fig.7.6 Dependence of number of holes drilled in Fe–0.8C (F-0008) materials on manganese sulphide addition and on drilling conditions. Alloy No. 1–6: data from Tab.7.4 [135,147,224].

Fig.7.7 (right) Dependence of number of holes drilled for steel F-0008 based on sponge iron powder (D-63) of density 7.0–7.1 g/cm^3 without and with 0.5% MnS addition on dependence on drill speed [135,147,224]. *Drilling test: HSS 3.5 mm drill, depth of hole 11.4 mm.*

At lower cutting speeds a higher number of holes drilled was attained but at the expense of longer time being necessary to drill the holes. Data listed in Tab.7.4 are graphically shown in Fig.7.6. As shown in Fig.7.7, the number of holes drilled decreased with increasing drill speed (>3000 rpm) more rapidly in materials with MnS addition.

Also in this case, the effect of manganese sulphide addition in relation to the iron powder grade and drilling conditions was similar to previous results. The machinability of both materials, without and with MnS addition, was affected by the drilling speed. Increasing drilling speed resulted in poorer machinability. The positive effect of MnS addition was increasing with decreasing drilling speed. It means that at a high drilling speed 0.5% MnS had only a minimal effect on machinability compared to those without MnS addition. It can therefore be concluded that the adhesive wear mechanism against which MnS is effective does not play a major role at high cutting speeds (higher temperature in the cutting zone).

7.1.5 Effect of non-sulphide machining aids

7.1.5.1 Effect of lead, bismuth and boron nitride

As stated in Ref.141, Fe–C steels modified by lead or bismuth addition exhibit essentially the same effect on thrust force at

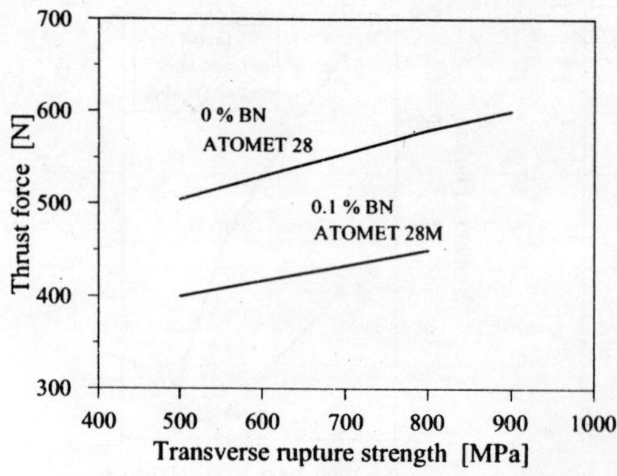

Fig.7.8 Relationship between drilling thrust force and transverse rupture strength for Fe–(0.6, 0.9)C steel based on atomised iron powder (ATOMET 28) with 0.1% BN addition [119]. *Drilling conditions: HSS 6.4 mm drill, drill speed 800 rpm, v_c = 16.1 m/min, feed 0.05 mm/rev., depth of hole 8 mm.*

densities in the range of 5.5 to 7.0 g/cm^3. In both cases the thrust force increased with the density increase, *i.e.* machinability becames poorer with decreasing porosity. The thrust force reduction due to lead or bismuth ranges from zero at the lowest densities to 20 to 50% at the highest densities. This makes lead and bismuth most attractive machining enhancers in high density parts. The addition of 0.25% of lead or bismuth was not adequate to reduce cutting forces in high carbon (~0.8%) steels; additions of 0.5% of these elements were effective, reducing the cutting forces in high carbon (0.8% C) steels at higher densities by about 25%. At lower carbon contents of ~0.45% C, additions as low as 0.25% Pb or Bi were sufficient to provide reduction in cutting forces for this carbon steel.

The effectivity of BN addition as a machining enhancer was confirmed by drilling tests, giving the relationship between drilling thrust force and transverse rupture strength of sintered Fe–C steels, as shown in Fig.7.8.

The addition of 0.1% BN in Fe–(0.6, 0.9)% graphite materials caused a decrease in thrust force by ~25% (~100 N) compared to those without BN addition. The thrust force in both materials was increasing with increasing transverse rupture strength in the tested range of 500 to 800 and/or 900 MPa. The course of both curves (without and with BN addition) in relation to the transverse rupture strength is parallel. It means that there is a relatively simple relationship between strength and thrust force in this case if regarded the mentioned decrease in thrust force by BN.

7.1.5.2 Effect of resin and oil impregnation

The effect of resin impregnation on the thrust force in drilling of

Tab.7.5 Effect of resin impregnation on thrust force in drilling of Fe–(0–0.8)C steels based on atomised iron powder 300M (0.12% Mn, 0.008% Pb, 0.018% S; Smith Inland). Sintering 30 min at 1120°C [141]. *Drilling conditions: HSS 6.3 mm drill, dril speed 1160 rpm, v_c = 23 m/min, feed 0.13 mm/rev., depth of hole 12.5 mm*

Carbon content [mass %]	Density [g/cm^3]		Thrust force [N]		Reduction in thrust force [%]
	Sintered	Impregnated	Sintered	Impregnated	
0	6.85	7.02	746	297	60
0.20	6.88	7.00	821	422	49
0.45	6.88	6.99	741	493	34
0.80	6.88	6.97	844	635	25

Remark: Reduction in drill thrust force due to the impregnation

Fe–C materials is listed in Tab.7.5; there are, however, no data about the type and chemical character of the resins used.

Impregnation provided a substantial reduction of 25 to 60% in thrust forces in all alloys. Reduction in thrust force steadily decreased as the carbon content of the alloy increased.

The drill thrust force was increasing with increasing carbon content more in as-impregnated alloys compared to as-sintered. The lowest drill thrust force was needed for impregnated plain iron compacts as seen in reduction in thrust force values.

To know in more detail the positive effect of resin impregnation on machinability of sintered steels, resin impregnated and unimpregnated Fe–C samples were prepared by the authors. In both materials, through holes 5 mm in diameter and 10 mm in depth were drilled and parallel M6 threads were tapped (manually). The microstructure of the machined (drilled) surface layers in both state is shown Fig.7.9 and that of the root of the thread in Fig.7.10.

The difference in deformation of the machined layers is clearly confirmed also by the measurement of the microhardness. The microhardness of the machined surface layer of the hole in impregnated material was 296 HV 0.01 and in the root of the thread 312 HV 0.01. In unimpregnated materials the corresponding microhardness values were 411 and 596 HV 0.01. The microhardness of the matrix (ferrite) in the distance 0.6 mm from the margin of the root was 135 HV 0.01. The difference in the deformation state of the threads between those tapped in impregnated and unimoregnated is clearly shown also at lower magnification, Fig.7.11.

This shows that in both drilling and tapping, a significantly lower deformation of the base matrix occurred in impregnated material compared to unimpregnated one. Because the pores were filled with resin, it was not possible to transfer the material into the pores. By

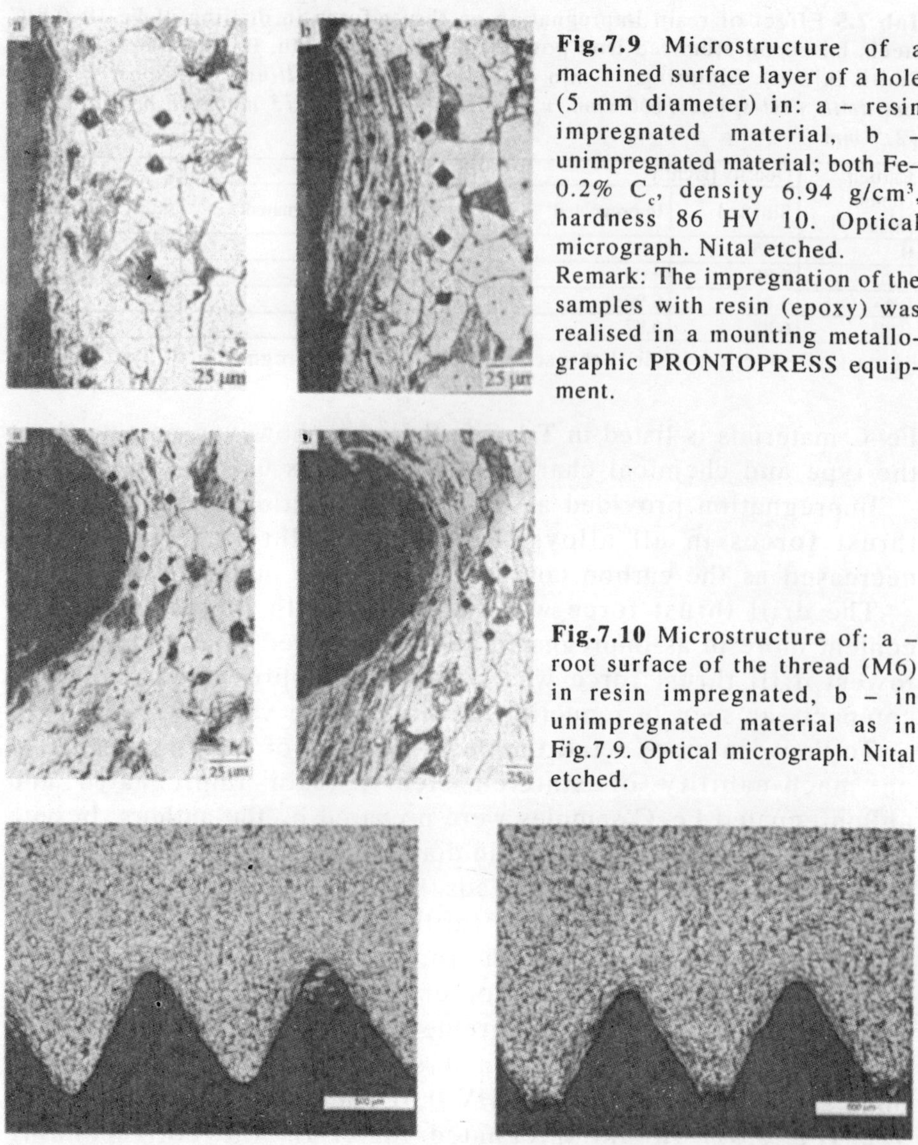

Fig.7.9 Microstructure of a machined surface layer of a hole (5 mm diameter) in: a – resin impregnated material, b – unimpregnated material; both Fe–0.2% C_c, density 6.94 g/cm^3, hardness 86 HV 10. Optical micrograph. Nital etched.
Remark: The impregnation of the samples with resin (epoxy) was realised in a mounting metallographic PRONTOPRESS equipment.

Fig.7.10 Microstructure of: a – root surface of the thread (M6) in resin impregnated, b – in unimpregnated material as in Fig.7.9. Optical micrograph. Nital etched.

Fig.7.11 M6 threads in: a – impregnated material, b – unimpregnated material as in Fig.7.9. Optical micrograph. Nital etched.

this the cutting forces could be lowered. The work-hardening was recorded to some extent (microhardness increase) also in impregnated material. Beside it is possible to presume (which was not investigated) a lubrication effect of resin, which at the high temperatures near the cutting edge becomes soft. This can be a further effect for improving the machinability of sintered steels by resin impregnation.

Resin and oil impregnation for improving the machinability of porous parts was regarded effective also in turning of sintered iron as confirmed in Fig.7.12.

Cutting resistance in turning of an impregnated PM iron part was minimal under the investigated conditions (100 N) and was not changed when varying the cutting speed up to 160 m/min. It is also lower than that for oil impregnated one. As shown a relatively high improvement in machinability in turning of sintered iron was attained also by oil impregnation.

In contrast to both oil and resin impregnated parts, only the cutting resistance of unimpregnated parts decreased with increasing cutting speed. Tool life, characterised by width of flank wear, was also extended by resin impregnation as shown in Fig.7.13. The flank wear increased with prolonged cutting time only in a small extent in contrast to the flank wear in turning of unimpregnated material. It means that with impregnated material the flank wear increases only slowly with prolonged cutting time.

7.1.6 Effect of machining aids on surface finish and tool wear

7.1.6.1 Effect of machining aids on surface finish

Surface finish is a very important criterion for machining especially in porous materials, and in some cases is used also as a

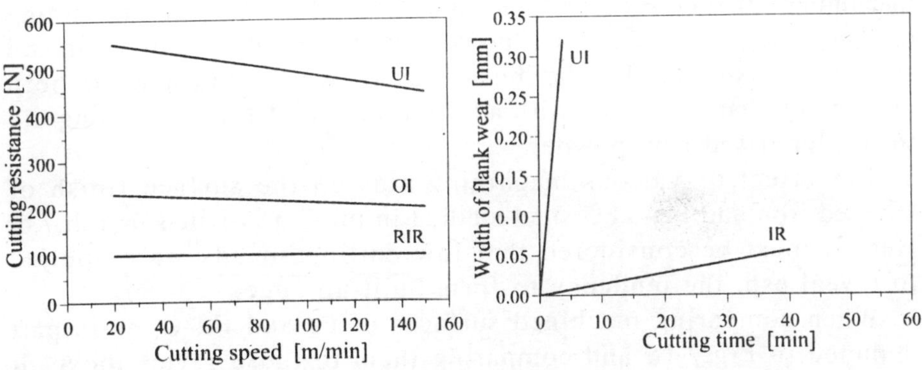

Fig.7.12 Relationship between cutting resistance and cutting speed in turning of sintered iron in dependence on oil and resin impregnation. Density 7.0 g/cm^3. UI – unimpregnated, OI – oil impregnated, RIR – resin impregnated (acc. to Ref. 265). *Cutting test: Cutting tool P20, depth of cut 1.0 mm, feed 0.1 mm/rev.*

Fig.7.13 (right) Effect of resin impregnation on tool life in turning (flank wear) of sintered iron in dependence on cutting time. Details as in Fig. 7.12. UI – unimpregnated, IR – impregnated with resinol [265].

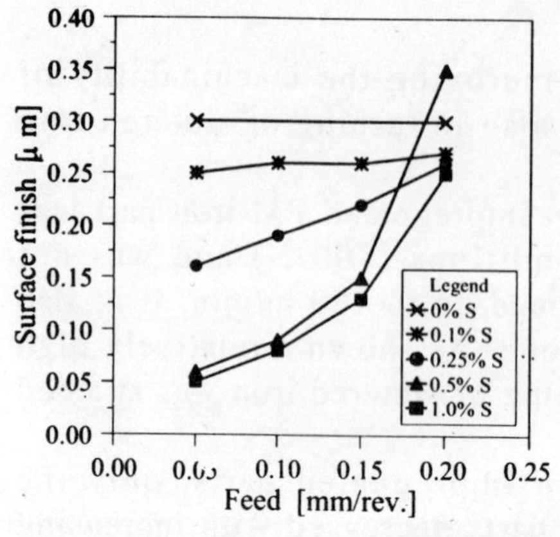

Fig.7.14 Surface finish *vs.* feed rate in turning of sintered iron for new carbide cutting edges in dependence on sulphur content added. Sponge iron powder (Ancor MH 100), density 5.8 or 6.2 g/cm^3 [107].

machinability index (see also 7.1.2.2). Surface finish is affected by all factors affecting the machinability of PM steels to varying extents including the cutting tools used. For improvement of the surface finish the sulphur effect was most closely investigated. Figure 7.14 shows the effect of sulphur addition on the surface finish in turning at various feeds of Fe–C steel.

Poor surface finish for all feed rates tested was exhibited by sintered iron without sulphur addition. The surface finish improved for increasing sulphur content of 0.1–1.0% at low feed rate up to 0.15 mm/rev., very good at 1.0% S but quite poor at 0.5% S at 0.2 mm/rev. All hydrogen containing sintering atmospheres caused some surface desulphurisation. Effect of 5% addition of a special machining aid consisting of 50% Cu and 50% Pb was tested in turning of Ancormet 101 + 2.5% Cu powder with regard to tool wear and surface finish [109]. The best results were obtained in both machining characteristics with addition of 5% of this machining aid to resulphurised iron powder.

The effect of various machining aids on the surface finish of sintered iron and Fe–0.8C steel studied in turning is listed in Tab.7.6. Here it must be considered that low cutting speeds were applied to reveal esp. the tendency to form built-up edges.

When comparing machined surfaces of Fe and Fe–C, as in part depicted in Fig.7.15 and comparing them to those given above, it stands out clearly that Pb is less effective here than both MnS and MoS_2 as shown in Fig.7.4. Fine Pb seems to be better than coarser one. In the case of Fe–C matrix, built-up edges seem to have formed more pronouncedly than in the case of plain Fe which is a somewhat unexpected result.

PbS is apparently more effective than Pb but, as in the case of

Tab.7.6 Properties and surface finish in turning of Fe and Fe–0.8C materials containing different machining aids. Bars 60×7×7 mm, compaction at 600 MPa, sintering 60 min at 1120°C in H_2; BN containing bars in $75N_2$–$25H_2$ atmosphere. Water atomised iron powder [251]. *Turning test: feed rate 0.15 mm/rev., depth of cut 0.1 mm, cutting speed approx. 16 m/min, dry cut, indexable inserts P20, uncoated*

a) Matrix Fe

Additive [mass %]	Density [g/cm^3]	Impact energy [J]	Roughness R_a [µm]	Additive [mass %]	Density [g/cm^3]	Impact energy [J]	Roughness R_a [µm]
0	7.25	>31.4	7...10	0.5 PbS	7.21	>47.6	7.2
0.3 MnS	7.22	>27.6	3.1...5.3	2.0 PbS	7.12	>31.2	2.3
0.5 MnS	7.21	>28.6	1.9...2.5	0.5 BN_f	7.21	11.5	**
1.0 MnS	7.18	18.7	1.3	0.5 BN_f*	7.29	>42.5	-
2.0 MnS	7.11	11.6	0.8	0.5 BN_c	7.04	9.2	**
2.0 MoS_2	7.06	18.3	1.5	0.5 BN_c*	7.14	>28.9	7.0
2.0 Pb	7.23	36.5	2.5	2.0 BN_c	6.85	3.9	>>

b) Matrix Fe-0.8% C

0	7.03	11.7	4.9...8.3	2 Pb	7.07	n.d.	2.1
0.5 PbS	7.06	8.9	8.0	0.5 BN_c	7.02	4.5	**
2.0 PbS	6.97	13.9	3.5	0.5 BN_c*	7.08	16.1	5.5
5.0 PbS	6.80	12.6	1.7	2.0 BN_c	6.76	3.6	5.0
2.0 graphite KS75***	6.69	1.0	4.2	2.0 graphite MFL***	6.77	4.7	4.85
2.0 graphite KS150***	6.42	1.9	3.1	-	-	-	-

Remark: *sintering at 1250 °C, **specimens too fragile, broke during machining; BN_f <1µm, BN_c – 50-180 µm, ***sintered at 1070°C

MnS, only above a given content; 0.5% PbS result in the same rugged surface as does 0.3% MnS while for 2% PbS the surface is rather smooth. For Fe–C matrix, once more the effect of the machining aid is less pronounced, smooth surfaces being obtained both for Fe and Fe–C only at rather high contents in the range of 5% PbS that are not feasible in practice.

Generally, however, compared to plain sintered iron or Fe–C material the additives used improved the surface quality with the notable exception of BN and coarse graphite grades, which, if the specimens could be turned at all, resulted in poor surface finish, apparently due to very low strength of the materials. (Here it must however be kept in mind that the BN contents commonly used for improving the machinability are significantly lower, usually <0.1%). BN remained stable during sintering and inhibited formation of stable sintering interparticle contacts resulting in lower impact strength. Fine BN also inhibits the dissolution of carbon in the iron matrix.

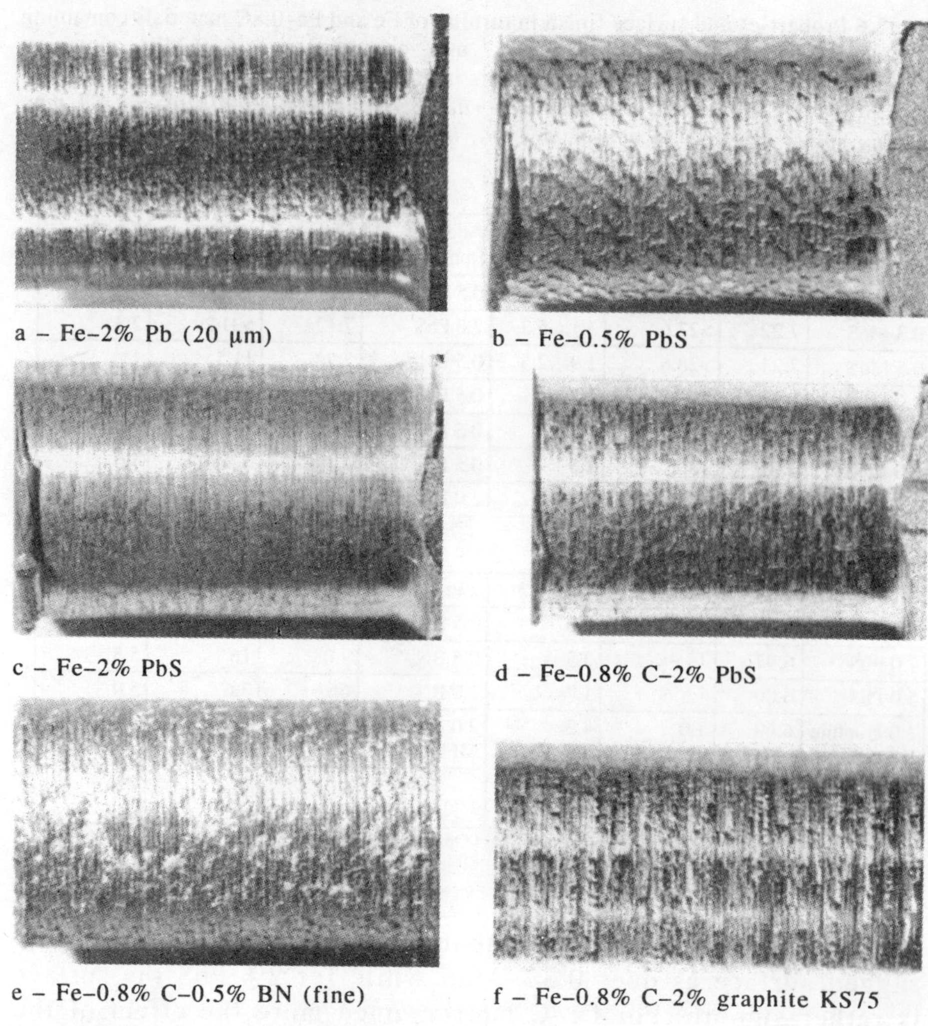

a – Fe–2% Pb (20 μm)

b – Fe–0.5% PbS

c – Fe–2% PbS

d – Fe–0.8% C–2% PbS

e – Fe–0.8% C–0.5% BN (fine)

f – Fe–0.8% C–2% graphite KS75

Fig.7.15 Machined surfaces of sintered iron and steel containing different machining aids [207]. (Courtesy of MIBA Vorchdorf). *Finish turning, hardmetal insert P20,* v_c *= 15 m/min, feed 0.15 mm/rev.*

In the case of materials containing free graphite, coarse graphite was added to the base mix of plain iron powder and fine natural graphite, and sintering was done at a temperature at which the fine graphite was readily dissolved but the coarse grades did not react. One moderately coarse grade (artificial grade KS75) and two coarse grades (natural grade MFL and artificial grade KS150) were used. A typical machined surface is shown in Fig.7.15f; the rather poor surface finish is evident, once more indicating that the real 'solid lubricants' as graphite or BN are less effective for improving the machinability than are MnS or $Fe_{1-x}S$.

In general, improvement in machinability was achieved at the expense of mechanical strength, as clearly visible from the materials containing MnS. Addition of 2% MoS_2 was comparable to 1% MnS, both with respect to surface finish and impact energy, however, it must be remined once more that in the machined samples there was no more MoS_2 present but rather sulphide $Fe_{1-x}S$ formed during cooling.

The impact tests showed that the inclusions incorporated into the materials by additives lower the mechanical strength to varying degrees. The amount of sulphur added is an important parameter and has to be considered when comparing the additives, since *e.g.* MoS_2 contains considerably more sulphur than, *e.g.* PbS [242].

According to Ref. 107, the best surface finish in turning exhibited the samples with 0.85 and 1.1% C (see Fig.5.18). Surface finish was not affected by feed in the range of 0.05 to 0.20 mm/rev.

As shown the machinability of sintered iron and of Fe–C materials, with the surface roughness being taken as a criterion for machinability, can be significantly improved by adding compounds such as sulphides or lead which are admixed as fine powders.

7.1.6.2 Effect of carbon, machining aids, and machining conditions on tool wear

The final criterion for a combined effect of all material and machining factors is tool wear and by this tool life time. In Refs. 107 and 152 it is shown that tool wear increases rapidly with the combined carbon content also in presence of sulphur, and the surface finish is consistently poor probably due to the relatively low density of the tested specimens. The effect of carbon content and machining time on tool wear in turning of Fe–C material is shown in Fig.7.16.

As shown the tool wear increased with increasing carbon content up to 0.85%, more at carbon contents above 0.5%. Tool wear increased with prolonged turning time and more also with increasing carbon content. A small decrease in tool wear was observed at 1.1% C. The lowest tool wear was reached in turning plain iron, without carbon.

Tool wear was lowered by addition of sulphur, *e.g.* with increasing sulphur content as shown in Fig.7.17. Tool wear decreased with increasing sulphur content independently of the carbon content. As shown, the lowest tool wear was reached for materials containing 0.5 and, especially, 1.0% S. To reach this

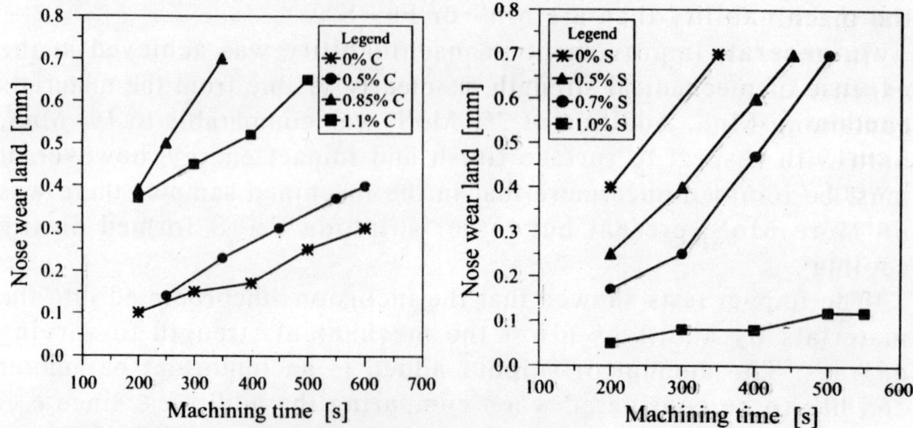

Fig.7.16 Tool wear (nose wear land) in turning of Fe–C material *vs.* carbon content and machining time (iron powder MH 100, density 6.5 g/cm^3) [107,152].
Fig.7.17 (right) Tool wear (nose wear land) *vs.* machining (turning) time for Fe–0.85C steel in dependence on sulphur content. Sponge iron powder (Ancor MH 100, density 5.8 or 6.2 g/cm^3) [107,152].

sulphur content at the surface of a part probably requires a practically hydrogen free sintering atmosphere to avoid the desulphurisation reaction that generates H_2S.

Tool wear in turning was markedly affected by sulphur additions, mainly at prolonged machining time. Minimum wear was attained with 0.5% S addition, independently of machining time. The effect of sulphur addition increased at long time machining. A carbon content of 0.85% caused a substantial increase of nose wear, except in the case of 1% S addition.

Sintered plain iron would be the best material with regard to abrasive tool wear, however, excessive galling occurs in drilling, tapping, etc., and the tool must be changed for this reason as a demonstration of the adhesive wear causing the galling of the tool. For higher strength structural material, the iron–carbon–sulphur mixes are, therefore, used. The best results in actual production were obtained for 0.5% sulphur and a combined carbon content of 0.35 to 0.5%.

7.1.7 Effect of machining aids on chip formation

In general, machining chips provide information on how the additives enhance the machinability. The chips with additives are much smoother than the baseline Fe–0.8C (300M, atomised iron powder, 0.02% S) chips. The 300M chips have many built-up edge (BUE) fragments deposited on them, which is rather surprising for this high carbon level. Built-up edge refers to the adhesion of a layer of work material on the cutting tool. BUE can cause also poor surface finish. The reduction of BUE fragments indicates that the additives reduce the chip-cutting tool adhesion. The additives increase chip curl and decrease chip thickness. Thinner chips indicate that they were formed at a higher shear angle. Different agents change the chip shape in different manner. Molybdenum disulphide, particularly, produced very small chips that are easily removed from the cutting area.

The size and the shape of the chips formed in machining depend on the workpiece properties and on cutting method and cutting conditions. In comparison to the chips formed in machining of wrought steels, the chips formed in machining of PM steels are in all cases short to very short and not uniform in shape, Fig.7.18. The chips formed in drilling of a soft steel are relatively thicker and more plain. The chips formed in drilling of a harder steel, Fig.7.18b, are thinner and screwy. The finest and more elongated are the chips formed in drilling of harder steel, Fig.7.18c. The chips formed in milling are relatively coarser without a predominant geometrical form, *i.e.* the size and shape of the chips is not uniform also in one machining operation.

7.1.8 Summary

- The effect of the tested machining aids on the compressibility of iron powders is minor or negligible.
- Manganese in small amounts (~0.5%) improved machinability also without sulphur addition. The optimum combination for machinability tested in drilling was attained with Fe–0.5% Mn–0.6% C steel.
- Less shrinkage was observed mainly with MnS addition independent on carbon content and compacting pressure (density). Sulphur and molybdenum disulphide cause swelling with increasing carbon content, apparently as a consequence of the liquid sulphide eutectic generated during sintering.
- Effect of sulphur, manganese sulphide and molybdenum disulphide on mechanical properties is not uniform and also depends on iron powder grade and density of the parts.
- The hardness of parts was increased by addition of molybdenum disulphide and sulphur compared to manganese sulphide addition. Transverse rupture strength of parts was increased with S, MnS and MoS_2 addition. For MoS_2 this can be explained by the matrix strengthening caused by dissolved Mo. Also in the pres-

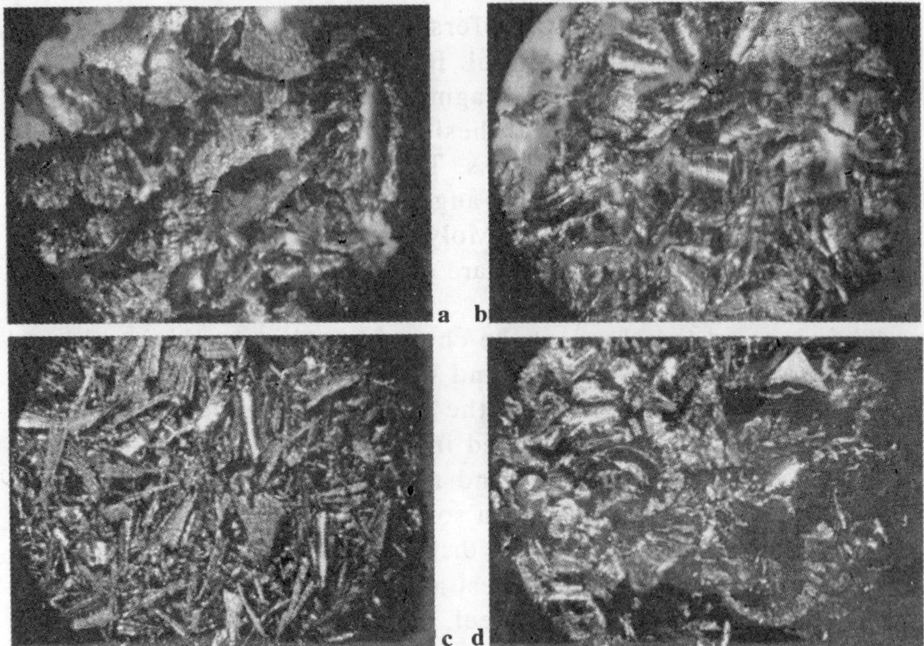

Fig.7.18 Typical size and shape of chips formed in machining of some PM steels. a – Fe–0.5Mo–0.1C_c, 65 HV 10; b – Fe–0.6C_c, 110 HV 10; c – Distaloy SE–0.5C_c, 202 HV 10; d – Fe–3Cr–1Mn–0.25C_c, 145 HV 10. a, b, c – drilling under identical conditions (v_c = 8 m/min), d – milling chips for chemical analysis, ×7.

ence of carbon the reactions between machining aids and matrix can occur.

- All machining aids (except BN and free graphite, at last in % concentrations) significantly improved the machinability of Fe and Fe–C materials. MnS additions improved the machinability of iron–carbon parts for all carbon contents in the range 0.3 to 0.9%. The optimum effect was attained at ~0.5% C.
- Minimal drill thrust force was achieved with 0.5% C and 0.5% MnS. Thrust force increased with higher carbon content. The reduction of thrust force due to MnS addition drops at carbon contents higher than 0.5%. Harder, stronger materials as shown need higher amounts of machining aids. The amount should be determined approximately in relation, *e.g.*, to the hardness of the material.
- In general, 0.5% addition of a machining aid seems to be a minimum considering some deviations in tested material properties and cutting conditions, and only in some cases is higher than 1%. The effect of some machining aid and amount should be always related to the required mechanical and other properties

of the parts.

- Better machinability was attained with steel based on sulphur prealloyed powders compared to admixed sulphur. The effect of sulphur prealloying on increase of mechanical properties and hardness of that steel was recorded.
- The effect of porosity on machinability of Fe–C parts without and with manganese containing aids was not uniform.
- The surface finish was affected by machining aids. Surface finish was significantly improved by sulphides. Also decomposed sulphides have a positive effect on the surface finish, indicating that $Fe_{1-x}S$ formed after sintering is also effective and not only the solid lubricant MoS_2. When surface finish was poor at low feed rates, it generally remained constant as the feed rate increased.
- Solid lubricants with a layered crystallographic structure such as graphite and hBN seem to be less effective machining aids than MnS or $Fe_{1-x}S$, resulting in poor surface finish if added in the % range. This seems to be at least partly linked to the adverse effect of these additives on the mechanical properties.
- Effect of iron powder grade on machinability was observed. More improvement in machinability was observed with S, MnS, and MoS_2 additions for atomised iron powder compared to sponge iron powder, but the effect was not uniform. The alloy based on sponge iron powder with 0.5% MnS showed a greater improvement in machinability for the slower speed and feed rates. The speed and feed rates tested were in general higher than those used in practice.
- The lubrication effect of a machining aid is higher at lower cutting speed and feed due to lower temperature in the cutting zone and due to relatively prolonged contact time with the cutting edge of the tool. Machining aids seem to be more effective against adhesive wear and built-up edges than against abrasive wear.
- Tool wear is a result of a combined effect of all material and processing factors taking part in a machining process. Tool wear increased with higher combined carbon content and decreased with increasing sulphur content. The best results in actual production were obtained for 0.5% sulphur and a combined carbon level of 0.35 to 0.5%.
- Resin and oil impregnation reduced markedly the thrust force in drilling and minimised the tool wear in turning, although the detrimental effect of higher carbon content is discernible also with impregnated materials. The better machinability of an impreg-

nated material could be related to lower deformation of the surface layers in drilled and tapped holes.

7.2 IRON–COPPER–CARBON STEEL

Copper containing sintered steels are very often used for production of sintered parts. The alloying with copper results in a strengthening effect but also in improvement of the machinability. The data concerning the effect of machining aids on mechanical and on other relevant properties and finally on machinability of iron–copper–(carbon) steel are also the basis for a more detailed knowledge about machinability of high strength steels. Figure 7.19 shows the characteristic ferritic–pearlitic microstructure of a Fe–2Cu–0.6C steel.

7.2.1 Effect of copper, manganese, and carbon

The effect of separate base alloying elements in PM steels, such as copper, manganese and carbon in adequate proportions, shows their mutual interaction affecting the machinability of Fe–Cu–C steels, Fig.7.20.

On the basis of pevious data concerning the effect of the mentioned single elements on the machinability of sintered iron it is possible to expect their effect also on improvement the machinability of copper alloy steels at adequate alloying.

As shown in Fig.7.21, a significant increase in the machinability of Fe–0.5C steel, measured as tool wear, can be reached with a copper addition above 5%. Parts requiring 0.85% C can be made machinable through the addition of 10% Cu in spite of very high sintered strength and hardness of the material (free copper).

As the copper amount increased, the composition of the material approached that of free machining steels. Copper in amounts of 2% caused a decrease in thrust force compared to 1% Cu which with

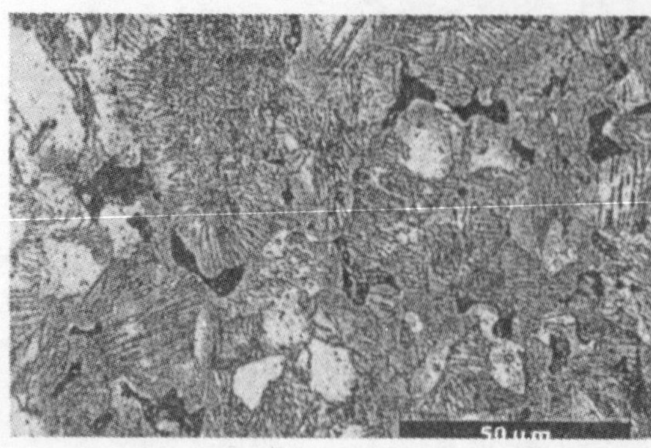

Fig.7.19 Microstructure of Fe–2% Cu–0.6% C steel. SC100.26 iron powder, sintering 30 min at 1120°C. Optical micrograph. Nital etched [235].

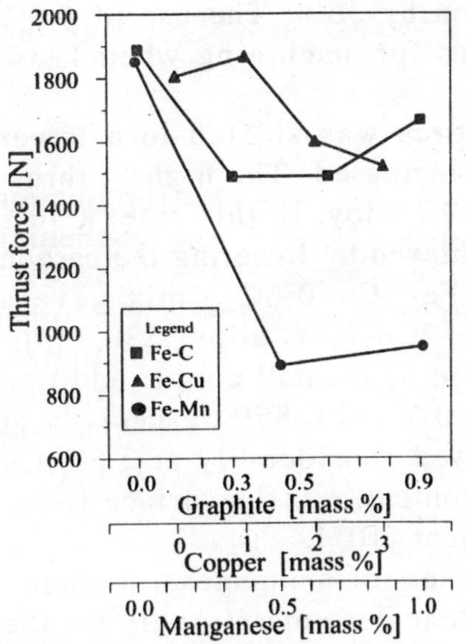

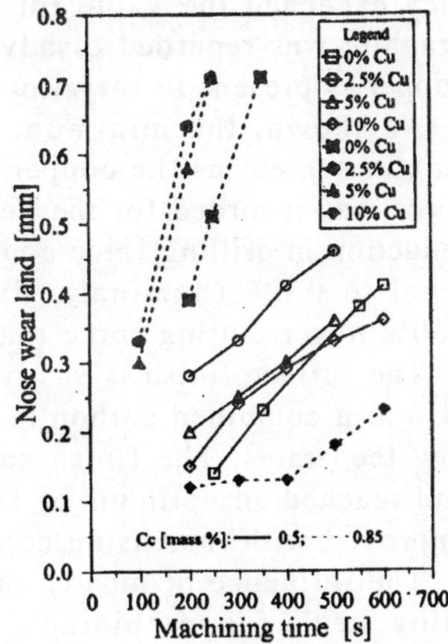

Fig.7.20 Effect of separate graphite, copper and manganese content on thrust force value in drilling for sintered iron (Atomet 28). Density 6.7 g/cm^3, sintering 30 min at 1120°C, endothermic gas (acc. to Ref. 108,230). *Drilling test: HSS 8 mm drill, drill speed 840 rpm, v_c = 21.1 m/min, feed 0.10 mm/rev., coolant – mineral oil.*

Fig.7.21 (right) Tool wear (nose wear land) in turning of sintered Fe–(0.5, 0.85)C steels without and with addition of copper vs. machining time (acc. to Ref. 152).

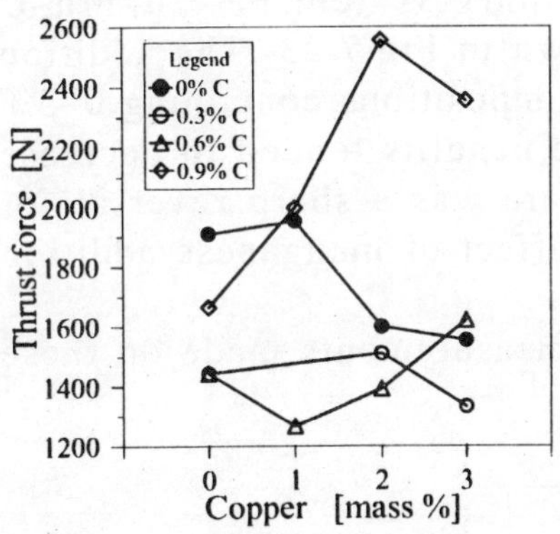

Fig.7.22 Influence of copper and carbon contents on thrust force values in drilling for Fe–Cu–C (graphite) mixes (acc. to Ref. 230).

further increasing copper content was changed only to a small extent. The combined effect of carbon and copper content on thrust force is shown in Fig.7.22, which shows in greater detail their contribution to the machinability of Fe–Cu–C mixes.

The lowest forces for Fe–Cu–C compositions were obtained with samples made from mixes containing 0.3–0.6% C_{nom} (graphite). With 0.9% graphite addition, the resulting samples showed substantially higher thrust forces, such that for 0.9% C plus 2% Cu they exceeded the value for plain iron by 36%. The use of 0.9% graphite was regarded disadvantageous for machining when 1–3% copper is present in the alloy [243].

As shown, the minimum thrust force was shifted to a lower carbon content as the copper content increased. The highest thrust force was recorded for the Fe–2Cu–0.9C alloy. In this case, a 40% reduction in drilling force could be achieved by lowering the carbon level to 0.6% (nominal). Even the Fe–3Cu–0.6C_{nom} mix gave a ~30% lower cutting force than the Fe–2Cu–0.9C alloy [230,233].

The surface finish is vastly improved by a small copper addition also at a combined carbon content of 0.5 and 0.85%, especially at low feed rates. The finish has improved considerably at 2.5% Cu and reached an optimum at 10% Cu content, *i.e.* the surface finish improved with increasing copper content [107].

The influence of mostly small amounts of manganese in copper alloy steels for machining enhancement depends strongly on the levels of other constituents. The addition of 0.5% Mn seems to be beneficial for all compositions containing 3% Cu and up to 0.6% C_{nom}. The effect of manganese is harmful at all copper levels for the 0.9% graphite mixes due to sinter hardening. The most unsatisfactory composition from the viewpoint of drilling force is the combination of Fe–2Cu–0.9C–0.5Mn [149].

The enhancing effect of 0.5% manganese addition on the change of the thrust force values for quaternary system Fe–Cu–Mn–C mixed steels was also proved, as shown in Fig.7.23. The addition of Mn was seen to be beneficial for compositions containing 0–3% Cu and 0–0.6% graphite, although the benefits tended to decrease as the graphite content increased. There was a sharp reversal for the 0.9% graphite content where the effect of manganese addition was detrimental at all copper levels.

According to Ref.10, thrust force measurements made on these

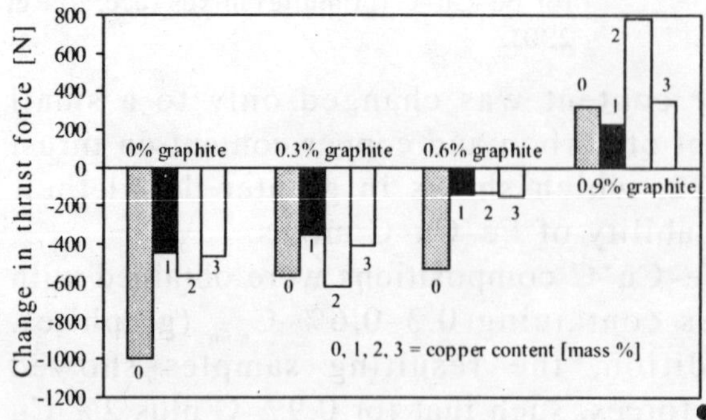

Fig.7.23 Influence of 0.5% manganese addition on changes in thrust force values in drilling for Fe–(0–3)Cu–(0–0.9)C mixed steels. Iron powder ATOMET 28 [230].

test-pieces containing various combinations of carbon, copper, and manganese confirmed the previous data. In binary iron–carbon, iron–copper, and iron–manganese systems, the drilling force was lowered by 21% for graphite additions of 0.3–0.6%, 13% for copper additions of 2–3%, and about by 50% for manganese additions of 0.5–1% compared to plain iron.

The addition of 0.5% Mn caused the reduction in thrust force by 40% compared to 0.5% graphite or ~2% Cu addition. The presented results limit the addition of the above mentioned alloying elements in quaternary systems in combination of up to 0.6% C, 0.5–1.0% Mn and 2–3% Cu for optimum machinability of Fe–Cu–C systems.

The mentioned additives, admixed in correct amounts, improve machinability without significantly changing the desired material properties. As follows from these data, the optimum amount of each of these elements should be added in binary systems but the influence of copper and carbon additions, especially on machinability, becomes more complex when these elements are added in combination. The influence of copper ranges from being beneficial in plain iron, Fig.7.20 and in low carbon compositions to being partially detrimental for 0.6% and 0.9% graphite additions, Fig.7.22. It follows from this that, not surprisingly, the relationships are more complicated for ternary and even more for quaternary systems.

Copper infiltrated steels are a special case in machining. Normally, copper infiltrating is used for bearings, with the copper content varying up to as high as 20%. In these steels it is possible to obtain the best machining conditions which can be defined as low tool wear, increased production rate, and better surface finish. It should be pointed out that the majority of machine parts manufactured from copper alloy steels do not fall into the category of copper infiltrated steel, but they have a composition in the range of 2.0 to 5.0% copper with carbon [152].

Phosphorus in copper- containing steels exerts a deteriorating effect on machinability. Extremely high tool wear in turning occurred in the presence of phosphorus (0.5% P) and further with increasing carbon content [107,152].

7.2.2 Effect of S, MnS and MoS_2

7.2.2.1 Effect of S, MnS and MoS_2 on mechanical properties of Fe–Cu–C steels

The effect of these machining aids on the dimensional change and transverse rupture strength of Fe–1.75Cu–0.8C (FC-0208) alloy steel based on two iron powder grades is listed in Tab.7.7.

These machining aids increased the expansion of FC-0208 samples, more for components based on atomised base powder compacts. The effect of sulphur is more pronounced than that of manganese sulphide or molybdenum disulphide at the same mass%, indicating that sulphur is the active aid in these systems. If the copper content is increased up to 10%, the sulphur addition will cause a large dimensional change after sintering [107]. In this case, the results somewhat contradict those obtained with Fe–C, especially regarding the sponge iron powder; also here, however, MnS exerts a significantly more adverse effect than S.

All three free-machining aids significantly reduced the transverse rupture strength of FC-0208 composition (up to 23% for atomised iron powder), except for MoS_2 in sponge iron powder base compacts. These results confirmed the possible effect of machining aids on strength. In this case it was an opposite effect as in Fe–C materials for which the hardness and TRS tended to improve after adding machining aids. It is therefore not possible to generalise individual test results without considering the alloy system.

The effect of various manganese sulphide contents on tensile strength of Fe–2Cu–(0.5, 0.8)C (FC–0205, FC-0208) steels of different porosity is shown in Fig.7.24.

As shown, the manganese sulphide addition caused slightly lower (<5%) tensile strength values in FC-0205 steel without significant change in relation to density. The tensile strength in both base materials was affected by the MnS addition when the density

Tab.7.7 Effect of S, MnS and MoS_2 additions on dimensional change (+Δ(l/l) and on transverse rupture strength (TRS) of sintered FC-0208 steel based on sponge (Pyron D-63) and atomised (Ancorsteel 1000) iron powder. Sintering 30 min at 1120°C, $90N_2$–$10H_2$ atmosphere [136]

Addition [mass %]	Iron powder (density)											
	Atomised (6.8 g/cm³)						Sponge (6.1 g/cm³)					
	Machining aid											
	S		MnS		MoS_2		S		MnS		MoS_2	
	Δl/l [%]	TRS [MPa]	Δl/l [%]	TRS [MPa]	Δl/l [%]	TRS [MPa]	Δl/l [%]	TRS [MPa]	Δl/l [%]	TRS [MPa]	Δl/l [%]	TRS [MPa]
0.00	0.20	924	0.20	931	0.24	938	0.27	572	0.31	581	0..31	565
0.25	0.87	786	0.24	897	0.70	807	0.55	552	0.35	542	0.54	569
0.50	0.87	752	0.31	862	0.68	772	0.56	538	0.36	555	0.44	593
0.75	0.80	731	0.34	841	0.71	717	0.55	554	0.38	523	0.44	572
1.00	0.90	731	0.35	834	0.75	717	0.56	551	0.38	509	0.41	593

exceeded 6.95–7.0 g/cm³. The decrease in the strength properties by increasing the MnS addition in steels of lower porosity was caused probably by the presence of this addition in the form of non-metallic inclusions in fracture (load bearing cross-section of the specimen in the tensile test) acting as initiators of dimple facets.

The data show that the mechanical properties of the copper steels prepared under these conditions were reduced by machining aids in comparison to Fe–C steel which shows the sensitivity of microstructure to all alloying and processing factors. The effect was more pronounced at a higher density level above 7.0 g/cm³. The TRS of the FC-0205-40 alloy also dropped by about 5% with MnS addition. No significant decrease in ductility or impact energy was found. The addition of 0.5% MnS caused a decrease in hardness by ~13 HRB.

7.2.2.2 Effect of MnS and MoS_2 on machinability of Fe–Cu–C steels

The copper containing steels without some machining aid show in machining an earlier drill failure, mostly due to an excessive wear of the flank. As listed in Tab.7.8, drill testing showed that the free-machining aids significantly improved the machinability of the FC-0208 steel, which effect also depended on the iron powder grade used and the density.

In general, the machinability of copper-containing steels with 0.8% C is regarded as very poor, regardless of density, as demonstrated by these results. In this case, the addition of sulphur produced the largest improvement in both sponge and atomised iron powder compositions. Molybdenum disulphide, although less effective

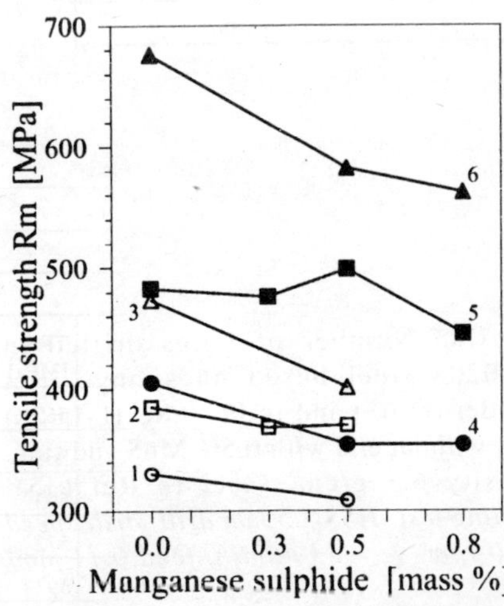

Fig.7.24 Effect of manganese sulphide content on tensile strength of FC-0205, FC-0208 steels. Sponge iron powder (D-63). 1 – FC-0205-35, 6.3 g/cm³, 57 HRB; 2 – FC-0205-40, 6.65 g/cm³, 74–61 HRB; 3 – FC-0205-45, 7.12–6.95 g/cm³, 62–74 HRB; 4 – FC-0208-40, 6.3 g/cm³, 69 HRB; 5 – FC-0208-50, 6.7 g/cm³, 79–76 HRB; 6 – FC-0208-60, 7.15 g/cm³, 88 HRB [104].

than sulphur, improved the drill life. In this case, manganese sulphide produced the smallest improvement in the drill life. The effect of the iron powder grade and, in this specific case, also of the density of the materials was manifested in machining and must be considered. It is the different effect of MnS addition on machinability of Fe–Cu–C compared to the Fe and Fe–C materials, as shown in Fig.7.25.

MnS additions enhanced the machinability of Fe–Cu–C materials, increasing the number of holes drilled by a factor of ~5. A significant reduction in thrust force in drilling by 37% and in torque by 30% was obtained by admixing of 0.2% S in form of MnS and 0.5% S in form of (FeMn)S to Fe–1.5Cu–0.6C (graphite) material (base atomised iron powder, 6.8 g/cm^3, 63 HRB). A further improvement can be achieved when the material contains fine homogeneously distributed (FeMn)S particles as a result of diffusion alloying with sulphur [222]. Using 0.5% high purity manganese sulphide in general increased by ~10% the number of holes drilled before drill failure compared to low purity manganese sulphide [241]. Sulphur addition contributes to decrease of the tool wear to

Tab.7.8 Effect of 0.5% machining aid addition (sulphur, manganese sulphide and molybdenum disulphide) on machinability (number of holes drilled) in drilling the Fe–2Cu–0.8C steel compacts based on sponge (Ancor MH 100) and atomised (Ancorsteel 1000) iron powder [136,137,242]. *Drilling test: HSS 3.5 mm drill diameter, drill speed 3000 rpm, v_c = 33 m/min, feed 0.08 mm/rev., depth of hole 12.4 mm*

Iron powder (density)	Machining aid			
	None	S	MnS	MoS_2
	Number of holes			
Sponge (6.1 g/cm^3)	2	608	81	108
Atomised (6.8 g/cm^3)	2	668	72	249

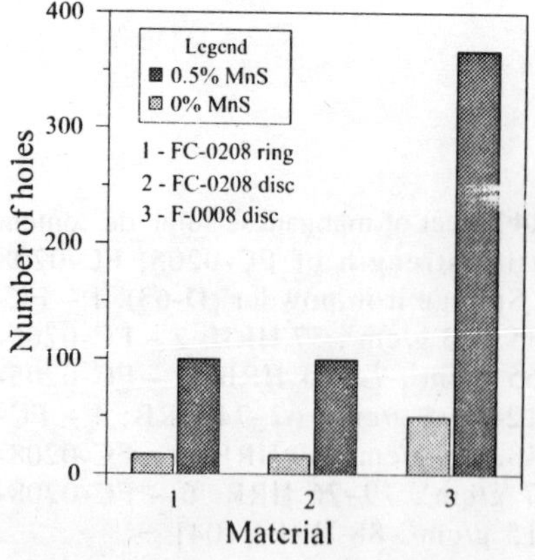

Fig.7.25 Number of holes drilled in FC-0208 steel based on sponge iron powder (D-63) and in Fe–0.8C (F-0008) steel without and with 0.5% MnS addition. Density 6.3 g/cm^3 (acc. to Ref.135). *Drilling test: HSS 3.5 mm drill, drill speed 3000 rpm, v_c = 33 m/min, feed 0.11 mm/rev.*

a greater extent in Fe–10% Cu compared to Fe–2.5% Cu alloy [107,241].

7.2.2.3 Effect of MnS and MnX

By combining various machining aids it is possible to expect further improvement in machinability in comparison to the case when a machining aid is used on its own. As an example, Tab.7.9 shows the combined and individual effects of MnS and MnX additions on the machinability of Fe–2Cu–0.8C (FC-0208) and of Fe–2Cu–0.5C (FC-0205) steel.

As in the previous case, very poor or almost negligible machinability was determined for FC-0208 steel without machining aid additions, with slightly better machinability obtained for FC-0205 alloys. When examining the effect of both aids on machinability, examination showed that there are also effects of the sintering atmosphere affecting the properties of the materials. For FC-0208 alloy, MnX used as a machining free-machining aid both on its own and in combination with up to 0.5% MnS improved machinability. In particular, the effect of MnS or MnX on the machinability of these materials is very hard to identify due to other factors affecting the final state. For those premixes based on sponge iron powder, the combination of MnS and MnX produced the longest drill life. The highest machinability of the investigated steels was obtained with the total (0.35 and 0.5)% (MnX + MnS) addition. As shown, the addition of 0.35% MnS or MnX on its own was

Tab.7.9 Combined effect of MnS and MnX additions on machinability (number of holes drilled) for FC-0208 and FC-0205 steels based on atomised (Ancorsteel 1000) and sponge (Ancor MH100) iron powder grades. Sintering: (1) – endothermic atmosphere, (2) – H_2–N_2 atmosphere. Density 6.8 g/cm³ [137]. *Drilling test: HSS 3.2 mm drill, drill speed 3000 rpm (v_c = 30 m/min), feed 0.15 mm/rev., depth of hole 12.7 mm*

Alloy	Iron powder grade	MnS/MnX addition [mass %]						
		0	0/0.35	0/0.50	0.10/0.25	0.15/0.35	0.35/0	0.50/0
		Number of holes						
FC-0208	Atomised (1)	2	220	186	91	134	107	312
	Atomised (2)	2	149	428	153	99	89	49
	Sponge (1)	2	23	55	108	157	64	42
	Sponge (2)	4	30	270	100	584	122	48
FC-0205	Atomised (1)	9	250	275	1400	473	356	498
	Atomised (2)	8	380	816	1400	1251	331	352
	Sponge (1)	9	62	120	111	416	54	152
	Sponge (2)	26	109	172	481	991	136	497

insufficiently effective. For FC-0205 alloys, better machinability was obtained with all combinations of the aids in comparison to FC-0208 alloy. The longest drill life was obtained with the premixes based on atomised iron powder with the MnS and MnX combination sintered in H_2–N_2 atmosphere rather than in the endothermic atmosphere.

Marked differences in machinability were observed for FC-0208 steels when using atomised iron powder as compared to sponge iron powder in combination with the carbon content. For FC-0208 steel made with atomised iron powder, a 0.5% MnS addition produced best drill life when sintered in an endothermic atmosphere. However, when sintered in $75H_2$–$25N_2$, 0.5% MnX produced best performance. For FC-0208 steels made with sponge iron powder, the combination of MnS and MnX produced the longest drill life under both sintering atmospheres [136]. The influence of the iron powder grade (sponge or atomised) together with that of the machining aid and the sintering atmosphere should be investigated further to find the physical reason.

7.2.3 Effect of sulphur prealloying

A substantial improvement in the machinability of Fe–2Cu–C steel was obtained by the use of sulphur prealloyed iron powder in combination with high graphite additions as listed in Tab.7.10. In this case, the effect of graphite addition is more visible in comparison to the data listed in Tab.7.3.

A marked increase in number of holes drilled was reached in sulphur prealloyed copper steels with high carbon content compared to those without sulphur addition. This shows perhaps a dominant role in machining of sulphur prealloying in high carbon copper alloy steels.

To reach optimum machinability enhancement by sulphur prealloying of a base iron powder, an adequate Mn/S ratio must be ensured. Tests were performed in drilling of samples (bushings) prepared from the sulphur prealloyed iron powders to which 2% of copper and 0.8% of graphite was added, Tab.7.11.

According to the measurement of axial drilling force for drilling one hole (3.2 mm in diameter and 9.5 mm in height), all alloys exhibited a relatively good machinability without larger differences between individual alloys (axial force 440–520 N) without any change up to 150 holes drilled when the test was finished. Larger Mn/S ratios improved the machinability of all Fe–S–Mn prealloyed materials. The results obtained by detailed microanalysis of built-

Tab.7.10 Effect of sulphur modification (prealloying) upon number of holes drilled for Fe–2Cu–(0.8–1.2)C steel. Density 6.85 g/cm³, N_2 –atmosphere [136]. *Drilling test: HSS 1 mm drill, drill speed 10000 rpm, v_c = 31 m/min, feed 0.012 mm/rev.*

Sulphur	Graphite [mass %]		
	0.8	1.0	1.2
	Number of holes		
Standard (no S)	<30	<30	<30
Prealloyed (modified)	>600	>600	350

Tab.7.11 Chemical composition of sulphur prealloyed powders (Domfer) in relation to the Mn/S ratio. Green density of the compacts 6.8 g/cm³, sintering 30 min at 1120°C, dissociated ammonia [271]

Powder grade	Composition [mass %]				
	Mn	S	Mn/S ratio	C	O
MP 37 HD 1	0.55	0.24	2.31	0.003	0.29
MP 37 HD 2	0.73	0.28	2.59	0.002	0.40
MP 37 HD 3	0.71	0.23	3.10	0.004	0.34
XP 73	0.40	0.20	2.00	0.013	0.20
MP 37 HDS	0.54	0.31	1.76	0.014	0.17

up edges on the drills and the chips showed that the oxidation of some amount of manganese in the material was beneficial to machinability.

Powders with relatively high Mn/S ratios had a lower compressibility. The composition 0.55% Mn, 0.24% S with Mn/S ratio of 2.31 appeared to be optimum for machinability of this alloy.

7.2.4 Effect of MnS and cutting conditions

The effect of a machining aid addition on machinability of some PM steels is closely related to the cutting conditions, in the case of drilling with drilling (cutting) speed and feed rate.

7.2.4.1 Effect of drill speed

The effect of drill speed on machinability of Fe–2Cu–0.8C (FC-0208) alloys with and without MnS addition is shown in Fig.7.26. As shown the number of holes drilled in the steel without MnS addition was small and was not affected by the drill speed. On the other hand, a marked increase in machinability was achieved with 0.5% MnS addition at 1000 rpm. With increasing drill speed to 3000 rpm, the effect of MnS addition decreased linearly to the level for the material without MnS addition. It can be assumed that due to increase of the temperature in the cutting zone at higher drilling speed the lubrication effect of MnS was chemically deteriorated.

The effect of drill speed on the machinability of FC-0208 and F-0008 alloy is shown in Fig.7.27. The difference in machinability between the Fe–C and Fe–Cu–C materials with MnS addition at low drill speed, in this case 1000 rpm, is negligible up to none. With MnS addition there is about an eight fold increase in the number of holes drilled in blanks FC-0208 when drilling at lower speed (1000 rpm). It was generally easier to drill a material at lower drill speeds, as shown before.

7.2.4.2 Effect of feed rate

The effect of feed rate on the machinability of FC-0208 steel is shown in Fig.7.28. Machinability improved with decreasing feed rate as was the case with the drill speed. Addition of 0.5% MnS shifted the machinability index vs. feed rate to higher values with respect to the curve for material without MnS. The increase in machinability with addition of MnS in relation to the machinability without MnS was constant, independent of the feed rate. At relatively high feed rate, *e.g.* 7.9 mm/s, it would be possible to increase the machinability from about 10 holes to almost ~230 holes when MnS was

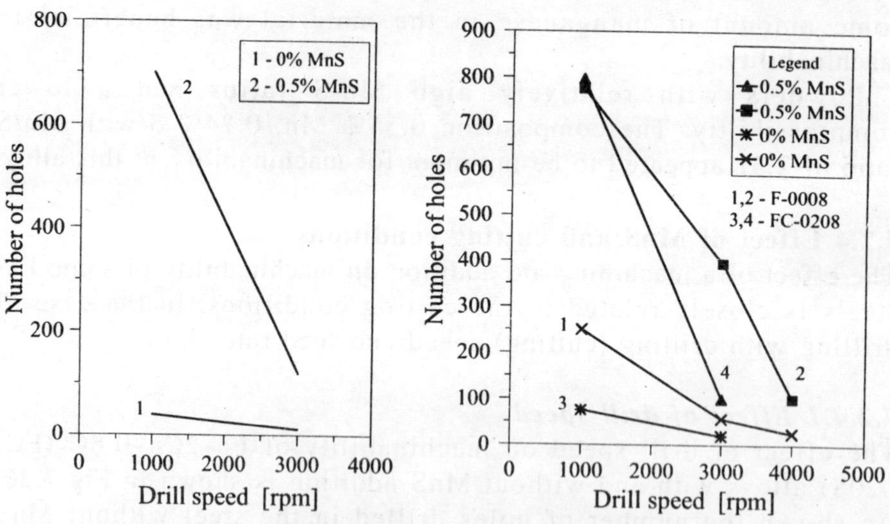

Fig.7.26 Relationship between the number of holes drilled in FC-0208 steel without and with 0.5% MnS addition (sponge iron powder) and drill speed. Density 7.0–7.1 g/cm^3 [224]. *Drilling test: HSS 3.5 mm drill, feed rate 7.9 mm/s, depth of hole 11.4 mm.*

Fig.7.27 (right) Number of holes drilled in FC-0208 and F-0008 materials without and with 0.5% MnS addition at various drill speeds. Density 6.4 g/cm^3, 0.75% C_c, hardness 56 to 60 HRB [147]. *Drill test: HSS 3.5 mm drill, feed rate 3.9 mm/s, depth of hole 25.4 mm.*

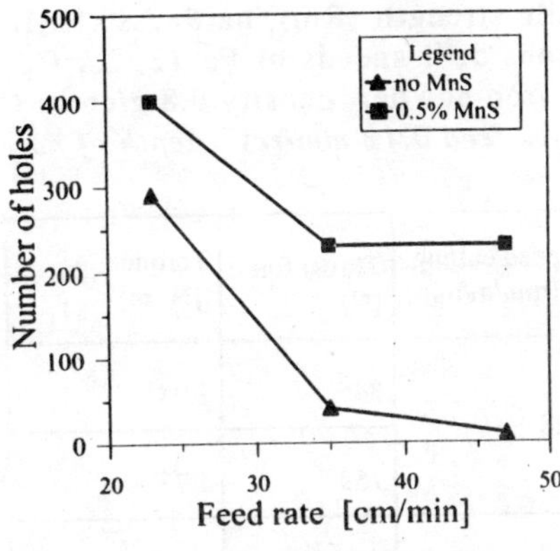

Fig.7.28 Effect of feed rate on the machinability (number of holes drilled) of FC-0208 compacts without and with 0.5% high-purity MnS addition [105]. *Drilling test: HSS 15 mm drill, drill speeds 2000, 3000, and 4000 rpm.*

added to the material.

The previous results are confirmed also by data listed in Tab.7.12 which show differences in the number of holes obtained under the same drilling conditions. By MnS addition a machinability increase in drilling of Fe–2Cu–0.8C material by about 5 times at higher cutting speed and by more than 10 times at lower drill speed was recorded. The lower machinability in 1.1% Cu containing steel reflects the effect of lower copper content on machinability.

7.2.5 Effect of non-sulphide machining aids

7.2.5.1 Effect of boron nitride

Boron nitride was shown to improve machinability also of copper alloy sintered steel parts like a free-machining aid. The effect of small additions of boron nitride on machinability in drilling of Fe–Cu–C and for comparison of Fe–C materials is listed in Tab.7.13.

Tab.7.12 Number of holes drilled without and with MnS addition in Fe–2Cu–0.8C (6.3 g/cm^3, endothermic atmosphere) and in Fe–1.1Cu–0.8C (0.7% Cc, density 6.2 g/cm^3, N_2+methane atmosphere) steels and machining parameters (sponge iron powder D-63). Hardness ~57 HRB [135]. *Drilling test: HSS 3.5 mm drill, depth of hole ~11 mm*

Steel	MnS [mass %]		Drill speed/cutting speed [rpm/m/min]	Feed rate [mm/s]
	0	0.5		
	Number of holes			
Fe-2Cu-0.8C	67	775	1000/11	5.9
	17	100	3000/33	3.9
	17	100	3000/33	5.9
Fe-1.1Cu-0.8C	15	85	3000/33	5.9

Tab.7.13 Effect of 0.1% BN addition on tensile strength (Rm), hardness (HRB) and thrust force and torque in drilling at various drill speeds of Fe–(2, 2.5)Cu–(0.55, 0.75)C and of Fe–0.6C steels (atomised iron powder, density 6.8 g/cm[3]); C – graphite [222]. *Drilling test: HSS 6.4 mm drill, feed 0.12 mm/rev., depth of hole 25 mm*

Steel	Rm [MPa]	HRB	Drill speed/cutting speed [rpm/m/min]	Thrust force [N]	Torque [N·m]
ATOMET 29- 0.6 % C (no BN)	252	53	1380/28	845	1.98
ATOMET 29M -0.6 % C (no BN)	249	46		752	1.73
ATOMET 29M-0.55 % C–2 % Cu – BN	364	63	2220/45	748	1.58
ATOMET 29M-0.75 % C–2.5 % Cu – BN	434	73		779	1.50

Remark: Suffix M – BN containing mix; BN content added not declared

A marked enhancing effect of BN on machinability of bearing caps manufactured using the material consisting of ATOMET 29M with 2.0% Cu and 0.6% graphite addition was confirmed in the large-scale production. At increasing copper addition to 2.5% and graphite to 0.8% in order to improve the strength, the machinability was not significantly deteriorated. Both machinability and strength requirements were met and the bearing caps manufactured from this material passed engine test at 140% of the service load [272].

It is shown by this that BN was also an efficient machinability enhancer in copper alloy steels as the cutting forces were reduced in the BN-containing materials. BN addition decreased the thrust force and torque to lower values compared to those for Fe–0.6% C material. Very stable cutting operation was observed at presence of BN. Also the detrimental effect of higher content of carbon and copper on machinability was minimised in presence of boron nitride. The effect of BN and graphite addition on drilling tool life (number of holes drilled) in Fe–C and Fe–2Cu–C alloys is shown in Fig.7.29.

Boron nitride addition caused an increase in drill life in all steels tested. A significant increase of 15 times in drill life was obtained in drilling of Fe–2Cu–0.9C_{nom} (graphite) material. The tool life increased to 74 holes when using BN addition compared to 5 holes without BN. For both Fe–C steels without and with BN addition the highest machinability was exhibited at 0.6% C.

The efficiency of boron nitride as a machinability enhancer was confirmed by the reduction of the cutting forces and by an increase

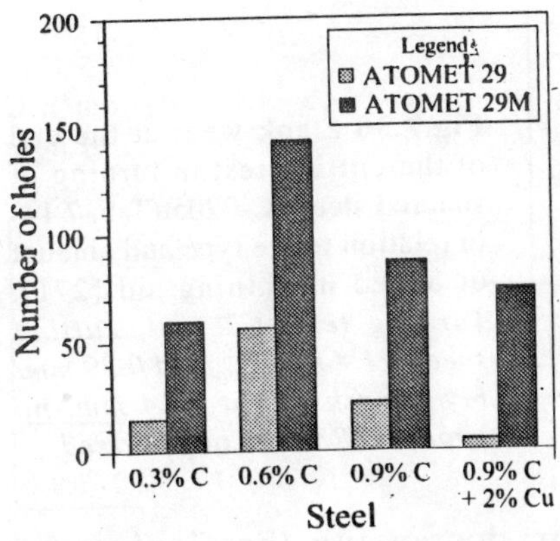

Fig.7.29 Number of holes drilled in Fe–(0.3, 0.6 and 0.9)C and Fe–2Cu–0.9C steels based on atomised iron powder (ATOMET 29) without and with BN (ATOMET 29M) addition. Alloy: 1 – Rm = 242 MPa, 44 HRB; 2 – Rm = 299 MPa, 57 HRB; 3 – Rm = 368 MPa, 72 HRB; 4 – Rm = 548 MPa, 82 HRB [119]. *Drilling test: HSS 6.4 mm drill, drill speed 800 rpm, v_c = 16 m/min, constant thrust force 685 N, depth of hole 25 mm.*

of the tool life, most pronouncedly in tested copper containing steel with 0.6% graphite addition.

A comparable effect of some machining aids in varying contents on machinability was tested in turning of FC-0205 alloy and with base sintered characteristics is listed in Tab.7.14.

The lower hardness of tested testpieces compared to TRS test bars was caused by lower effective cooling rate of the testpieces due to larger mass and cross section at the given cooling rate in the furnace. The TRS values should therefore be taken only as a qualitative indicator for the effect of machining aids. However, it stands out clearly that machining aids affected hardness and transverse rupture strength of the steel.

The effect of machining aids on the machinability in turning of this steel is shown in Fig.7.30. All free machining additives extended tool life and reduced tool wear significantly. When cutting FC-0205 material with no additive, flank wear increased progressively throughout the test until the tool failed after 94 cuts. The addition of 0.1% BN produced the lowest flank wear after 100

Tab.7.14 Base sintered characteristics of FC-0205 steel in dependence on machining aid type and addition (MnS, sulphur prealloyed, BN). Density 6.63–6.69 g/cm^3, sintering 30 min at 1120°C, $90N_2$–$10H_2$ atmosphere (acc. to Ref. 273)

Addition [mass %]	Testpiece (Ø 44.5/25.4 x 25.4 mm)			TRS test bar	
	C [mass %]	S [mass %]	HRB	MPa	HRB
None	0.58	0.007	55.3	844	69.6
0.35MnS	0.58	0.11	53.7	841	69.5
0.50MnS	0.57	0.18	52.1	803	68.7
0.14S*	0.50	0.13	47.7	709	64.0
0.35S*	0.57	0.30	54.7	718	66.3
0.1BN	0.55	0.012	49.2	750	65.2
0.2BN	0.58	0.008	50.8	764	65.3

Remark: *sulphur prealloying (resulphurized iron powder), hexagonal BN

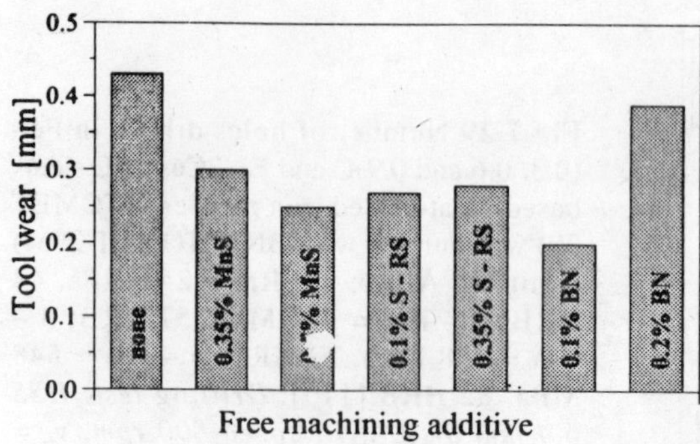

Fig.7.30 Flank wear at the end of the cutting test in turning of sintered steel FC-0205 (Tab. 7.14) in relation to the type and amount of added machining aid [271]. *Turning test: C2 tool, cutting speed 91.4 m/min, feed 0.29 mm/rev., depth of cut 1.14 mm, no coolant. RS – resulphurised.*

cuts. Sulphides were more effective in decreasing the flank wear than 0.20% BN addition. There was little difference between the different sulphide additions in flank wear. However, increasing the sulphur content from 0.14 to 0.35% had no significant influence on flank wear. Introducing a machining aid tends to reduce the cutting and radial forces but has less effect upon feed force. Turning FC-0205 with 0.1% BN required a low cutting force that was significantly lower than those required with 0.2% BN addition. Effect of a machining aid on chip form especially in the resulphurised powder material was also observed. As shown, the effect of the type and amount of machining aid – 0.35/0.5% MnS, 0.14/0.35% S, resulphurised; 0.1/0.2% BN – on surface finish was also clearly different [273].

7.2.5.2 Effect of retained graphite

Graphite is a base hardening element in production of steels in powder metallurgy, but it can partly contribute as a solid lubricant in powder mixes and as a machinability enhancer with good lubrication properties during machining, which effect is well known from grey cast iron (grey, nodular).

One of the methods for retaining free graphite in sintered materials is the production of sulphur prealloyed iron powders (>0.1% S) and to add graphite to this powder in a higher amount compared to graphite addition for improving mechanical properties. It is stated that the sulphur is likely to segregate on the surface to retain graphite in the sintered steel after sintering [274], apparently by inihibiting graphite dissolution.

The machinability of Fe–2Cu–C steel samples prepared by this way compared to those based on iron powder without prealloying with sulphur is shown in Fig.7.31. As follows from Ref. 274, the enhancement effect of free graphite in the steel made from sulphur-

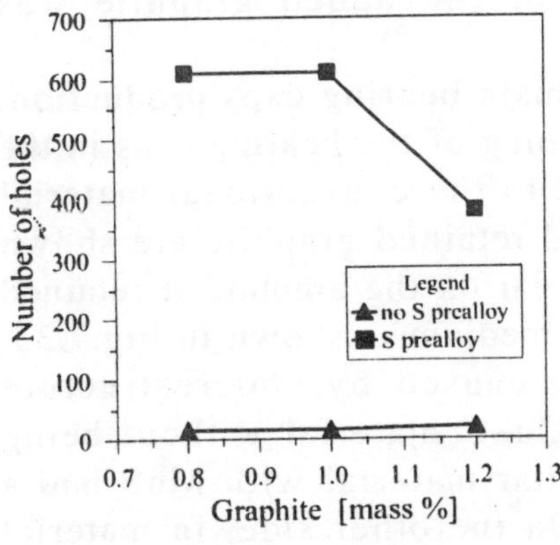

Fig.7.31 Effect of graphite addition on machinability (number of holes drilled) of Fe–2Cu–(0.8–1.2)% graphite materials, based on atomised iron powder without (0.02S, 0.15Mn) and with 0.1% S prealloying (0.06Mn). Sintering 1130°C, 20 min, N_2 – atmosphere, density 6.85 g/cm³ [274]. *Drilling test: HSS 1 mm drill, drill speed 10000 rpm, v_c = 31 m/min, feed 0.012 mm/rev., through hole depth 10 mm.*

containing material was confirmed by 600 holes drilled with 0.8 and 1.0% graphite addition (disolved carbon content 0.50–0.65%) and 350 with 1.2% graphite addition (combined carbon content 0.7%). The number of holes drilled in the material made from powder without prealloyed sulphur was less than 20. The 0.1 to 0.3% retained (free) graphite content in the pore sites of sintered steel is considered for the main factor for improving machinability. It was supposed that the adsorption of sulphur on the surface of iron powder reduces the carburisation rate and part of graphite remains in the pore sites of sintered steels after sintering.

The tensile strength of a tested Fe–2Cu–X graphite steel with 1.2% graphite addition based on sulphur prealloyed powder was 390 MPa, and without prealloying 400 MPa with 1% graphite addition. The conventional Fe–2Cu–0.6C steel (SC1200.26) has a tensile strength ~490 MPa [77]. The materials with lower addition of graphite like 1.2% for retaining a part of it in free form have significantly lower strength. It is the dilemma – enhanced machinability or high mechanical properties.

The other mechanism of inhibiting the diffusion of graphite into iron powder is using boron oxide (or BN, see chapter 6.1.6.1). Green compacts containing graphite were sintered at <700°C to remove the lubricant, then painted with an aqueous solution of 8% boron oxide (B_2O_3) penetrating inside the compacts. The compacts were dried (crystals of boron oxide remained in the pores and gaped the powders). In subsequent sintering at a temperature lower than the diffusion temperature of carbon (<600°C), boron oxide became glassy covering the graphite surfaces and obstructing the contact of graphite and iron powder. This process was followed by conventional sintering (two-stage sintering). The amount of retained graphite increased as the amount of the added graphite was increased.

This method was applied also for main bearing caps production. Machinability results in continuous turning of the bearing caps in the total length of cut 7000 m compared to the conventional material and with MnS addition and developed retained graphite are shown in Fig.7.32. The dependence of tool wear on the amount of retained graphite as a machining enhancement medium is shown in Fig.7.33. Large differences in hardness were caused by microstructure modification. The conventional Fe–Cu–C material without being permeated with boron oxide and similar material with MnS had a pearlitic microstructure (67 HRB). On the other side, in material where boron oxide was permeated therein from the surface, softer ferrite grains were observed.

The cutting resistance of the material with retained graphite was

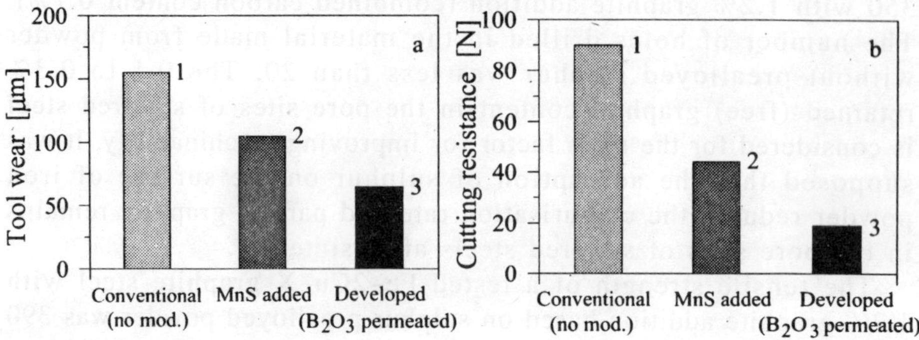

Fig.7.32 a – Tool wear (flank wear) and b – Cutting resistance in turning of bearing cups (Fe–1.5Cu–1.0% graphite) without and with MnS addition and with retained graphite (B_2O_3 method). Density 6.72 g/cm^3, hardness 67 HRB for conventional and with MnS material and 28 HRB for B_2O_3 material. 1 – conventional (no modification), 2 – 0.5% MnS addition, 3 – retained graphite [203,275]. *Cutting test: CBN cutting tool, cutting speed 180 m/min, feed 0.1 mm/rev., depth of cut 0.15 mm.*

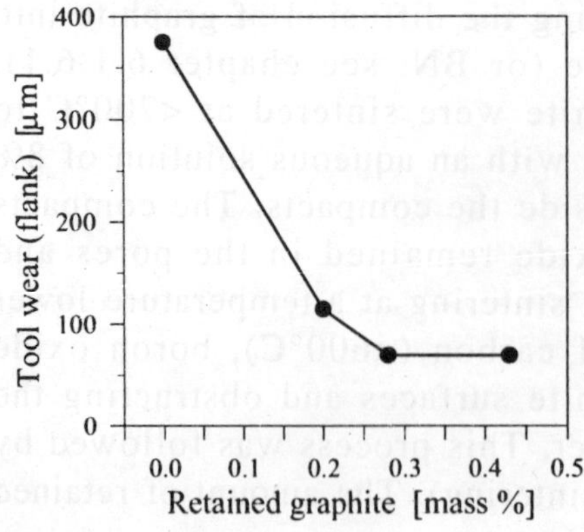

Fig.7.33 Tool wear in turning (flank wear) as in Fig.7.32 in dependence on amount of retained graphite [275].

lowered and the tool wear was greatly decreased. Low hardness (120 HV) was a proof that the diffusion of added graphite was obstructed. The tool life was prolonged by about 5 times compared to that of a conventional non-modified material and by about 2 times to that of the MnS-added material.

7.2.5.3 Effect of enstatite

In the group of oxide-base materials used for machinability enhancement (see Tab.6.1), there is a group of mineral powders having a slipping property – such as talc and enstatite, classified in a magnesium metasilicate base iron group. The effect of enstatite addition on machinability was tested on valve guide and high strength material with the properties listed in Tab.7.15.

The mechanical properties of investigated materials were not affected by the addition of 0.6 and 0.3% of enstatite, and of 0.5% MnS, respectively.

The effect of enstatite and MnS addition on machinability was tested in reaming tests of the valve guides material (Cu alloy steel) and separately in turning of a high strength steel. The inner diameter of 500 samples of valve guides was continuously worked by a carbide reamer. The corresponding flank wear on the tool and the dimensional transition of the inner diameter as the machinability indices are shown in Figs.7.34 and 7.35. Minimum wear and minimum changes in the machined inner diameter of valve guides were attained with enstatite. The wear of the reamer in reaming of valve guides with enstatite addition was the lowest. Enstatite had very good chip breaking action and had good chip discharging properties.

7.2.5.4 Effect of resin impregnation

The machinability in drilling of FC-0208 steel based on atomised iron

Tab.7.15 Base properties of valve guide material (Fe–4.5Cu–0.25P–0.5Sn–2.0C) and of high strength material (Fe–4Ni–1.5Cu–0.5Mo–0.5C–graphite, Distaloy SE) with addition of enstatite. Sintering 60 min at 1130°C, N_2-base atmosphere [233]

Material	Enstatite [mass %]	Density [g/cm^3]	UTS [MPa]	Hardness HRB	KC [J]	Fatigue strength [MPa]
Valve guide material	0.6	6.6	380	77	4	75
High strength material*	0.3	7.0	750	98	20	205

*See chapter 7.4.2.3

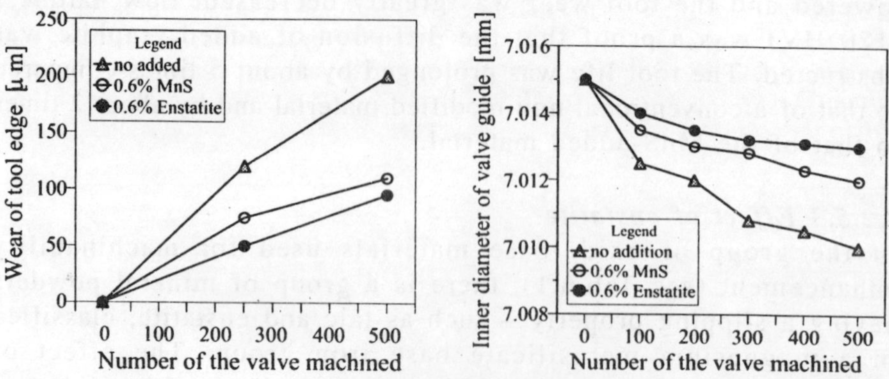

Fig.7.34 Effect of enstatite and MnS addition on tool edge wear in reaming of valve guides (Ø6.4→7 mm) (Tab.7.15) [208]. *Cutting test: Carbide reamer, 960 rpm, feed 0.2 mm/rev., depth of cut 0.3 mm, water solubility lubricant.*
Fig.7.35 (right) Effect of enstatite and MnS addition on the inner diameter of the valve guides. Cutting test as in Fig.7.34 [208].

powder was improved by resin impregnation, using an anaerobic impregnation technique. The axial force and torque in drilling resin-impregnated material was lower than when drilling conventional unimpregnated PM parts. A reduction of drilling forces by up to 75% was attained [109]. PM impregnated material showed less surface roughness variation as compared to conventional material, when subjected to different machining conditions in drilling [166]. The impregnated samples also showed generally improved fatigue strength and wear resistance [276].

Table 7.16 presents the data showing the comparative effect of MnS addition and resin impregnation on the machinability of FC-0208 and FC-0205 steels with diverse densities. The poorest machinability (relatively high thrust force, lowest number of holes drilled) was shown by the FC-0205 material without machining aids with the lowest density. In contrast the FC-0208 steel had the poorest machinability at the highest density of 7.2 g/cm^3 without MnS addition. An extremely large difference was recorded in drilling between FC-0208 and FC-0205 alloys at densities of 6.7 and 7.1 g/cm^3. The results show the deteriorating effect of higher carbon content on machinability in the FC-0208 compacts. 0.5% MnS caused a marked increase in the number of holes drilled and a decrease in thrust force. The 0.8% MnS addition was not effective in decrease of the thrust force for FC-0208 material except for the lowest density. It is not in agreement with number of holes drilled. Resin impregnation decreased the thrust force for both steels,

Tab.7.16 Machinability results in drilling (thrust force Fz, number of holes drilled – N.h.) for FC-0208 and FC-0205 steels with varying densities without and with high purity MnS addition and of resin impregnated. Atomised iron powder. Sintering 30 min at 1120°C, $85N_2$–15 d.a. atmosphere [277]. *Drilling test: HSS 9.53 mm drill, speed 1250 rpm, v_c = 37.4 m/min, feed 0.23 mm/rev., depth of blind hole 25.4 mm, no coolant*

FC-0208								
Density	No addition		0.5 % MnS		0.8 % MnS		Resin impregnation	
[g/cm³]	F_z [N]	N.h.	F_z [N]	N.h.	F_z [N]	N.h.	F_z [N]	N.h.
6.3	2632	8	1980	27	1572	23	655	32
6.7	2709	32	1887	192	1985	189	288	192
7.2	2957	11	1390	192	1931	192	-	-
FC-0205								
6.3	2238	64	1687	80	-	-	-	-
6.7	2047	192	1479	192	-	-	786	192
7.1	2149	192	1314	192	-	-	-	-

Remark: Test stopped after 192 holes drilled without drill failure

mainly at a density of 6.7 g/cm³ also in comparison to the MnS addition.

7.2.5.5 *Effect of calcium sulphide*

Reaction sintering can be a method for the formation of a very fine machining aid in the form of calcium sulphide in the matrix (see 6.1.4.2). This method was tested in machining valve guide compacts. The parts compacted from a Fe–Cu–C powder mix with the mentioned addition were sintered at 1000°C. High machinability of the compacts was verified by reaming trials, using a single flute carbide reamer, at a feed rate of 430 mm/min, a speed of 2400 rpm and a coolant pressure of 1 MPa. The stock removal was 0.6 mm on the diameter. 5000 components were machined with one tool [261]. Data for machining these parts without $CaCO_3/MoS_2$ for comparison are missing.

7.2.6 Summary

- Copper alone and in combination with graphite is considered to be a machinability enhancer to a limited extent. Optimum reduction in cutting forces was achieved with Fe–(2, 3) Cu–0.6% graphite mixes.
- Surface finish was improved with increasing copper content.
- The addition of 0.5% manganese seems to be beneficial for compositions containing 3% Cu and up to 0.6% graphite. The effect of manganese is harmful at all copper levels for mixes containing 0.9% graphite.
- For optimum machinability in quaternary system the alloying was

limited up to 0.6% C, 0.5–1.0% Mn and 2–3% Cu.

- All machining aids presented substantially improved machinability with special differences, *e.g.* sulphur, manganese sulphide or molybdenum disulphide, MnX and non-sulphide additives. Their use should be adapted for the respective material and machining conditions. This allows the right selection of copper alloyed material with a machining aid for special PM applications that require extensive machining.
- However, 0.5% MnS provided optimum results. 0.3% MnS or MnX was not sufficient and 0.8% MnS showed no further improvement in machinability. The 0.5% MnS addition reduced generally the thrust forces by 30% compared to the same composition without MnS addition.
- From the practical point of view, the 0.5% MnS (higher purity) addition seems to be effective in most cases.
- The sulphur prealloyed powders were superior in machinability to the FC-0208 base steel with admixed MnS.
- Boron nitride was an effective additive to improve machinability of carbon–copper steels, as confirmed by drilling thrust force and tool life evaluation.
- Retained graphite (control of sintering) and enstatite can be effective for special materials and machining conditions. The use of enstatite significantly improved effectiveness in turning of sintered parts and caused a decrease in tool wear. The mechanical properties of materials with retained graphite are lower compared when only combined carbon is present. It is a compromise between the machinability and mechanical properties.
- Resin impregnation was in general found to be the most effective method for improving machinability, even better than 0.5% MnS addition. For the Fe–Cu–C materials, resin impregnation reduced the axial force by 70%, when combined with 0.5% MnS the reduction was 75%.
- Lower drilling speed and feed rate contributed to the technical effectiveness of machining, especially when MnS was added.
- The adverse effect of some machining additives on the mechanical properties has to be considered.
- Sintering atmosphere (endothermic, hydrogen, hydrogen–nitrogen) in combination with carbon and with machining aid type must be considered. Better machinability was attained when sintering in cracked ammonia compared to endothermic gas.
- Some effect of the base iron powder type (atomised, sponge) was observed in all cases. Higher machinability was reached

with atomised base iron powder with respect to cutting process applied. The reduced (sponge) iron powders used for the low density samples resulted in the poorest machinability. The difference in machinability of sintered steels based on atomised and reduced powders are often mentioned. Better machinability was attained in some cases with the reduced iron powder. The causes of it may be the differences in density and/or in powder source related to manufacuring method (reduced, atomised iron powder).

- The lowest density material with 0.5 and 0.8% C had the poorest machinability. In some cases, the same results were obtained in some cases in Fe–C materials, in other cases, the opposite was found.
- At the same density and processing conditions the Fe–2% Cu–0.5% C steels were more machinable than those with 0.9% carbon.

7.3 NICKEL ALLOYED STEELS

Nickel is a very ductile metal, which does not undergo transformations in its basic fcc crystal structure up to its melting point and in general is more difficult to machine compared to other metals. This holds also for many nickel alloyed steels. Because most high strength PM steels are based on alloying with nickel, and since the machinability of newly developed nickel-less steels was not investigated so far, for nickel containing steels methods are needed to diminish the deteriorating effect of nickel on machinability.

In terms of the adverse effect on machinability, it is important to consider the higher concentration of nickel in interparticle necks areas between the starting iron powder particles. Ni alloyed sintered steels are commonly prepared from mixed or diffusion bonded, *i.e.* chemically heterogeneous powders, and during sintering the active diffusion of nickel occurs mainly in the small contact areas (due to higher dislocation density) than in the bulk material, esp. in the grain cores, the diffusivity of Ni in austenite being very slow. This also results in the formation of a nickel martensitic microstructure in localised areas. This type of martensite, which appears white after etching, is only slightly harder than ferrite. On the other hand, hard martensite shells form around the starting iron particles in combination with carbon, whereas the particle cores remain softer. This is a great problem for the machinability of nickel alloy steels. It must be noted that also wrought steels containing nickel exhibit poor machinability.

The machinability of Ni-containing steels is much poorer, in general about 8-10 times lower tool life being observed as compared to Cu-containing steels with similar mechanical properties. Nickel-containing steels have many negative factors to machinability based on structure, *e.g.* higher hardness, increased toughness and segregation of pores among grain boundaries [149]. The reason for the poorer machinability of Ni-containing steels is caused by mentioned structural heterogeneity in relation to the carbon content with a small amount of complex carbides $(Fe, Ni)_3C$ [138]. Finally one reason for poor machinability of Ni-alloy steels can also be found in the interaction of the cutting tool and the chip (friction pair) as shown in Fig.3.29.

With increasing nickel addition, the effect of carbon also becomes stronger in relation to the mechanical properties. This is because the amount of martensite in nickel alloy steels is considerably higher already after cooling in a sintering furnace in comparison to a nickel-free carbon steel. In regard to the fact that most of PM steels, especially high strength steels, are alloyed with different amounts of nickel, its effect on machinability of these steels through the microstructure formed will be perhaps a long-term task for investigation.

The effect of the machining aids on machinability of sintered low-alloyed nickel steels is, therefore, very sensitively joined with the heterogeneity and with proportion of particular phases in the microstructure, and knowledge gained at machinability of Fe-Ni-C steel can be useful also for the Distaloy type steels. A characteristic heterogeneous microstructure of a sintered mixed nickel steel compact is shown in Figs.7.36 and 7.37.

7.3.1 Effect of machining aids

7.3.1.1 Effect of MnS on mechanical properties

The effect of a machining aid on mechanical properties of a PM steel can be approximately linked to the machinability. When the mechanical properties of a steel are lowered by the machining addition, machining is easier and when the mechanical properties are higher, machinability deteriorates. Only a negligible effect of a machining aid on the mechanical properties, or none at all, is desirable.

As stated in Ref. 104, a negligible effect of 0.5% MnS addition on the mechanical properties was recorded for prealloyed Fe-2Ni-0.5C (FL-4205), FL-4205+1.5%Ni and Fe-2Ni-0.5Mo-0.5C (FL-4605) steels. This is confirmed also by the data in Tab.7.17 giving nearly equal mechanical properties for FN-0205 steel without and

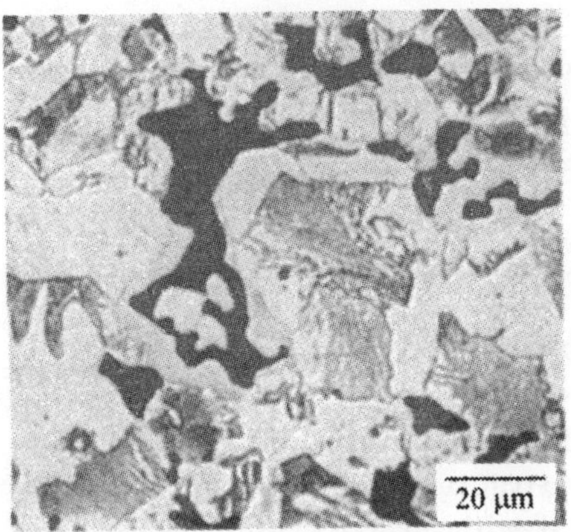

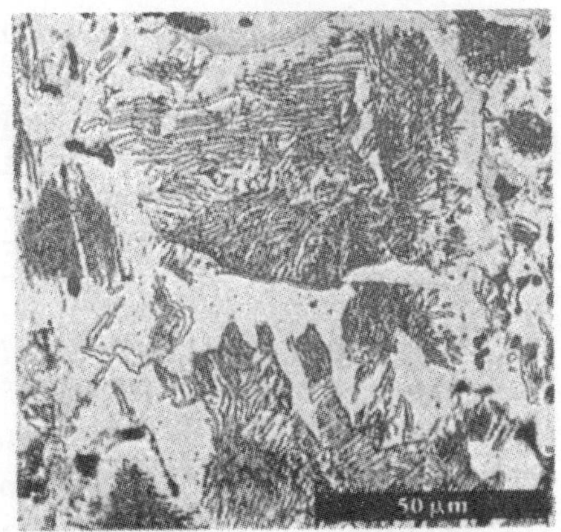

Fig.7.36 Microstructure of Fe–2Ni–0.5C steel. Compacted, sintering 30 min at 1120°C. Optical micrograph. Nital etched [238].
Fig.7.37 (right) Microstructure of Fe–3Ni–0.6C steel, sintering 30 min at 1120°C, density 7.10 g/cm^3, NC 100.24 iron powder. Optical micrograph. Nital etched [235].

with 0.5% MnS addition independently on density. The difference in the results was in the range ±(3–5)%.

Figure 7.38 shows that the effect of the manganese sulphide addition on the tensile strength of FN-0205 and FN-0208 alloys is not in all cases negligible. At lower densities, there is no significant difference between the properties of the compacts alloyed with MnS and those without MnS addition. MnS addition lowers the tensile strength by 15–20% only for the highest density parts (alloy 7,8) [244]. This could be caused by the higher fraction of MnS particles situated in the sintered interparticle necks acting as non-metallic inclusions during fracture. Hardly any effect of MnS addition on hardness was observed. Probably, the small differences in the microstructure of the steel and by this surely in the mechanical properties also should be studied in more detail.

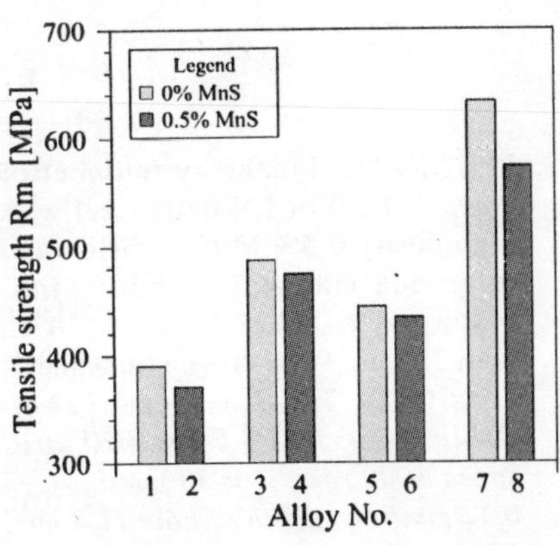

Fig.7.38 Effect of 0.5% high-purity MnS addition on tensile strength of FN-0205 and FN-0208 alloys. 1– FN-0205-25, sintered density (s.d.) 6.98 g/cm^3, 69 HRB; 2 – as 1, s.d. 6.92 g/cm^3, 64 HRB; 3 – FN-0205-30, s.d. 7.23 g/cm^3, 77 HRB; 4 – as in 3, 77 HRB; 5 – FN-0208-35, s.d. 6.89 g/cm^3, 74 HRB; 6 – as in 5; 7 – FN-0208-45, s.d. 7.31 g/cm^3, 92 HRB; 8 – as in 7 (acc. to Ref. 104).

Tab.7.17 Mechanical properties of FN-0205 steel in relation to the density. Atomised iron powder. Sintering 30 min at 1125°C, dissociated ammonia (acc. to Ref. 244)

Density [g/cm³]	Rm [MPa]	Rp0.2 [MPa]	HRB	KC [J]
6.4	251	183	43	5
6.8	340	216	51	14
7.2	474	250	59	27
7.5	562	292	67	45

7.3.1.2 Effect o, MnS on machinability

The effect of MnS addition on machinability of nickel containing steels must be linked to the sintering temperature and by this to its effect on the formation of the microstructure as shown in Fig.7.39. The machinability of the base material without MnS addition dropped from an average of 75 holes for the compacts sintered at lower temperature to an average of 50 holes for the high temperature sintered compacts, probably due to the change in the proportion of the nickel martensite area. The addition of 0.5% MnS improved significantly the machinability of compacts sintered at either temperature, in both cases around 500 holes being attained. Machinability of the alloy was not strongly affected by the density, at least not in the range of 6.8 to 7.5 g/cm³.

Comparable data about the effect of MnS addition on tensile strength, hardness and machinability (by drilling) of FN-0205 and F-0008 steel compacts without and with MnS addition are listed in Tab.7.18. Here the unexpectedly positive effect of MnS addition on the machinability of these materials is evident. In this case a decrease in tensile strength and hardness with MnS addition was recorded. The 0.5% MnS addition improved machinability very significantly in both alloys, and the number of holes drilled (3500) in Ni-alloyed steel and also in Fe–C steel (4200) must be regarded as very high. This effect was not observed in previous results. The

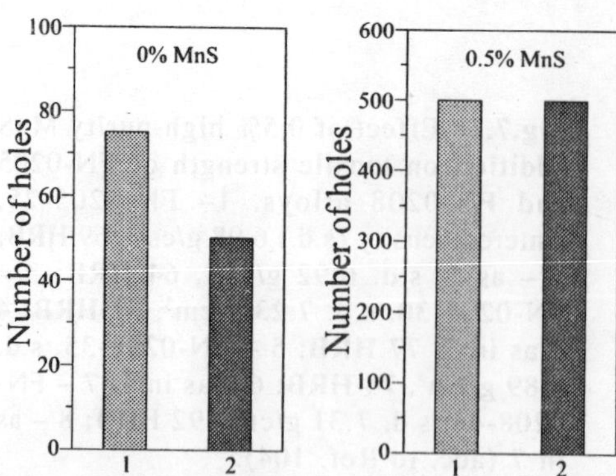

Fig.7.39 Machinability (number of holes drilled) of FN-0208 steel with high purity 0.5% MnS addition and reference material without MnS. Sintering: 1 – 1125°C, 2 – 1250°C, both 30 min, $90N_2$–$10H_2$ atmosphere, density of 7.2–7.3 g/cm³ [244]. *Drilling test: HSS 3.5 mm drill, drill speed 3000 rpm, v_c = 33 m/min, feed 0.12 mm/rev., depth of hole 11.4 mm.*

Tab.7.18 Effect of MnS addition on tensile strength (Rm), hardness (HRB) and machinability (number of holes drilled) of FN-0205 and F-0008 steels (atomised iron powder): density 7.05 g/cm³ [224]. *Drilling test: HSS 3.5 mm drill, drill speed 4000 rpm, v_c = 44 m/min, feed 0.12 mm/rev., depth of hole 11.4 mm*

MnS [mass %]	FN-0205 Rm [MPa]	HRB	Number of holes FN-0205	F-0008
0	428	65	100	152
0.5	373	60	3500	4200

machinability results with the MnS addition cannot be explained by minor differences in tensile strength and hardness values. A more detailed investigation of all factors taking part in the cutting process is necessary, *i.e.* of the material properties including a very thorough analysis mainly of the microstructure and of the machining conditions used. The result of this investigation could be a contribution for improving the machinability of steels containing not only nickel but also other alloy elements.

Improvement in machinability with (0.5, 1.0)% MnS addition was recorded when drilling the prealloyed Fe–2Ni–0.5Mo–0.25Mn–0.0C (Astaloy A) compacts (density 6.8 g/cm³) sintered at 1120°C for 30 min in an endothermic atmosphere (Rm = 250 MPa) [11]. The 81 holes were drilled with 0.5% MnS and 96 holes with 1.0% MnS against 38 holes without MnS addition (increase in the number of holes drilled was 113% for 0.5% MnS addition and 153% for 1.0% MnS addition) [148].

Some differences in the machinability of prealloyed and mixed steels also without machining aid addition are expected and can be explained by corresponding differences in the homogeneity/heterogeneity of the microstructure at the same density and in the mechanical properties. Mixed steels exhibit mostly higher elongation due to the presence of relative higher portion of ferrite in the microstructure and prealloyed powder-based materials higher hardness due to more homogeneous microstructure. Therefore, the mixed steels should be better machinable.

7.3.1.3 *Effect of resin impregnation*

The comparison of the effects of MnS addition and resin impregnation, respectively, on the machinability of FN-0205 and FN-0208 steels is listed in Tab.7.19. In connection with these results not only a relatively small difference in thrust force must be noted compared to the previous results but also a high difference in machinability (number of holes drilled). Without MnS addition, a significantly higher machinability in the number of holes drilled at

Tab.7.19 Machinability (thrust force, number of holes drilled) in FN-0205 and FN-0208 steels without and with 0.5% MnS addition and resin impregnated [277]. *Drilling test: HSS 9.53 mm, drill speed 1250 rpm, v_c = 37 m/min, feed 0.23 mm/rev., blind hole depth 25.4 mm in the specimen – thickness 27 mm*

Alloy	Density [g/cm^3]	0 % MnS		0.5 % MnS		Resin impregnated	
		Thrust force [N]	Number of holes	Thrust force [N]	Number of holes	Thrust force [N]	Number of holes
FN-0205	6.9	2134	171	1350	192	-	-
	7.2	2158	>192*	1252	192	-	-
FN-0208	6.9	2904	20	1638	192	679	>192*
	7.3	2794	27	1563	160	-	-

Remark: *Test stopped after 192 holes drilled without drill failure in all tests

a lower difference in thrust force values was attained in FN-0205 compared to FN-0208 steel without MnS addition. This result can be attributed only to the carbon content. The thrust force in machining of materials with MnS addition decreased by more than 30% for FN-0205 steel and by ~40% for FN-0208 steel. The difference in density had a negligible effect on machinability.

A marked improvement in machinability, measured as decrease in the thrust force values, was determined as a result of resin impregnation. The thrust force decreased by more than 70% compared to the material without MnS addition and by about 60% when MnS was added.

7.3.2 Dual-phase nickel steel

The term dual-phase refers to a class of steels that can be processed by intercritical heat treatment to obtain a dispersion of martensite in the ferrite matrix. The problem of these steels is to create a microstructure consisting of two proportionally different phases while, at the same time, to produce a high strength sintered steel easier to machine than current PM steels of corresponding properties.

An example of such a material can be the Fe–8.8% Ni alloy steel prepared by two methods. In the first one, nickel was added to the iron powder as a master alloy (water atomised powder – 22.2% Ni, 0.15% Mo) and in the second one a powder mixture was prepared from elemental powders. In both cases (0.1, 0.3, 0.5, 0.8)% graphite was admixed. The test pieces were pressed at 600 MPa and sintered at 1200°C for 30 min in pure hydrogen followed by furnace cooling. To create the dual-phase microstructure, sub-zero treatment was carried out at –100°C in liquid nitrogen for 6

h. This treatment increased the microhardness of Ni-rich austenite from 210 HV in the as-sintered condition to 616 HV without any change in the bainite regions and in strength. This treatment increased the apparent hardness of the samples with 0.8% C from ~170 to 250 HV. The sub-zero treatment converted soft machinable austenite into harder wear-resistant martensite.

Therefore, in terms of machinability, the Fe–8.8Ni–(0.1–0.8)C dual-phase steels were easy to machine in the as-sintered condition and hardly machinable in the condition after sub-zero temperature treatment [278]. The quantitative data about easy machining of as-sintered dual phase, which are lacking, could be a contribution to the machining of nickel alloy steels.

7.3.3 Summary

- Additions of up to 0.5% MnS had a measurable adverse effect on the mechanical properties of the tested nickel-containing steels, mainly with higher density materials. This addition of MnS reduced tensile strength by more than 10%. Increasing the MnS content to 0.8% was even more detrimental to the highest density material, though less significant at lower densities of 6.4–6.7 g/cm^3.
- These conclusions regarding the effect of MnS on the mechanical properties of mixed nickel alloy steels are also valid for nickel prealloyed Fe–0.5Ni–0.6Mo (FL-4205) and Fe–1.8Ni–0.55Mo (FL-4605) steels and also with further admixing of nickel reaching a tensile strength of >600 MPa and hardness up to ~90 HRB.
- The Fe–2Ni–0.5C steels without and with MnS or other machining aids were consistently more machinable than those containing 0.8% C. The number of holes drilled increased when the carbon content decreased from 0.8% to 0.5%. The effect of carbon content on machinability was dominant.
- For both carbon levels, there was no significant difference in the machinability of compacts with a sintered density of 6.9 to 7.2/7.3 g/cm^3, taking the number of holes drilled as a criterion. For the steels with a higher density (~7.2 g/cm^3) and a high carbon content (0.8%), the 0.5% MnS addition was effective.
- The addition of 0.5% MnS was effective in improving machinability, particularly in FN-C208 and less in FN-0205, when taking the thrust force values as a criterion.
- Resin impregnation was most effective in decreasing thrust force, providing a 75% reduction for FN-0208 steel.
- For the tested nickel-containing steels, relatively large differ-

ences in the machinability results from different test series measured under the same or similar cutting conditions were recorded.

7.4 DIFFUSION ALLOYED STEELS

Diffusion alloyed steels are a group of steels that have been used successfully in the production of structural parts for high strength and high precision applications because of their better technical properties and their overall ease of manufacture.

These steels are alloyed in varying combinations and amounts with nickel, copper, and molybdenum. They are prepared as 'non segregable mixes', exhibiting similar compactibility and sintered microstructures as the respective mixes, with appropriate amounts of graphite. According to the base powder used in their preparation, diffusion bonded type alloys (frequently known under Höganäs designation 'Distaloy') can be divided roughly into two groups. The first group is formed by those alloys prepared on the basis of plain iron powders (mixed powder systems) and the second one consists of the alloys prepared on the basis of prealloyed Fe-Mo powders (hybrid material systems).

According to this classification, the first group incorporates: a) medium alloyed Fe–1.75Ni–1.5Cu–0.5Mo steels based on sponge iron powder (Distaloy SA) and on atomised iron powder (Distaloy AB), and b) higher alloyed Fe–4Ni–1.5Cu–0.5Mo steels based on sponge iron powder (Distaloy SE) and on atomised iron powder (Distaloy AE). The base powder mixes differ in compressibility depending on the base iron powder grade. At suitable carbon contents, even the as-sintered materials exhibit high strength and hardness and are also suitable for heat treatment. A typical microstructure is shown in Fig.7.40a.

The second group of diffusion alloyed steels is formed by those produced by adding the following elements to the prealloyed Fe-1.5Mo powder by diffusion bonding:

- 2% Ni (Distaloy DC-1, DC – dimensional control). This material is designed to achieve very high dimensional accuracy. With admixed graphite, it ensures high as-sintered strength, and the microstructure (Fig.7.40b) is formed with a relatively large amount of bainite and some martensite, at least at low cooling rates in a common belt furnace.
- 2% Cu (Distaloy DH-1, DH – direct hardening). This material is designed mainly for sinter hardening and the microstructure is then hard, martensitic-bainitic (Fig.7.41a).

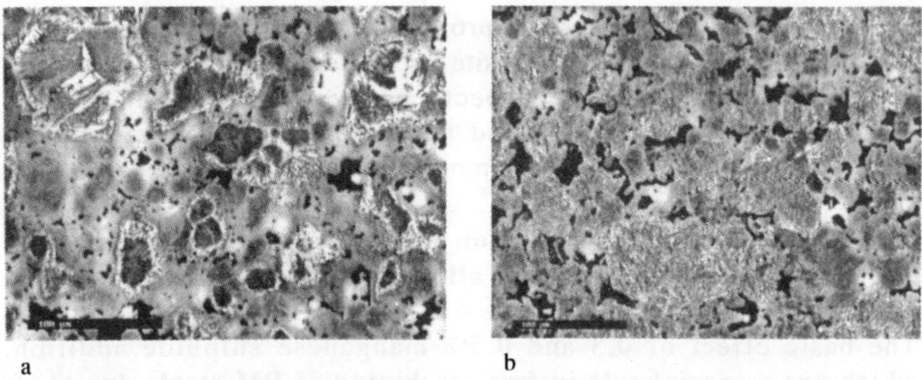

Fig.7.40 Microstructure of as-sintered: a – Distaloy AE-0.5% C steel, b – Distaloy DC-1-0.5% C steel (hybrid system) (Courtesy of Höganäs AB). ×150

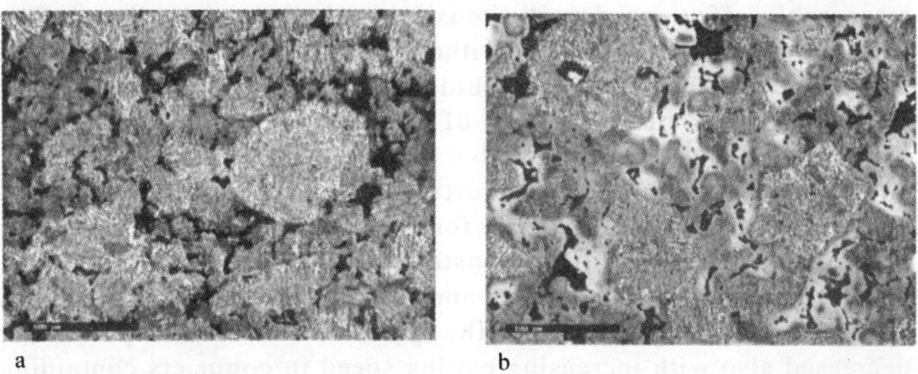

Fig.7.41 Microstructures of as-sintered: a – Distaloy DH-1-0.5% C steel, b – Distaloy HP-1–0.5% C steel (both hybrid systems) (Courtesy of Höganäs AB). ×150.

- 4% Ni (Distaloy HP-1, HP – high performance). Because of the high nickel content, in this material, when admixed with graphite, the microstructure consists of martensite and bainite with approx. 2–3% of retained austenite after common belt furnace sintering (Fig.7.41b). A tensile strength of ~1000 MPa is achievable at 7.0 g/cm^3.

An alternative to the Distaloy DC and Distaloy DH materials are alloys based on prealloyed Fe–0.85Mo (Ancorsteel 85 HP) powder with diffusion bonding of 2% Cu or 2% Ni.

The main characteristics of these alloys, *e.g.* combination from medium up to high alloying with elements with different solubility in iron with carbon, result in high strength and hardness exceeding ~200 HV 10 which level is in general regarded as the practical limit for the machinability of a sintered material. The microstructure of Distaloy type alloys in dependence on alloying and on the cooling rate is characterised in various proportion by ferrite and bainite, or by prevailing bainite and martensite up to martensite and bainite with retained austenite. These aspects are very important for the machinability of the steels and for machinability improvement methods compared to steels of more simply alloyed steels.

7.4.1 Effect of machining aids on properties and machinability of Fe–1.8Ni–Cu–Mo diffusion alloyed steels

7.4.1.1 Effect of MnS

The basic effect of 0.3 and 0.5% manganese sulphide addition, which has a special role in free machining of PM steels due to its very broad use, on properties and machinability of Distaloy SA alloy is listed in Tab.7.20. The manganese sulphide addition had no measurable effect on density, tensile strength, and hardness. The machinability of the material without MnS addition is regarded as very poor. The manganese sulphide additions caused a significant increase in the machinability of the material compared to the material without MnS addition, but without a larger difference when comparing 0.3% and 0.5% additions. The effect of 0.5% MnS addition on tool forces – thrust force and torque in relation to the cutting speed – is clearly demonstrated in Fig.7.42.

In this case, both torque and thrust force increased with increasing cutting speed in the plain Distaloy compacts but decreased also with increasing cutting speed in compacts containing 0.5% MnS, the torque more so. Increasing feed rate had little effect on surface finish and roundness. Generally, as the feed rate and speed increased the quality of the edge of the hole, particularly at the exit, deteriorated.

7.4.1.2 Effect of various machining aids on mechanical properties and machinability

As shown before, sulphur, manganese sulphide, molybdenum disulphide, selenium and tellurium are often regarded as the best machinability enhancing additives also for high strength materials.

Tab.7.20 Density, tensile strength (Rm), hardness (HV 5), and machinability (number of holes drilled) of Distaloy SA-0.6% graphite without and with 0.3 and 0.5% MnS addition [148]. *Drilling test: HSS 3.5 mm drill, drill speed 3000 rpm, v_c = 33 m/min, feed 0.08 mm/rev., depth of hole 12 mm*

MnS [mass %]	Density [g/cm^3]	Rm [MPa]	HV 5	Number of holes
0	6.92	624	191	13
0.3	6.91	623	190	100
0.5	6.91	622	191	113

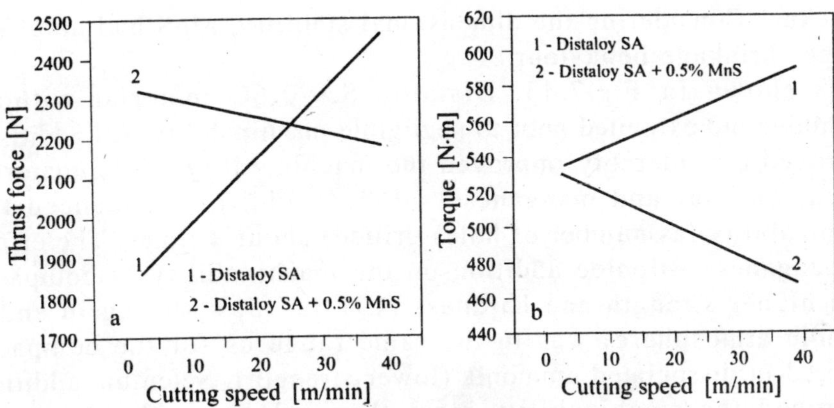

Fig.7.42 Least squares line fits of variation of: a – thrust force and b – torque with cutting speed in drilling of Distaloy SA-(X)C alloy without and with MnS addition. Sintering 20 min at 1150°C, endothermic atmosphere; density 6.2 and 6.6 g/cm³ [279]. *Drilling test: HSS 9.5 drill, drill speed 273 rpm, v_c = 8.1 m/nin, feed 0.1mm/rev., drill point angle 118°, drill helix angle 29°. (XC – carbon content not declared).*

Their effect on the tensile strength and dimensional change of Distaloy SA-0.6C alloy is shown in Tab.7.21.

This example of the influence of different types and contents of machinability enhancing additives on the properties of an alloy steel clearly demonstrates the complexity of the problem. These substances have undesirable side effects on the properties of the alloy. Sulphur, selenium and tellurium lower the tensile strength, especially tellurium by 16 to 25% (the worst) compared to the value without the machining aid, whereas MnS and MoS_2 have a much weaker effect in this respect. In particular, low MnS contents up to 0.5% do not obviously decrease the strength properties at all as shown also before. Sintering of the compacts in an endothermic atmosphere caused an increase in strength, possibly through some

Tab.7.21 Tensile strength (Rm) and dimensional change of Distaloy SA-0.6% graphite in dependence on S, Se (elemental), tellurium (elemental), MnS and MoS_2 addition. Compaction 600 MPa, sintering 1120°C, 30 min, dissociated ammonia or endothermic atmosphere, density 6.9–7.0 g/cm³ [148,209]

Machining aid [mass %]													
No	S		Se		Te			MnS					MoS_2
0	0.15	0.25	0.3	0.5	0.05	0.15	0.25	*0	*0.3	*0.5	0.5	0.85	0.8
Rm [MPa]													
590	550	520	560	530	490	460	440	624	623	622	590	580	580
Dimensional change [% linear]													
0.02	0.02	0.40	0.16	0.22	0.23	0.30	0.32	-	-	0.01	0.01	0.01	0.28

Remark: *sintering in endothermic atmosphere, hardness 191 HV 5

C pickup. Considering the dimensional stability, MnS had no effect on the shrinkage behaviour.

As shown in Fig.7.43, Distaloy SA–0.6C material without machining aid exhibited poor to negligible machinability. All additives improved considerably improved the machinability of the material tested. Sulphur and manganese sulphide additions enhanced the machinability (as number of holes drilled) about 8 times. The effect of manganese sulphide addition on the machinability of compacts with higher strength and hardness (191 HV 5, sintering in endothermic atmosphere) was in the same range as for the compacts sintered in dissociated ammonia (lower strength). Selenium addition improved the machinability more than sulphur and manganese sulphide additions. Molybdenum disulphide addition improved the machinability from 15 to 235 holes for compacts sintered in cracked ammonia. Tellurium addition had the highest enhancing (~15 times) effect on machinability. This can also be related to the lowest strength of the steel. Apparently a compromise between the mechanical properties required and the machinability of a material has to be found.

7.4.1.3 Effect of copper and MnS addition on machinability

In previous results, poor to very poor machinability of Distaloy SA-C alloy steel was shown. Regarding the previously mentioned positive effect of copper on the machinability of some PM steels, the effect of additional admixing 2 and 4% of copper on the machinability of Distaloy SA and AB with 0.4 and 0.8% graphite was investigated (total copper content of 3.5 or 5.5%; total content

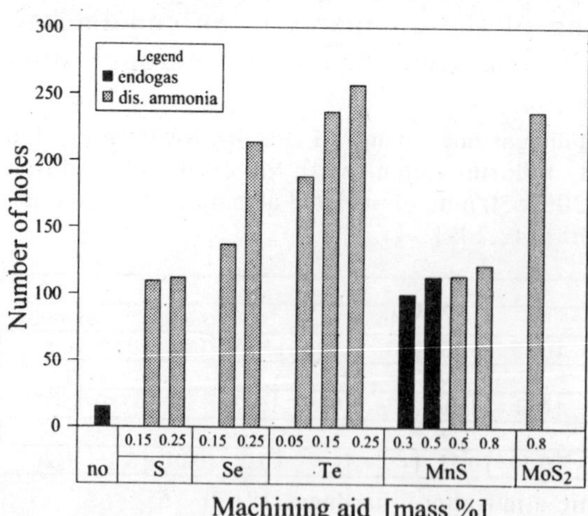

Fig.7.43 Machinability (number of holes drilled) of Distaloy SA-0.6C (graphite) alloy without and with addition of S, Se, Te, MnS and MoS_2 [148,209]. *Drilling test: HSS 3.5 mm drill, drill speed 3000 rpm, v_c = 33m/min, feed 0.08 mm/rev., depth of hole 12 mm (All compacts drilled as-sintered).*

of alloying elements 5.75 or 7.75% in alloy). 0.2 and 0.5% manganese sulphide additions were involved in the machining testing of such steels. Hardness and porosity values of the tested alloys are listed in Tab.7.22.

The characteristic high hardness of the alloys was the result of density and of carbon and copper content. Regarding the high alloying and carbon content of the materials, the detailed analysis of the microstructure showed a percentage of the major microstructure constituents which can provided a contribution to better understanding the machinability of the alloys, Tab.7.23. Ferrite was present in all materials tested. Increasing carbon and copper content decreased the proportion of ferrite and increased the proportion of bainite and martensite (in spite of the tricky characterisation of bainite by image analysis, its proportion may be a contribution to better characterisation of the microtructure in relation to the machinability of a material). Increasing carbon content from 0.4 to 0.8% increased the proportion of martensite in the microstructure, mostly at the expense of ferrite. The proportion of pearlite was not affected by the carbon and copper content. These microstructural characteristics controlled the machinability of the materials.

The specimens (bushing Ø65/35.5 × 20 mm) prepared from the above-mentioned powder alloys, were subjected to turning, drilling, and tapping tests. The main results attained in these extensive cutting tests of the alloys are presented below.

Turning. Turning test: TiC + TiN coated carbide inserts (P25) (see chapter 4), cutting speed 78 m/min to 200 m/min, feed 0.1 mm/rev., depth of cut 0.5 mm.

- Increase in carbon content of 0.4 to 0.8% increased cutting forces.
- Increasing additional copper content from 2 to 4% in Distaloy AB with 0.8% carbon decreased the cutting forces by about 15-

Tab.7.22 Hardness and porosity of Distaloy SA and Distaloy AB alloys with 0.4 and 0.8% C and with 2 or 4% copper addition. Sintering 45 min at 1120°C, endothermic atmosphere [139]

Alloy	Distaloy SA+2 % Cu		Distaloy AB+4 % Cu	
Compacting pressure [MPa]	500	700	500	700
Carbon [mass %]	Hardness HV 10			
0.4	162	205	183	223
0.8	217	256	224	253
Porosity [%]	18.1	16.6	14.3	12.8

Remark: Distaloy SA based on sponge iron powder, Distaloy AB based on atomised iron powder

Tab.7.23 Percentage of major microstructural constituents in random areas in Distaloy SA and AB materials tested [139]

Carbon [mass %]	0.4		0.8	
Alloy grade	Distaloy SA	Distaloy AB	Distaloy SA	Distaloy AB
	Proportion of microstructure constituents [%]			
Ferrite	19	15	9	7
Pearlite	49	50	46	49
Martensite	16	20	29	32
Porosity [%]	16	15	16	14

Remark: all values ± 1 to ± 3%; bainite not determined

30%. The Distaloy SA materials showed better machinability, through lower cutting forces than equivalent Distaloy AB materials with a possible effect of different porosity.

- The material with higher porosity gave a longer tool life and lower cutting forces.
- In the machined surfaces and in the chips, a higher microhardness in specific microstructural constituents was measured compared to the bulk material. MnS-containing materials showed less marked densification.
- The longest tool life was recorded at a cutting speed of 100 m/min. At 200 m/min, there was a large drop in the insert life when the depth of cut was increased to 0.5 mm.
- A manganese sulphide addition of 0.2 and 0.5% improved tool life for the Distaloy SA materials. In the machining of Distaloy AB materials, tool life was not affected by the addition of manganese sulphide if lubrication was not used. In such a case the tool life was very short anyhow.

Drilling. Drilling test: HSS 5 mm drill (P1, P4, TiN coated, see chapter 4), drill speed 663, 1268 and 1786 rpm, v_c = 10, 20 and 30 m/min, feeds 0.10, 0.15 and 0.20 mm/rev., through hole depth 20 mm.

- In drilling Distaloy AB materials, about 78 holes were drilled at a drilling speed of 30 m/min and a feed rate of 0.15 mm/rev. with TiN-coated drill when compared to Distaloy SA (12 holes).
- Increasing the density had no effect in low copper and carbon compacts.
- Increasing the additional copper content from 2 to 4% had no effect on the lower density low carbon materials, but reduced the drill life in material in combination with high carbon content and higher density materials also under lubrication conditions.
- Drill lives were significantly reduced (100→8 holes) when the carbon content in Distaloy AB was raised from 0.4 to 0.8%. High carbon content materials could be best drilled with the P4

drill bit.

- The addition of manganese sulphide to the high carbon low copper materials improved drill performance at a feed rate of 0.15 mm/rev. in Distaloy AB, but reduced the tool life in Distaloy SA materials.
- Increasing the drilling speed from 10 m/min to 30 m/min at feed 0.15 mm/rev. had no effect on material Distaloy AB (0.4% C, 2% Cu, compaction 700 MPa) and P1 drills; however, if the feed was increased to 0.20 mm/rev. there was a reduction in drill lifetime at the high speed for material Distaloy AB (0.4%C, 2% Cu, compaction 500 MPa). The use of a feed rate of 0.20 mm/rev. at 30 m/min (highest cutting conditions used) reduced the drill lifetime compared to a feed of 0.15 mm/rev. at the same speed for a range of tested materials.
- The use of lubrication during the drilling process increased drill lives.
- Materials with a hardness lower than 200 HV 10 were easily drilled under the test conditions used (drilling was stopped at 100 holes drilled). The high hardness materials required P4 drill bit.
- Distaloy AB materials were easily drilled compared to Distaloy SA materials.
- The results attained clearly show that the relatively hard materials, if possible, should be machined with hardmetal tools.

Tapping. Tapping test conditions: predrilled holes 5 mm diameter, tap M6, 15 mm deep, tapping (spindle) speeds 530, 795 and 1061 rpm, v_c = 10, 15 and 20 m/min, tapping fluid.

- The optimum speed for tapping with the TiN-coated straight fluted tap performed at 20 m/min, for the TiN-coated with spiral flute at 10 m/min, and for the HSS (spiral flute) at 15 m/min. The best condition was the one with lubrication. The addition of MnS had no effect on Distaloy AB materials unless a TiN-coated tool was used, and the effect was only sligthly better for Distaloy SA material probably due to the densification of the machined surface of the hole manifested in an increase of microhardness.
- Lowering the carbon content or increasing the copper content was more favourable than MnS addition. When copper was added to 0.8% carbon alloys, the tapping torque was reduced.

It was proved that the copper had an enhancing effect on machinability in three basic machining processes of high hardness Distaloy SA and AB type alloys. The base iron powder grade used for the preparation of the alloys (sponge SC100.26 for Distaloy SA, atomised ASC100.29 for Distaloy AB) showed different behavior in turning compared to drilling (wear mechanisms in the cutting process). Manganese sulphide addition increased the beneficial effect of copper. A close relationship was shown between the repective properties of the tested materials, machining methods and conditions including tools (wear mechanism). On the basis also of these results it can be supposed that in machining of materials of 'very high hardness' the dominant mechanism is probably abrasive wear and therefore, the contribution of MnS to better machinability is lower than that of decreasing the hardness of a material.

7.4.2 Effect of machining aids on properties and machinability of Fe–4Ni–Cu–Mo diffusion alloyed steels

7.4.2.1 Effect of MnS

The effect of manganese sulphide addition on mechanical properties and machinability of Distaloy SE diffusion alloy steels is listed in Tab.7.24.

There is no appreciable difference between the strength of the compacts without and with 0.5% high-purity MnS addition. The same conclusion was also valid for impact strength and fracture toughness.

The machinability of the compacts with no carbon was improved from about 20 holes without MnS to 990 holes with 0.5% MnS addition. In compacts with carbon, machinability improved from 77 holes without a machining aid to about 1600 holes with 0.5% MnS. In this case, manganese sulphide improved the machinability of the Distaloy SE compacts even in the presence of 0.5% carbon in the same proportion. It is necessary to note the high machinability of

Tab.7.24 Tensile strength (Rm), hardness (HRB), density, and machinability (number of holes drilled) of Distaloy SE-(0, 0.5)% C steel compacts without and with 0.5% high purity MnS. Sintering 40 min at 1135°C, and at 1125°C, $90N_2$–$10H_2$ atmosphere. (acc. to Ref. 105,224,244) *Drilling test: HSS 3.5 mm drill, drill speed 2000 rpm, v_c = 22 m/min for alloy 1 and 2, 4000 rpm, v_c = 44 m/min for alloys 3 and 4, feed 0.12 mm/rev., depth of hole 11.4 mm*

MnS/C [mass %]	Density [g/cm^3]	Rm [MPa]	HRB	Number of holes	Ref.
0/0	7.0	425	61	20	181
0.5/0		405	60	990	
0/0.5	7.2	*1205 (HT)	105	*77 (AS)	105,
0.5/0.5		*1200 (HT)	105	*1600 (AS)	203

Remark: *alloys heat treated (HT) (840°C, oil quenching); all materials drilled only as-sintered (AS), (the scale for HRB hardness measurement is up to 100 only)

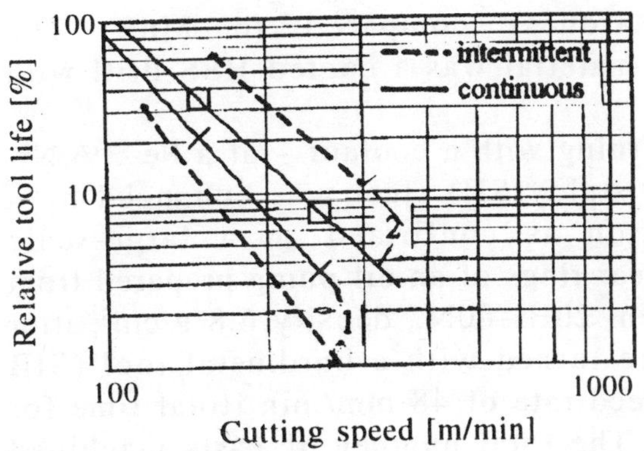

Fig.7.44 Relative tool life in continuous and intermittent cutting of Distaloy AE-0.5C alloy. Density 7.0 g/cm^3. 1 – 0% MnS, 2 – 0.5% MnS (acc. Ref. 280). *Turning test: PVD-TiN coated hardmetal (CNMG 120408), cutting speed 200 m/ min, feed 0.1 mm/rev., criterion 0.3 mm flank wear.*

these alloys with MnS addition without and with 0.5% carbon at a relatively minimal difference in density.

Figure 7.44 shows the strong effect of 0.5% MnS addition in enhancement in continuous and intermittent face turning of Distaloy AE-0.5% C alloy due to reduced strain in the shear plane. For intermittent turning as compared to continuous one, an insert with a more ductile core is to be selected in order to withstand force oscillations that occur during the operation.

Figure 7.45 shows relative tool life in drilling for plain iron and Fe–2Cu–0.5C material without and with 0.5% MnS and for Fe–2Cu-0.5C and Distaloy AE–0.5C material without and with 0.3% MnX addition. When drilling Distaloy AE-0.8C, the addition of 0.5% MnS caused only a minor improvement in machinabiliy, while the addition of 0.3% MnX caused a significant increase in drill life. It means that MnX is apparently a more effective machining aid for

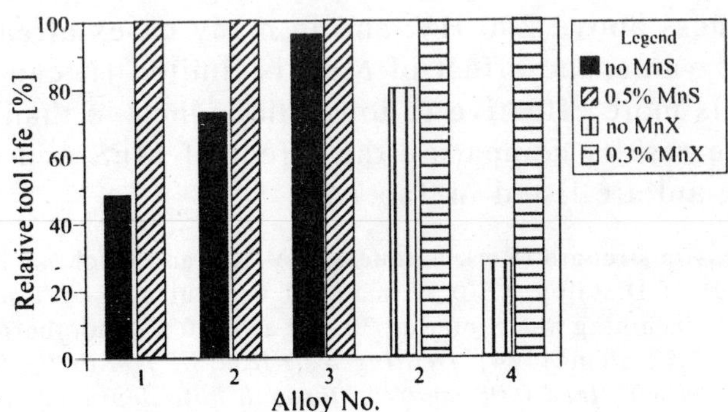

Fig.7.45 Effect of 0.3% MnX addition compared to 0.5% MnS addition on relative tool life in drilling of: 1 – plain iron (ASC100.29), 2 – Fe–2Cu–0.5C, 3 – Distaloy AE-0.5C, 4 – Distaloy AE–0.8C material [acc. to Ref.280]. *Drilling test: HSS 4 mm drill (point angle 118°), depth of hole 10 mm, dry drilling test.*

materials with higher hardness compared to MnS. The recommended drill for this material was a coated HSS drill with lubrication [280].

The machinability – in turning with a coolant – of a Fe–3% Ni-2.4% Cu–0.3% graphite (Distaloy SH) steel was enhanced by a manganese sulphide addition as confirmed in a large-scale production test. The parts, gear rings of an oil pump prepared from this steel (1200°C for 45 min, $20H_2$–$80N_2$, density 6.8 g/cm^3, after coining 6.9–6.8 g/cm^3) were turned with a hardmetal tool (SIB 2103D-T) at 800 rpm at a feed rate of 48 mm/min (total time for machining of a part 30 s). The total number of parts machined without MnS addition per tool was 1085 and with 0.3% MnS addition 2250. The addition of 0.3% MnS increased the tool life by about 100% [148].

7.4.2.2 Effect of 'new' machining aid

The development of some new machining aids with improved effect is a part of many studies concerned with the machinability of PM parts. In this way, the development of new additives can be a contribution to the further and substantially higher improvement in the machinability of high strength Distaloy type steels (see 6.1.3.3). In Ref. 233 the effect of a 'new' aid with the density a half that of iron by mass (appr. 2% by volume), without further data being given, on the machinability of Distaloy AE-0.5C steel, which seems to have the poorest machinability in this group of materials, was tested in drilling compared to MnS addition. The aim was to develop an additive which should give to the high strength materials a satisfactory machinability because in high strength PM materials with hardness above 200 HV, and in many cases already markedly below this value, the effect of MnS is limited (it can be supposed that MnS is more effective to lower the adhesive than the abrasive wear). The results comparing the effect of MnS and of the 'new' machining aid are listed in Tab.7.25.

Tab.7.25 Tensile strength (Rm), hardness (HV 10), and machinability (number of holes drilled) of Distaloy AE-0.5C material without and with addition of MnS and of 'new' machining aid. Sintering 30 min at 1120°C, endothermic atmosphere, density 7.08–7.12 g/cm^3 [154]. *Drilling test: HSS 3.5 mm drill, drill speed 3000 rpm, v_c = 33 m/min, feed 0.08 mm/rev., through hole depth 12 mm, no coolant*

Addition	Rm [MPa]	HV 10	Number of holes
0%	750	200	5
0.5% MnS	715	195	25
0.3% 'new'	740	200	74

The shrinkage caused by the machining aids tested was about 0.4% higher compared to those without aids. The hardness was not affected by these machining aids. A decrease in tensile strength from 750 MPa to 715 MPa with MnS and to 740 MPa with the 'new' additive was determined.

The machinability of a material with 0.5% MnS addition was almost 5 times better than that of the base material without MnS addition (5→25 holes drilled). Material with the addition of 'new' machining aid had a machinability 15 times that of a material without a machining aid (5→74 holes) and 3 times that of the material with 0.5% MnS addition. The high hardness and tensile strength of the material with the 'new' machining aid should be noted.

The results indicate that when machining high strength steels manganese sulphide cannot withstand the pressure all the times and coating by MnS is broken in some places in the cutting zone. As stated in Ref.154 this 'new' machining aid forms a stronger coating than MnS and thus withstand higher pressures at cutting.

7.4.2.3 Effect of enstatite on machinability

The effect of enstatite and MnS addition on machinability in turning of Distaloy SE-0.5C steel (properties listed in Tab.7.15) is shown in Fig.7.46.

In the case of MnS and mainly 0.3% enstatite addition, the tool life was markedly increased and stable improved surface finish was attained for long cutting distances. Enstatite, as in other alloys, having low cutting resistance, caused a lowering of the shearing energy by the chip breaking along with the lubricating action between the tool and chips. When the material without a machining

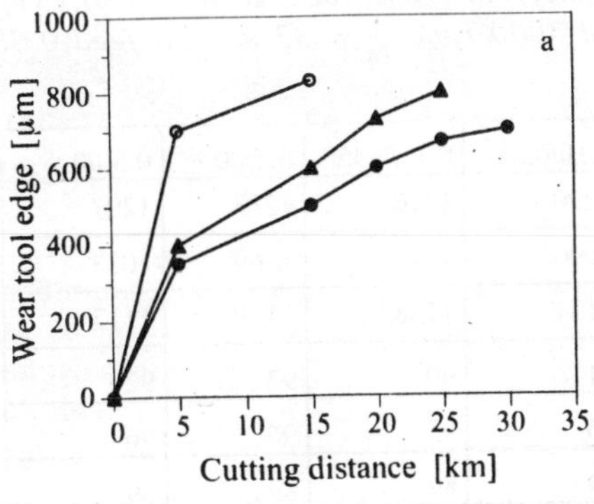

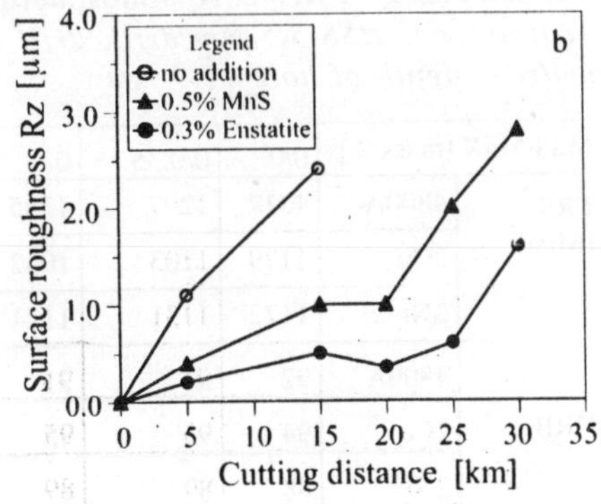

Fig.7.46 Effect of enstatite and of MnS addition on machinability in turning: a – tool edge wear, b – surface roughness of Distaloy SE-0.5C steel in dependence on cutting distance [208]. *Turning test: CBN tool, cutting speed 120 m/min, feed 0.2 mm/rev., depth of cut 0.3 mm, water soluble lubricant.*

aid was turned, the chipping on the tool edge occurred after about 10 km cutting distance.

7.4.2.4 Effect of MnS and MnX on properties and machinability of hybrid steels based on prealloyed Fe–0.85Mo powder

The typical example of high strength and high hardness steels are those based on prealloyed powders with admixed or diffusion bonded elements, *e.g.* copper or nickel. Not surprisingly, these steels are characterised by poor machinability. In such case an improvement in machinability can be attained by adding machining aids or by the use of some special heat treatment or cutting method.

For improving the machinability of Fe–0.85Mo (Ancorsteel 85HP) steel with admixed copper or nickel the effect of MnS and MnX addition in various proportions was tested. The samples were cold compacted, and for comparison, also samples from Distaloy 4800A (Distaloy SE) powder with appropriate graphite addition were prepared. The transverse rupture strength, hardness, and machinability of the investigated alloys is listed in Tab.7.26.

As shown, the free-machining aids used had a minor effect upon hardness and also as stated on dimensional changes. The transverse rupture strength of the Ancorsteel 85HP alloy with 2% Cu or 2% Ni was reduced by 6–13% by adding machining aids. The data suggest that 0.5% addition of either may reduce the TRS of these

Tab.7.26 Combined effects of MnS and MnX additives on transverse rupture strength (TRS), hardness (HRB) and machinability (number of holes drilled) of Distaloy 4800A-0.5C (4800A), Ancorsteel 85HP-0.9C (2Cu) and Ancorsteel 85HP-0.5C (2Ni) alloy compacts (carbon – graphite admixed). Compacting 600 MPa, sintering 30 min at 1120°C, 75H_2–25N_2 atmosphere, density 7.0 g/cm^3 (acc. to Ref. 136,137). *Drilling test: HSS 3.5 mm drill, drill speed 3000 rpm; v_c = 33 m/min, feed 0.08 mm/rev., depth of hole 12.7 mm*

MnS/MnX [mass %]		0/0	0/0.35	0/0.50	0.10/0.25	0.15/0.35	0.35/0	0.50/0
TRS [MPa]	4800A	1317	1297	1255	1290	1290	1283	1297
	2Cu	1179	1103	1062	1090	1076	1069	1028
	2Ni	1172	1131	1124	1131	1138	1159	1117
HRB	4800A	92	91	91	91	90	92	92
	2Cu	94	95	95	96	95	96	96
	2Ni	88	89	89	89	89	89	89
Number of holes	4800A	3	17	40	61	12	7	14
	2Cu	2	72	80	188	122	74	37
	2Ni	1	1	2	2	3	4	2

materials. The presence of 0.35% of an aid did not significantly reduced TRS.

All three alloys without machining aids can be described as non-machinable. The machinability of Distaloy 4800A steel increased with the addition of both machining aids in an amount 0.35–0.5% MnX and mainly in combination of MnS and MnX (0.1MnS-0.25MnX).

The Ancorsteel 85HP–2Ni–0.5C alloy can be classified as non-machinable also with an addition of the machining aids tested. Once more, nickel has to be defined as an element deteriorating the machinability of PM steels.

As shown, the highest improvement in machinability by addition of MnS/MnX was attained in the Fe-0.85Mo-0.5C alloy with addition of 2% copper, which in this case acts as an element increasing machinability.

To improve the machinability of Ancorsteel 85HP–2Ni–0.5C alloy, the effect of annealing was examined. This alloy exhibited a high resistance to tempering–annealing (150–870°C) which had only a slight effect on its microstructure and properties. The simple tempering and annealing cycles up to 650–730°C produced less improvement in machinability than tool improvements. Annealing did not produced the desired coarse pearlite, nor did it increase drill life. Annealing at 870°C in dependence on cooling rate has changed the failure mode from abrasive to adhesive. The drill appeared to 'stick' to the workpiece when cutting the annealed test pieces. Not surprisingly, the number of holes drilled in as-sintered compacts was 3 and in as-tempered at 450°C (optimum) only 14.

All tests performed also showed that the individual improvements of the tool, *e.g.* high speed steel grade, flute shape, or coating, had relatively little effect upon the drill life under the test conditions employed. However, when combined, the drill life was increased significantly. The highest drill life was recorded for TiN coated high cobalt HSS drills with a split point [136,137,242].

7.4.3 Summary

a) Distaloy type steels

- The machinability of diffusion alloyed sintered steels, considering some differences in composition presented by both Fe-1.75Ni–1.5Cu–0.5Mo–C (Distaloy SA) and Fe–4Ni–1.5Cu-0.5Mo–C (Distaloy SE) alloys, is poor to negligible. Improvement in machinability of these alloys by some suitable methods must be related to their poor basic machinability.

- The effects of S, MnS, MoS_2, MnX, Se,´Te, 'new' aid and enstatite as machining aids in various amounts, the effect of carbon content and especially of copper addition and density on the machinability of these alloys were tested. The results proved that machining of Ni-containing diffusion alloyed steels is more difficult compared to the materials with a simpler composition and less complex microstructure.
- The machining aids investigated affected the dimensional changes and mechanical properties of the materials to a different extent, the changes were mostly negligible, especially with MnS addition. Tellurium addition caused the relatively largest decrease in mechanical properties and increase in the dimensions compared to other machining aids.
- For some materials, large scatter in the machinability results under nominally equal cutting conditions was determined.
- The carbon content was found to be decisive for machinability. Materials with 0.4% carbon were more easily machined than materials with 0.8% C. Increasing the carbon content from 0.4 to 0.8% increased cutting forces in turning and resulted also in a significant decrease in machinability by drilling. High carbon materials could be best drilled with the P4 drill bit.
- All machining aids tested improved the machinability to varying degrees, and the effect of a single agent or method for improving the machinability must be considered in relation to the actual composition, processing conditions and, therefore to the actual mechanical and microstructure properties of the machined material.
- Manganese sulphide additions of 0.3 or 0.5%, esp. as high purity grade, caused a marked increase in the machinability of Distaloy SA base alloy, *e.g.* 13→113 holes drilled under constant conditions. The effect of MnS was the same for compacts sintered in an endothermic atmosphere (higher strength) and in dissociated ammonia.
- Addition of 0.5% MnS caused a decrease in the thrust force and torque in Distaloy SA alloy, also with increasing cutting speed. Increasing feed rate had little effect on surface finish.
- Manganese sulphide addition had no effect in tapping. Lubrication was effective, in contrast.
- Selenium and tellurium increased the machinability more than MnS and S. Tellurium improved machinability to a larger extent compared to other machining aids, but at the expense of mechanical properties, as shown above.

- Molybdenum disulphide addition (0.8%) increased the machinability to the level of that attained with Se and Te (0.25%) addition but at higher tensile strength.

The effect of additional 2 and 4% copper addition on the machinability in turning, drilling and tapping was tested for Distaloy SA and AB alloys with 0.4 and 0.8% C, and the following results were obtained:

- Cu addition caused a significant increase in hardness.
- Copper addition improved machinability in both processes.
- Lowering the carbon content or increasing the copper content was more favourable than the MnS addition. Increasing copper content to 4% with 0.8% carbon decreased the cutting forces in turning, but reduced the drill life in higher density materials. When copper was added to the 0.8% carbon alloys the tapping torque was reduced.
- At higher porosity of the compacts the tool life in turning was longer.
- Longest tool life in turning was achieved at a cutting speed of 100 m/min compared to higher cutting speed.
- At higher cutting speeds in drilling, *e.g.* at 30 m/min, a lower feed has to be used and vice versa.
- Cutting conditions that resulted in long tool life also caused a low surface deformation of the material in the cutting zone. Wet machining conditions reduce deformation of material.
- An effect of the base iron powder grade used for preparation of diffusion alloyed powders on machinability was recorded.
- The Distaloy SA (base sponge iron powder) material was turned more easily than Distaloy AB (base atomised iron powder) but the other way round in drilling. This shows that the base iron powder grade used for the preparation of a Distaloy material affects not only the mechanical properties but also, depending on the type of machining operation (drilling, turning), the machinability.
- The optimum machinability of Distaloy AE alloy in drilling was achieved with 0.5% MnS addition and 0.5% carbon content.
- The machinability of Distaloy AE material was better with 0.3% 'new' agent and with 0.3% enstatite compared to 0.5% MnS addition.

b) Cold compacted hybrid sintered steels with machining aids

- Both, MnS and MnX machining aids significantly improved the machinability (drill life) for Ancorsteel 85 HP–2Cu–0.5C alloy. Copper addition was effective in improving the machinability of the alloy compared to the Distaloy 4800A and Ancorsteel 85 HP-2Ni–0.5C alloy.

- A 0.35% total addition of the machining aids (MnS+MnX) was close to optimum under the conditions tested for the Distaloy 4800A and Ancorsteel 85HP-2Cu alloys. The combination of both free-machining aids was more efficient than single additives alone.
- No effect of single or combined MnS and MnX addition on the machinability of Ancorsteel 85HP–2Ni–0.5C alloy was recorded. This can be considered as a proof for the adverse effect of nickel on machinability of PM steels.
- Machining aids are less successful in higher strength PM molybdenum nickel steels.
- The results attained for the machinability of these alloys demonstrated a possible enhancing effect of machining aids and especially of alloying on the machinability of steels with high strength and high hardness. The reason is found in the sensitive microstructural characteristics of the materials.
- The heat treatment (annealing) experiments revealed a gap in knowledge about sintered steels.

Further intensive testing will provide a clearer assessment of the machinabiliy of diffusion alloyed materials. The potential problem of chemical reaction involving manganese sulphide and titanium nitride coatings also needs to be addressed.

The machinability of Distaloy PM compacts varies considerably with the specific alloy composition and processing conditions. It was shown that a considerable variation in machinability could be achieved with small changes in alloy composition. For example, changing the Distaloy type from SA to AB, or compacting the material at 500 MPa instead of 700 MPa, will improve the machinability of the component; however the mechanical properties have to be sufficient. Changes in the copper and carbon content may often be more beneficial than adding manganese sulphide or other machining aids.

7.5 CHROMIUM, MANGANESE, CHROMIUM–MANGANESE, AND SILICON ALLOYED STEELS

With respect to an increased interest in chromium, chromium-molybdenum, and also manganese, chromium–manganese and chromium–manganese–molybdenum low-alloy steels for production of structural parts, the data about their machinability will gain importance [47]. These are the steels alloyed with the elements of high affinity for oxygen. Iron–silicon steels belong also to this group of materials. These steels form a new group of high strength steels with many possibilities to affect the final properties in order to meet special requirements for some parts, *e.g.* through heat treatment processes. Therefore, the machinability of these steels should be investigated in parallel to the processing parameters and mechanical properties. A short overview of tensile strength and hardness of some alloys based on chromium prealloyed powders and of some manganese steels is shown in the following figures for different carbon levels and manufacturing routes. Here it stands out clearly for Cr-steels that with increasing carbon content and increasing density of the materials, strength and hardness increase, Figs.7.47 and 7.48; the sintering temperature is less relevant since higher sintering results in lower combined carbon content due to more effective deoxidation (see chapter 2.42; 5.13). The relationship between carbon content and mechanical properties is also evident with more complex alloy systems, as shown for Cr–Mo–X alloyed steels in Figs.7.49 and 7.50.

Figure 7.51 shows the tensile strength of Fe–(2–4)Mn–(0.5–1.0)C alloys based on SC100.26 iron powder. New effective combinations of mechanical and toughness properties can be expected in the hybrid Fe–Cr–Mn–Mo–C steels which however will exhibit perhaps poor machinability under conventional conditions [49]. There is hardly any knowledge about the machining of sintered silicon steels, due to their limited use mainly as soft magnetic materials [75].

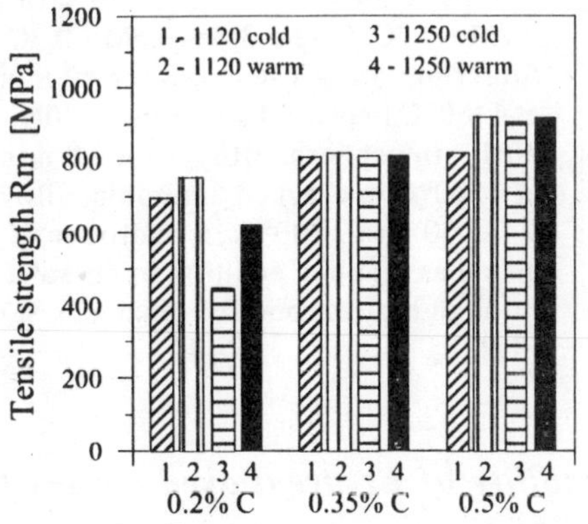

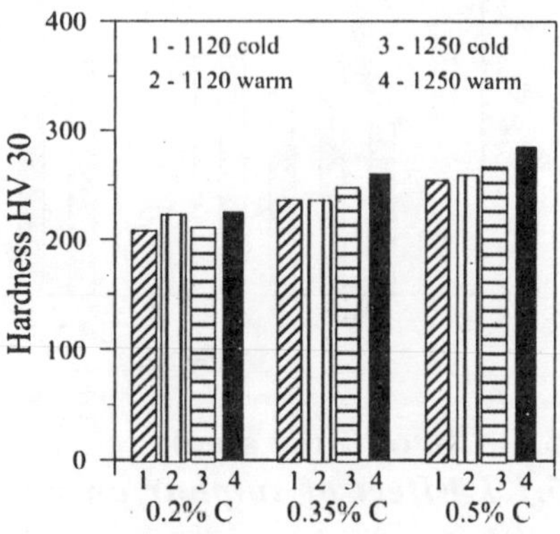

Fig.7.47 Tensile strength of Cr-steels based on Astaloy CrM powder in dependence on manufacturing route (cold and warm compaction at 700 MPa, sintering temperature) and carbon content (graphite) [281].

Fig.7.48 (right) Hardness of Cr-steels based on Astaloy CrM powder as in Fig. 7.47 [281].

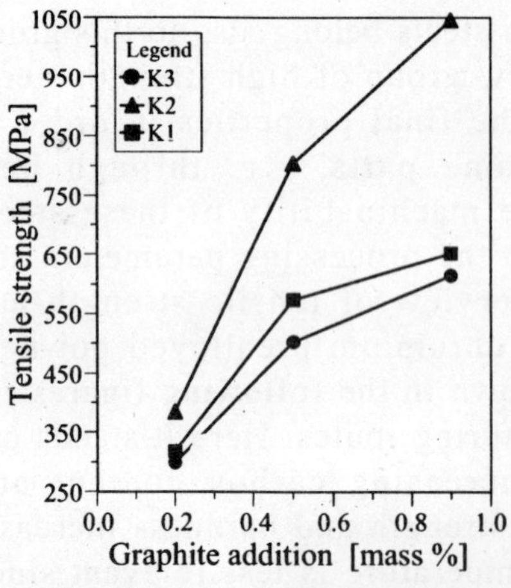

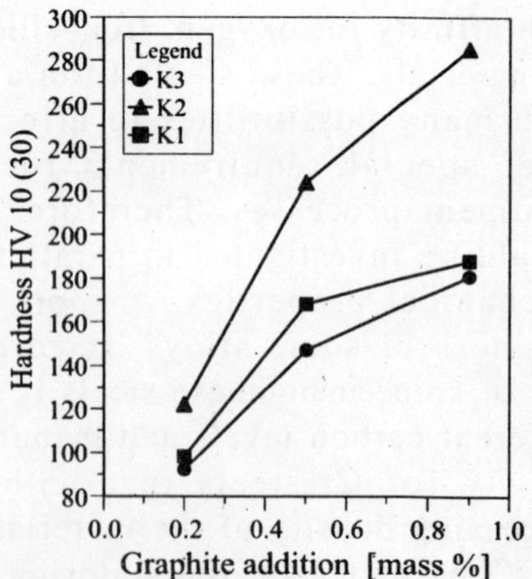

Fig.7.49 Tensile strength of Fe–Cr–Mo–XC alloy steels based on: K1 – Fe–1Cr-0.3Mo–0.3V (KIP 103V powder), K2 – Fe–3Cr–0.3Mo–0.3V (30CRV powder), K3 – Fe–1Cr–0.7Mn–0.2Mo (4100V powder) prealloyed vacuum annealed powder in dependence on graphite addition (C_c: 0.15–0.19% for 0.2% graphite, 0.28–0.30% for 0.5% graphite, 0.58–0.60% for 0.9% graphite). Compaction at 690 MPa, sintering 60 min at 1120°C in dissociated ammonia [282].

Fig.7.50 (right) Hardness of Fe–Cr–Mo–XC alloy steels as in Fig.7.49 [282].

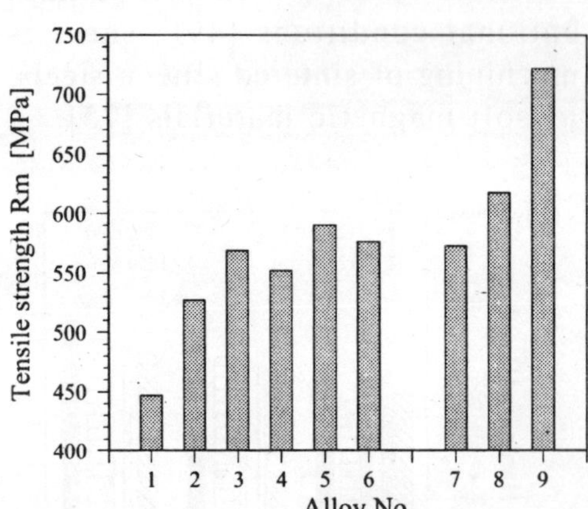

Fig.7.51 Tensile strength of sintered Fe-Mn-(Mo)-C steel in dependence on sintering conditions (as-sintered density of 6.8 to 6.93 g/cm³). Alloy: 1 – Fe-2Mn-0.4C; 2 – Fe-2.5Mn-0.7C; 3 – Fe-2.5Mn-1C; 4 – Fe-3Mn-0.7C; 5 – Fe-3Mn-1C; 6 – Fe-4Mn-0.4C; 7 – Fe-2.5Mn-0.7C; 8 – Fe-3Mn–0.2Mo-0.7C; 9 – Fe-3Mn–0.5Mo-0.5C. Sintering: alloy 1–6 and 8 for 40 min at 1180°C industrial sintering in $70N_2$-$30H_2$ atmosphere, alloy 7 for 60 min at 1200°C in cracked ammonia, alloy 9 for 60 min 1200°C in hydrogen (C added as graphite addition in part also through high carbon ferromanganese) [39].

7.5.1 Chromium steels

7.5.1.1 Effect of sulphur on machinability of Cr prealloyed sintered steel

The machinability (in drilling only) of a prealloyed Fe–1Cr–0.7Mn-0.2Mo–C steel (AISI 4100, oil atomised powder) was tested with various sulphur contents. Sulphur was added to/into the powder by three methods:

- prealloying; manganese content was kept constant in starting powder,
- prealloying; manganese content was increased in proportion to increasing sulphur content,
- admixing MnS powder.

The effect of sulphur content added to the base prealloyed manganese containing powder (starting 0.78% Mn) and the addition method on machinability (number of holes) in this steel is shown in Fig.7.52.

Machinability in drilling was improved by increasing the sulphur content from 0.05%. The relatively largest increase in the number of holes drilled was attained in tested sulphur prealloyed alloys. Increasing the sulphur content up to 0.3% increased the number of holes drilled by a factor of 10. A small difference in machinability was found when sulphur was added by other methods.

A characteristic marked drop in the number of holes drilled was recorded with increasing drilling speed in the range of 700 to 1500 rpm, with minor or no effect of sulphur addition. The highest machinability, *i.e.* the maximum number of holes drilled, was reached at the lowest drill speed of 700 rpm. When testing the sulphur-prealloyed material in turning, it was possible to increase the cutting speed from 60 to 105 m/min with increasing sulphur content.

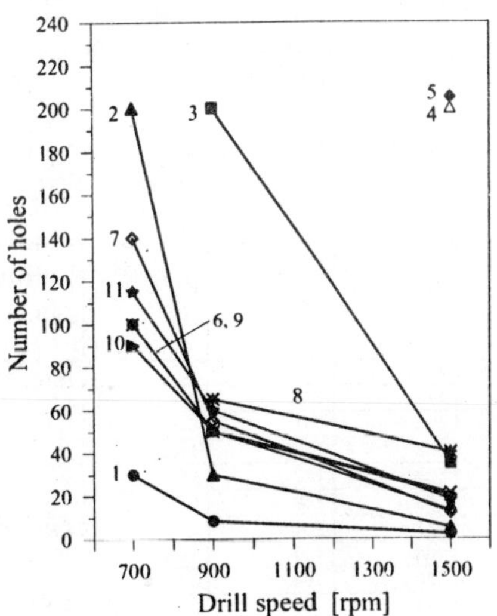

Legend		
Plot	S [mass %]	Method of addition
1	0.02	S-prealloying
2	0.05	
3	0.10	
4	0.20	
5	0.30	
6	0.10	S+MnS
7	0.20	
8	0.30	
9	0.10	MnS admixing
10	0.20	
11	0.30	

Fig.7.52 Machinability (number of holes drilled) in SAE 4100 (oil atomised) as a function of sulphur content, addition method, and of drill speed. Density 7.0 g/cm³, hardness 80–88 HRB, tensile strength 550–640 MPa (acc. to Ref. 283). Drilling test: HSS 10 mm drill, feed 0.15 mm/rev., blind hole depth 25 mm.

Prealloying with 0.05% S was recommended as a compromise between the machinability and the mechanical properties of the sintered steels based on chromium prealloyed powder.

7.5.2 Manganese steel

Sintered manganese steel parts are produced only in a very small number of cases. For this reason, the machinability of manganese steel, except for manganese additions not exceeding ~0.5% in combination with sulphur for machinability enhancement as mentioned, was not investigated. In terms of the possible use of manganese as an alloying element for high strength steel production, the basic mechanical properties and machinability test results of some mixed manganese steels without machining aids are listed in Tab.7.27. The effect of the iron powder grade and carbon content on the characteristics of materials was also investigated.

The length of holes drilled was chosen as machining index because in some alloys it was not possible to drill even one single hole through in the samples with 10 mm thickness.

The machinability of these steels was affected by all investigated factors, *e.g.* by manganese and carbon content and iron powder grade. The 1% and 2% Mn steels with the carbon content in this

Tab.7.27 Density, tensile strength (Rm), hardness (HV), impact energy (KC) and machinability (length of holes drilled – L) of Fe–Mn–C alloys in dependence on iron powder grade and carbon content. Manganese carrier – high carbon ferromanganese (76% Mn, 6.5% C), compacting 600 MPa, sintering 60 min at 1120°C, dissociated ammonia. Carbon content: alloy 1–6 combined, 7–15 added with high carbon ferromanganese [153]. *Drilling test: HSS 3 mm drill, drill speed 850 rpm, v_c = 8 m/min, constant thrust force 333 N, test sample – rectangular impact test bar (10 × 10 × 55 mm)*

Alloy No.	Iron powder grade	Mn added [mass %]	C [%]	Density [g/cm^3]	Rm [MPa]	HV 10	KC [J]	L [mm]
*1	SC100.26	2	0.30-0.35	6.41	447	105	9.8	127
*2		3		6.36	472	128	8.4	1.0
*3		4		6.30	492	151	4.5	1.0
4		2		6.81	462	128	16.2	26
5		3		6.75	529	142	10.5	5
6		4		6.71	563	169	6.7	1.0
7		1	0.08	6.82	337	96	26.7	454
8		2	0.16	6.88	404	137	19.1	355
9		4	0.32	6.65	520	193	4.0	2.0
10	NC100.24	1	0.08	6.81	334	104	23.3	425
11		2	0.16	6.68	384	142	19.0	376
12		4	0.32	6.65	503	179	5.7	1.0
13	ASC100.29	1	0.08	6.86	312	91	31.3	560
14		2	0.16	6.74	386	133	24.8	472
15		4	0.32	6.68	514	162	5.4	1.0

Remark: *compaction 400 MPa, L = number of holes drilled × 10 mm

case of up to 0.16% can be regarded as machinable, and the 4Mn steel with a carbon content of ~0.30% was not machinable. In this case, a relatively better machinability was recorded for the steels based on atomised powder grades as compared to reduced iron powder grades. A large difference in the machinability of the manganese steel in relation to the manganese and carbon content cannot be directly related to some other property of these steels because the differences in machinability were much larger than those in the properties shown. Relatively high homogeneity of the microstructure is characteristic for tested manganese steel based on sponge and atomised iron, Fig.7.53. Some coarser iron particles in the core of the samples are ferritic.

Despite the relatively poor machinability of manganese steels, the gears (see Fig.2.38) manufactured from Fe–3Mn–0.5Mo–0.3C steel, were machined under the conditions used for wrought steel gears.

In relation to the machinability of manganese steel the role of manganese as an element forming manganese sulphides used as a machining aid has to be remembered. In Ref. 282 a method for improving the machinability of PM steels called 'Mn content iron powder + MoS_2' technique, is described. According to this, the dissolution of MoS_2 occurs during sintering and sulphur from MoS_2 reacts with manganese in the base matrix. By this new MnS particles, more than 50% of them <3 μm in size, are formed. The

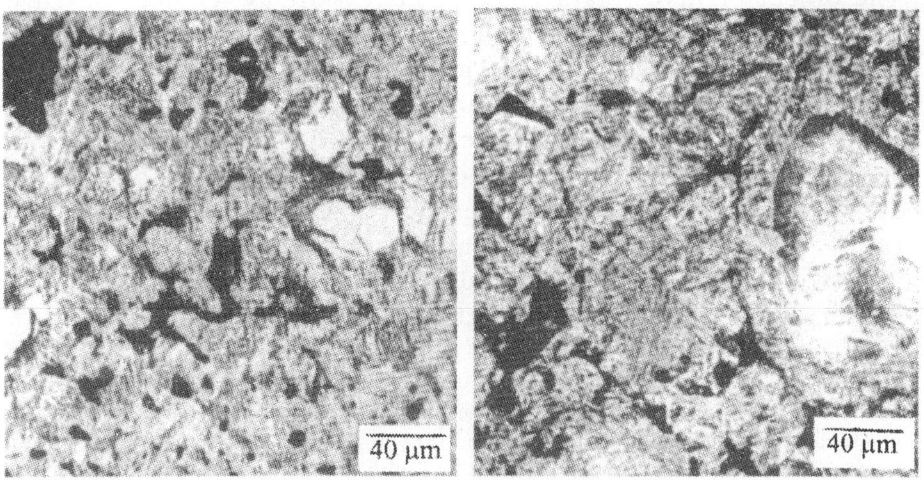

Fig.7.53 Microstructures of sintered Fe-4Mn-0.25C steel based on: a – sponge iron powder NC100.24; b – atomised iron powder ASC100.29; high carbon ferromanganese. Compaction 600 MPa, sintering 60 min at 1120°C, dissociated ammonia. Optical micrographs. Nital etched.

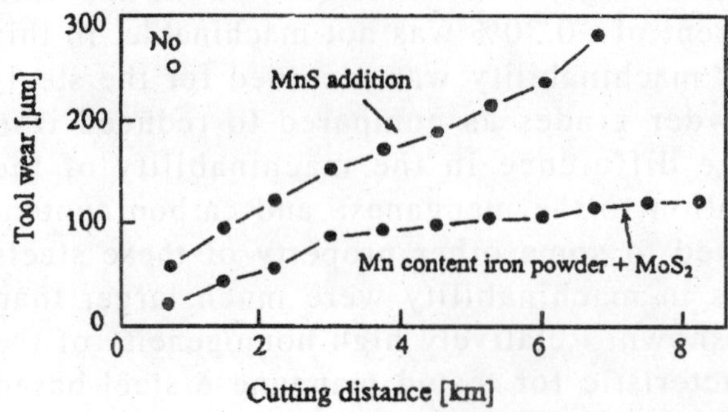

Fig.7.54 Machinability of Fe-1Mo-1Ni-0.5Cr-0.5Mn-1% graphite steel ('Mn content iron powder + MoS_2 method') in turning (valve seat insert). Green density 6.7 g/cm^3, sintering 60 min at 1180°C, dissociated ammonia [284].

finer distribution of the finer particles contributes more to improvement of machinability compared to MnS added directly to the powder mix. The effect of this method was tested in machinability (turning) of Fe–1Mo–1Ni–0.5Cr–0.5Mn steel with an addition of 0.5% MoS_2 compared to a material with 0.5% MnS and without any machining aid. The results are shown in Fig. 7.54. Moreover, the S supply source to obtain this phenomenon is not limited to MoS_2 powder. It was assumed that this method can be used for machinability improvement of high strength steels, and by this including manganese steels.

7.5.3 Chromium-manganese steel

The object of investigation was a Fe–Cr–Mo–V–C steel based on a chromium prealloyed vacuum annealed powder with the addition of manganese. The mechanical properties and machinability of the tested materials are listed in Tab.7.28.

In relation to the tensile strength and hardness, the chromium steel exhibited lower machinability compared to Fe–(1, 2)–Mn steel as was expected due to the presence of Cr-carbides and nitrides (after sintering in dissociated ammonia). A decrease in machinability of the steel was observed with manganese addition but at a higher increase in tensile strength.

Tab.7.28 Density, tensile strength (Rm), hardness (HV), impact energy (KC) and machinability (length of the holes drilled – L) of Fe–1Cr–0.3Mo–0.3V–0.25C steel (base 103V prealloyed powder) without and with manganese addition (high carbon ferromanganese) [153]. Drilling test: as in Tab.7.27

Mn added [mass %]	Density [g/cm^3]	Rm [MPa]	HV10	KC [J]	L [mm]
0	7.02	375	145	18.0	174
1	7.00	498	170	17.3	160
1.5	7.00	580	178	15.0	114

7.5.4 Iron-silicon steel

For an electromagnetic brake for the asynchronous electric motors the sintered bodies, core (∅130×30 mm, mass 2880 g) and armature (∅130×10 mm, mass 770 g) were prepared in the form of discs by compacting and sintering. The compacts were sintered at 1250°C for 2 h in vacuum, and the shrinkage, which is typical for the Fe-Si system, was 8% in height and 5% in diameter

The prototype parts (30 pcs.) for the measurement of magnetic properties with final dimensions, Fig.7.55 and 7.56, were machined by turning, boring (grooves for coils), tapping (M6) and threading (M36×1.5). The mass of removed material by machining from the part – core was 770 g. The coarse-grained ferrite microstructure was homogeneous. The machining of 30 parts with hardmetal tools was performed without detrimental flank wear. The machined parts were annealed at 650°C for 30 min in hydrogen to relieve the stresses formed by machining. After this annealing the magnetic induction of the parts was increased compared to the as-machined state [285]. The characteristic ferritic microstructure of Fe–Si sintered steel is shown in Fig.7.57.

7.5.5 Summary

- Machinability in drilling of as-sintered prealloyed low chromium steel was increased by sulphur addition in amount of 0.3% by ~10 times and the cutting rate in turning could be increased from 60 to 105 m/min.

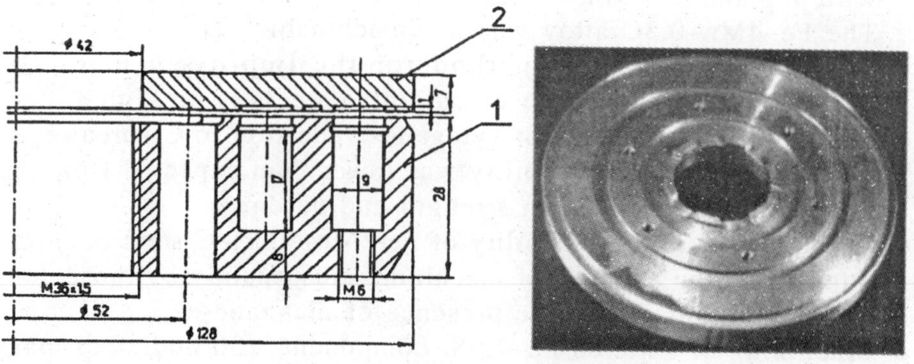

Fig.7.55 Dimensions (mm) of the core (1) and of the armature (2) produced from mixed sintered Fe-4.5Si steel (sponge iron powder + silicon in form of Fe90Si - ferrosilicon). Sintering 2 h at 1250°C, vacuum, density 7.24 g/cm³ (rel. density 93.6%), hardness 205 HB.

Fig.7.56 (right) Photography of the armature of a sintered electromagnetic brake.

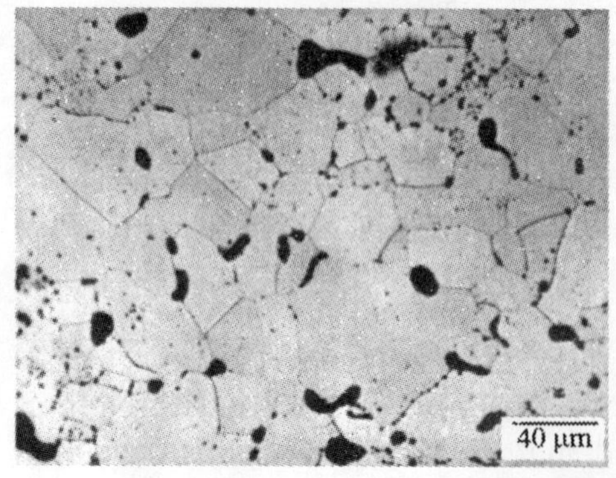

Fig.7.57 Microstructure of Fe–4.5Si steel as in 6.55. Optical micrograph. Nital etched.

- In sulphur prealloyed low chromium powder, 0.05% S was recommended as the best.
- A marked effect of drilling speed on the number of holes drilled in this steel was determined. Lower cutting speeds were recommended (~700 rpm) independently of sulphur content.
- Machinability of mixed manganese steel without machining aids was affected by iron powder grade, and manganese and carbon content. Higher machinability in drilling was exhibited by specimens based on atomised iron powder compared to sponge iron powder grades.
- In the investigated range, the optimum machinability was found with 1% and 2% Mn steels with carbon contents up to 0.16%. The Fe–4Mn–0.3C alloy was not machinable. The 'Mn content iron powder + MoS_2' method for the improvement of the machinability of manganese containing steels is presented.
- The deteriorating effect of manganese addition to Cr-prealloyed powder steel on machinability was lower than expected from the corresponding increase in strength and hardness.
- Improvement in machinability of Fe–(Mn)–(Cr)–C steel could be attained by the addition of machining aids, mainly by those containing sulphur due to the presence of manganese.
- Machining of sintered Fe–4.5Si components (30 bodies of magnetic brake) in turning, boring, tapping and threading was performed with hardmetal tools without significant flank wear.

7.6 IRON–PHOSPHORUS STEEL

In wrought steels, phosphorus causes embrittlement by grain boundary segregation. Consequently, the phosphorus concentration is limited to the lowest possible value. On the other hand, phosphorus belongs to the group of the cheapest alloying elements with the highest hardening effect in iron. In sintered steel, P is a widely used sintering activator, as stated in chapter 2.5.2.4.

The machinability of Fe–P–(C) alloys depends on the material characteristics as for other steels, with the exception of the inadequate relation between the strength and elongation in phosphorus-alloyed steels compared to those alloyed with other elements, *i.e.* by relatively high strength with relatively high ductility which prevents the very simple 'correlation' as higher strength – poorer machinability. Figure 7.58 shows the characteristic microstructure of Fe–P steel.

High strength properties are achieved by further addition of carbon, copper and/or nickel used for the production of structural parts. Carbon-free Fe–P alloys are employed usually as soft magnetic materials.

7.6.1 Effect of copper and nickel

For testing the machinability of P-containing steels the effect of other strengthening additions can be of significant importance. Among those, copper, in some other alloys enhancing the machinability and nickel, deteriorating machinability, can be representative. Both copper and nickel cause strengthening of Fe-P steels. Their effect in amounts of 1 to 4% on density and hardness of Fe–(0.3,0.6)P–(0,0.5,0.9)C (PASC30, PASC60) is listed in Tabs.7.29 and 7.30.

Copper addition caused a decrease in density and an increase in hardness. The heterogeneity in the microstructure of PASC30 alloy without carbon in both cases can be concluded from the scatter in the hardness measured with low load (HV 0.1).

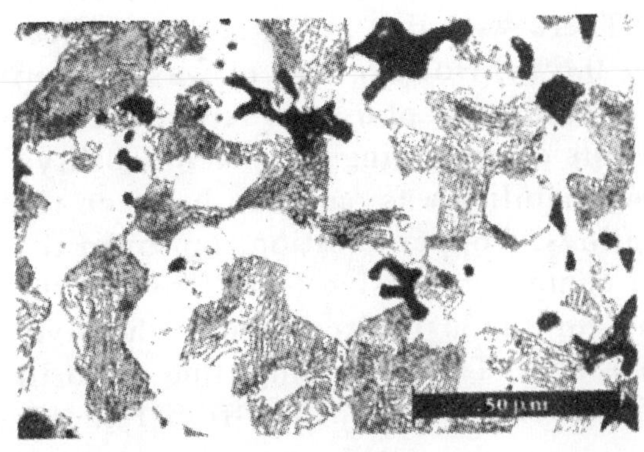

Fig.7.58 Microstructure of Fe–0.45% P–0.5% C (base ASC100.29 iron powder) steel. Sintering 30 min at 1120°C, density 6.98 g/cm^3 [235].

Tab.7.29 Density (ρ) and hardness (HV) values of PASC30 and PASC60 alloys with addition of (1 to 4)% copper and (0,0.5,1.0)% carbon. Compacting 600 MPa, sintering 30 min at 1120°C, exothermic atmosphere (acc. to Ref. 149)

Copper [mass %]	Alloy									
	*PASC30		PASC30-0.5C		PASC30-1.0C		PASC60-0.5C		PASC60-1.0C	
	ρ [g/cm³]	HV 0.1	ρ [g/cm³]	HV 5	ρ [g/cm³]	HV 5	ρ [g/cm³]	HV 5	ρ [g/cm³]	HV 5
1	7.00	293-319	6.99	180	7.00	192	7.08	220	7.10	230
2	6.99	297-330	6.87	187	6.91	199	6.92	235	6.93	241
3	6.90	300-341	6.81	193	6.81	205	6.83	243	6.90	252
4	6.81	304-350	6.75	208	6.78	290	6.80	257	6.85	260

Remark: *austenitisation at 900°C, 60 min, air cooled

Tab.7.30 Density (ρ) and hardness (HV) values of PASC30 and PASC60 alloys with addition of (1 to 4)% nickel and (0,0.5,1.0)% carbon. Compacting at 600 MPa, sintering 30 min at 1120°C, exothermic atmosphere (acc. to Ref. 149)

Ni [mass %]	Alloy									
	*PASC30		PASC30-0.5C		PASC30-1.0C		PASC60-0.5C		PASC60-1.0C	
	ρ [g/cm³]	HV 0.1	ρ [g/cm³]	HV 5	ρ [g/cm³]	HV 5	ρ [g/cm³]	HV 5	ρ [g/cm³]	HV 5
1	7.13	290-350	7.10	170	7.19	178	7.25	190	7.27	200
2	7.21	310-359	7.18	185	7.24	190	7.28	222	7.32	228
3	7.27	320-368	7.23	210	7.28	231	7.31	236	7.38	248
4	7.32	334-370	7.24	228	7.31	240	7.38	248	7.45	265

Remark: *austenitisation at 900° C, 60 min, air cooled

An increase in density with minor changes in hardness in Fe-P alloy with addition of nickel is characteristic for these alloys also in comparison to copper addition.

The effect of copper and/or nickel addition as strengthening elements for the mentioned PASC30 and PASC60 alloys on their machinability in drilling is shown in Figs.7.59 and 7.60.

The drill life was lowest for alloys without carbon but with 1% of copper or nickel addition. There was also no significant effect of increased P-content (0.3 or 0.6%). However, there was a small but conclusive effect of an increase of phosphorus and carbon content from 0.5 to 1.0% towards deteriorating the machinability.

A marked difference in machinability was recorded between the Fe–P–Cu–C and Fe–P–Ni–C alloys. Copper addition improved the machinability. With the copper content increasing from 1 to 4%, the machinability of Fe–P–C alloy continuously decreased to the level of Fe–P alloy, at increasing hardness. The machining time per drill to failure decreased with copper content increasing from 1 to 4%

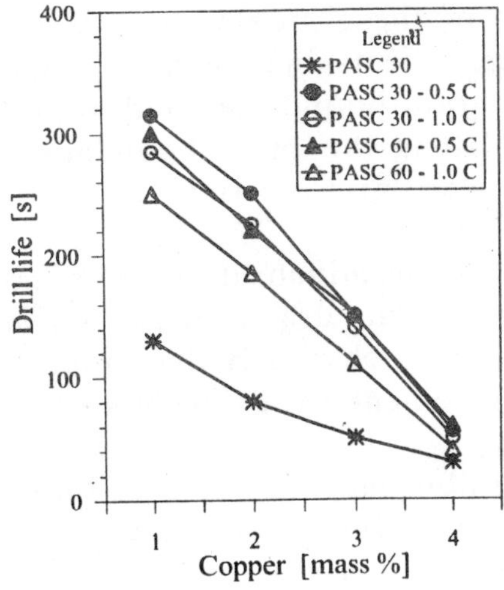

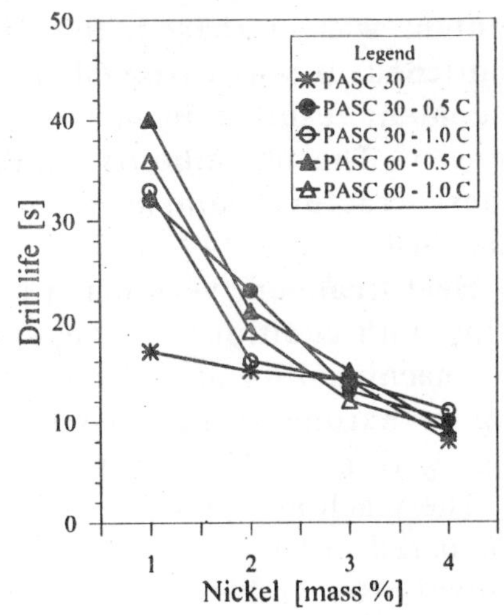

Fig.7.59 Effect of copper addition on machinability of PASC30-C and PASC60-C alloys (time per drill to failure) – properties are shown in Tab.7.29 (acc. to Ref. 149). *Drilling test: HSS 1.6 mm drill, drill speed 740 rpm, v_c = 3.7 m/min, feed = 0.075 mm/rev. Criterion – time taken per drill up to failure.*
Fig.7.60 (right) Effect of nickel addition on machinability of PASC30-C and PASC60-C alloys (time per drill to failure) – properties are shown in Tab.7.30. (acc. to Ref. 149). *Drilling test conditions as in Fig.7.59.*

in both PASC30 and PASC60–0.5C steel compacts. The increase of P-content from 0.3 to 0.6% had no effect. In contrast the effect of increasing C-content from 0.5 to 1.0% on the decrease of machinability was significant in the investigated range.

When the machinability of PNC 45 steel, determined as the number of holes drilled, was 20, that of steels with 0.6% C was 10, and that of steels with 0.3% C and 2% Cu was 41. This also confirmed the positive effect of copper addition at low carbon content on the machinability of P-containing steels [149,286]. The carbon content has a greater effect on the machinability of these steels.

Increasing the P content does not give any conclusive result for the machinability characteristics of sintered steels.

In the case of nickel, a sharp decrease from 1% Ni in the tool life up to 2% Ni addition and a gradual decrease in the tool life by 75% with the Ni-content increasing up to 4% at a hardness increase of ~30% was recorded, Fig.7.60. The machinability of Fe–P–C alloys with an addition of nickel can be regarded as very poor, compared to those with copper addition. When the C- and P-content was increased, the tool life of these steels was greatly shortened. It was assumed that in these steels increasing Ni- and P-content resulted in more clear-up grain boundaries and higher density. The phosphorus segregated along the grain boundaries. These factors consume more power giving rise to poor machinability.

Heat treatment very much lowered the machinability of PASC30 alloy with additions of copper or nickel. This drastic reduction in the machinability of as-heat treated steels is obviously due to much higher hardness and possible embrittlement by grain boundary segregation of phosphorus in this case.

The machined phosphorus steels exhibit improved surface finish compared to Fe–C material which improved further with increasing phosphorus and even more with higher carbon contents. The phosphorus-bearing materials gave the best surface finish over the entire span of P-contents used. A positive effect of phosphorus on the finish of inner holes turned in a stator of Fe–(1.2–1.5)P (soft magnetic alloy; hardness 120–140 HB, porosity ~15%, cutting speed of 120–130 m/min, cut depth 1 mm and fine 0.05 mm) was observed; while the finish was shiny as polished [287].

7.6.2 Effect of manganese sulphide

The effect of carbon and manganese sulphide addition on the mechanical properties of PNC45 alloy is listed in Tab.7.31 and on machinability is shown in Fig.7.61.

The addition of 0.5% MnS significantly improved the machinability of all steels. Addition of 0.3% MnS increased the number of holes drilled in PNC45 alloy by about 8–10 times. The poorer machinability of PNC45+0.6C compared to those materials without carbon corresponds to previous results which show some relationship to the hardness. Manganese sulphide additions improved

Tab.7.31 Density (ρ), tensile strength (Rm), elongation (A) and hardness (HV 10) of PNC45 steel with carbon and manganese sulphide additions. Compaction 589 MPa, sintering 30 min at 1120°C, endothermic atmosphere (acc. to Ref.148)

Alloy No.	Fe-0.45P+X [mass %]	ρ [g/cm^3]	Rm [MPa]	A [%]	HV 10
1	0	6.90	361	9.1	121
2	0.3 MnS	6.90	358	9.0	122
4	0.6C	6.82	442	4.9	145
5	0.6C, 0.3MnS	6.81	439	5.1	144

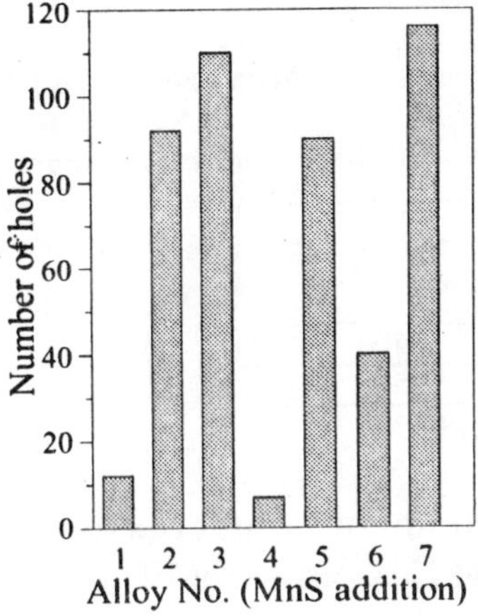

Fig.7.61 Effect of (0.3, 0.5)% manganese sulphide addition on machinability (number of holes drilled) of PNC45 steel with carbon and copper alloying. Bar - addition: 1 – none; 2 – 0.3% MnS; 3 – 0.5% MnS; 4 – 0.6% C; 5 – 0.6% C+0.3% Mn; 6 – 0.3% C+2% Cu; 7 – 0.3% C+2% Cu+0.3% MnS (acc. to Ref. 148). *Drilling test: HSS 3.5 mm drill, drill speed 3000 rpm, v_c = 33 m/min, feed 0.2 mm/rev., depth of hole 12 mm.*

the machinability of PNC45+0.6C about 15 times. The manganese sulphide also substantially improved the machinability of Fe–0.45P–2Cu–0.5C alloy (No.7) to the level of C- and Cu-free materials (No.2,3).

Compared to the previous results, a larger improvement in machinability in drilling of Fe–0.45P (A45P) alloy (Rm = 350 MPa, 54 HRB) was attained by 0.5% manganese sulphide addition. At drilling tests (v_c = 44 m/min), without MnS 20 holes and with MnS addition 840 holes 11.4 mm deep were drilled [224]. Similarly, in the Fe–0.45P (A1000B) alloy (density 7.0 g/cm^3) when drilling at v_c = 33 m/min, 151 holes without MnS and again more than 804 holes with 0.5% MnS addition were drilled [135]. The differences in the machinability of Fe–P material without and with MnS addition listed in the references cannot be explained by small deviations in the machining conditions but must be caused by the material properties.

A comparison of the effect of low- and high purity manganese sulphide additions and of the sintering temperature on the machinability of Fe–0.45P soft magnetic steel is listed in Tab.7.32. A small difference in the strength properties of compacts in dependence upon the purity of manganese sulphide was determined. When low purity MnS was used, the elongation dropped by 30% (15 *vs* 10%) and impact energy by ~20% compared to the compacts with no or with high purity, possibly through MnS agglomeration.

In this case, a very large improvement in machinability also with low purity MnS addition in comparison to previous results was recorded. Further improvement by ~100% in number of holes drilled was attained by the use of high purity manganese sulphide. The

Tab.7.32 Effect of 0.5% low and high purity manganese sulphide addition and of the sintering temperature on machinability (number of holes drilled) in Fe–0.45P alloy. Sintering in $90N_2–10H_2$ atmosphere. Tensile strength 400 MPa and hardness 50 HRB for compacts sintered at 1125°C [244]. *Drilling test: HSS 3.5 mm drill, drll speed 4000 rpm, v_c = 44 m/min, feed 0.12 mm/rev., depth of hole 11.4 mm*

Sintering temperature [°C]	Density [g/cm³]	MnS [mass %]		
		0	0.5 (l. p.)	0.5 (h. p.)
		Number of holes		
1125	7.2	38	1740	3500
1250	7.4	16	1460	3500

Remark: l. p. – low purity, h. p. – high purity

sintering of Fe–P steel at 1250°C resulted in a decrease in machinability in comparison to sintering at 1125°C in spite of lower porosity. Generally, the machinability of compacts sintered at either temperature was improved markedly by the use of manganese sulphide and especially a further significant increase by the use of high purity manganese sulphide.

When comparing drill tests (HSS 9.5 mm drill, drill speed 273 rpm) of plain iron (NC100.24) and of Fe–0.45P (PNC45) powder compacts sintered at 1150°C in an exothermic atmosphere it was determined that: a) hole integrity improved as drill speeds and feed rates were reduced, b) at equivalent density levels in PNC compacts a little improvement in hole integrity compared to NC100.24 compacts was obtained, c) drills with a 'standard' helix angle of 29° and point angle of 110...125° produced optimum consistencies in hole geometries [279].

7.6.3 Summary

- Poor machinability was exhibited by Fe–(0.3,0.45,0.5) P steels containing carbon.
- Manganese sulphide addition (0.3,0.5)% caused a significant improvement in the machinability of the tested materials, without and with carbon.
- High purity manganese sulphide addition had a stronger effect on machinability compared to low purity manganese sulphide.
- Large differences in machinability were recorded when testing under (nominally) equal conditions.
- Additional alloying of P-containing steels with copper and nickel for strengthening resulted in significantly different machinability. The machinability of Fe–P–Cu–C alloys was substantially higher, by ~10 times, compared to those alloyed with nickel.
- The surface finish was improved with phosphorus and even more with further addition of carbon.

7.7 STAINLESS STEELS

The requirements for stainless steels have aesthetic (colour and shine) and technical reasons and, therefore, are broader compared to low alloyed structural steels. They are:

- corrosion and oxidation resistance,
- mechanical properties, *e.g.* ductility,
- hot strength,
- wear resistance,
- magnetic and antimagnetic properties,
- hardenability,
- machinability.

In principle, the machinability of stainless steel parts cannot be as good as that of those parts with a lower alloy content due to high contents of alloying elements up to 30%, mainly of chromium and nickel. This feature is well known also from machining of wrought stainless steels, especially austenitic grades, which cause great problems due to considerable work hardening and tendency to form very long chips.

The stainless steels with their special microstructure and mechanical properties compared to conventional structural steels are used as PM material for the production of many structural parts which require secondary machining operations. Stainless steels are generally much more difficult to machine than might be expected from their low hardness which is mainly due to their high alloying with elements which deteriorated also the machinability of low alloyed structural steels.

Powder metallurgy procedures are used for producing austenitic chromium–nickel steels (*e.g.* AISI 304 and 316), ferritic chromium (*e.g.* AISI 430) and martensitic steels (*e.g.* AISI 410). The preparation of duplex stainless steels is of great interest for research and, consequently, their machinability should also be an interesting subject. The compressibility of stainless steel powder mixtures is rather low due to high nickel and chromium contents. Compacting at 600 MPa with an addition of 1% lubricant usually results in a density of the parts of 6.4–6.6 g/cm^3.

In order to increase compressibility of stainless steel powders to ~6.9-7.1 g/cm^3 and by this also the properties the preparation of duplex stainless steel by different methods is investigated. The final bi-phase microstructure in dependence on processing parameters will become the new object for machining [288].

Thus, relatively high porosity of the parts is characteristic which must also be considered with regard to machinability. Also the low thermal conductivity of stainless steel parts as a consequence of high alloying and low density is a further factor impairing the machinability.

The difficulties in machining of sintered stainless steels compared to, for example, low alloy structural steels are caused by the following factors:

- high alloying mainly with chromium and nickel,
- lower thermal conductivity,
- high work hardening rates,
- higher porosity,
- occasionally: sintering atmosphere containing nitrogen.

Both high alloying and porosity reduce the thermal conductivity of the material, particularly in highly porous, lower density parts. Consequently, the heat formed in the cutting zone rapidly increases the temperature of the workpiece–tool cutting pair, resulting in a marked reduction in tool life.

When sintering, the lubricant must be completely burnt out, and the sintering atmosphere should have a low carbon activity. The composition of the sintering atmosphere strongly affects the final properties and, consequently, the machinability of PM stainless steels. Most PM stainless steel parts are sintered in hydrogen and in vacuum and less frequently in nitrogen containing atmospheres at lower or higher temperatures. The effect of carbon and/or nitrogen must be eliminated to attain real stainless steel quality.

When sintering a stainless compact in a nitrogen-containing atmosphere (dissociated ammonia, mixed nitrogen–hydrogen atmosphere), nitrogen diffuses into the matrix. Depending on the reactive pore surface of the material, alloying, sintering temperature, and cooling rate, the nitrogen content in sintered compacts increases from about 0.026 to 0.72% [161], 0.03 to 0.2% [289] or 0.014% to 1.22% [290] when compared to hydrogen sintered materials, especially if cooling was too slow. Due to high affinity for chromium, adsorbed and diffused nitrogen forms complex nitrides and when some carbon content is present then nitrogen forms complex carbonitrides. The microhardness of the nitrides and carbonitrides is ~1200–1400 HV which creates highly abrasive conditions for the cutting edge, especially HSS drills.

Sufficiently low carbon and oxygen contents, together with their effect on the microstructure (these factors are essential for satisfactory corrosion resistance), can decrease the deteriorating effect of nitrogen in the atmosphere.

Martensitic grades, *e.g.* AISI 410, are soft and easily worked after sintering in hydrogen or in vacuum. Martensitic hardening during sintering can be ensured by sintering in nitrogen with an addition of either 0.1% carbon (sintering in hydrogen) or 0.2% carbon (vacuum sintering). Sintering should be carried out in a high purity reducing atmosphere (most efficiently hydrogen). In Ref.126 it is claimed that sintering in argon or in vacuum preventing reoxidation of open pore surfaces during cooling is not secured. Their high strength after deformationn and high work hardening rates also contribute to poor machinability. The austenitic stainless steels have high strain hardening exponents, so that cold-working rapidly increases their hardness and strength. Of particular importance is the spread between the yield and tensile strengths, which is much greater for austenitic stainless steel materials than for low or medium-carbon ferritic steels. Consequently, larger deformation is required for a stainless steel to reach its tensile strength than for low-carbon steels. This causes 'guminess' in machining which results in a built-up edge on the tool and poor surface finish.

As in other PM steels, the machining aids are also used for improving the machining of PM stainless steels. In terms of more complicated basic factors impairing the machinability of PM stainless steels it is necessary to ensure greater improvement by the use of machining aids and other methods. Their effectiveness is determined by the general requirement to improve – also economically – the machinability of these steels while not (too much) adversely affect the mechanical properties of these steels.

In this connection, it is necessary to stress once more the difficult machining of wrought stainless steels. Before machining, ferritic and martensitic steels are usually soft-annealed whereas in the case of austenitic stainless steels nothing can be done to prevent work hardening.

The comparative evaluation of wrought and PM stainless steels is possible due to equal chemical composition. For further comparison, Table 7.33 gives the machining rating of several wrought stainless steels, with the machinability of free-machining AISI B 1112 steel (max. 0.13% C, 0.16–0.23% S) taken as a standard, 100%. Grades 304 and 316 are about the same, and only the martensitic and precipitation hardenable grades are rated as having poor machinability.

Comparative data concerning the machinability of 303 and 304 wrought stainless steels with some structural steels is listed in Tab. 7.34.

Tab.7.33 Machinability rating of wrought stainless steels [289]

Material	B 1112	303	304	316	410	440 C	17-4PH
Rating [%]	100	140	95	102	94	65	55

Tab.7.34 Machinability (average drilling time for drilling 35 holes) of wrought 304 (density 7.93 g/cm³) and of 303 and 304 PM stainless (density ~6.55 g/cm³, sintering at 1218°C, dissociated ammonia) and of some PM structural steels [289]. *Drilling test: HSS 6.35 mm drill, drill speed 480 rpm, v_c = 9.6 m/min, constant drilling force 780 N*

Material	Wrought 304	PM 303 N	PM 304 N	PM 304 L	F-000-10	FC-0208-40	FX-1008-50
Hardness	80 HRB	60 HRB	60 HRB	55 HRB	55 HRB	70 HRB	80 HRB
Drilling time [s]	84±12	37±2	82±21	44±6	47±5	55±5	47±3

Remark: FX-1008 copper infiltrated

Despite their lower density, PM stainless steels could be drilled faster compared to wrought steel. Low density iron, copper steel, and copper infiltrated steel all could be drilled quite fast with consistent drilling times. The free-machining grade of PM austenitic stainless steel (303) drilled faster than any of the PM materials compared and the drilling was better than copper infiltrated steel which is generally regarded as a readily machinable PM material. Higher nitrogen content in 304 N (2050 ppm) steel increased the hardness and lengthened the drilling time compared to 304 L (280 ppm) steel [289].

7.7.1 Austenitic stainless steel

7.7.1.1 Effect of density and sintering temperature and atmosphere

Sintering conditions, mainly the sintering atmosphere, play a more important role in the machining of stainless steels compared to other structural alloyed steels.

Effect of density. As mentioned and shown in chapter 5.2.3 (Fig.5.11–5.13), density is more relevant for the machinability of sintered stainless steel compared to other sintered iron base materials. Independently of other factors, the lower density of PM stainless steel material is the first factor that must be considered in the machinability process. The drilling of a stainless steel of low density (relative density ~70%) occurs with higher thrust force due to high work hardening rates compared to one of higher density material. The average drill flank temperatures are also significantly higher at the higher cutting speed and increased with the hole depth, as expected. This is then the reason for the lower number of holes drilled at the higher cutting speed.

Effect of sintering temperature and atmosphere. The effect of sintering temperature and especially of the atmosphere on the machinability of some widely used austenitic PM stainless steels is listed in Tab.7.35.

The results shown indicate that raising the sintering temperature reduced the required drilling time by about 15%. This was found for all austenitic grades evaluated, and the results were quite consistent. The 303 grades were the best at both temperatures and could be drilled faster with more uniformity.

In all cases, holes could be drilled faster in the material sintered in hydrogen. In a nitrogen-containing atmosphere, the amount of nitrogen absorbed at higher temperature was smaller (~250 ppm) than that at lower temperature (1500–2000 ppm). This explains the improvement in machinability as the sintering temperature increased and was the reason for poorer machinability due to the supposed larger amount of nitrides formed at the lower sintering temperature. These main results were also confirmed when the average thrust force at a constant feed rate (40 mm/min, 800 rpm) in the drilling force was measured as the machining index [243,289].

In contrast to the 'interrupted' cutting theory of the effect of porosity on machinability, metallographic analysis showed that the newly generated surface at drilling was hardened and densified below the cutting edge and along the walls of the holes. The depth of the work hardened layer was almost always greater than the depth of cut. The depth of densification increased with porosity and tool wear. Thus, the drill cutting edges did not cut the porous base material but rather contacted the densified and work hardened

Tab.7.35 Effect of sintering temperature and atmosphere on machinability (average drilling time for one hole) of austenitic PM stainless steels. Density ~6.50 g/cm³ [289]. *Drilling test: HSS 6.35 mm drill, drill speed 480 rpm, v_c = 9.6 m/min, constant thrust force 780 N, through hole depth 25 mm*

Sintering temperature [°C]	Material					
	303 N	303 LSCN	304 N	304 LSCN	316 N	316 LSCN
	Drilling time [s]					
1120, d. a.	44±4	43±4	92±12	119±15	-	-
1220, d. a.	37±2	37±2	82±21	92±18	76±20	87±14
1220, H_2	33±2	30±3	44±6	-	-	-

Remark: d.a. – dissociated ammonia, LSC material - extra corrosion resistant grades alloyed with copper and tin. The average drilling time for 304 wrought steel was 84±12 and for 316 steel was 62±8 s

material. At low densities, cracks existed in the top layer of the material at the bottom of the hole and were connected to the porosity below the surface. This could easily cause material fracture during drilling [142].

7.7.1.2 Effect of MnS and sintering atmosphere

The effect of machining aids on machinability of stainless steels is more closely related to the sintering conditions compared to other alloys. The effect of dissociated ammonia and of N_2+CH_4 atmosphere, both containing nitrogen but in markedly different amounts, and of manganese sulphide addition on machinability of 316L steel is listed in Tab.7.36. When comparing the effect of both atmospheres, it may be seen that the higher number of holes was drilled in materials sintered in dissociated ammonia with lower nitrogen content compared to the nitrogen–methane atmosphere. A large difference in the number of holes drilled was found for materials containing MnS and sintered in dissociated ammonia and nitrogen–methane atmosphere, respectively. This difference in the number of holes drilled (>1760 vs. 352), especially in stainless steel, cannot be explained only by the formation of larger amounts of hard compounds in the latter atmosphere.

Table 7.37 gives the optimum cutting speed and feed, and Fig.7.62 shows in more detail the optimum cutting speeds for 316LHC material as a function of the sintering atmosphere and manganese sulphide addition.

The effect of the sintering atmosphere and of manganese sulphide addition on the cutting conditions in the drilling of this steel was also clearly demonstrated in this case. Higher speeds and feeds could be used for drilling the materials sintered in hydrogen, especially with the addition of 0.5% MnS. Tool life increased with increasing MnS addition but it is necessary to consider the deteriorating effect of increasing cutting speed. Drilling speed

Tab.7.36 Effect of sintering atmosphere containing nitrogen on machinability (number of holes drilled) of 316L stainless steel without and with 0.5% MnS addition [161]. *Drilling test: HSS 3.5 mm drill, drill speed 700 rpm, v_c = 7.9 m/min, feed 0.1 mm/rev., depth of hole 11.4 mm*

Density [g/cm^3]	Sintering temperature/time [°C/min]	Atmosphere	0 % MnS	0.5 % MnS	Ref.
			Number of holes		
6.9	1135/25	Dis. ammonia	~50	>1760	31
6.6	1120/30	N_2+CH_4	23	352	25

Remark: hardness 70 HRB for compacts sintered in dissociated ammonia (d. a.)

Tab.7.37 Optimum cutting conditions for drilling of sintered 316LHC stainless steel sintered in hydrogen and in nitrogen–hydrogen atmosphere, respectively, without and with MnS addition [161]

Cutting conditions	Sintering atmosphere			
	H_2		$70N_2$-$30H_2$	
	no MnS	0.5 % MnS	no MnS	0.5 % MnS
Cutting speed [m/min]	8.6	14.9	5.5	10.2
Feed [mm/rev.]	0.25	0.25	0.15	0.20

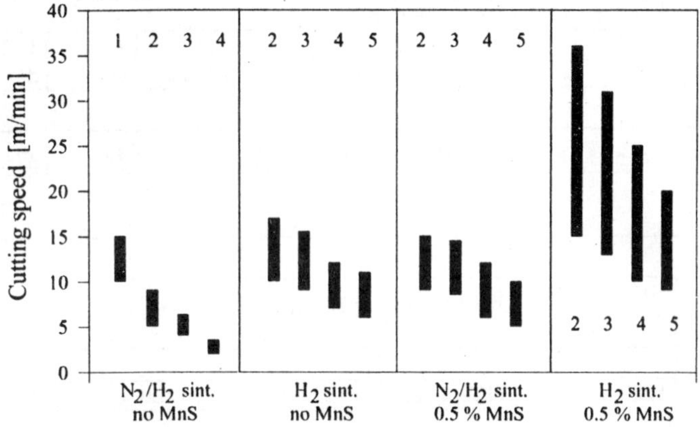

Fig.7.62 Optimum cutting speeds for drilling of 316L steel without and with 0.5% MnS addition, sintered in hydrogen and in nitrogen–hydrogen atmosphere, at different feeds (10% ranges) [289]. Feed: *1 – 0.05 mm/rev., 2 – 0.10 mm/rev., 3 – 0.15 mm/rev., 4 - 0.20 mm/rev., 5 – 0.25 mm/rev.*

decreased with increasing feed, especially in the material with the highest machinability. With increasing feed the temperature in the cutting zone increased, resulting in less lubricating effect of the MnS addition for machinability enhancement, and thus the tool wear increased.

7.7.1.3 Effect of MnS, BN and sintering atmosphere

The machinability of 304 and 316 wrought and of 316, 303 and 304 PM steel grades sintered in dissociated ammonia and/or in hydrogen at two temperatures without and with MnS and BN addition is listed in Tab.7.38.

304 wrought grade is characterized by poor machinability. 316 and 304 PM grades sintered in dissociated ammonia are less machinable compared to wrought grades. On the other side, 303 304 PM grades sintered in hydrogen exhibited significantly better machinability.

Tab.7.38 Effect of sintering temperature and atmosphere on the machinability (average time for drilling one hole) of PM stainless 316, 303 and 304 grades without and with MnS and BN addition and of wrought 304 and 316 grades. Density: 6.55 g/cm³ for 303 N, 6.52 g/cm³ for 304 N, 6.59 gcm³ for 316 N (acc. to Ref. 289). *Drilling test: HSS 6.35 mm drill diameter, drill speed 480 rpm, v_c = 9.6 m/min, constant thrust force 780 N, blind hole depth 25.4 mm*

a)

Wrought		PM								
304	316	316 N	316 LSCN	303 N	303 N	303 LSCN	303 LSCN	303 L	303 LSCN	303 LSC
		Sintering [°C] for 32-48 min								
		1220 d. a.	1220 d. a.	1220 d. a.	1120 d. a.	1220 d. a.	1120 d. a.	1220 H_2	1220 d. a.	1220 H_2
Drilling time [s]										
84±12	62±8	76±20	87±14	37±2	44±4	37±2	43±3	33±2	37±2	30±3

b) continued

PM					
304 N	304 LSCN	304 LSCN	304 L	304 N + 0.5MnS	304 N + 0.5BN
Sintering [°C] for 32-48 min					
1220, d. a.	1220, d. a.	1120, d. a.	1220, H_2	1220, d. a.	1220, d. a.
Drilling time [s]					
82±21	92±18	119±15	44±6	45±8	91±30

Remark: d. a. – dissociated ammonia, LSCN – extra corrosion resistant with copper and tin alloying

Raising the sintering temperature from 1120°C to 1220°C reduced the required drilling time by about 10%. This was explained by the smaller amount of nitrogen absorbed at the higher sintering temperature.

When sintering in hydrogen, a small increase in machinability was attained compared to the H–N atmosphere and in all materials the holes were drilled faster with greater uniformity in drilling times. The 304 grades showed the largest difference in drilling in relation to both sintering atmospheres. The combination of sintering in hydrogen and MnS addition resulted in the best machinability, *i.e.* in improved tool life [161]. The material without MnS sintered in nitrogen-hydrogen was in general most difficult to machine.

Different hardness of the materials was determined, 116 HB after sintering in hydrogen and 152 HB after sintering in the nitrogen-containing atmosphere.

The MnS addition was effective for drilling of 304 N steel

sintered in dissociated ammonia but this mix was still not as good as the PM 303 grade. The difference in machinability between 304 and 303 grades is large. The highest machinability in the investigated range was exhibited by the 303 grade sintered, either in dissociated ammonia or hydrogen.

The 304 material with an addition of 0.25% graphite could not be machined. These results correspond to the presumption of the presence of chromium carbide precipitates at the grain boundaries formed during slow cooling. This resulted in an extremely poor machinability, and in tool failure through abrasive wear.

The addition of up to 1.5% selenium also caused a decrease in thrust and torque in drilling 304 and 303 stainless steels. The presence of this additive resulted in a reduction in the mechanical properties and in part dimensions during sintering [202].

7.7.1.4 Effect of sulphurisation and resin impregnation

The effect of both machining aids on machinability of PM 303, 304 and 316 grades in the drilling test is listed in Tab.7.39. The sulphurised alloy 303 grade was a little better in machining than the 304 or 316 grades, but still all were very poor up to not machinable. The different sintering conditions in this case had no effect on

Tab.7.39 Machinability data (average number of holes drilled and thrust force) of sulphurized 303, 304 and 316 PM steels sintered under various conditions. Some materials resin impregnated (anaerobic impregnant) [145]. *Drilling test: HSS 9.5 mm drill, drill speed 500 rpm, v_c = 15 m/min, feed 0.23 mm/rev., no coolant, blind hole depth 25.4 mm*

Steel/sintering temperature [°C]													
303 N1	303 N2	303 N2	304 L	304 L	304 N1	304 N2	304 N2	316 L	316 L+ imp.	316 N1	316 N1+ imp.	316 N2	316 N2+ imp.
1120	1315	1315	1290	1290	1120	1315	1315	1290	1290	1120	1120	1315	1315
Density [g/cm³]													
6.5	6.5	6.9	6.5	6.8	6.4	6.4	6.9	6.5	6.5	6.4	6.4	6.5	6.5
Number of holes													
37	45	47	3	4	2	1	2	20	192	1	192	4	192
Thrust force [N]													
558	514	539	671	647	777	904	681	542	121	773	166	729	128

Remark:.Sintering time and atmosphere: N1 – 32 min, d.a.; N_2 – 48 min, 90% d.a.+10% N_2; L – 45 min; imp. - resin impregnation; testing finished at 192 holes drilled without drill failure

machinability in spite of high temperature sintering; higher density had also no measurable effect on machining behavior. The sulphurization did not contribute in this case to the markedly enhanced machinability of tested materials.

Only resin impregnated samples achieved 192 drilled holes when drilling was stopped at the lowest thrust force. Resin impregnation was more effective in machinability improvement of the materials when compared to the sintering temperature and resulphurisation effect. A relatively larger difference was determined when machinability was measured in the thrust force. The highest thrust force in the tested range was determined for 304 alloy sintered at 1315°C in the H–N atmosphere [145].

The relationships between the thrust force and the number of holes drilled is shown in Fig.7.63. It is clearly shown, that in this case, the sulphurisation of the steels, a sintering temperature increase by 200°C and different amounts of nitrogen in the atmosphere have only a small effect on their machinability. The number of holes drilled increased with lowering the thrust force.

7.7.1.5 Effect of waterglass, boron nitride and talc

Soda waterglass (Na_2SiO_3 – 70.6%) (waterglass) and talc increase the number of available free-machining aids and were used as machining aids in tests in comparison to BN for austenitic 304 L grade.

The machinability tests were carried out in turning of 304 L steel specimens (600 MPa, bushing Ø 63/20 × 63 mm, sintering at 1200°C in vacuum) without and with machining aids. The tests were performed by turning with different cutting tools at different cutting speeds. The tool life in minutes, determined when the flank wear reached 0.2 mm, was used as the machining index. The effect of

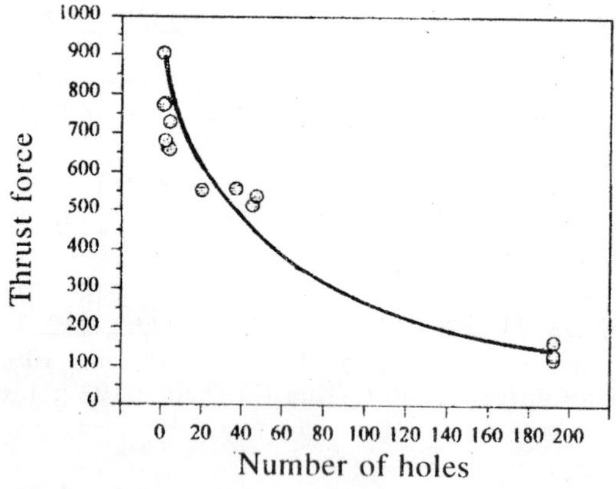

Fig.7.63 Relationships between thrust force and number of holes drilled for steels listed in Tab.7.39.

waterglass addition on the machinability is shown in Fig.7.64.

In these tests, the effect of the tool material in relation to the type of machining aid was clearly visible. When waterglass was added, the tool life was improved most markedly by 3% addition of waterglass for K10 and P20 tools. In contrast, the cermet tool life became progressively longer with increasing waterglass addition.

When adding 1% BN, the maximum tool life was obtained for all three cutting tools, with the tool life of K10 showing the largest improvement (10 times). A ceramic tool tested was not suitable for cutting the sintered austenitic stainless steels due to early chipping during cutting.

When the waterglass addition was combined with adding up to 3% BN, a significant improvement in the K10 tool life was achieved (90 min at 150 m/min cutting speed). Figure 7.65 shows the complex effect of optimum waterglass, BN and talc additions on the tool life in turning of PM 304 L for different cutting tools.

As regards the effect of waterglass and talc on machinability, on the basis of good results it is assumed that the waterglass or talc adheres to the relief face at the tool/workpiece interface and may lubricate the worn surface of the tool. The addition of talc was less effective in comparison to waterglass.

The real tool life was lower with increasing cutting speed and improved mainly at a lower speed (100 m/min). The minimum difference in the effect of machining aid additions was recorded for the cermet cutting tool. The best results with all machining aids tested were obtained with cermet tool for the cutting speed range 100-200 m/min.

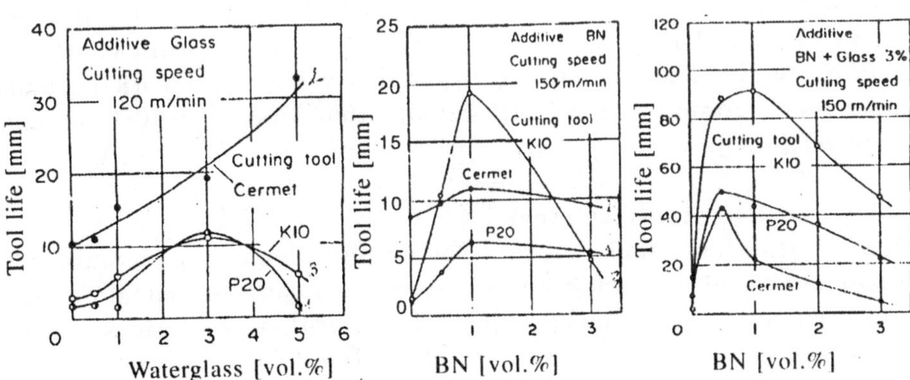

Fig.7.64 Dependence of tool life in turning of PM 304 L steel on addition of: waterglass (1–5)%, boron nitride (1–3)%, and 3% waterglass+(1–3)% BN, all for hardmetal cutting tool P20 and K10 and cermet, feed 0.05 mm/rev., depth of cut 1.0 mm [291].

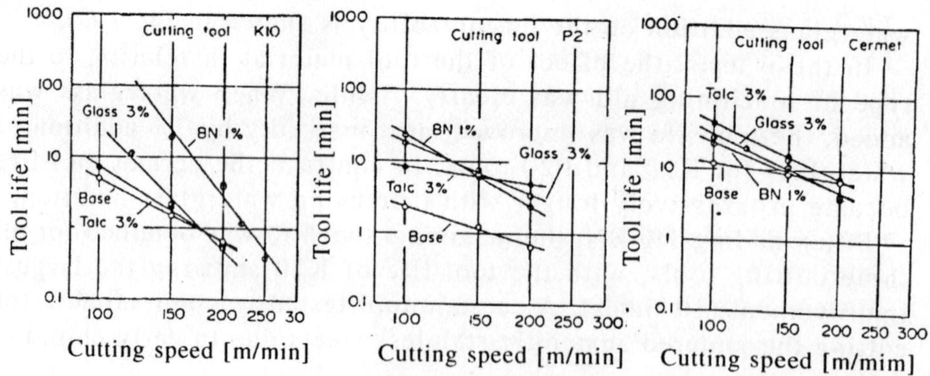

Fig.7.65 Dependence of tool life in turning of base PM 304 L steel without and with addition of boron nitride (1%), waterglass (3%) and talc (3%) addition for cutting tool: K10, P20 and cermet on cutting speed [291].

Here the effect of cutting tool material on machinability of PM 304 L steel is shown also in relation to the machining aid used; this should be taken into account also in machining of other PM alloy steels for all machining processes. It is a consequence of different strength of adhesion of the cutting tool material to the workpiece material in the contact zone and of the machining aid forming a lubricant (friction).

When drilling PM grades 304 L it was also found that the material with a density of 70% resulted in much higher wear rates than the material with 90% density. When measuring micro-hardness in the fully dense region at the bottom below the tip of the drill, it was noted that the low density materials hardened to 490 HK, while 90% dense material hardened to only 300 HK. Since the drill is always cutting through the almost fully dense work hardened material, the drill life was determined by the hardness of the workpiece. It appears that the low density material work hardens more when compressed to full density than does the 90% density material [129].

7.7.2 Ferritic stainless steel

Ferritic stainless steel powders for structural parts are commonly water atomised and must be subjected to a subsequent annealing treatment to provide acceptable compressibility. Ferritic stainless steels (base Fe–17Cr–1Mo–0.7Si–0.25Mn) do not contain nickel. Nevertheless, compared to austenitic stainless steel powders, ferritic grades are more difficult to compact to higher densities. Therefore, in most cases parts are produced that are compacted at 600 MPa with a density of 6.5–6.6 g/cm^3.

7.7.2.1 Effect of MnS and sintering atmosphere

The effect of sintering atmosphere and cutting speed on the tool life at drilling of 430LHC ferritic stainless steel without and with 0.5% MnS addition is shown in Fig.7.66. The worst machinability was recorded for the alloy without MnS addition and sintered in the N–H atmosphere. The 0.5% MnS addition improved the tool life more markedly than the change of the atmosphere. The compacts sintered in hydrogen (hardness 97 HB) with 0.5% MnS addition showed the highest tool life, and the optimum cutting speeds were increased by a factor of 2–4.

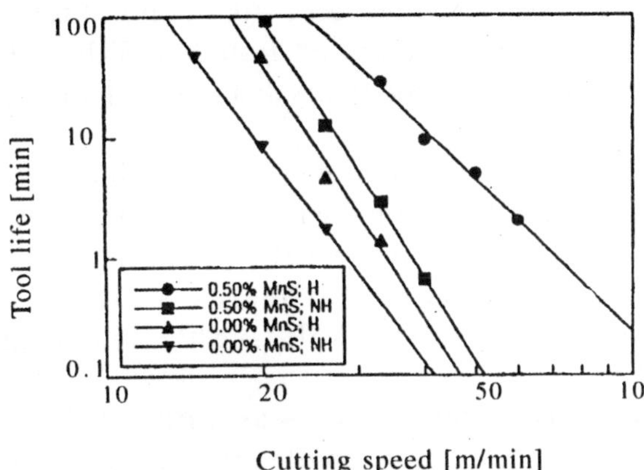

Fig.7.66 Effect of sintering atmosphere and of 0.5% MnS addition on drill life (min) in drilling of stainless steel 430LHC for varying cutting speed. Sintering 30 min at 1280°C, hydrogen or $70N_2$-$30H_2$ atmosphere (hardness 145 HB). Density 6.5–6.6 g/cm^3 [290]. *Drilling test: HSS 5.0 mm drill, drill speed 0-12.000 rpm changed, feed 0.1 mm/rev., blind hole depth 7.5 mm.*

Results obtained for the machinability of sintered stainless steel 430LHC with varying MnS addition are shown in Fig.7.67; machinability improved with increasing MnS contents after hydrogen sintering.

The tool life continuously improved by raising the MnS content. The differences beyond 0.5% MnS are, however, minimal, especially for higher drilling speeds. An addition of 0.50–0.75%

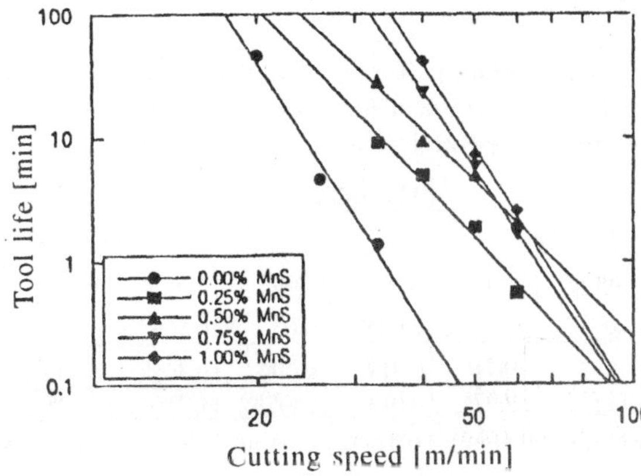

Fig.7.67 Influence of MnS addition (0–1.0)% on tool life (min) when drilling compacts of 430LHC steel sintered in hydrogen [290]. *Drilling test conditions as in Fig. 6.66.*

MnS improved the drill performance to a higher degree than it was degraded by sintering atmosphere effects. As shown, higher speeds reduced the tool life by between one and two orders of magnitude in both cases tested.

7.7.3 Effect of carbon and nitrogen

A marked effect on machinability is exerted by the composition of the sintering atmosphere. This effect is linked to the carbon and mainly the nitrogen content in the materials when sintering was carried out in nitrogen/hydrogen. The composition of the sintering atmosphere affects the final carbon, oxygen, and nitrogen contents in both austenitic and ferritic stainless steels as listed in Tab.7.40 for the mentioned elements after sintering in hydrogen and in the N_2–H_2 atmosphere.

The high oxygen content listed results more from its increase caused by the oxidation of the chips formed rather than from its content in the as-sintered workpiece material. High nitrogen contentin materials sintered in N–H atmospheres prove the formation of chromium nitrides in these steels during the sintering process.

The as-sintered density of the specimens was only slightly affected by the sintering atmosphere. The synthetic atmosphere with its nitriding effect which is also observed in, for example, dissociated ammonia, increased the hardness by about 50 HB to 145 HB, compared to 97 HB for the hydrogen sintered ferritic steel. Due to higher hardness, chip formation required higher plastic deformation work necessary and, consequently, more heat was generated in the cutting zone. The relatively poorer machinability of materials sintered in N–H atmospheres is mainly the effect of Cr-nitride inclusions.

Tab.7.40 Carbon, oxygen, and nitrogen content of 316 L and 430LHC stainless steel taken from chips. Sintering: 1280°C for 30 min either in pure hydrogen or in a mix atmosphere of N_2–H_2 (according to Ref.161 and 290)

Alloy	316L				430LHC			
Atmosphere	H_2		$70N_2$–$30H_2$		H_2		$70N_2$–$30H_2$	
Analysis area	Surface	Bulk	Surface	Bulk	Surface	Bulk	Surface	Bulk
C [mass %]	0.059	0.042	0.078	0.052	0.025	0.032	0.105	0.054
O [mass %]	0.193	0.225	0.247	0.274	0.219	0.302	0.305	0.300
N [mass %]	0.026	0.032	0.717	0.657	0.014	0.020	1.216	1.038

Remark: Chips for chemical analysis taken from a surface of about 0.3 mm thickness and from bulk chips

Under all conditions, higher contents of C, O, and N were found in the subsurface layer compared to the bulk of the compacts. The measured content of nitrogen and, similarly, of oxygen in alloy 430LHC was ~60% higher compared to 316L alloy. Depending on the manganese sulphide addition in the range of 0 to 1%, the sulphur content in both alloys increased from 0.02–0.04% to ~ 0.3%, as mentioned previously [225].

Increased nitrogen content in stainless steels, therefore, clearly shows on the formation of hard chromium nitrides (nitrogen–pearlitic microstructure) resulting in abrasive wear of the cutting edges and, consequently, in shorter tool life.

7.7.4 Effect of drill material and geometry on machining of wrought stainless steels

In following are listed some results about the effect of drill material and geometry on machinability of a wrought stainless steel. The investigation of the effect of tool material and its geometry on machining also of PM steels, especailly of stainless steels, is surely one of the ways to improve their machinability. Performance results achieved in drilling of a wrought stainless steel with the drills of different materials and with different geometry are listed in Tab. 7.41.

Firstly, the effect of TiN coating on HSS drill with modified geometry can be noted. Secondly, a significant increase in machinability of the mentioned stainless steel by the use of hardmetal TiN coated drill with a point angle of 130° with a convex cutting lip compared to HSS drills was achieved.

Tab.7.41 Performance comparison of the effect of drill type on machinability of AMG 1.3 (EN9) stainless steel. *Drill 8 mm, 20 mm long through holes, lubrication emulsion 5%* [292]. *(The characteristics of the drills used (DIN 1897/BS 328/ISO 235): A120 (DIN 1897 RN) – HSS standard helix, 118° up to 2.9 mm and over 13 mm 135°, surface treatment bright below 1 mm. blue 1 mm; A520 (DIN 1897 R) – HSS TiN, special point angle 130°, thinned point convex cutting lip; R520 (DIN 1897 R) – micrograin hardmetal TiN, special point angle 130°, convex cutting lip)*

Drill type	Drill speed [rpm]	Feed [mm/rev.]	Penetration [mm/min]	Number of holes cut in 30 s	Number of holes cut in 30 min
A120 (HSS)	1074	0.16	172	4	258
A520 (HSS coated)	1592	0.28	446	11	669
R520 (HM)	3580	0.33	1181	30	1773

7.7.5 Summary

- The austenitic 316L grade steel showed the poorest machinability of the investigated materials under the given cutting conditions. The machinability of the 303 grade is the highest compared to the 304 and other austenitic grades.
- The machinability of the ferritic steel is better than that of the austenitic one due to lower amounts of alloy elements, especially due to the absence of nickel.
- Sintering in hydrogen results in markedly better machinability compared to sintering in a nitrogen containing atmosphere, mainly in ferritic grades.
- The material without MnS sintered in nitrogen–hydrogen was most difficult to machine. This was accompanied by a general increase in the nitrogen content which was higher in ferritic compared to austenitic alloys.
- The drilling time was shorter with materials sintered at higher temperatures (1280°C) compared to, for example, 1120°C. However, the effect of the sintering temperature on the machinability of the tested alloys was not uniform.
- The addition of 0.5% MnS resulted in optimum improvement in the machinability of the tested materials. The combination of hydrogen sintering and MnS addition resulted in the relatively highest machinability.
- The absolutely highest machinability was recorded with resin-impregnated materials in comparison to those after resulphurisation and/or MnS addition. The impregnated materials also showed greatly reduced roughness variations.
- Effective additives tested for improving the tool life in turning of 304 grade included waterglass, boron nitride, and talc. The combined addition of boron nitride and soda waterglass to the austenitic stainless steel showed 300 times longer tool life than without the addition.
- The effect of the relative density on machinability in the tested range was not uniform.
- The optimum cutting speed and feed in drilling and turning must be defined in relation to the base material composition and the properties, to the machining aid addition and the sintering atmosphere and temperature, and finally in relation to the productivity. Relatively low cutting speeds and feeds contributed to higher machinability of the tested alloys.
- The effect of the tool material (hardmetal, cermet) on machinability in turning of the 304 grade was determined in re-

lation to the machining aid type. The optimum tool type should be chosen in dependence on workpiece composition and machining method applied.

7.8 COMPARING THE MACHINABILITY OF VARIOUS STEELS UNDER DIFFERENT CUTTING CONDITIONS

The machinability of PM steels can be examined in more detail by comparing the results attained under equal machining conditions for materials with different composition and mechanical and microstructural properties. The other method could be testing a material with defined properties with tools from different material and especially varying geometry, also under different cutting speed and feed, thus obtaining a 'machinability profile' for a given material. Usually, the results obtained in machining by one cutting method cannot be used for another cutting method without some adaptation.

If the former approach is taken, differences in the machining results can be attributed at least roughly to the mechanical and microstructural properties of specific materials because the cutting tool and cutting conditions are constant in such a case. The relationships between machinability and wear and the mechanical properties for some materials may provide a contribution for a more detailed analysis of these alloys.

In this chapter, therefore some results are shown that have been gained in machining of various PM steels under one defined test condition, *e.g.* the results about the effect of manganese sulphide addition on machinability of steels of different properties, and possible relationships between the machinability, wear, and mechanical properties of the workpieces are presented. These results are completed by selected data showing the best machinability attained under conditions presented before. This could contribute to a short overwiew on some problems associated with the machining of PM steels.

7.8.1 Drilling test with constant thrust force

The results presented so far have shown that most research work in machinability of PM steels has been done by drilling test. The drilling test is a technically relatively simple machining method for testing sintered steels which can be also easily instrumented for measurement of relevant parameters. Drilling with constant thrust force is affected only by the workpiece material being machined, except of drill material and geometry. In the following data are listed about the machinability of structural steels of different chemical composition, also with MnX addition. The common factor in this case is constant thrust force.

The materials tested by this method were Fe–C steels based on sponge and atomised iron powders, Fe–Cu–C, and Distaloy SA and SE alloys compacted and sintered under the same conditions.

7.8.1.1 Basic properties and machinability

The basic characteristics of Fe–C, Fe–Cu–C, and of Distaloy SA and SE steels with 0.3 and 0.7% graphite, without and with MnX as machining aid, as selected for drill testing with constant thrust force are listed in Tab.7.42.

The effect of the base iron powder grade (sponge, atomised) in Fe–C alloys on the mechanical properties is clearly discernible. With the materials based on atomized iron powder, higher tensile strength, hardness and mainly impact energy values were obtained.

The effect of MnX addition in the Fe–C steels differed from that in the Distaloy type materials. The MnX addition in the Fe–C alloys increased the strength and hardness of the material but lowered these properties in Distaloy SA, and can be supposed that the same effect would be also in Distaloy SE alloy. The impact energy was lowered by MnX in all cases. This observation suggests that MnX

Tab.7.42 Carbon content (C_c), density (ρ), tensile strength (Rm), hardness (HV 10), unnotched impact energy (KC) and length (L) of holes drilled in investigated steels. Code: S – sponge iron powder (SC100.26), A – atomised iron powder (ASC100.29), SA - Distaloy SA, SE - Distaloy SE, MnX-0.5% addition. Compaction 600 MPa, sintering 1120°C, 30 min, cracked ammonia [153]

Ser. No.	Alloy	C_c [mass %]	ρ [g/cm^3]	Rm [MPa]	*HV 10	KC [J]	L [mm]
1	S-0.3C	0.33	6.97	189	76	20.8	2490
2	S-0.3C-MnX	0.29	6.82	214	82	14.3	2100
3	S-0.7C	0.61	6.94	234	110	13.6	1390
4	S-0.7C-MnX	0.63	6.84	248	111	11.4	210
5	A-0.3C	0.29	7.04	220	74	26.5	1210
6	A-0.3C-MnX	0.32	6.99	229	80	21.5	1040
7	A-0.7C	0.56	7.08	253	105	22.2	280
8	A-0.7C-MnX	0.59	7.02	247	107	21.4	410
9	S-2Cu-0.7C	0.61	6.87	365	143	16.5	110
10	SA-0.3C	0.28	6.96	533	158	22.5	1180
11	SA-0.3C-MnX	0.31	6.87	496	169	18.9	1460
12	SA-0.7C	0.55	6.94	598	191	15.5	45
13	SA-0.7C-MnX	0.67	6.92	555	182	14.6	14
14	SE-0.3C-MnX	0.29	6.90	572	187	20.1	53
15	SE-0.7C-MnX	0.53	6.89	623	202	19.5	1

Remark: *measured on cross-section of tested bars

in Fe–C materials under given sintering conditions tends to dissolve, at least in the presence of carbon and harden the material. This possibility was not mentioned previously; there is mainly the question of the thermal stability of MnX machining aid in steel. This hardening effect was not observed with alloyed matrix.

The machinability test results for these steels are graphically shown in Fig.7.68. Drilling tests were perfomed on the shear faces of rectangular impact test bars (10 × 10 × 55 mm), perpendicular to the pressing direction, which corresponds to usuall drilling of parts in practice. The machinability rating was determined by taking the total length of holes in *mm* drilled in a material under constant thrust force until drill failure occured. Drilling was continued on all identically produced specimens until the end point was reached.

It showed that the spread of machinability data between the tested materials was considerably larger than the range in tensile strength, hardness or impact energy values. This is the result of higher sensitivity of the workpiece material to the failure type of the workpiece in machining compared to the fracture mechanism when testing the mechanical properties.

The machinability of Fe–C steels based on atomised iron powder was more than 50% lower when compared to that of the materials based on sponge iron powder, *i.e.* quite the reverse of the mechanical properties. A significant decrease in machinability was

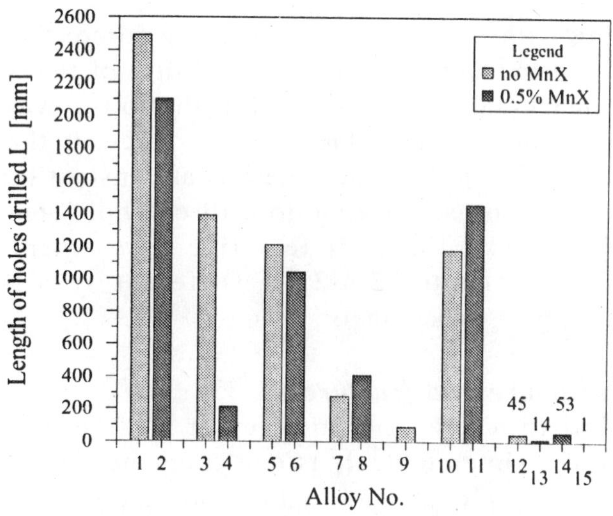

Fig.7.68 Machinability (the length of holes drilled – *L*) for steels listed in Tab.7.42 [153]. *Drilling test: HSS 3 mm drill, drill speed 850 rpm,* v_c *= 8 m/min, constant thrust force 333 N. Number of holes drilled = L/10.*

caused by an increase of the carbon content from 0.3 to 0.7%, especially in steels based on the atomised iron powder.

The effect of MnX as a machining aid on the machinability was not uniform. MnX additions in Fe–C materials impaired machinability and as expected, improved it in Distaloy type steels. As mentioned, an opposite effect of MnX was determined for tensile strength and hardness. The explanation of different effect of MnX on Fe–C and on Distaloy type materials is missing. The hardening effe : of MoX in Fe–C material indicates that it is not CaF_2 as stated in [209].

The machinability of the Fe–Cu–C alloy (No. 9) was similar to that described previously for the copper alloy steels (chapter 7.2).

Very poor up to negligible machinability in the investigated range was recorded for Distaloy SA–0.7C and Distaloy SE–(0.3, 0.7)C alloys (L = 45→1 mm) despite the fact that the hardness of the alloys was rather moderate, in the range from 148 to 176 HV 10. Considering the differences in tensile strength and hardness, the machinability of Distaloy SA–0.3C was better than that of some Fe–C steels.

The Distaloy SE–0.53C_c –MnX material (No. 15) was drilled also with a HSCo 3 mm drill under cutting conditions used. Also in this case only 1.5 mm depth of hole was drilled before drill failure. It means that this type of material cannot be drilled under the conditions applied here, perhaps coated hardmetal drills would be the best.

When drilling with a constant axial drilling force, the feed rate adjusted itself approximately to the machinability of the workpiece. The lower the machinability of a material the lower was the feed rate as a consequence of the increased resistance of the material to machining. By drilling dfferent materials at constant thrust force it would be possible to use the time to drill one hole (feed rate) in each material to make the first tentative assessment of their machinability (see chapter 3.4.2.1). Different results can be obtained in turning of these steels.

7.8.1.2 Microstructure and fracture

The basic microstructure and to a lesser degree the fracture characteristics of a sintered steel, as one component of the cutting pair affect the cutting process and by this cutting tool life. All these characteristics should be considered as part of a dynamic process, *i.e.* they change during the cutting process, forming a microstructure with new physical and metallurgical properties as shown

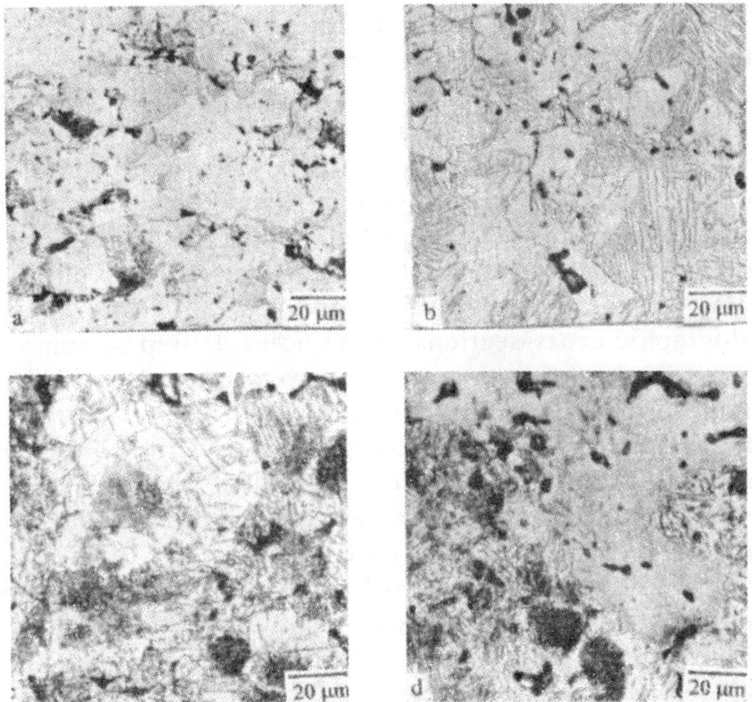

Fig.7.69 Microstructure of: a – Fe–0.29C_c steel (No. 2); b – Fe–0.56C_c steel (No. 7); c – Distaloy SA–0.55C_c (No.12); d – Distaloy SE–0.53C_c (No.15, see Tab.7.42). Optical micrograph. Nital etched.

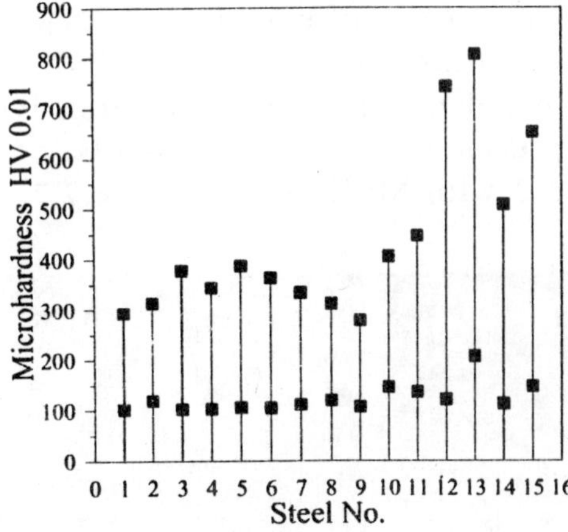

Fig.7.70 Microhardness value range (HV 0.01) of microstructural constituents in the steels listed in Tab.7.42.

and discussed before. The microstructures of some investigated materials as listed in Tab.7.42, from the best down to the poorest machinability, are shown in Fig.7.69.

The microhardness values of the microstructural constituents give a more detailed view on the heterogeneity of workpiece material properties and by this way on its machinability, Fig.7.70.

Ferrite was present in the microstructure of all alloys. According to the microhardness values, the microstructures of the tested alloys formed two groups: ferritic–pearlitic ones and ferritic–bainitic–martensitic ones for Distaloy type steels. The wide range of the microhardness values in all alloys confirmed the heterogeneity of the microstructure shown in the micrographs. In machining, these microstructural constituents come into contact with the cutting tool edge during a single rotation of the tool.

Metallographic cross-sections of the holes drilled in some alloys also showed in this case that the freshly machined surface of the hole was deformed, with the extent of deformation being higher when drilling a softer material, Fig.7.71. Consequently, the microhardness values measured in undeformed materials, Fig.7.70, do not correspond to the microhardness of the deformed layer which in reality is relevant for the cutting process, Tab.7.42, and, therefore, the former can be regarded only as general information about the properties of a workpiece. According to the sections, with the harder material a more uniform surface finish of the hole according to sections was observed.

The fracture surface of a material shows more clearly the presence of undissolved particles (non-metallic inclusions) in a material which affect to varying degrees not only the mechanical properties but also the machinability. These particles may be particles of a machining aid and/or non-metallic inclusions. They always come into contact with the cutting tool. Figures 7.72–7.74 show fracture surfaces of some tested materials after the tensile test. They can be regarded as representatives of these steels.

The fracture of the material No.2 with MnX addition, Fig.7.72,

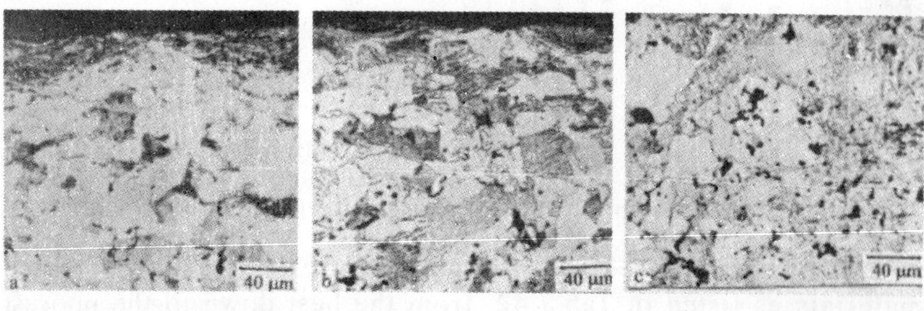

Fig.7.71 Cross-section of a hole surface drilled in: a – Fe–0.29C_c steel (No. 5); b – Fe–0.56C_c steel (No.7); c – Distaloy SA-0.28 C_c steel (No. 10) (see Tab.7.42). Optical micrograph. Nital etched.

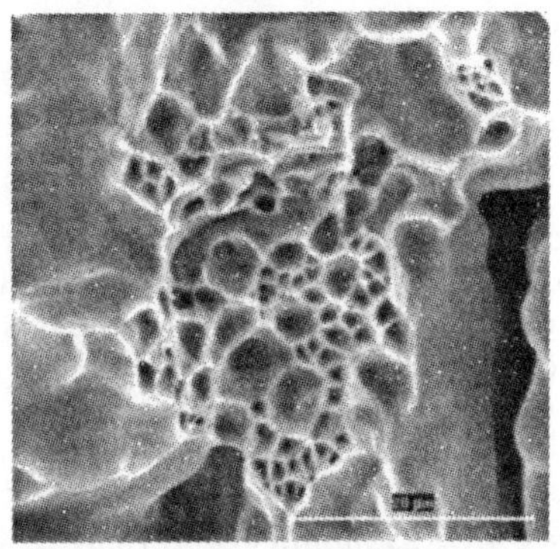

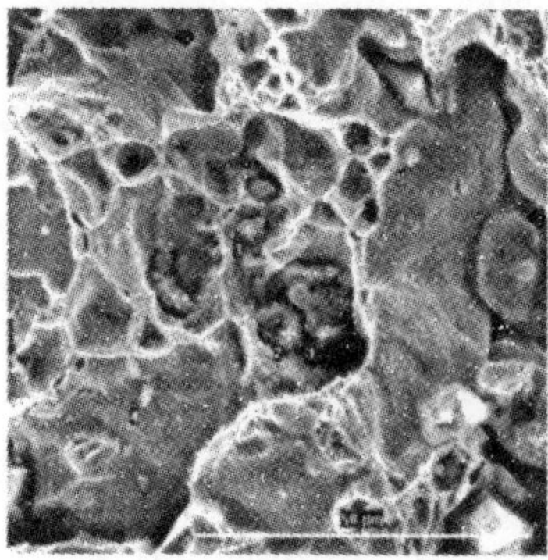

Fig.7.72 Tensile fracture surface of Fe–0.29C_c steel (alloy No. 2, Tab.7.42). SEM.
Fig.7.73 (right) Tensile fracture surface of Distaloy SA–0.67C_c steel (alloy No. 13, Tab.7.42). SEM.

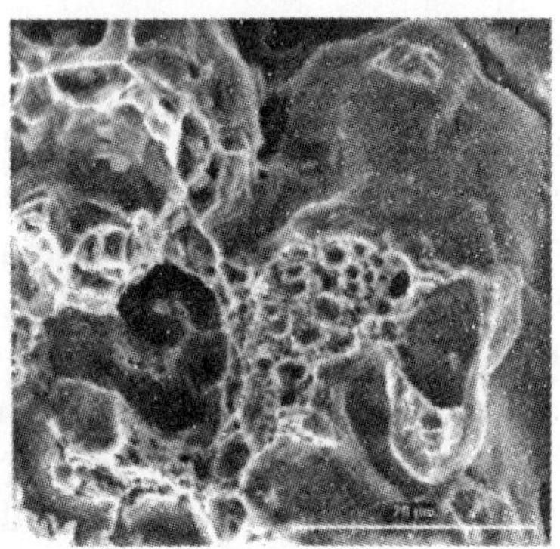

Fig.7.74 Fracture tensile surface of Distaloy SE-0.29 C_c steel (alloy No. 14) (see Tab.7.42), SEM.

was ductile with dimple morphology and without inclusions in the dimples. The pore surfaces were clean, *i.e.* also without particles of the machining aid (MnX). It can be concluded from it that the MnX dissolved during sintering. The fracture surface of the Distaloy SA–0.7C with MnX addition showed dimple facettes with inclusions and some very fine particles on free pore surfaces, which can be MnX particles. Microanalysis of a particle confirmed the presence of calcium, as shown before in Fig.6.18. This could confirm the presence of MnX particles if it is accepted that MnX is CaF_2. The fracture of Distaloy SE–0.3C material with MnX addition was similar with the particles on the pore surfaces. Some brittle transgranular cleavage facets in Distaloy SE–0.7C material were recorded.

7.8.1.3 Effect of drilling speed at constant thrust force on machinability of Distaloy SA steel

The samples of Distaloy SA–0.7 graphite steel (Tab.7.42, No. 12)

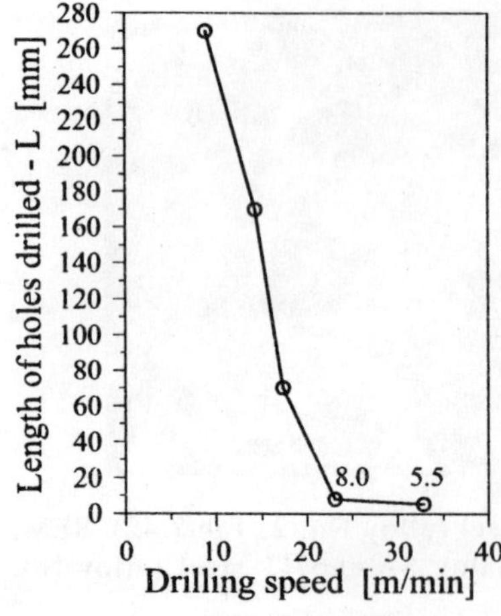

Fig.7.75 Machinability (length of holes drilled – L) of Fe–1.75Ni–1.5Cu–0.5Mo–0.55C_c steel (properties Tab.7.42, alloy No.12) in dependence on drilling speed. *Drilling conditions: HSS 3 mm drill, drilling speed: v_c = 8.9 m/min (950 rpm); v_c = 14.3 m/min (1520 rpm), v_c = 17.3 m/min (1840 rpm); v_c = 23 m/min (2440 rpm); v_c = 32.8 m/min (3480 rpm); constant thrust force 200 N, specimen – impact test bars 10 × 10 × 55 mm. (Data – courtesy of K. Vasilko).*

were tested in drilling under various drilling speed with lower constant thrust force for comparison to the results shown for this material in Fig.7.68. The results attained are shown in Fig.7.75.

Drilling with a lower thrust force of 200 N compared to previous 333 N (Fig.7.68) improved machinability (45→270 mm – length of holes) of this steel. In contrast, increasing drilling speed in the range of 8.9 to 23 m/min caused a significant decrease of the machinability (270→8 mm – length of holes). It shows that beside the cutting speed and feed the thrust force value becomes an important variable in drilling PM steels for the estimation of optimum drilling conditions for a steel with the defined properties.

7.8.1.4 Drill failure appearance

The loss of the cutting efficiency of a drill manifests in the flank wear, which is often measured and evaluated as machining index.

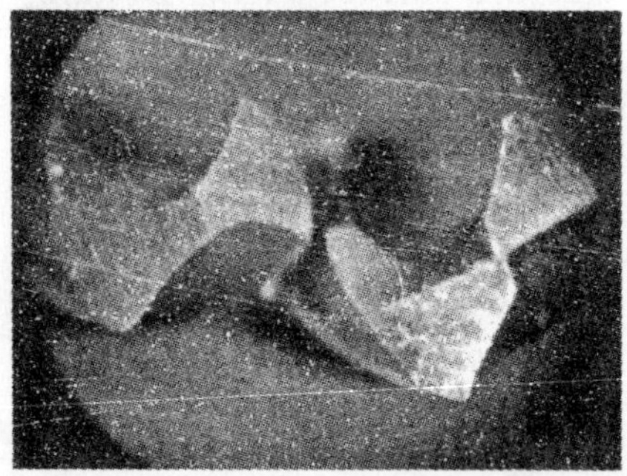

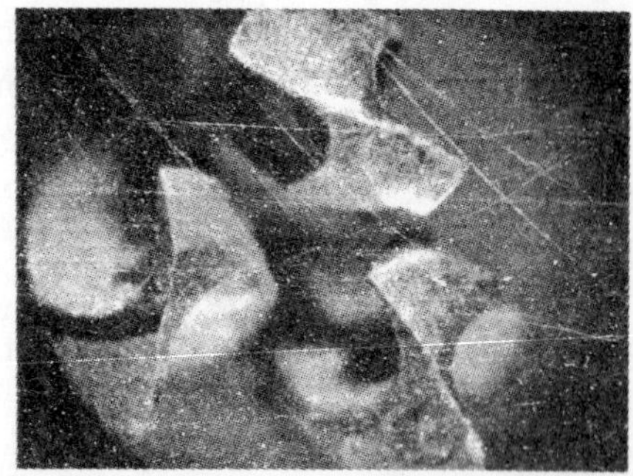

Fig.7.76 View on drill point – cutting edges of a new standard HSS 3 mm drill (left) and of HSCo 3 mm drill with modified geometry (right).

Fig.7.77 (right) View on three failed cutting edges of HSS 3 mm drills after drilling the Distaloy based materials No. 12–14 (Tab.7.42).

In Fig.7.76 are shown the drill points of a new standard HSS 3 mm drill (left) and of a HSSCo 3 mm drill (right, modified spiral geometry) with cutting and chisel edges. The differencies in the geometry of these drills are visible.

As mentioned before, the Distaloy SE–0.7C–MnX material was also unsuccesfully drilled with this HSSCo drill.

Figure 7.77 shows three standard HSS 3 mm drills with failed drill points after drilling with the steels No. 12–14 (Tab.7.42) with a constant thrust force. The wear of the cutting edges is large. In this case, the failure of these drills is not exactly measurable, *e.g.* as flank wear, because at the drling with constant thrust force the feed is stopped itself but the drill can further rotate and by this the wear of a drill is larger. Figure 7.78 shows the failured drill point of a HSS drill after drilling the holes 174 mm long in the Cr-alloyed steel. In this case, the abrasive wear of the drill was significantly larger compared to previous drills (chromium nitrides) with the failed flute.

Figure 7.79 shows a cross-section (see Fig.3.44) of the last

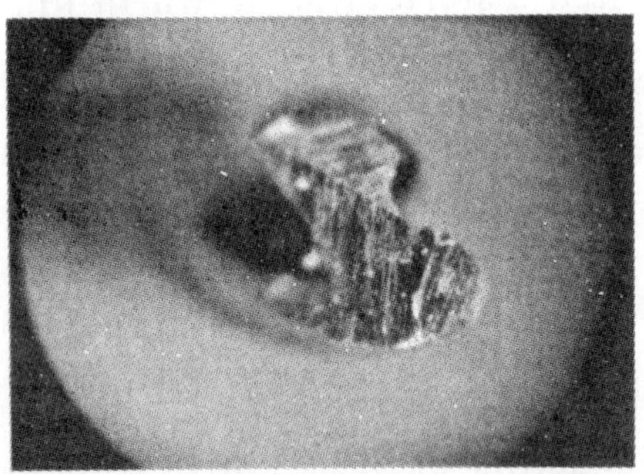

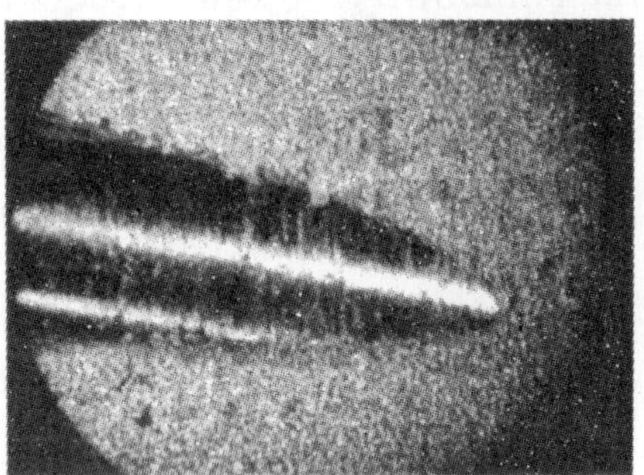

Fig.7.78 View of the failed cutting edges of a standard HSS 3 mm drill after drilling Cr-steel (Tab.7.28).

Fig.7.79 (right) Section of the last unfinished cuspated hole in Fe–0.63C_c–MnX steel (material No.4, Table 7.42).

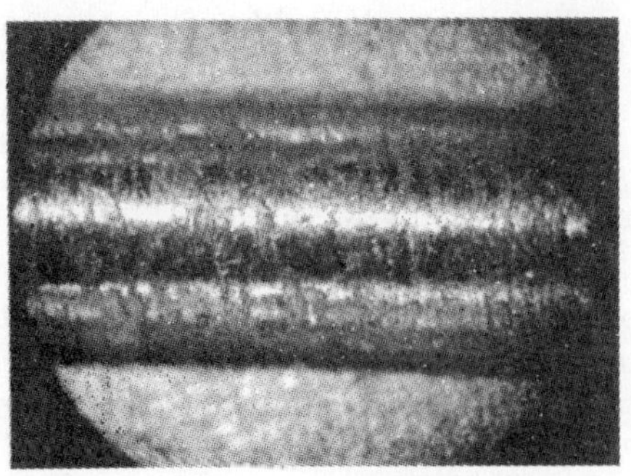

Fig.7.80 Drilled surface hole in Fe–0.3C–MnX (material No.2, Tab.7.42). Drilling conditions as in Fig.7.68.

unfinished hole at the test due to drill failure in material Fe–0.63C_c–MnX steel. Due to abrasive wear, the drill bit wear was cuspated (flute worn out). It proves that also in materials without 'hard' phases in the initial state, and of course in, *e.g.* Cr-alloyed steels, under some cutting conditions abrasive wear of the cutting tool occurs. The surface of a through hole drilled in this material is shown in Fig.7.80.

7.8.2 Effect of some machining aids on machinability of various alloys

Here, results will be presented that were obtained when comparing the machinability of various materials without and with MnS addition, which is considered in many cases as the best machining aid for most sintered steels if chemical purity is taken into account, and also various materials with addition of other machining aids. It may contribute to better knowledge of the effectiveness of individual machining aids on machinability of various matrials.

The results concerning the effect of 0.5% MnS addition on machinability of some frequently tested materials is shown in Fig. 7.81 The tensile strength of the tested Fe–C and QMP 4401 (Fe–0.1Mn–0.8Mo) alloys was ~220 MPa and that of the alloyed materials 380–410 MPa, and the corresponding hardness 18–20 HRB and 54–69 HRB, respectively. It is necessary to note the large differences in machinability of F–0008, FC–0208 and FN–0205 steels without and with MnS addition which do not correspond with some previous results. The highest number of holes drilled can be regarded as encouraging.

A more detailed view on the machinability of these alloys without

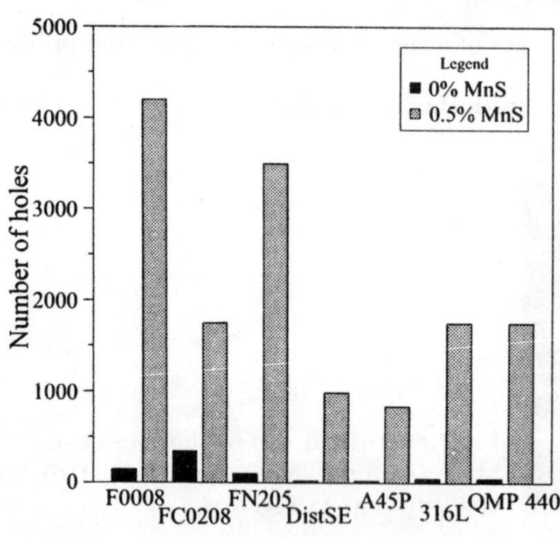

Fig.7.81 Machinability (number of holes drilled) of various materials without and with 0.5% MnS addition (acc. to Ref. 224). *Drilling test: for the iron based compacts (density ~7.0 g/cm³) – HSS 3.5 mm drill, drill speed 4000 rpm, v_c = 44 m/min, feed 0.12 mm/rev.; for stainless steel compacts (density 6.9 g/cm³) – drill speed 700 rpm, v_c = 7.7 m/min, feed 0.1 mm/rev., depth of hole 11.4 mm for both material groups.*

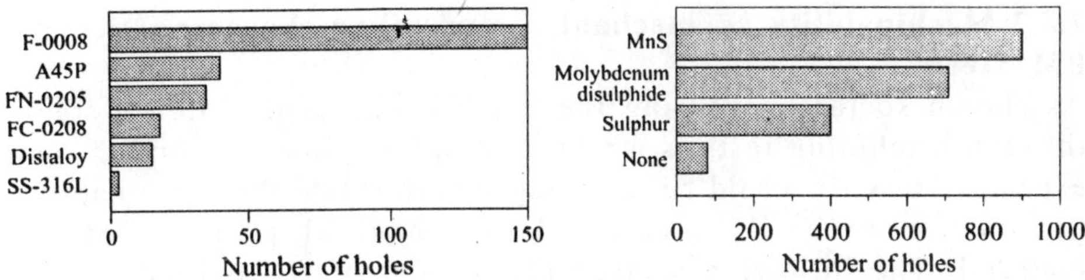

Fig.7.82 Machinability (number of holes drilled) of some materials without machining aid. Density 7.2 g/cm³ [105,293]. *Drilling test: HSS 14.2 mm drill, drill speed 4000 rpm, v_c = 178 m/min, feed 0.8 mm/s.*

Fig.7.83 (right) Machinability (number of holes drilled) in F-0008 steel with various machining aids (0.5%) [105,293]. *Drilling test: HSS 14.2 mm drill, drill speed 4000 rpm, v_c = 178 m/min.*

machining aid is shown in Fig.7.81, considering the difference in relation to Fig.7.82. The machinability of all materials without MnS addition was low (20–25 holes drilled) except FC-0208 alloy (300 holes drilled). Only three holes were drilled in stainless steel. These results can be related approximately with uniform apparent hardness and tensile strength.

The effect of the mostly used machining aids on the machinability of Fe–C (F-0008) steel is shown in Fig.7.83.

Manganese sulphide (purity not defined) provided the relatively highest improvement in machinability of these alloys, followed by molybdenum disulphide with the minimum effect on the mechanical properties and dimensional change of the parts. It can be concluded that MnS was most stable at the sintering temperatures of ferrous alloys and did not interact with the matrix of a large number of ferrous alloys. This is a somewhat broader view showing that the presence of manganese sulphide and of other sulphur-based agents on the tool surface helped to considerably reduce the friction between the tool surface and the workpiece when compared to other machining agents. Due to larger differences in machinability with various machining aids, and, especially, with MnS addition when compared to the differences of the mechanical properties of the compared steels, one can assume that the lubricating effect of the aids also depends on the main characteristics of the workpiece and the tool material. The high enhancing effect of 0.5% MnS addition shown is not standard for all grades of sintered steels.

7.8.3 Machinability vs. mechanical and other characteristics of PM steels

As shown so far, measuring the machinability of sintered steels through machining tests is a rather complex, time consuming, and expensive task. It would be advantageous to relate the machinability to some more easily measurable mechanical property of the material. The following section gives most of the published data showing whether any of the mechanical properties of a material is sufficient for defining its machinability and whether such a determination is possible with sufficient accuracy.

In wrought steel, tensile strength and primarily hardness are the main properties used to predict machinability. The most simple dependence is the relationship between hardness and the carbon content (regarding alloying). Increasing the carbon content increases the hardness which in turn reduces the machinability of wrought steel, at least as a first approximation.

Therefore, in PM parts machining it also is a practical tendency to find a possible relation between the machinability of a material and its one or more mechanical properties or other material characteristics.

7.8.3.1 Machinability vs. hardness

Hardness is easy to measure and could also be an interesting property of a PM material in relation to machinability. In porous sintered steels, the apparent hardness includes also the porosity and to some extent also the heterogeneity of the microstructure which both affect negatively the cutting process of PM steels. In the following, data will be presented on the relationships between apparent hardness and tensile strength of some sintered steels and their machinability. It will be possible to deduce from these results to which extent these material properties can be/or not related to machinability.

The apparent hardness values of PM steels show only a very general correlation to the machinability – steels of lower hardness are commonly more machinable than steels of higher hardness with a poor accuracy to the higher hardness level. More reliable data on the machinability of PM materials (whichever index is used for characterization), on one side, and on apparent hardness, on the other side, will enable an assessment of their relationships and its range for a practical prediction of the machinability.

For PM materials this relationship can be used to a limited range if the amount of ferrite present in the microstructure is taken into

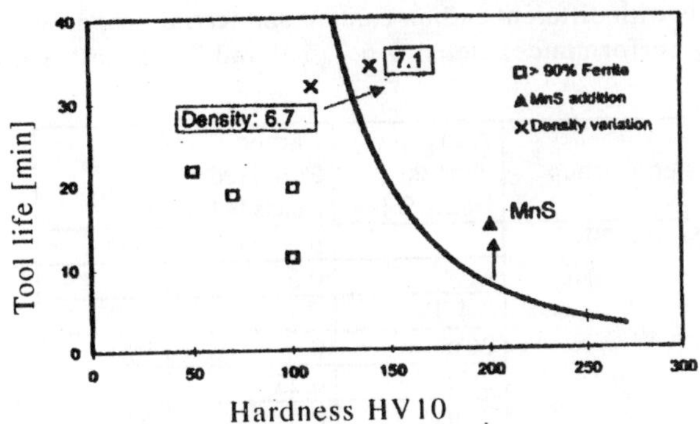

Fig.7.84 Tool (GC1015) life vs. hardness in face turning for Fe–C materials with high amount of ferrite, of various density and with MnS addition [118,157,280]. *Turning conditions: surface cuting speed 200 m/min, feed 0.1 mm/rev., tool life criterion 0.3 mm flank wear.*

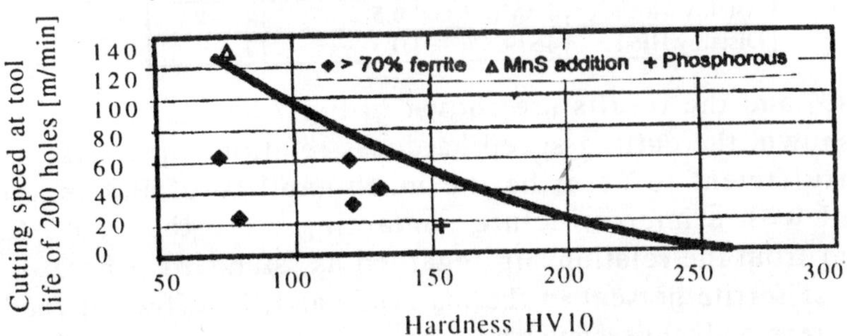

Fig.7.85 General relationship between cutting speed and hardness with a tool life of 200 holes drilled in sintered iron based materials [118]. *Drilling test: HSS 4 mm drill, tool life of 200 holes drilled, feed 0.06 mm/rev., depth of hole 10 mm; tool life criterion was total failure of the drill.*

account, and as shown in Fig.7.84 also the MnS addition and density level. The materials with more than 90% ferrite lie outside the common hardness/tool life criterion used in machining of wrought steels, and therefore this relation is only of limited use for plain sintered iron.

A large scatter in the correlation between tool life in turning and hardness including factors, is shown in the figure. The tool life–hardness curves denotes only very approximately a possible course of this relationship regarding also the density, ferrite proportion and MnS addition.

With the aim of proving a possible relationship between drill life and hardness, several groups of PM materials (plain iron, diffusion alloyed, prealloyed, hybrid) with varying density levels were subjected to drilling performance tests. The materials are listed in

Tab.7.43 PM steels with different carbon content and ferrite proportion in microstructure tested in drilling performance. Density 6.7, 7.0 and 7.3 g/cm³, sintering 30 min at 1120°C [118]

Material group	Alloy element [mass %]	Carbon Premixed [mass %]	Ferrite [%]
ASC100.29	-	0	100
	-	0.5	55
	0.45P	0	100
	2Cu	0	100
	2Cu	0.25	80
	2Cu	0.5	30
	2Cu	0.8	1
Distaloy AE (4 Ni–1.5Cu–0.5Mo)		0	90
		0.25	35
		0.5	8
		0.8	2
Astaloy Mo	1.5Mo	0.5	-
Distaloy HP-1	4Ni–2Cu	0.5	1

Tab.7.43, and the results are shown in Fig.7.85.

As shown, the cutting speed/hardness relationship is also in this case insufficient to describe the machinability of PM materials. Material with a microstructure containing more than 70% ferrite diverged from the relationship regarded as 'standard', therefore the amount of ferrite present in the materials and the effect of feed rate for different drill types must be taken into account when describing the machinability. As a rule of thumb, ferrite in amounts of ~90% can be taken to indicate decreased cutting performance, relative to the hardness correlation.

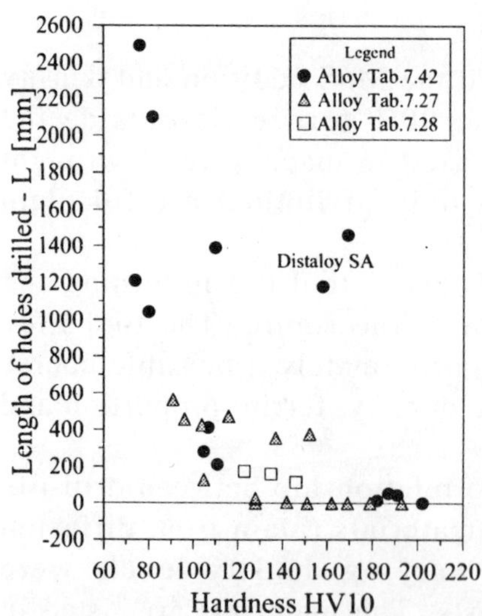

Fig.7.86 Length of holes drilled in drilling test with constant thrust force (as criterion for machinability) vs. apparent hardness for the materials listed in Tab.7.27, 7.28, 7.42.

Figure 7.86 shows the relationship between the machinability and the apparent hardness of steels drilled with constant thrust force. The tested materials cover a wide compositional and hardness range. As shown, three areas in this relationship between the machinability and the apparent hardness can be defined for the tested alloys. The results show a marked decrease in machinability with increasing hardness of the alloys up to ~110 HV10. In this case, the machinability of the materials can be classified by the apparent hardness only in the range of ferritic–pearlitic (Fe–C) steels with different proportions ferrite/pearlite in the microstructure. In the hardness range from ~100 to 200 HV 10 with heterogeneous microstructures, the length of drilled holes reached only up to 400 mm, *i.e.* there was only a small change in machinability with increasing hardness. It means that the hardness is also in this case an insuffcient characteristic for these materials in relation to the machinability tested under equal conditions. The reason for this must be found in the microstructural character of the alloy steels and by this in the 'machinability' of the individual microstructural components and their proportion in the structure. The machinability of Distaloy SA–0.3C (158 and 169 HV 10) greatly differed from this relationship, Fig.7.68. The steels with a hardness of ~130 to 200 HV 10 have been classified as less and as non-machinable.

The relationships between the thrust force in drilling, the number

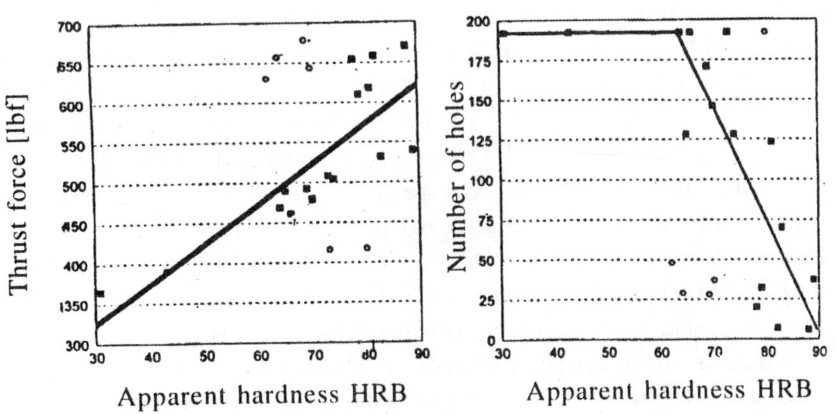

Fig.7.87 Machinability (drilling, thrust force) vs. hardness for various PM steels (F-0005-8, FC-0205-8, FN-0205-8), density 6.7–7.0 g/cm³ [145]. Conversion: 1 lbf = 4.44 N.

Fig.7.88 (right) Number of holes drilled vs. hardness for various PM steels as in Fig.7.87. Drilling stopped at 192 holes maximum drilled [145].

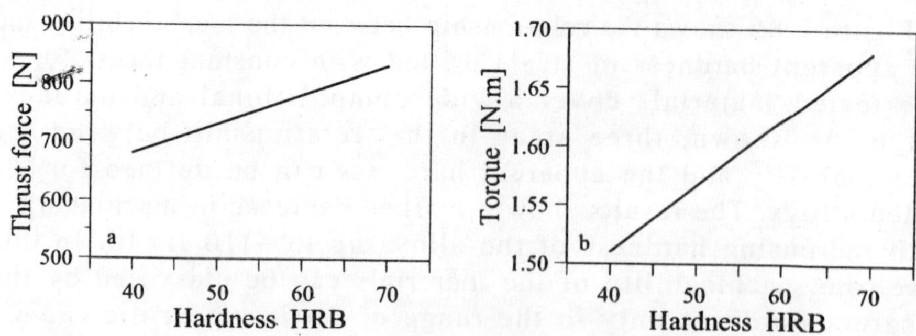

Fig.7.89 Relationship between hardness and: a – thrust force, b – torque for Fe-2Cu-0.55C with BN addition (ATOMET 29 M) for the density range 6.3 to 6.8 g/cm³ [222].

of holes drilled, and the hardness shown for other alloys in Figs.7.87 and 7.88 accomplish the previous results.

The shown scatter of the recorded data must be taken into account. The drilling force increased rapidly when the hardness increased beyond the 70–75 HRB (approx. 130–140 HV10) level. It was also noted that the low alloy steels FL-4205 and FL-4605 were less machinable than predicted for the hardness range of 62–72 HRB, and the Distaloy FD-0205 and FD-0405 were more readily machined than would be expected from the graph. Similarly, Fig. 7.88 indicates that according to the number of holes drilled in relation to hardness, the machinability starts to decrease at ~70 HRB.

Linear relationships of thrust force and torque to the hardness were gained when tests results only of one material were taken, Fig.7.89. It could be concluded from these results that as a machining index, the number of holes drilled seems to be more sensitive to the hardness of the workpiece material (a larger scatter in the results due to the heterogeneity of the microstructure) compared to the thrust force and torque which indicate the linear relationship. A more detailed view on the relation between hardness and machinability can be made only throught the testing a larger number of materials of various composition and properties but also in such cases a sufficiently general relationship will not probably be found due to known heterogeneities in the microstructure characteristics for PM steels.

The linear relationships recorded between the thrust force and torque as machining indices and the hardness can be explained by the deformation behaviour of the material in the cutting zone. As the density and hardness of the material increased, the thrust force and torque increased linearly. At low density, the cutting forces were low because the excessive work hardening induced cracks in the material and made it brittle. As the density increased, the material became stronger and the extent of work hardened layer was reduced. At high density, densification under the tool was minimised and the hardness of the densified layer became independent of the density of the part.

In connection with the investigation of a possible relation between the machinability and apparent hardness of a material, it must be noted that the results attained by the measurement of the hardness by different main methods (Brinell, Vickers, Rockwell) are not fully comparable. Also for the presented apparent hardness Vickers and Rockwell B values of tested materials in this book is not a sufficient comparison (there are data only for the relation between the Vickers and Rockwell B hardness for wrought steel) (see Appendix). For this reason, it also is proposed to measure the apparent hardness especially of sintered materials by the 'universal hardness' method. The method is based on the measurement and recording, continuosly the acting force and the correspoding penetration depth of the indentor. As stated Vickers hardness is less sensitive to processing. In contrast to this, by means of the 'universal hardness test' it is also possible to detect the different type of response of materials based on sponge and atomised iron powder. The automated measurement of 'universal hardness' should enable to distinguish between materials sintered once or twice at 1120°C [294].

According to this encouraging information it can be expected that the measurement of the 'universal hardness' will contribute to the more exact relation between the machinability and the hardness as the easily measurable material property.

Regarding the possible relation between the machinability and apparent hardness of PM steels it is necessary to take into account the differences in the processes occurring at the hardness measurement and machining. Hardness measurement occurs at relatively 'longer' time in relation to the small affected area of the microstructure and can be called 'the static' process. In contrast machining can be called 'the dynamic' process because it is relatively very fast in relation to the size of the microstructural grains, affects significantly a larger area of the microstructure, and causes the side effects as an increase in the temperature and the deformation of the material.

7.8.3.2 Machinability vs. wear

Friction in the tool/workpiece contact zone, classified by tool wear, is a dominant factor in machining. It is an effect of the workpiece properties, including machining aids on one side and on the other side of the tool material and geometry under the increased temperature generated under the cutting conditions used. 'Wear resistance', like 'machinability', is not a physical property of a material but the manifestation of the failure resistance under specific loading (friction) conditions, *i.e.* a system, not a material, behaviour. For wrought structural steels, wear resistance is found in most cases to be directly related to the hardness.

To determine the possible relation between the machinability and dry wear of one of the friction pairs in the cutting process, *i.e.* the workpiece, the samples used for drilling tests were also tested for dry wear. The relationship between the machinability (lenght of holes drilled) and dry wear of the alloys is shown in Fig.7.90.

A linear dependence between the machinability in drilling with a constant thrust force and the dry wear of the investigated materials is indicated, with a scatter in the range of lower wear values. The machinability decreased with decreasing wear of the materials. This means that the alloys with poor machinability exhibited low dry wear. When testing these samples to seizure, the seizure of this friction pair occured at the load ~1600 N and the temperature in the friction zone increased from 50–60°C to ~250°C [68].

The use of a counterpart test disc for the friction test that consists of the tool material could contribute to the study of the

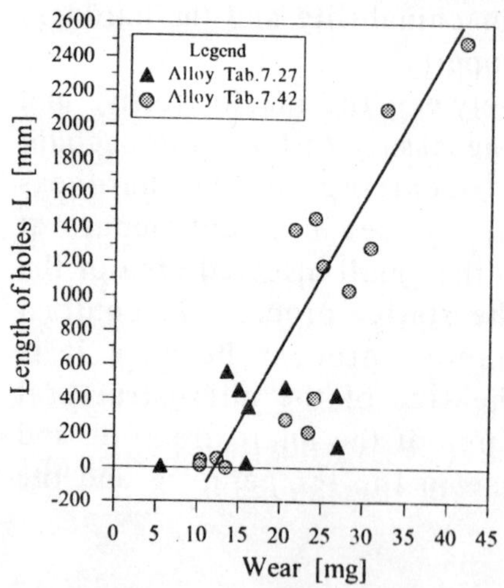

Fig.7.90 Length of the holes drilled – *L* vs. dry wear of steels listed in Tab.7.27, 7.28, 7.42 [68,69,153]. *Wear test: disc – block (test sample) method, load 100 N, sliding distance 1000 m, speed 0.5 m/s, test sample – impact test bar.*

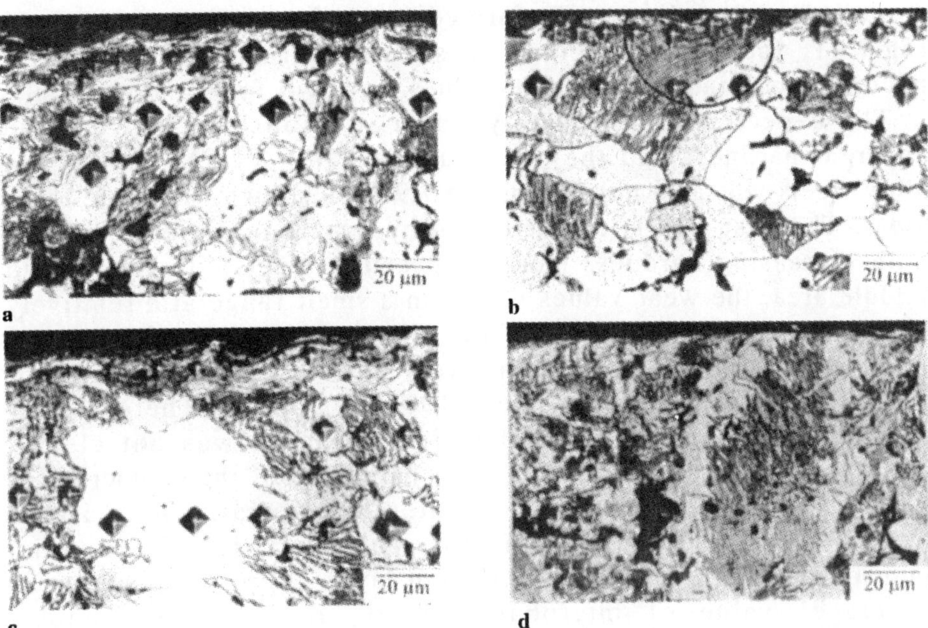

Fig.7.91 Cross-section of the friction track in the alloy: a – Fe–0.7C, *w* = 41.9 mg, 324 HV 0.01; b – Fe–0.7C–0.5MnX, *w* = 23.4 mg, 342 HV 0.01; c – Fe–2Cu–0.7C, *w* = 18.4 mg, 477 HV 0.01; d – Distaloy SA-0.7C, *w* = 10.3 mg, 515 HV 0.01 [69] (for materials see Tab.7.42); *w* – dry wear (cumulative mass loss); (friction test conditions in Fig.7.90). Optical micrograph. Nital etched.

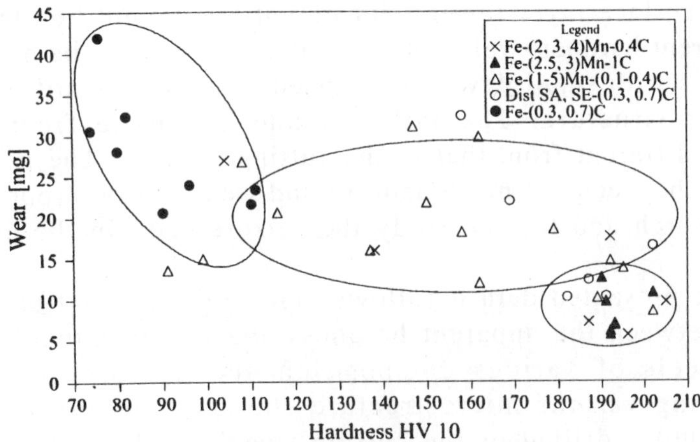

Fig.7.92 Dependence of dry wear on apparent hardness of materials tested for machinability. (Tab.7.27, 7.28, 7.42). Wear test conditions as in Fig.7.90.

mechanism of tool wear under the temperature generated as a consequence of friction in the cutting zone. The tests can be performed up to seizure of the friction (cutting) pair.

During dry wear testing of the PM samples a similar work

hardened surface layer was formed as in the cutting process. In Fig.7.91, the microstructure of surface deformed layers with microhardness indentations and values is shown for some materials.

The expected approximately linear relationships between the apparent hardness and the dry wear resistance of PM carbon and alloy steels was also not found, Fig.7.92. As shown, there are three areas. The first area is characterised by a marked decrease in wear with increasing hardness (ferritic–pearlitic steels). In the second middle area, the wear values varied in a small range at a relatively large difference in hardness of the alloys. The third area shows a rapid decrease in wear indicating very poor to no machinability of the alloys at the hardness about 190–200 HV 10. The effect of MnX addition on the wear of tested materials was not clearly identified, probably due to high work-hardening of the contact layer.

The apparent hardness of a sintered material, which is the nearest property to the machinability characteristics of a material considering the deformation of the cut layer, expresses however an 'average' value of microhardness of single microstructural constituents deformed by the indentor, including pores. The effect of porosity on the cutting process, as mentioned in chapter 5.21, must be taken into account. The cut occurs in a work hardened deformed layer without pores, according to the deformation theory explaining the effect of pores in the cutting process. The hardness of the new surface layer formed during cutting cannot be identical to the apparent hardness value of the material. This can be accepted for all alloys with small deviations caused by particular differences in the microstructure. The failure mode at tensile fracture is completely different from that in the cutting process. The possible relationship between the machinability and wear can be proper only for the research and for the study the seizure state in the cutting process.

From the presented data it follows that no direct relation could be found between the apparent hardness and the machinability of sintered steels of various composition and by this also with corresponding various microstructure character, taking *e.g.* the number of holes drilled or the cutting speed for drilling a given number of holes or for turning. The presented data do not give sufficiently accurate information on the machinability and hardness of PM steels because no correlation exists.

Regarding all various processing and physico-metallurgical factors, which define the microstructure homogeneity/heterogeneity and properties of sintered steels, it can be said that no general direct relation between the apparent hardness an machinability of PM steels can be identified.

An approximate relationship between the number (length) of holes drilled and the hardness was established only for material of defined simple microstructure, *i.e.* in this case in Fe–C alloys. The material heterogeneity with different proportion of microstructural constituents from ferrite to martensite and by this with hardness in a wide range, is very difficult to describe by one material property or by a general machinability index. In Ref. 295 it was also stated that all such data as mentioned for the characterisation the machinability are valid only for materials by the same porosity of the workpiece and the need for enhancing machining theory in the areas of workpiece microstructure was stressed.

7.8.4 High machinability rating PM steels and conditions

Most of the hitherto listed data about the machinability of various PM steels, obtained mainly by drilling, showed a large scatter of the results independently of the machining index. Table 7.44 lists the materials and cutting conditions with the highest number of holes attained drilled each under given drilling conditions in different PM

Tab.7.44 Selected data showing the largest number of holes drilled in some PM steels with different machining aids attained under different drilling conditions (listed in the Tables above) and corresponding total length of holes drilled

Material	Density [g/cm³]	Machining aid [mass %]	Drilling conditions			Number of holes	Length of holes [mm]	Table
			Drill dia. [mm]	rpm	Cutting speed [m/min]			
Fe-0.5C (at.)	6.8	0.5MnS	3.5	3000	33	890	11300	7.2
Fe-0.5C (at.)	6.8	0.5MoS₂	3.5	3000	33	718	9100	72
MP 36S	6.6	S prealloy	6.35	2300	46	200	5000	7.3
Fe-0.8C (sp.)	6.3	0.5MnS	3.5	1000	11	725	8000	7.4
Fe-0.8C (at.)	7.2	0.5MnS	3.5	4000	44	2514	28000	7.4
FC-0208 (sp.)	6.1	0.5S	3.5	3000	33	608	7500	7.8
FC-0208 (at.)	6.8	0.5S	3.5	3000	33	668	8200	7.8
FC-0205 (at.)	6.8	0.10MnS +0.25MnX	3.2	3000	30	1400	17800	7.9
FC-0205 (at.)	6.8	0.15MnS +0.35MnX	3.2	3000	30	1251	15900	7.9
Fe-2Cu-0.8C (sp.)	6.3	0.5MnS	3.5	1000	11	775	8500	7.12
F-0008 (at.)	7.05	0.5MnS	3.5	4000	44	4200	48000	7.18
FN-0205	7.05	0.5MnS	3.5	4000	44	3500	40000	7.18
Dist. SE-0.5C	7.2	*0.5MnS	3.5	2000	22	1600	18200	7.24
**Dist.AB-0.5C	7.1	0.5MnS	6.0	1000	18.8	>300	6000	7.50
Ancorsteel 85HP-2Cu-0.5C	7.0	0.10MnS +0.25MnX	3.5	3000	33	188	2400	7.26
Fe-0.45P	7.2	0.5MnS	3.5	4000	44	1740	19600	7.32
Fe-0.45P	7.2	*0.5MnS	3.5	4000	44	3500	40000	7.32
316L	6.9	0.5MnS	3.5	700	7.9	>1760	8600	7.36
316L	6.5	IMP	9.5	500	15	>192	4900	7.39

Remark: at. – atomised iron powder, sp. – sponge iron powder, *high purity, **tempered, IMP – resin impregnated

steels. The shown large number and length of holes drilled, shows that in the machining of PM steels results were attained which significantly exceed other ones attained under equal or similar cutting conditions. The reason for this was not investigated. These impressive results can be very interesting for practice and, especially, for a more detailed analysis of materials properties and cutting conditions, including drill type and geometry used. They can be accepted as indicating a potential for a more significant general improvement in the machinability of PM steels with a smaller scatter in the results. It may be presumed that in these machining tests it was not possible to characterise all factors with sufficient accuracy. However, it may be assumed that if such satisfactory results have been experimentally obtained once it should be possible to attain their reproducibility.

7.8.5 Summary

- Drilling with constant axial drilling (thrust) force showed to be a simple and effective method for testing the machinability of PM steels, with the number or length of holes drilled as machinability index. The use of the standard impact test bars enables more simple preparation of the drilling test and facilitates combination with mechanical testing of a material.
- Fe–0.3C materials were easier to drill compared to Fe–0.7C materials.
- The Fe–C materials based on sponge iron powder were easier to machine compared to those based on atomised iron powder.
- Distaloy SA–0.7C and Distaloy SE–(0.3, 0.7)C exhibited poor machinability.
- In contrast, Distaloy SA–0.3C achieved better machinability than even some Fe–C materials.
- 0.5% MnX addition showed ambiguous results. MnX addition in Fe–C materials caused a decrease in machinability and an increase in tensile strength and hardness, but vice versa in Distaloy type steels.
- The similarity of the deformation of machined surface layers and of those subjected to dry friction wear tests was demonstrated.
- When comparing the effect of the most commonly used machining aids on the machinability, the highest improvement was caused by manganese sulphide, followed by molybdenum disulphide and sulphur addition.
- The effect of a machining aid is markedly affected by the base alloying of a material and thus by its properties. The highest

improvement was attained in Fe–C materials.

- No direct relation was found between hardness, and machinability (as number of holes drilled) when data for various alloys and machining conditions were used.
- The microstructure should be more carefully investigated in relation to machinability of a material, as shown when comparing steels with a relatively wide hardness range.
- In a smaller range of materials, a linear relationship between the number of holes drilled and thrust force and torque and hardness was demonstrated.
- A approximately linear relation between the dry wear rate of a material and its machinability was indicated – the lower the wear the lower the machinability.
- Relatively large differences in the results of machinability tests performed under equal conditions were observed.
- The highest machinability results (maximum number of holes drilled) gained in some works exceed by some orders of magnitude other results attained under similar or (nominally) equal drilling conditions. A more detailed investigation is necessary here; in any case, however, these results can be regarded as an encouraging contribution for improving the machinabiliy of PM steels in drilling since they show that positive machining results are possible if the appropriate parameters have been defined.

7.9 STANDARDIZING THE MACHINABILITY OF PM STEELS

To obtain data for a broad range of PM steels in machining under controlled identical conditions which would enable standardizing their machinability, nearly 100 sintered steels materials were subjected to the drilling test.

In this test series, average machinability values were taken for specimens with 6.7 and 7.0 g/cm^3 density levels. Iron–carbon steels, copper as well as nickel alloyed ones, in prealloyed and diffusion alloyed variants, without and with MnS anddition as well as resin impregnated ones were involved in the test programme. The materials are listed here in the ascending order by the number of holes drilled, Tab.7.45. Once 192 holes have been reached (which was set as the test limit), the list continues by decreasing drilling force values. Thus, in this table the very poorly machinable materials are at the top of the list (least number of the holes and highest drilling force). In the MPIF Standard 35 [296] only corresponding 'Rating' values are listed. In Tab.7.45 these are complemented by the corresponding number of holes drilled, cutting

Tab.7.45 Standard machinability rating of PM steels [145,277,296]. *Drilling test: 9.5 mm drill, drill speed 1250 rpm, v_c =37.2 m/min, feed 0.23 mm/rev., blind hole depth 25.4 mm, failure criteria – drill breakage or cutting edge wear >0.38 mm, no coolant*

Material	Number of holes drilled	Ax. drill (thrust) force [lbf]	[N]	Cutting edge wear [mm]	Rating
FD-0208	6.5	664	2948		5
FN-0208	22	648	2977		18
FC-0208	27	624	2771		22
FL-605	28.5	667	2961		24
FL-4205	42.5	637	2828		35
FD-0405	53.5	535	2373		44
FD-0208+0.5MnS	67.5	394	1439		55
FD-0405+0.5MnS	81	364	1616		66
AISI 1045 (wrought)	122	479	2127	0.21	100
FL-4405	128	496	2202	0.21	105
FLN-4205	134.5	548	2433		110
FLN-4205+0.5MnS	139	292	1296	0.41	114
FN-0205	178	494	2193	0.41	146
F-0008	192	488	2167	0.28	153
FC-0205	192	469	2082	0.36	157
FD-0205	192	416	1847	0.18	160
FC-0208+0.5MnS	192	391	1736	0.30	168
FL-4205+0.5MnS	192	365	1718	0.33	171
F-0005	192	377	1674	0.41	177
FN-0208+0.5MnS	192	336	1638	0.33	181
FL-4605+0.5MnS	192	305	1621	0.33	184
FC-0205+0.5MnS	192	336	1492	0.20	201
FN-0205+0.5MnS	192	305	1354	0.38	220
FD-0205+0.5MnS	192	304	1351	0.38	220
F-0008+0.5MnS	192	300	1332	0.33	222
F-0005+0.5MnS	192	231	1023	0.28	263
FC-0208+IMP	192	193	857	0.20	286
FC-0208+0.5MnS+IMP	192	161	715	0.18	305
FN-0208+IMP	192	153	679	0.13	310
F-0008+IMP	192	140	622	0.08	317
FC-0205+IMP	192	123	546	0.20	328

Remark: IMP – resin impregnated: 1 lbf = 4.44 N

edge wear, and axial drilling force values (as machining indices) from Refs. 145 and 277.

The results show that no relation was found between the indices. On the basis of these extensive machining tests performed under identical conditions it may be concluded that the number of holes drilled is preferred as a machining index for rating the machinability of PM steels. However, more than half of the steels tested reached 192 holes, and the test was then stopped. The higher the number of holes drilled in PM steels, the more machinable was the steel, and, therefore the higher was the machinability rating. According to these data the cutting edge wear data provide no meaningful values for the least machinable materials. The axial drilling (thrust) force in relation to the number of holes drilled can

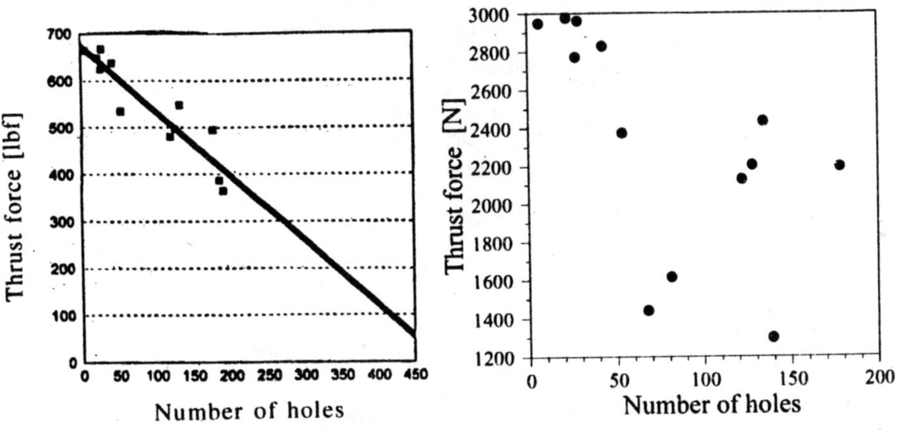

Fig.7.93 Machinability correlation between drilling (thrust) force (Fz) in *lbf* and the number of holes drillled [145,277]. Data from Tab.7.45. *Drilling test: HSS 9.53 mm drill, drill speed 1250 rpm,* v_c *= 37.4 m/min, feed 0.23 mm/rev., blind hole depth 25.4 mm.*
Fig.7.94 (right) Machinability correlation between drilling (thrust) force (Fz) in *N* and the number of holes drillled up to drill failure. Data from Tab.7.45.

also be considered as a possible machinability index. Thus, a compromise was made in the form of a graphical linear relationship between the drilling force and the number of holes drilled for the steels where the test data stopped at 192 holes, Fig.7.93. Figure 7.94 illustrates the equal machinability correlation with the thrust force values in *N*.

From the Fig.7.94 one can derive only the tendency that higher number of holes was drilled at lower thrust force. According Fig. 7.93 a linear relationship was found and used to extrapolate an 'estimated number of holes drilled for those materials where the test data stopped at 192 holes'. The presumption that these materials (18 from 30 tested) would be drilled minimum in the equal scatter of number of holes drilled as those which reached less than 190 holes is very important.

$$\text{Estimated number of holes drilled} = \\ = 488.76 - 0.725 \text{ lbf (drilling force).}$$

A further approach for developing a machinability rating system for PM steels followed the method used for wrought steel using the wrought AISI 1045 steel as reference material. The machinability rating was determined from the following equation (the number of holes for materials with >192 holes was determined according to

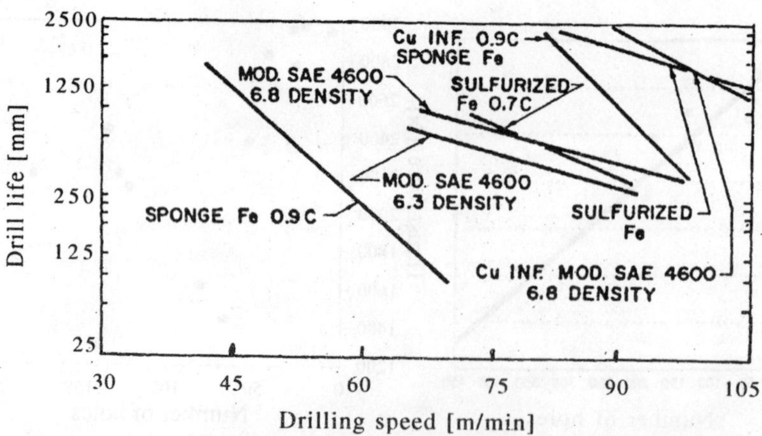

Fig.7.95 Machinability of some PM steels [134]. Sulphurized – 0.04% S, Inf. – copper infiltrated.

the previous equation):

$$\text{Machinability rating} = \frac{\text{Number of holes drilled in PM steel}}{\text{Number of holes drilled in 1045 steel}} \times 100$$

For example, according to this equation, the normalised machinability rating for FN-0205 alloy is:

$$178/122 \times 100 = 146.$$

According to these data, sintered FN-0205 steel should be better machinable than medium carbon wrought steel.

For PM steels, the higher the number of holes drilled, the more machinable the steel is and, therefore, the higher is the machinability rating. The rating based on the results with a large scatter has only very informative value for comparison the machinability of PM steels of various compositions. The force measurements alone cannot always be used to predict tool wear in machining [144,155]. These results present a broader view on machinability in drilling of 30 PM steels, which represent the most widely used basic compositions, machined under identical conditions. In spite of it, no technically exact quantitative comparison of the results obtained in drilling of a material with those presented here is possible. Designers and PM part producers can better relate to machinability expressed in terms of the number of holes drilled before drill failure, as also follows from these results [255].

A partial summary comparing the relation between the drill life and drilling speed of some sintered Fe–C and low alloy steels (sulphurised, infiltrated, various density) is shown also in Fig.7.95.

A large increase in drill life at an increase of drilling speed was caused by copper infiltration for Fe–0.9C and SAE 4600 steel. The sulphurisation of Fe–0.9C steel resulted in an increase in drill life on the level of SAE 4600. A minor increase in drill life was attained the sulphurised Fe–0.7C steel compared with the previous case.

7.10 SPECIAL PROCESSING AND MACHINING ROUTES FOR HIGH STRENGTH-HIGH HARDNESS PM STEELS

From the data presented so far it follows that the machinability of differently sintered high strength, high hardness alloy steels with additions of various types of machining agents can be characterised as low to poor. This restricts the wider use of these steels for the production of highly stressed complex shaped parts. In addition to this, higher alloying is most effective in the heat treated state of the materials which treatment makes them virtually unmachinable. This is the reason for the development and application of some PM processing steps with special attention to facilitate the machining of these steels and/or for testing the cutting conditions and cutting tool and geometry for such parts.

In this connection, it is important to note again that the machining conditions of a PM part are closely linked to the overall history of production and the processing steps used up to the final properties

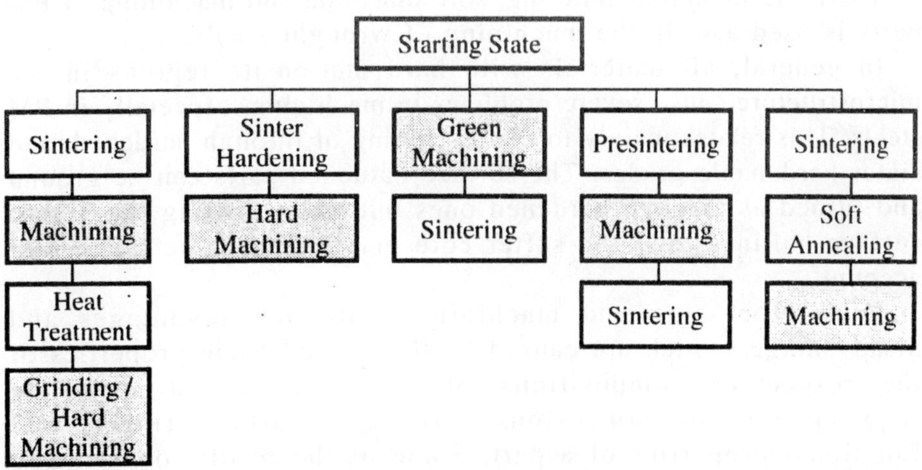

Fig.7.96 Main production routes and machining methods for high strength and high hardness PM steels.

of a sintered part with very complex shape.The main production routes, including hardening processes with corresponding machining methods applicable for high strength, high hardness steels, which are hardly machinable, are shown in Fig.7.96.

The first route is the most conventional one. If necessary a supplementary grinding or hard machining step can be applied after heat treatment (through hardening, surface hardening) of parts with an apparent hardness of 50–60 HRC.

The second possible route is simpler in processing steps which includes sinter hardening as a special PM process followed by hard machining. The resulting microstructure of a material formed by cooling down the parts straight from the sintering temperature with an increased cooling rate is commonly martensite or martensite and bainite with the microhardness in the martensite regions exceeding 650 HV. The apparent hardness in dependence on alloying, carbon content and cooling rate can be about 40 HRC. These main material characteristics decide about the option of the optimum hard machining conditions to realise the required machining operation on a part.

The third processing route, green machining with subsequent sintering, or sinter hardening, again a special PM processing route, seems to be the simplest but is very closely linked to the basic powder properties and the compacting process.

The similar route is characterised by presintering of the parts (to obtain the necessary strength) and machining with subsequent final sintering/sinter hardening.

The route including sintering, soft annealing and machining of PM parts is used also in the machining of wrought steels.

In general, all materials with hard martensite regions in the microstructure cause severe problems in machining, especially in PM steels. This relates mainly to the machining of through hardened, and sinter hardenable grades. The surface hardened parts can be ground and turned as through hardened ones but when drilling the higher surface hardness and the softer core material must be taken into account.

Each processing and machining route has advantages and disadvantages which are caused by the special basic properties of the respective compositions or alloys combined with the requirements for dimensional accuracy, surface finish, and functional properties of a part. Some of the results obtained in machining of these materials show the possibilities and limitations of each method. The results indicate in general the problems why it is not possible to machine all grades of high strength steels as of wrought steel even under modified processing and cutting conditions.

7.10.1 Green machining

Green machining is a PM process aimed at reducing the machining problems of PM parts by machining them in the green, *e.g.*, as-compacted, condition prior to sintering: The lack of strong interparticle bonding results in low cutting forces. The technique is known from the processing of ceramics. Green machining offers one of the alternative routes for the production of high strength PM steels but at present mainly in combination with warm compaction technology. The green strength of conventional metal powder compacts is too low to withstand the cutting and clamping forces employed in machining operations. The green strength of warm compacted parts is two to four times higher than that of cold compacted parts. The cold compaction technology with polymeric lubricants increasing the density and the green strength is another way, thus facilitating successful green machining of pressed parts. In any case, the green strength limitations also restrict the handling of the parts and require a careful adaptation of the cutting parameters and tool geometry to the green compact properties.

The first condition for successful green machining of a powder compact part is satisfactory green strength of a compact being machined attainable through interparticle bonds which enable the clamping of parts in jaws. The optimum state for green machining of a part cannot be characterised by the apparent hardness value. The cutting forces and moments are reduced in green machining approximately to one eighth of the values required for conventional machining. The decrease in machining forces makes it possible to reduce the clamping forces and to fix the green parts without damage. Most green machining operations are applicated to warm compacted parts with high density and by this with high green strength. Green machining can be applied for drilling, tapping, turning, milling and other cutting methods [209].

In green machining, the cutting parameters and tool geometry are mainly responsible for the quality (surface finish) obtained. For drilling, a more favourable drill with a 118° split point geometry (consistent quality over a broad feed range) compared to a standard drill with 118° point angle (large sizes of breakouts) is recommended. The surface finish quality which is of major importance is affected by the stiffness of the tools and the sharpness of the cutting edges. For optimum results using uncoated HM drills (higher stiffness compared to HSS drills) is the best. The use of uncoated tools is recommended since each coating rounds the cutting edge and reduces sharpness, and green compacts are very sensitive to rounded edges since the local geometrical conditions in the shearing plane change from a positive to negative tool geometry [165].

The second condition is minimum dimensional change of the green machined part during sintering. It means that the dimensional changes which occur during sintering must not exceed the required dimensional tolerances of a green machined hole, tap, thread or an undercut. The dimensional changes of a part caused by sintering depend on powder grade, density and composition and thus the selection of the starting powder is an important factor or the proper production of a part that which is to be green machined. Due to not quite regular shrinkage of all dimensions of a shaped part, differences, *e.g.* in entering and outlet diameter of a hole or thread may result which has to be considered.

The process of green machining (especially cutting resistance) is also affected by the microstructure of the base and alloying powder particles in a green compact. The microstructure of all green parts subjected to green machining is mixed and consists of the base powder (plain iron or prealloyed powder), additionally added alloying elements, graphite, lubricant, and/or binder type which can also act as basic green machining enhancers reducing friction.

Warm compaction aims at increasing the density of parts and, in combination with alloying and carbon content, to achieve as-sintered high strength properties of alloys, but from the beginning a special access to their machinability is required. Beside warm compaction, conventional cold compaction is used for preparation of high strength steel parts based on hybrid Distaloy type alloys with an appropriate binder. The parts produced by these compaction methods are subjected to green machining, at the same time an addition of some machining aid to the powder can be an interesting method for improving the green machinability. The green strength commonly increases with specific surface area of the powder, *i.e.* with surface irregularities of the powder (which however results in lower green density).

Basic preconditions for green machining are high green strength (>20 MPa) of the material which is achieved by warm compaction of Densmix powders or cold compaction of a powder mix with addition of special polymeric or other lubricants, and cutting parameters. There are three different methods to increase green strength, which are:

- addition of chemical adhesive,
- increasing green density,
- increasing particle irregularities (specific surface area).

The way to increase green density of a part is also to remove all lubricant from the powder mix and then to use die wall lubrication during compaction or special binder. Normally, die lubrication is manifested more in a green strength increase than in a green density increase.

The advantage of green machining are lower cutting forces and lower tool wear expressed in lower overhead costs, and the disadvantages are limited stability of the green compacts due to some internal defects and a lower product quality (surface, breakouts) regarding also a sintered state [165].

7.10.1.1 Green machining of warm compacted materials

As an example for green machining in drilling, tapping and turning, Tab.7.46 lists some of the results obtained under different cutting conditions with warm compacted alloyed powders.

The effect of cutting tool material (HSS, HM – stiffness) with its geometry in green machining on cutting forces and surface quality requires special attention with the aim to eliminate an additional machining step after sintering. Figure 7.97 shows the

Tab.7.46 Chemical composition, compacting pressure, green density and green strength of warm compacted (Densmix, Ancordense) compacts subjected to green machining. *Drilling test: HSS steam treated 5 mm drill, hole depth 12 mm, point angle 90°, 118°, 130°, cutting speed of 40, 80, 120 m/min, feed of 0.025, 0.05, 0.1, 0.2 mm/rev.; tapping test: HSS M6 straight flute-spiral point, cutting speed 10, 20, 40, 60 m/min. Face intermittent turning test: hardmetal uncoated and coated (TiN, Ti(C,N) and HSS TiN coated cutting tools, cutting speed 100–600 m/min, feed 0.0125–0.2 mm/rev., depth of cut 0.0125–0.2 mm*

Iron powder grade	Composition [mass %]					Compacting pressure [MPa]	Green density [g/cm³]	**Green strength [MPa]	Ref.
	*C	Mo	Cu	Ni	Cr				
Astaloy 85 Mo	0.3	0.85	-	-	-	800	7.31	20-21	65, 267, 297
Distaloy DH1	0.5	1.5	2.0	-	-		7.28		
Distaloy AE	0.5	0.5	1.5	4.0	-	700	7.30	27	65, 298, 299
						540	7.15	25	
Astaloy CrM	0.5	0.5	-	-	3.0	800	7.15	30	
						650	7.00	27	
Ancorsteel 85HP	0.4	0.85	-	2.0	-	-	7.33	33	115, 136

Remark: *as added graphite, **in bending, Distaloy AE for comparison. Apparent hardness of green compacts was in the range of 83 to 96 HV 5 and microhardness of base prealloyed powder particles of 146 to 156 HV 0.01 [65].

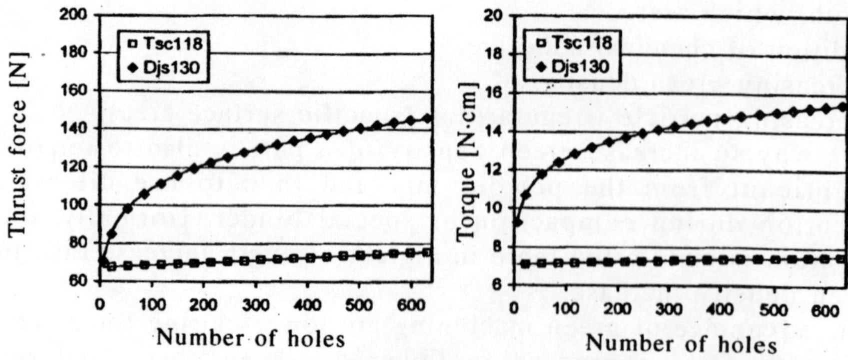

Fig.7.97 Thrust force and torque in green drilling of warm compacted Fe–3Cr–0.5Mo (Astaloy CrM)–0.5% graphite compacts vs. number of holes drilled [65]. *Drilling test: HSS (Djs130, 130°, brightened and hardmetal (Tsc118, 118°, K10) 5 mm drill, depth of hole 12 mm, cutting speed 80 m/min, feed 0.05 mm.*

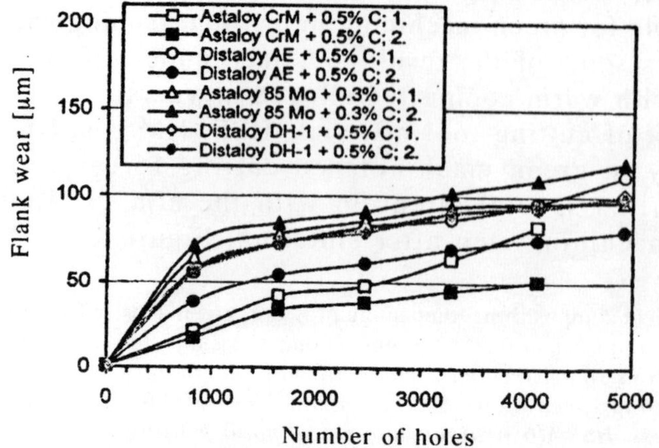

Fig.7.98 Width of the flank wear at the major cutting edges as a function of the number of holes drilled with an HM drill (Tsc118) in warm compacted green compacts from powders of different composition [65].

thrust force and torque as a function of the number of holes green drilled in warm compacted Fe–3% Cr–0.5% Mo (Astaloy CrM) powder compacts which is the subject of present special research interests. Beside the chemical composition and density, the alloying method is also an important factor with respect to green machining properties.

As shown, green machining reacts very sensitively to a change of the tool material (HSS, HM) and also of tool geometry which occurs with increasing tool wear. The thrust force and torque in green drilling with a HSS drill increased with increasing number of hole drilled. When drilling with a hardmetal drill, in contrast, these forces were constant (due to less wear) which seems to be more advantageous for green drilling. The mean size of breakout at entering and outlet edges (~0.3 mm for both) and, similarly, the surface finish quality were independent of the number of holes drilled with a hardmetal drill.

When comparing drilling of as-sintered and as-green warm compacted parts (compacting pressure 550 MPa) the thrust force and torque values are 80–90% lower in green drilling [300]. As shown in Fig.7.98, the wear of a tool depends on the number of holes drilled which which varies depending on the properties of the base powder mix.

The highest increase in wear was caused in drilling approximately up to 800 holes drilled for all compositions, *i.e.* a typical run-in period is observed. A smaller uniform wear increase of the drills was recorded in drilling over 800 holes up to 4000–5000 holes drilled. The minimum wear in the investigated range was recorded for CrM-steel. The material being drilled causes larger differences in wear than does varying green strength as listed in Tab.7.46. It follows from this that the effect of the powder particle properties (*e.g.* difference in microhardness values of the alloy particles) is much greater than, for example, integral properties of the compact such as macrohardness, density, and/or green bending strength.

It was also stated in Ref. 301 that good green machinability in spite of low green strength was exhibited by the cold compacted compositions Fe–3% Cu–1% graphite and Fe–1% graphite having radial crushing strength of about 60 MPa.

Very good conditions for additional green machining were found for the warm flow [15,16] and high energy compacted parts [18,19] exhibiting high density and green strength.

7.10.1.2 Green machining of parts with improved bonding strength

When high density and, consequently, high strength of a sintered part – which is best achievable by warm compaction of an appropriate powder – is not the deciding property, it is tried to attain the high green strength needed for machining also by cold compaction. The methods are based on the addition of some binders which increase the bonding between the powder particles.

Polymeric lubricant. The high green strength of a compact needed for machining can be obtained by the use of a polymeric lubricant (HGS). For parts with a density of ~6.8 g/cm³ this lubricant can provide a high green strength of ~48 MPa at a compacting temperature of ~55°C by subsequent curing.

The process in turning the grooves in timing sprockets produced from a Fe–1.8Ni–0.55Mo (Atomet 4601)+2.5% Cu+0.9% graphite powder mix as well as 0.65% HGS lubricant was tested. The sprockets were compacted to a density of 6.7–6.8 g/cm³. The best results were attained in green turning the sprockets when turning at 900 rpm and with a relatively high feed rate of 0.20 mm/rev. A substantially longer tool life was obtained with the process compared to as-sintered machining, and no burrs formed along the machined edges in green machining of cured parts. Satisfactory green machining was also achieved in semi-production trials of parts of FN-0205 alloy [302].

Chemical adhesive lubricant. As stated in Ref. 303, in order to increase the green strength of the parts, a chemical adhesive lubricant designated GS-Lube was added to the base powder mix (Distaloy AE+0.5% graphite). Premixes with common lubricant Amide Wax were prepared. The pinion gears were compacted to 7.0 (560 MPa) and 7.25 g/cm³ (900 MPa). The parts with GS-Lube were compacted at elevated temperature (55°C). The turning with CNC milling machine of green pinion gears, Fig.7.99, was tested with cutting speeds of 400–600 m/min at feeds of 0.005–0.01 mm/rev. (the green machining operation was carried out in 4 steps). The critical steps in the turning operation occurred when the insert entered and left (with possibly most damage on the teeth) the pinion gear. Cracks and breakouts were found on the surface of the outlet

Fig.7.99 Green turned pinion gear (22.3 mm outer diameter, 11.6 mm height) [303].

on both materials compacted to a green density of 7.0 g/cm³. The green machining of parts with a green density of 7.25 g/cm³ resulted in no cracks or damage in the material containing GS-Lube. In contrast the parts with Amide Wax had some cracks and breakouts on the teeth also at the density of 7.25 g/cm^3. The use of the GS-Lube lubricant increased the green strength of the parts (by curing) as required for green machining and, on the basis of the turning test results, was considered fully successful for this purpose.

7.10.2 Heat treatment processes and machinability

7.10.2.1 Soft annealing and presintering

These widely different heat treatments are discussed together here since they have been jointly described in the same publications. Both soft annealing and presintering heat treatment routes are applicable to high strength sintered steels and lead to improved machinability. However, the optimum method in both cases depends on a number factors which cannot be simplified.

Soft annealing is applied to improve the machinability of wrought steels and, therefore, should be also applicable to high strength sintered steels represented by, for example, Distaloy type alloys. This procedure was performed including an additional low-temperature heat treatment- partial hardening after machining in order to restore the original as-sintered properties, with strength and hardness exceeding the values acceptable for machinability.

Presintering treatment is a well suited process for the sintered components which without this treatment must be subjected to hard machining, such as, for example, sinter hardenable materials. Presintering is performed at a temperature sufficient to attain the required strength for withstanding the machining operation without damaging the parts; frequently it can be combined with the delubrication treatment.

Soft annealing and presintering. In Ref. 304 there is a description of the influence of soft annealing and presintering compared to sintering on the machinability of Fe–4Ni–1.5Cu–0.5Mo–0.5C (Distaloy AE) and Fe–8Ni–1Mo–0.5C (Distaloy AG), which can be accepted for mostly known high alloy, high hardness steels.

The heat treatment processes (compaction at 600 MPa) used for testing the effect on machinability of the above mentioned alloys were performed by the following processing routes:

- *presintering*: 900°C, 30 min, $95N_2$–$5H_2$ atmosphere for Distaloy AE and AG followed by machining and sintering 30 min at 1120°C in endothermic atmosphere + CH_4 (Distaloy AE) and 60 min at 1150°C in $95N_2$–$5H_2$ atmosphere (Distaloy AG),
- *annealing* (soft): sintering as at previous sintering route, anneal-

ing 60 min at 500°C, N_2 atmosphere, machining – both alloys (hardness 200 HV 10 Distaloy AE and 250 HV 20 Distaloy AG), followed by partial hardening 60 min at 750°C, N_2 (Distaloy AE, hardness 230 HV 10) or 60 min at 725°C, N_2 (Distaloy AG, hardness 310 HV 20),

– *sintering*: as at the presintering route followed by final machining.

As claimed by the author, soft annealing at 500°C provided the lowest tensile strength and hardness of both materials. (Soft annealing at 500°C is not usual and the changes in the microstructures which affected the decrease in mechanical properties of the materials were not sufficiently demonstrated).

A further heat treatment in the range of 725–750°C (partial

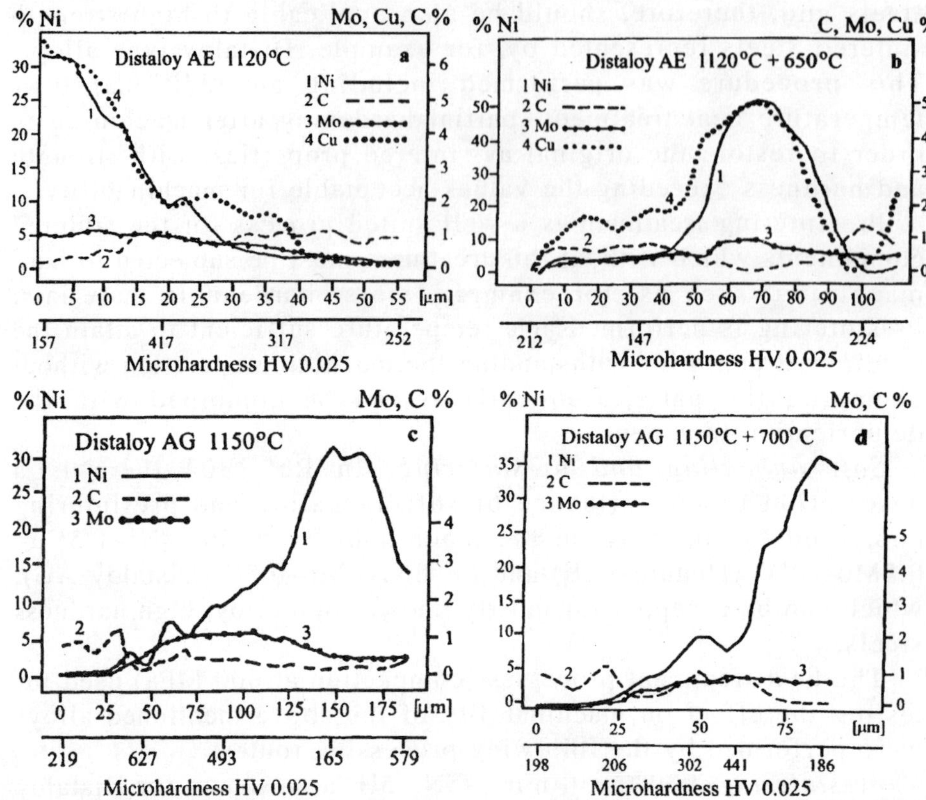

Fig.7.100 Linear microprobe analyses of sintered Distaloy AE and Distaloy AG alloys. Distaloy AE: a – sintering 1120°C, b – annealing 650°C; Distaloy AG: c – sintering 1150°C, d – annealing 700°C [114,304,305].

hardening) had to recapture the as-sintered properties. The tensile strength of Distaloy AE alloy soft annealed at 500°C was 620 MPa and that of Distaloy AG 730 MPa and after partial hardening 730 and 980 MPa.

It was further claimed that the non uniform distribution of nickel and copper which, on their own, have individual effect on machinability and in combination a more marked effect, was lowered by annealing when compared with the as-sintered condition, as shown in linear microprobe analyses of both alloys in Fig.7.100 completed with the corresponding microhardness values.

The microhardness values show more schematically the changes in

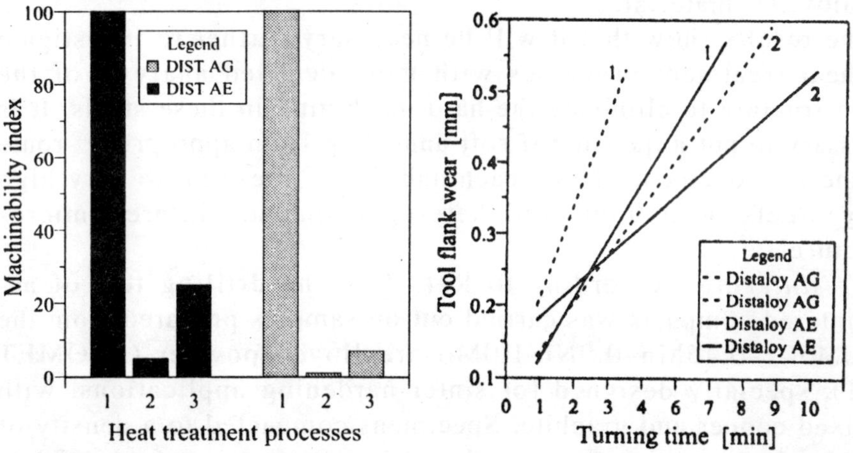

Fig.7.101 Machinability (machinability index) in drilling of Distaloy AE and Distaloy AG after three different heat treatment processes applied. 1 – presintering (900°C), 2 – sintering (1120 or 1150°C), 3 – soft annealing (500°C) [acc. to Ref.114,305]. *Drilling test: HSS 3.5 mm drill, drill speed 3000 rpm, v_c = 33 m/min, feed 235 mm/min, depth of hole 12 mm, criterion totally worn drill.*

Fig.7.102 (right) Machinability (tool flank wear) in turning of Distaloy AE and Distaloy AG. 1 – as-sintered, 2 – as-soft annealed [acc. to Ref. 114 and 305]. *Turning test: Carboloy 883 tool bit (cermet CT 515), speed 100 m/min, feed 0.2 mm/rev., cut depth 0.4 mm, criterion 0.4 flank wear or 15 min.*

microstructures of both alloys by annealing with regard to machining. The decrease in microhardness values by the reduction of the nickel and copper concentration heterogeneity in the microstructures was found in different levels by annealing of Distaloy AE and of Distaloy AG (although is is not clear how a low temperature annealing treatment should result in better homogenization than sintering – according to literature data the diffusivity of Ni in austenite at 1120°C is almost three orders of magnitude faster than at 650°C).

The machinability of these steels was tested in drilling, and the results are shown in Fig.7.101 and for turning in Fig.7.102.

The machinability in drilling of both alloys was highest as-presintered. Soft annealing (500°C) caused a signicantly lower machinability compared with as-presintered, relatively more in Distaloy AE steel. The machinability of Distaloy AG after soft annealing was at the same level as that of as-sintered Distaloy AE material. The same results were obtained in the turning test. Considering the large shrinkage of the high-Ni Distaloy AG at sintering, the presintering process was not advantageous for this alloy due to dimensional reasons but it can be applied efficiently to Distaloy AE material.

The results show that it will be necessary further to investigate the heat treatment processes with more detailed analysis of the microstructure to eliminate the hard machining in these steels. It is necessary to put a question if soft annealing is an appropriate route for the improvement of the machinability of these up to very high alloy steels with a marked heterogeneous as-sintered microstructure.

Presintering. According to Ref. 306, the drilling test of as-presintered compacts was carried out on samples prepared from the Fe–0.45Cr–0.45Mn–0.9Ni–1.0Mo prealloyed powder (ATOMET 4701), specially designed for sinter hardening applications, with admixed copper and graphite. Specimens compacted to a density of 6.8 g/cm^3 were presintered in the temperature range from 670 to 970°C for 28 min in a $90N_2$–$10H_2$ atmosphere and tested in drilling; the thrust force was measured, Fig.7.103.

As shown, the thrust force in drilling presintered specimens increased with the pre-sintering temperature. For the thrust force,

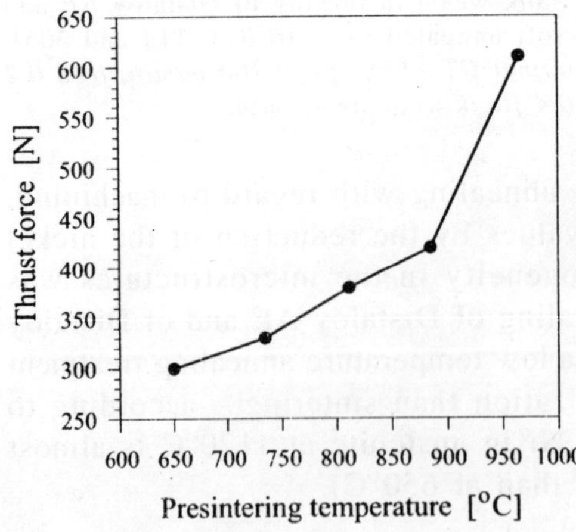

Fig.7.103 Effect of presintering temperature on the average thrust force in drilling of Fe–0.45Mn–0.45Cr–0.90Ni–1.0Mo (prealloyed ATOMET 4701)+2% Cu+0.9% graphite alloy, green density 6.8 g/cm^3 [306]. *Drilling test: black oxide coated HSS 6.35 mm drill with a helix angle of 118°, drill speed 2220 rpm, v_c = 44 m/min, feed 0.12 mm/rev. Fifteen holes were drilled to a depth of 11.2 mm for each test condition.*

the rate of increase was linear in the temperature range from 650 to 890°C with a steep increase from 400 to 620 N at the temperature of 970°C while for torque, which was also measured, it was almost linear throughout the full temperature range studied (0.6–1.1 N·m). This is a consequence of an increase of the mechanical strength and hardness of the material with increasing temperature due to the formation of metallic bonds between the powder particles in the compact and the diffusion of graphite and copper into the ferrous matrix.

The optimum presintering temperature for drilling this alloy was found to be in the range from 810 to 890°C. At 650 and 730°C, the presence of soft ferrite grains promoted the densification of the layer under the surface of the cut as the drill penetrated the material. The structure in the presintering temperature range from 810 to 890°C minimised the densification of the layer under the surface cut. At a temperature of 970°C, the presence of hard martensite and bainite, as a consequence of virtually complete carbon dissolution [307], rapidly led to the destruction of the cutting edge of the tool. The annealing treatment at 870°C changed the failure mode of the HSS drill. The drill appeared to adhere to the workpiece and map rather than overheat. The optimum microstructure which enabled a smooth cutting surface of the drilled hole was composed of pearlite.

Generally, presintering is a technique that enhances the machinability without too much additional cost (esp. if combined with delubrication), and the material can be adjusted for optimum machinability simply by varying the presintering temperature. Similar to green machining of warm compacted materials, the dimensional change during sintering has also to be considered here.

7.10.2.2 Hard turning of sinter hardenable steels

Sinter hardening is a post-sintering treatment which is integrated in the sintering process and can be approximately associated with a through hardening process but avoiding thermal stresses and distortion.

Tab.7.47 Chemical composition of sinter hardenable powders used (4200 series acc. to MPIF standard 35, Domfer) [65,265]

Powder grade	Content of alloy elements [mass %]						
	Mn	Ni	Mo	S	Si	P	Fe
MP52	0.40	0.45	0.70	0.02	0.09	0.02	Bal.
MP52FM	0.70	0.45	0.70	0.25	0.09	0.02	Bal.

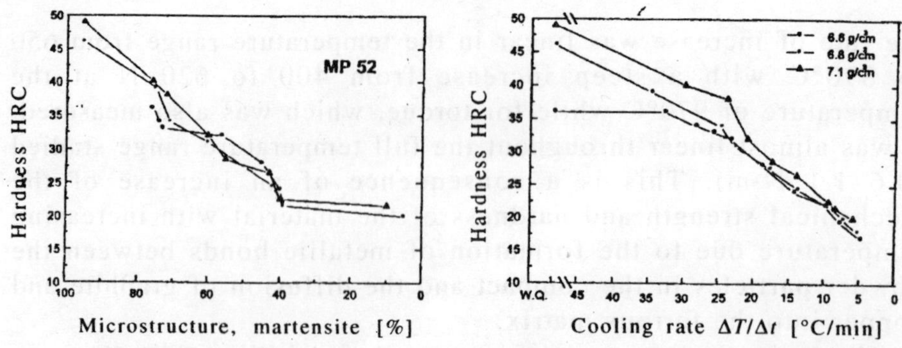

Fig.7.104 Apparent hardness of MP52–2Cu–0.65C_c alloy with varying density as a function of: left – the proportion of martensite in the microstructure, right – the cooling rate between 550 and 350°C [265].

The manganese and sulphur prealloyed sinter hardenable powders may contribute to improved machinability of such steels. Two types of sinter hardenable powders were selected for testing in turning and drilling, and these grades MP52 and MP52FM (the latter with Mn and S) are used to show the effect of sulphur prealloying on the machinability (the mean MnS particle size is 5 μm) The chemical composition of the powders used is listed in Tab.7.47. The powders, with admixed 2% Cu and 0.75% graphite were compacted to a density of 6.8 g/cm^3 (rectangular bars and discs) and sintered in 25H_2–75N_2 atmosphere (final as-sintered hardness 27 HRC, C_c = 0.65%).

Sinter hardening was simulated by austenitising the samples at 980°C and cooling in nitrogen at different rates. Figure 7.104 shows the required cooling rate of MP52 and MP52FM steels in the critical temperature range to obtain a given microstructure and apparent hardness. The highest hardness of the parts was attained at 100% martensite in the microstructure after water quenching. At a cooling rate of ~30°C/min, the parts exhibited a hardness of about 40 HRC with 80% martensite in the microstructure. The effect of the density on the hardness decreased with lower cooling rate.

Machining tests of the steels were performed in turning and in drilling. Figure 7.105 shows the relation between the thrust force in drilling and the number of holes drilled in the tested steels for various cutting speeds. High axial thrust values are typical when drilling sinter hardenable steels compared to as-sintered ones. Significantly, sulphur prealloying caused a decrease in the thrust force and an increase in the number of holes drilled in MP52FM steel, compared to MP52 steel.

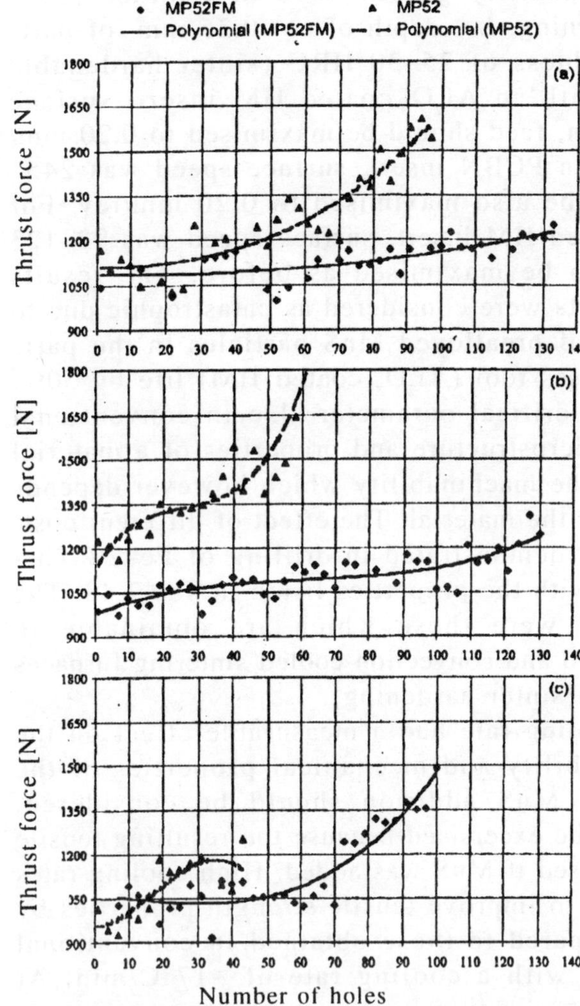

Fig.7.105 Evolution of axial (thrust) force in drilling of MP52 and MP52FM steel samples. Drill speed [m/min]: a – 61, b – 75, c – 83 (acc. to Ref. 308). *Drilling test: Uncoated hardmetal 6.75 mm drill, constant feed 0.10 mm/rev., depth of hole 10 mm.*

For MP52FM steel, the cutting speed in drilling with uncoated HM drill should be ~76 m/min and for MP52 should be ~53 m/min using a feed rate 0.10 mm/rev.

Turning tests of the steels were carried out with hardmetal inserts coated with alumina, alumina ceramic inserts, and TiAlN coated PCBN tools at cutting speeds of 123 to 308 m/min and feeds of 0.10 to 0.20 mm/rev. respectively. Good hardness and chemical stability at high temperature as well as oxidation resistance were the selection criteria for the cutting tools. The average width of the flank wear surface of a cutting edge 0.38 mm was the criterion for the tool life (typical criterion for tool indexing while performing finishing operation in turning).

The following machinability guides were determined for a finishing operation in turning, *i.e.* depth of cut 0.25 mm, of parts having an apparent hardness of 25–30 HRC (sinter hardenable steels). For MP52FM with an Al_2O_3 coated HM insert, surface speed was 154–185 m/min, feed should be maximised to 0.20 mm/rev. For MP52FM with an PCBN insert, surface speed was 245–277 m/min, feed should be also maximised to 0.20 mm/rev. For MP52 with an Al_2O_3 coated HM insert, surface speed was 92–123 m/min, feed should also be maximised as before. The results obtained with Al_2O_3-inserts were considered as catastrophic due to chipping. The presence of prealloyed MnS particles in the parts made of MP52FM increases tool (Al_2O_3 coated HM) life by 50%.

The cooling rate is a critical parameter also in conventional sintering affecting the microstructure and properties of a material with resulting effect on the machinability which however depends on the specific features of the material. The effect of different post-sintering cooling rates is demonstrated in drilling of Fe–2% Cu–0.9% C (FC-0208) steel with the properties listed in Tab.7.48. The cooling rates employed were those which are obtainable in conventional water jacketed and convection cooled sintering furnaces (lower than those used at sinter hardening).

The post-sintering cooling rate had a measurable effect on the microstructure, machinability and mechanical properties of the material. The effect of MnS addition should be considered. However, caution should be excersised because the resulting tensile strength properties decreased if MnS was added. High cooling rates (~40°C/min) can be used to improve tensile strength properties by approximately 15% compared to those obtained in conventional water jacketed furnaces with a cooling rate of ~17°C/min. At

Tab.7.48 Mechanical properties (Rp0.2, Rm, HRB) of FC-0208 (atomised Kobelco 300 MA powder)+0.5% MnS steel in dependence on the cooling rate. Sintering in endothermic atmosphere. Density ~6.61 g/cm³, elongation ~2.3% (no effect of the cooling rate). Cooling rate treatment: 1 – controlled tempering, water jacket, 2 – conventional water jacket, 3 – atmosphere convection cooling [266]. *Drilling test: TiN coated 19 mm hardmetal drill, drill speed 145 rpm, v_c = 8.7 m/min, cutting oil, drilled part thickness 25 mm. Criterion: drill motor load, surface finish, hole entry and exit size*

Treatment	Cooling rate [°C/min]	Rp0.2 [MPa]	Rm [MPa]	HRB
1	8.75	253	337	68
2	16.7	348	394	73
3	39.8	497	503	84

~9°C/min cooling rate, coarse lamellar pearlite and divorced eutectic carbides, at 16.7°C/min fine lamellar pearlite, and at ~40°C/min unresolved pearlite can be formed in the microstructure of the material tested. Usually lower cooling rates, achievable using a temperature-controlled water jacketed cooling zone can be used to improve the machinability of the parts at the expense of strength properties. In any case, assessing the effect of the cooling rate on the microstructure of a part in terms of machinability must always be related to the mass of the singular part.

The lowest drill motor load of 3.1 kW for drilling (single drill) was required after treatment 1 of the material (lowest hardness). The treatment 2 caused an increase in drill motor load to 3.33 kW and the treatment 3 to 3.51 kW (highest hardness of the material). A reduced cooling rate (~9°C/min), resulting in the lowest hardness as compared with the faster cooling rates, can be used to improve significantly the machinability at small differences in surface finish. An increase cooling rate (~40°C/min) can be used to improve the mechanical properties. The question is if the resultant difference is economically interesting.

In terms of the hole size (entry and exit side) consistency and variability for these three treatments were not significantly different. However, treatment 2 was actually closer to the nominal hole size.

7.10.2.3 Machinability of hybrid warm compacted steels based on prealloyed Fe–1.5Mo powder

Machining tests in drilling and tapping of warm compacted and sintered Distaloy DH-1 compared to Astaloy 85 Mo and Distaloy AE steels, all with graphite addition and without machining additives, show the limits in machining for such a group of materials. The effect of the cooling rate was also studied. The chemical composition, density and hardness of investigated materials are listed in Tab.7.49.

Tab.7.49 Chemical composition, density and hardness of warm compacted (Densmix) and sintered Distaloy DH-1-C, Astaloy 85 Mo–C and Distaloy AE-C steels (reference material). Compacting pressure 800 MPa, disc 80 mm dia. and 15 mm height, sintering 1120°C, 95N_2–5N_2 atmosphere (acc. Ref.309,310). *Drilling and tapping test: HSS and fine grained HM 5.0 mm drill, HSS M6 tap, drill speed from 0 to 8000 rev./min, feed 0.1 mm/rev., blind hole depth 10 mm*

Powder grade/cooling rate	Composition [mass %]				Density	Hardness
	C	Mo	Cu	Ni	[g/cm³]	HV 5
Astaloy 85 Mo+0.3 %C/1 K/s	0.3	0.85	-	-	7.32	175
Distaloy DH-1+.0.5 %C/1 K/s	0.5	1.50	2.0	-	7.18	267
Distaloy DH-1+.5 %C/3 K/s	0.5	1.50	2.0	-	7.18	337
Distaloy AE+0.5 %C/1 K/s	0.5	0.50	1.5	4.0	6.99	232

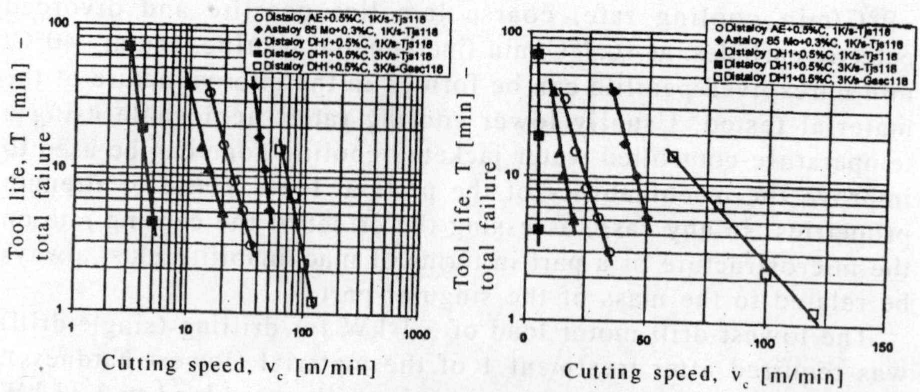

Fig.7.106 Tool life in drilling of warm compacted materials and Distaloy AE+ 0.5% C. *Dry drilling, feed 0.1 mm/rev.* According to the Taylor model. Tjs 118 – HSS standard drill, Gasc 118 – HM drill [309,310].
Fig.7.107 (right) According to the experimental model [309,310].

Figure 7.106 shows the Taylor tool life lines and Fig.7.107 the exponential tool life in drilling these materials. When using HSS drills, the machining characteristics of warm compacted materials deviated significantly from those of the reference material Distaloy AE+0.5%C. With the HSS drill, only extremely low cutting speeds (3.7 m/min) lead to some results with the material Distaloy DH1–0.5C/3 K/s, which is not economically acceptable. When drilling this material with HM drill, cutting speed increased from 3.7 m/min (HSS drill) to about 53 m/min. It follows from it that the mentioned hardest material can be drilled only using HM drills.

When drilling under wet conditions, the cutting speed for Astaloy 85 Mo was more than 60% higher. The influence of cooling and lubrication on the drill performance was more pronounced with Distaloy DH1–0.5C/1 K/s. When using a standard HSS drill, the drilling of Astaloy 85 Mo–0.3C/1 K/s, is most economical under dry condition.

The tapping of Astaloy 85Mo–0.3C/1 K/s, and the drilling and tapping of Distaloy DH1–0.5C/1 K/s, can be done only under wet conditions.

The hardest material Distaloy DH1–0.5C/3 K/s, cannot be drilled or tapped with standard HSS tools under dry nor under wet condition. However, the results of a dry drilling test with a fine grained HM drill can be accepted as very promising for this material, even if the hole quality was not always as good as that obtained with the HSS drills.

7.10.2.4 Hard turning of through hardened and tempered steels

The face hard turning test was performed with water quenched

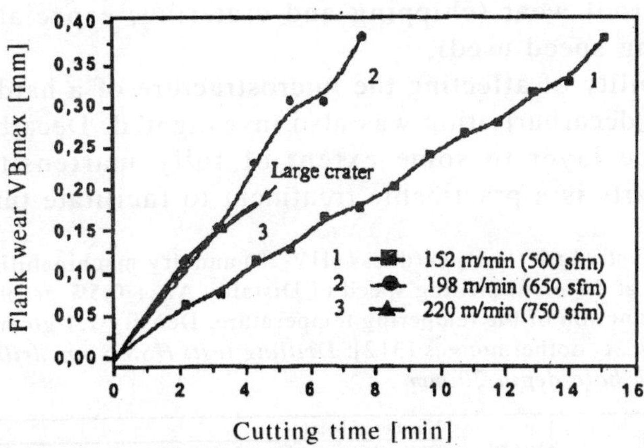

Fig.7.108 Tool life (flank wear) of PCBN inserts in hard turning of MP52FM+ 2% Cu+0.75% graphite steel as a function of surface speed and cutting time (*feed rate 0.20 mm/rev., depth of cut 0.25 mm, dry cutting*). Density 6.8 g/cm^3, sintering at 1120°C in $10H_2$–$90N_2$ atmosphere and heat treated for maximum hardness [311].

MP52FM alloy (Tab.7.47) of a untempered fully martensitic microstructure (~60 HRC) in dry conditions with constant feed of 0.20 mm/rev. and a depth of cut 0.25 mm using different tools (PCBN, cermet and coated hardmetal tools).

The tool life in hard turning of MP52FM alloy in dependence on cutting time for PCBN inserts shows the effectiveness of the process, and the results obtained with a cermet tool used to remove the decarburised layer, Fig.7.108.

Heat treatment of PM steels by through hardening followed by tempering is a route for production of PM parts with highest mechanical properties and better machinability compared to the untempered, fully martensitic, *i.e.* as-quenched, state (60 HRC). The beneficial effect of MnS formed in these steels based on sulphur prealloyed powder on the machinability was confirmed. In the test a loss in lubrication behaviour of MnS at high cutting speeds was observed.

It was stated that PCBN tools were the only ones that could be used for cost effective machining of PM parts with fully martensitic microstructure. As shown the optimum cutting speed was ~150 m/min depth of cut of 0.25 mm with a feed rate of 0.20 mm/rev., while at higher cutting speed a significant decrease in cutting tool life was observed. Increasing this cutting speed by 30% to 198 m/min led to a marked decrease in tool life. It was shown that high cutting speeds can lead to the formation of a diffusion layer on the surface of the tool, formed by oxidized manganese and silicon, accelarating tool wear (chipping and cratering, especially at the highest cutting speed used).

The possibility of affecting the microstructure of a hard PM part by controlled decarburisation was also investigated. Decarburisation of the surface layer to some extent of fully martensitic sinter hardenable parts is a practicable treatment to facilitate turning but

Tab.7.50 Tensile strength (Rm), hardness (HV 20) and dry machinability (number of holes drilled) at different drilling speed of Distaloy AB + 0.5% graphite + 0.5% MnS steel as a function of the tempering temperature. Density 7.1 g/cm³, sintering 30 min at 1120°C, ε dothermic gas [312]. *Drilling test: HSS 6 mm drill, feed 0.05 mm/rev., through hole depth 20 mm*

Tempering temperature [°C]	Rm [MPa]	HV 20	Number of holes Drill/cutting speed [rpm/m/min]]		
			500/9.4	1000/18.8	1500/28.2
250	970	432	57	19	251
300	940	326	*300	*300	250
350	910	320	*300	*300	226

Remark: *Test stopped after 300 holes drilled without drill failure. Retained austenite was less than 2 %. As-sintered Rm = 570 MPa, 172 HV 20

it is very tricky to obtain reproducible thickness of the decarburised layer in all parts and in different batches. This treatment cannot be effective in drilling (soft surface layer and hard core).

In an investigation of the fatigue properties of Fe–1.75Ni–1.5Cu–0.5Mo–0.5C (Distaloy AB) material, the machinability of as-heat treated (860°C/60°C/oil) and tempered was also tested. Tensile strength, hardness, and machinability (test results in drilling) of the material are listed in Tab.7.50.

These results compared to those obtained in machining of hard PM steels, as, *e.g.* Distaloy DH-1 and other, are surprising. The results show another view on the machinability of high hardness steels, *e.g.* that heat treated and tempered PM steels – in this case diffusion alloyed steels with hardness markedly above 200 HV10 – are machinable better than as-sintered ones.

A significant effect of the tempering temperature was recorded. The machinability of the samples tempered at 250°C is lower than for those tempered at 300 and 350°C especially for the low cutting speeds. It follows from it that even at a hardness corresponding to the tempering temperature of 300°C the machinability in drilling (HSS drill, 1000 rpm) and the fatigue strength of this material should be very acceptable. Effect of cutting speed is clear.

The comparison between machinability in drilling and endurance limit of this material is shown in Fig.7.109 as a function of the tempering temperature. The figure also shows that optimum

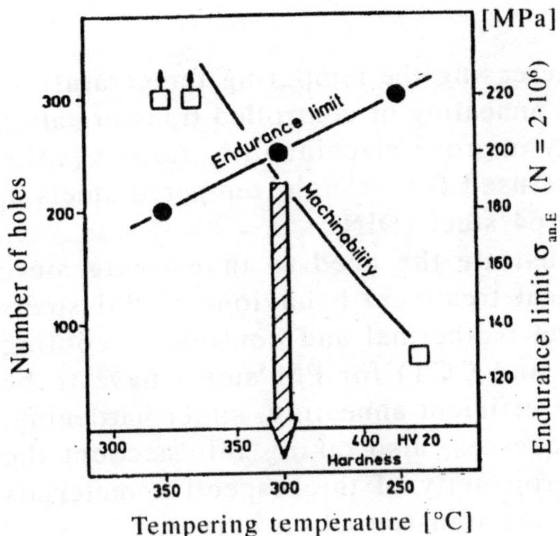

Fig.7.109 Machinability (number of holes drilled) and endurance limit for heat treated Distaloy AB+ 0.5% graphite+0.5% MnS steel as a function of hardness and tempering temperature (see Tab.7.50) [312]. *Drilling test: drill speed 500 rpm, v_c = 9.4 m/min, no coolant.*

machinability and fatigue strength of the steel are obtained at a tempering temperature of 300°C.

It must be noted that also wrought carbon steels nearly always have better machinability than alloy steels of comparable carbon content and hardness. Steels hardened and tempered to hardness levels above 300 HB are an exception to this observation; under such conditions, alloy steels have superior machinability, which is usually attributed to, first, the higher tempering temperature required to temper an alloy steel to a specified hardness level and, second, non-uniformity of microstructure due to limited hardenability in carbon steels. The same effect on the machinability was observed in different PM alloy steels.

The microhardness values suggested that a sort of precipitation hardening may have occurred in the pearlite and bainite areas of the microstructure. This may account for the increase in yield strength and loss of ductility with increasing tempering temperature. In contrast, the nickel-rich areas show a slight reduction in microhardness on tempering. Considering also the microstructure heterogeneity it is possible to presume that the hardness increase was caused by the transformation of the retained austenite to martensite ('300°C brittleness').

The experimental results indicated in general that the heat treatments employed did not sufficiently change the microstructure in relation to the machinability of PM steels. The tempering temperature range of 150–700°C should be investigated. It cannot be expected that common tempering conditions exist for all high alloy PM steels.

It can be expected that increasing the tempering temperature to 650–700°C, *i.e.* to subcritical annealing or controlled transformation annealing after sintering, may improve machinability further [137]. This is the tempering process used for wrought tempered steels at ~540–680°C, *e.g.* for 42CrMo4 steel (DIN).

The published results illustrate the need to investigate more generally the fundamental heat treatment behaviour of PM steels in relation to machining. The isothermal and continuous cooling transformation curves (TTT and CCT) for PM steels have to be established in order to define efficient annealing, sinter hardening, and other heat treatment processes, also taking into account the chemical and structural heterogeneity of the respective materials which cannot be reflected in the transformation curves.

Machining of steam treated parts. For this investigation the premixes formed by atomised iron powder with addition of (0, 0.5, 1.1)% MnS as well as of (0.3, 0.7)% CaF_2 (common valve seats lubricant) were compacted to sample 40 mm in diameter (density 6.4 g/cm^3, sintering 20 min at 1120°C in $90N_2$–$10H_2$ atmosphere) and steam treated at 530°C. The samples were turned with TiC cermet insert at cutting speed 250 m/min at feed 0.08 mm/rev. and a depth of cut 0.5 mm.

It was stated that both CaF_2 and MnS reduced built-up when the samples were machined as-sintered. CaF_2 addition degraded the surface finish of sintered samples while this was not in the case of steam treated parts. When machined steam treated parts, the behaviour of both additives was significantly different. In the case of MnS the sharp edge of the tool tip was maintained. CaF_2 was effective in eliminating metal built-up when machining as-sintered parts but not for steam treated parts. MnS prevented built-up for both as-sintered and as-steam treated parts. Steam treated parts showed consistently superior finish compared to only as-sintered parts. However, further investigation of the effect of CaF_2 on machinability of PM steels is needed [313].

7.10.3 Summary

a) *Green machining*

- Warm compacted materials with a green density of about 7.3 g/cm^3 considering the limitations (dimensional change) in properties of parts in sintered state can be green machined if the machining is performed with suitable cutting parameters and tool material and geometry.

- The high green strength of a compact necessary for green machining can also be obtained by the cold compaction of premixes with addition of a polymeric lubricant.
- Successful green machining is a function of both green strength of the material (>20 MPa) and of base physical–metallurgical properties of the powder (specific surface area, hardness, compressibility).
- Significantly lower cutting forces are required and lower tool wear is encountered compared to other machining methods.
- The cutting conditions should be adapted to the powder composition and green strength including density.
- Tool material and geometry and tool wear depend on the base powder and finally affect the integrity of the machined surface. Both cutting forces and surface finish can be improved by changes of drill type and profile. The choice of tool will be a compromise between low cutting forces, surface finish, and tolerances.
- Since in green machining the geometry is defined in the green state, dimensional changes during sintering affect the machined dimensions; this is particularly noticeable with drilling and tapping where only fixed dimensions – bore diameters, threads – can be machined.
- The use of a special lubricant (HGS) or an adhesive lubricant (GS-Lube) provides also sufficient strength for clamping and green machining of parts cold compacted to the density ~6.8 g/cm^3. Surface finish and edge integrity of machined parts can be improved if a curing process is done before green machining.

In drilling the best results in terms of drilling force and surface finish were achieved with a HSS-135° split point drill. Low thrust force is an important factor in green drilling.

- Drills with geometry improving the chip removal from the hole to be drilled, which is a key factor in green drilling, should be preferred.
- High cutting speeds and low feeds are the best also for improving surface finish.
- Breakouts are characteristic for green drilling; the outlet edges are usually slightly larger than those at the entering edges; the best overall results in the breakout sizes were obtained with a conical 90° point angle drill at the low depth of cut of 0.025 mm and a feed of 0.05 mm/rev.
- The best results at the entering edges were obtained with the

drill with a short overall length.

The effect of graphite addition on the green machining of parts should be investigated, considering the effect of retained graphite on machinability (see chapter 7.2.5.2). The results obtained with standard tools can be exceeded by using adapted tool material and geometry. The best results were obtained at drilling with sharp uncoated hardmetal drills (high stiffness).

In tapping, none of the threads produced complied with the specifications (ISO 1502), independently of the material.

- Tapping was optimised with spiral point geometry taps.

In turning, the criterion was edge integrity in relation to the cutting speed, feed, depth of cut, structural integrity, and surface quality. The critical moments at this process are when the cutting edge enters the material, and when it leaves it, which causes damages in the shape, *i.e.* breakouts of the edges of the slots.

- A sharp uncoated cutting edge geometry and highly positive rake angle were essential regarding the quality criteria for the machined surface of a part required.
- The cutting process was realisable in the cutting speed range between 100 and 600 m/min with HM insert (OCGT11T302FN-27, Plansee) when a feed of 0.025 mm/rev. was used at the depth of cut of 0.0125 mm/rev.

b) *Heat treatment processes*

- For high strength and high hardness steels, machining can be performed after suitable heat treatment.
- Soft annealing at 500°C followed by partial hardening at 700–725°C was slightly effective in machining compared with the sintered state.
- Presintering in the temperature range of 810 to 890°C seems to be most effective for machining of sinter hardenable materials. Also here, however, dimensional change during sintering affects the machined dimensions. The pearlitic microstructure was the best for machining.
- For sinter hardened steels, the cooling rate has a dominant effect on machinability. The optimum cooling rate attainable in production furnaces must be chosen in relation to the mechanical properties required and the machinability of the material.
- Sinter hardened steels could be drilled with hardmetal drills.
- Hard turning of through hardened and sinter hardened steels was possible only with PCBN inserts.
- Sulphur prealloyed powders and/or MnS addition in sinter

hardenable steels resulted in machinability enhancement.

- Tempering of through hardened Distaloy type alloy steels increased significantly the machinability in drilling with HSS drills compared to the as-sintered state. The optimum tempering temperature in this case can be determined as 300°C at cutting speed of 18.8 m/min. However, the effect of tempering depends on the composition and microstructure of a material, and no general conditions can be given.

7.11 MACHINING OF POWDER FORGED STEELS

Powder forging (PF), also called 'sinter forging', is a process for manufacturing virtually fully dense parts especially for high dynamic loading conditions. Powder forging enlarges the range of powder metallurgy parts into the region with properties even exceeding those of wrought steels. Basically, the almost complete elimination of pores diminishes greatly their known deteriorating effect on the properties of the parts. In such case, high impact and fracture toughness, fatigue behaviour and structural homogeneity are mandatory properties of the parts. If speaking of 'fully' dense parts, it implies forged parts with typical porosity ~2.0–1.0%. In powder forged parts the porosity in the mentioned range, as 'residual' porosity, is concentrated in the surface layer only, resulting from local cooling of the preheated compact by contact with the cooler forging die before and during the forging stroke. In this chapter more details are presented regarding the PF technology and PF parts properties compared to the conventional sintered materials described above since this technology can be regarded to offer wide perspectives for powder metallurgy development. Since commonly neither the dimensions, contours nor the surface finish of the as-PF parts fulfill the requirements for the final properties of a structural part, machining has an important role in the production flow sheet. The most important factor affecting positively the machinability of PF parts is 'full' density and a microstructure that is more homogeneous in mixed steels compared to pressed and sintered ones up to a fully homogeneous microstructure in prealloyed powder materials, with a generally lower amount of alloying elements compared to sintered steels.

Powder forging, as an additional technological step to the base pressing and sintering technique, requires special equipment and instrumentation, and, therefore, is effective mainly for the production of parts with properties that cannot be attained by other methods used in powder metallurgy for the densification of parts. The optimum solution are automatic forging lines [314]. Manual powder forging of the sintered preforms with or without graphite (emulsion) coating and heating in a protective atmosphere can also be performed, *e.g.* in research and development.

Two different PF technologies for 'net' shape parts have been developed, *i.e* drop (upset) forging or 'powder forging', mainly with lateral material flow, and 'hot repressing', with deformation mainly in the *z* axis (the preform profile is the same as the profile of the final part, all densification is in the pressing direction), *i.e.* as densification [110,315,316]. The flow sheet for both variants of powder forging is shown in Fig.7.110.

The decision between these two methods depends on the loading conditions of a part in practice, *e.g.* if the surface layer or the total cross section of a part is highly loaded. The first method is used for the parts requiring 'full' density with regard to service properties. The second method is employed for parts with porosity usually <5% but which due to large size up to some kilograms are usually produced by powder forging (lower pressure for the compaction of preforms and for forging). Examples of such parts are, *e.g.* the parts for agricultural machines and similar components. The objective of powder forging is to reduce extensive machining of parts which cannot be effectively produced by the known and new 'press and sinter' PM routes. By powder forging, excellent mechanical properties can be combined with wide geometric variety and the highest mass and dimensional accuracy compared with

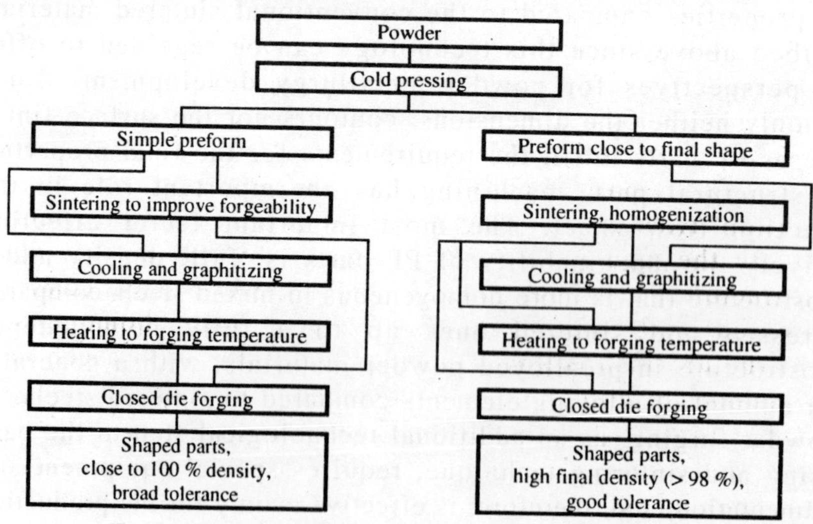

Fig.7.110 Flow sheet for both variants of powder forging.

wrought steel forging [317].

Powder forging has its position in the production of many parts for engine transmissions, as, for example synchroniser rings, differential pinion gears, but also for gears in general, rocker arms for setting braking force, sprocket wheels for cranes, and many others. A frequently described example for PF is the production of connecting rods, in which case considerable machining is necessary [21]. This part is produced at present also by warm compaction [318]. The production of roller bearing rings by powder forging can be regarded as a special case due to the special loading mode of these parts, and therefore it will be discussed here in more detail [22,23,319].

Powder forging is a compaction process in a closed die, acting on the height of a part, and therefore high precision in the mass of the preform is necessary (weighing of each sintered preform). Furthermore the forging for full density is not necessary in all cases. Therefore powder forging to 98–99% density is used for parts working under high surface stresses. The problem with the surface layer containing residual porosity is eliminated by the already minimum machining and the part becomes fully dense for the application.

As mentioned before, the main difference between conventional powder metallurgy compaction methods for the fabrication of the highly dense parts (~7.5 g/cm^3), including hot pressing methods, and powder forging lies in the deformation process of the preform at forging of a sintered preform in a closed die. Two factors are acting during the powder forging process, *i.e.* dynamic force and high temperature. The lateral material flow (lateral yielding) in the die cavity during powder forging to an extent of ~10% in a heated (200–300°C) closed die is a condition to close the pores under shear stresses. The preform must have a smaller width or diameter than the die and larger height, the density should be about 6.5 g/cm^3. These are the most important factors for the achievement of 'full' density required for highest dynamic properties of a final component. The hot repressed parts do not attain as high dynamic properties.

In general the powders used for forging could not – at least not yet – be made as exact copies of the wrought steels they are supposed to replace only due to some restrictions in powder production; however, this is not necessarily desirable. It is advantageous and realisable to prepare powders with a composition that results in optimum properties of the parts on a realistic economical level. For powder forging, both mixed and prealloyed powder grades are used, depending on individual technical and economical decisions. The positive experiences with the preparation and powder forging of water atomised low alloyed nickel, chromium and molybdenum alloyed steel powders give a perspective for a broader use of powders of similar composition as wrought steels.

The oxygen content in atomised powders used for powder forging is <0.20%, in some cases <0.15–0.10%, which is a good basis for further processing. The microstructural homogeneity of powder forged parts based on prealloyed powders is the reason for their high mechanical (mainly toughness and fatigue) properties, *e.g.* for contact fatigue endurance and fracture toughness.

The production of parts by powder forging has established its position in PM industry but growth is limited at present due to different reasons. However, new developments in powder forging can be expected for the large-scale production of parts simpler in shape and larger in size and mass which cannot be produced by conventional powder metallurgy routes especially for highly dynamically loaded parts.

7.11.1 Mechanical properties of powder forged Cr-prealloyed steel

When speaking about the machining of powder forged steels, at first their mechanical, especially toughness properties must be determined and shown. The following data concerning the mechanical properties of PF low-alloyed Cr-steel should be taken only as examples of application in heavy duty parts. At present, *e.g.* Astaloy CrM, Astaloy CrL and/or KIP 103V and 30CRV (all nickelless) powder grades are designed for such applications.

Since the large-scale production of the low chromium prealloyed powders is an industrial standard at present, the following data will show the properties of these PF steels attainable with AISI 4100 powder. Figure 7.111 shows the strength properties of powder forged Fe–1Cr–0.7Mn–0.2Mo (SAE 4100) steel samples with addition of graphite as a function of the forging temperature.

Hardness was increased not only with increasing carbon content but also with increasing forging temperature. After forging at 1050°C and cooling in air, the hardness (HV 30) was 315 for S4, 349 for S6 and 356 for S8 steel.

The effect of forging temperature on the notch toughness of PF steels as in Fig.7.111 on the tensile properties is shown in Fig.7.112. The results clearly demonstrate that for high toughness PF steels the

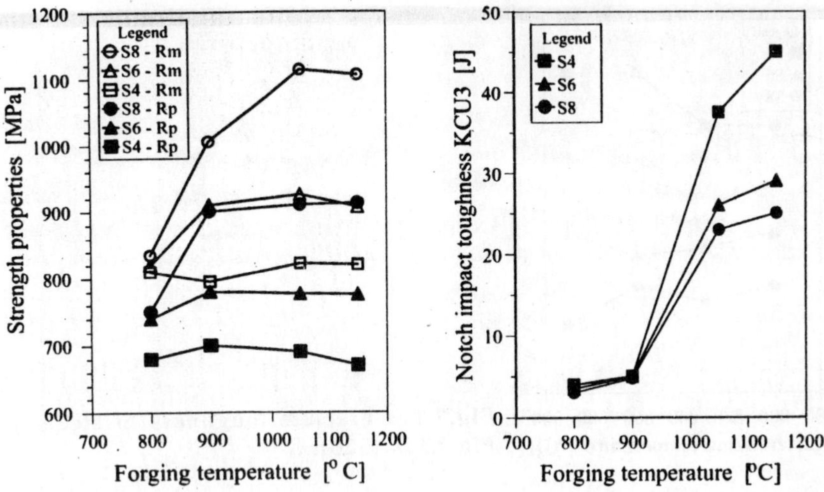

Fig.7.111 Tensile and yield strength of powder forged steel (prealloyed powder Sumiron 4100S) as a function of forging temperature and graphite addition: S4 – 0.4%, S6 – 0.6%, S8 – 0.8 mass %. Combined carbon content after forging at 1150°C for S4 = 0.32%, for S6 = 0.44% and for S8 = 0.63% [320].
Fig.7.112 (right) Effect of forging temperature on notch toughness of PF materials as in Fig.7.111 [320].

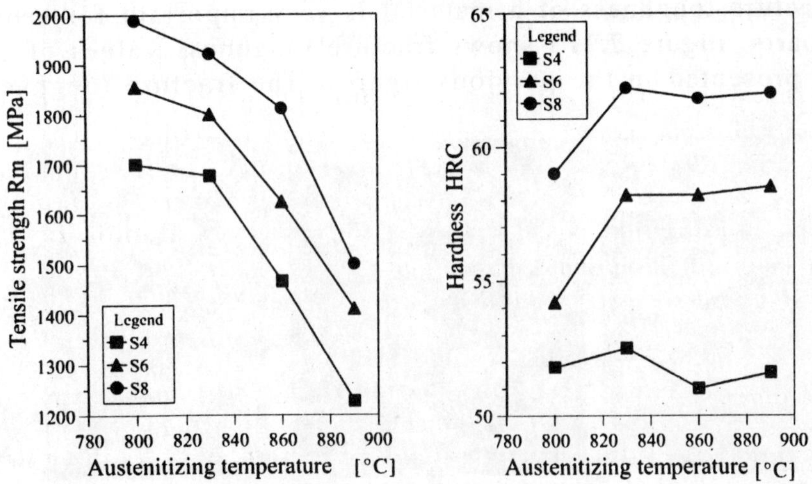

Fig.7.113 Quenched and tempered tensile strength values of powder forged Fe–1Cr–0.7Mn–0.2Mo steel (Sumiron 4100S) as in Fig.7.111 as a function of the austenitizing temperature. Density 7.83 g/cm^3. Tempering: 165°C, 1 h, air [320].
Fig.7.114 (right) Quenched and tempered hardness of steels as in Fig.7.113 [320].

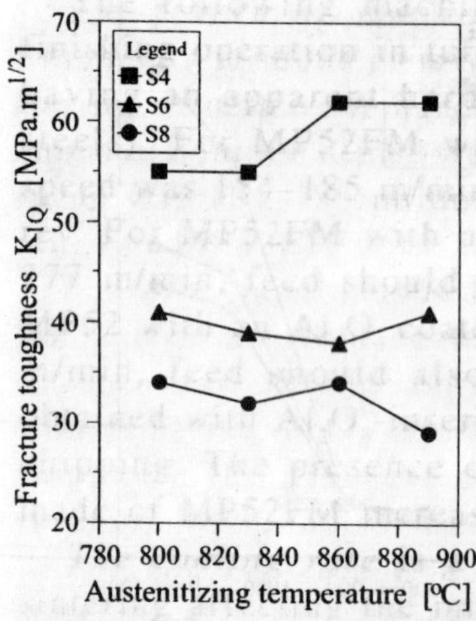

Fig.7.115 Fracture toughness of steels as in Fig.7.114 [320].

forging temperature should be above ~1050°C.

In most cases, the as-heat treated properties of parts are critical for application in the case of an optimum composition and adequate processing methods being selected. Figures 7.113 and 7.114 show the as-heat treated tensile strength and hardness values of powder forged steel in dependence on austenitising temperature. The high hardness values, above 60 HRC at 0.63% C, must be noted. Also, the fracture toughness of a material is very important for heavy duty parts. Figure 7.115 shows fracture toughness values of the steels presented in the previous figures. The fracture toughness

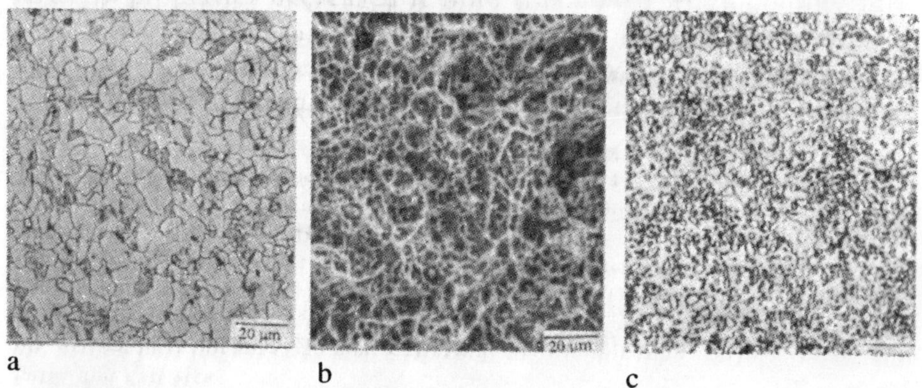

Fig.7.116 a – as-forged microstructure of Fe–1Cr–0.7Mn–0.2Mo–0.2C_c (Sumiron 4100S powder) samples. Forging temperature 1050°C. Optical micrograph. Nital etched. b – Tensile fracture surface of a PF sample as in Fig.7.113. Sintering 1 h at 1200°C, H_2, forging temperature 1150°C, SEM. c – As-spheroidized microstructure of Fe–1Cr–0.7Mn–0.2Mo–0.62C_c (Sumiron 4100S powder) samples. Forging temperature 1050°C, spherodizing 4 h at 750°C in nitrogen. Optical micrograph. Murakami etched.

values are higher than those required for wrought bearing steel (min. 21 MPa·m$^{1/2}$) (SAE 52100, DIN 100Cr6) [321].

The characteristic fine-grained and homogeneous microstructure of as-forged steel (S4, Fig.7.111) is presented in Figs.7.116. The fracture surface of this steel has an uniform fine ductile dimple morphology as shown in Fig.7.117. The microstructure of the PF part as shown here is isotropic without any texture, clusters of non-metallic inclusions, or carbides. Figure 7.116c confirms the homogeneous distribution of fine carbides in the as-spheroidized microstructure of powder forged Fe–Mn–Cr–Mo–C steel (see Fig.6.19).

7.11.2 Machining of powder forged steel in general

Machining of PF components is in many cases an essential part of the manufacturing process. This is most widely acknowledged for PF connecting rods, which are manufactured in large numbers in several countries, Fig.7.117. The total production for 2002 has been estimated to be well above 20 million pieces [315]. For powder metallurgy conrods, regardless if they have been pressed and sintered or powder forged [316] machining inevitably involves turning the small and the large bore as well as drilling the oil hole and drilling and tapping the threaded holes for the bolts, while separating the conrod bearing cap is usually done by fracture splitting [322]. Usually, the conrods are also flat ground to thickness as a first step to give a precise basis for the subsequent machining processes, and the bores are honed to improve roundness and surface finish [280]. Therefore, the machinability of PF steels is an item of high practical relevance, a common measure being addition of about 0.5% MnS to the starting powders.

Although they are almost fully dense, PF steels differ in

Fig.7.117 Powder forged connecting rods for passenger cars (Courtesy of Mashad Powder Metallurgy Co., Mashad, Iran).

Ser. No.	Cutting conditions			Material		
	v_c [m/min]	f [mm]	a_c [mm]	A	B	C
1	30	0.1	0.5			
2	120	0.1	0.5			
3	30	0.5	0.5			
4	120	0.4	0.5			
5	30	0.1	1.5			
6	120	0.1	1.5			
7	30	0.4	1.5			
8	120	0.4	1.5			

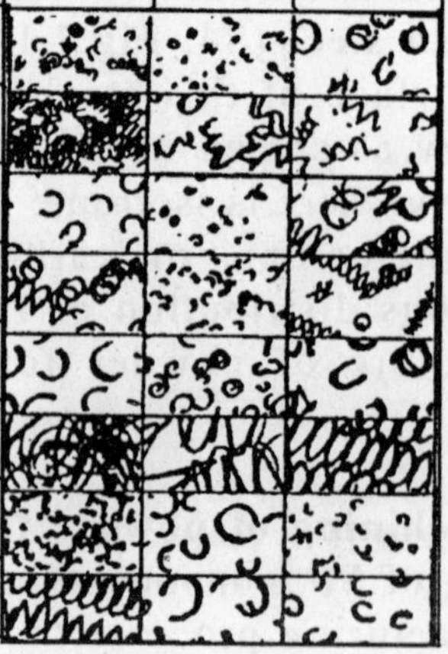

Fig.7.118 Shape of the chips formed in turning of: A – forged wrought bearing steel (AISI 52100, DIN 100 Cr6), B – PF Fe–0.32C–0.7Mn–1.0C–0.2Mo (Sumiron 4100S) steel, C – PF Fe–0.33C–1.9Ni–0.5Mo–0.3Mn (Astaloy A) steel [155]. Hardmetal cutting tool P20, v_c – cutting speed, f – feed, a_c – depth of cut.

machinability from wrought steels. If formation of the chip is regarded the wrought steel workpieces must be called 'long chipping' while the powder forged workpieces can be called 'short chipping'. Figure 7.118 shows the shape of the chips formed during turning of wrought bearing steel (SAE 52100) and two grades of powder forged steels under different cutting conditions. The shape and size of chips formed in turning the PF steels confirm their better machinability compared to the approximately equivalent wrought steel.

The principal difference between PF and wrought steel forgings with regard to machining are the surface properties. The wrought steel forging surface is covered by a thick layer of scale. The allowance of material in forged wrought steel parts for machining has to be large, also due to relatively large decarburisation at heating to forging temperature. The surface of a PF part is usually covered by a layer of graphite, to protect the part from oxidation during heating to forging temperature and to contribute to lubrication of the die during the forging process. Conventional deburring processes are realisable if necessary [323].

For assessing the machinability the mentioned limited amount of porosity, the composition, the compacting and sintering conditions and forging conditions (forging temperature, die temperature) must be considered. The microstructural heterogeneity of powder forged steels from mixed powders is lower compared to as-sintered, but remains noticeable also after additional annealing (dynamic recrystallization).

Drilling, threading, tapping, turning, reaming, and processes for improving dimensional accuracy and surface quality also in the as-heat treated state are the most frequently needed machining processes in powder forged parts. The machining of powder forged parts occurs in two areas:

- unavoidable and necessary for all PF parts regardless of shape,
- additional machining like in wrought forged parts (drilling, threading, undercuts, etc.).

The unavoidable is limited machining of outer surfaces of a part by turning to achieve metallic appearance and base dimensional tolerances and surface finish (machining of the graphitised surface layer) for further heat treatment and machining processes (of course, if considerable machining is necessary to attain the base tolerances this adversely affects the competitiveness of PF). If the outer surface of a part is not functional for the respective application and the preforms were not coated with a graphite emulsion before heating, the surface is not usually sufficiently clean without additional machining (possible oxidation in cooling).

Machining of powder forged parts (removal of surface residual porosity, shaping of inner bores, undercuts, contours of random shape and dimensions, drilling of holes etc.) is done using tools both with geometrically defined and/or undefined geometry [324].

The machining of powder forged materials does not differ markedly from that of conventional wrought steels when regarding some differences in properties and alloying as mentioned above. With some modifications they can be put also through the existing machining lines designed for parts from wrought steel to be replaced, but also through special lines dersigned for the given part and operations. There are not many special data considering the machinability of PF parts because their producers do not encounter significant problems.

In general, powder forged steels based on commercially available powders with specifications similar to, *e.g.* SAE 4600, SAE 4100, SAE 4140 and SAE 1045 can be machined as well as their wrought equivalents or better, provided that they have been processed correctly, and that mainly the tool geometry was properly adapted if necessary. In such case the powder forged parts can be machined in the same line as wrought steel.

In principle, there are three possible approaches for improving the machinability if necessary:

- addition of a suitable machining aid,
- the material should be subjected to special heat treatment prior to machining (annealing; spheroidisation in rare cases),
- modification of the machining conditions including tool material and geometry.

All three approaches have their benefits and drawbacks, and especially the high hardness of PF parts cooled in air must be regarded (*e.g.* 315–356 HV 30 for materials in Fig.7.111 forged at 1050°C).

The first approach includes mainly addition of MnS in a lower amount compared to sintered steels or the use of sulphur prealloy powders which seems to be the simplest method.

The second approach must be related to the base characteristics of an as-forged part and to the machining methods and surface quality required.

The third approach has disadvantages only in those cases when the modification requires changes in the base adjustment, *e.g.* of an automatic machining line. The small-scale forging of parts occurs without larger problems with regard to machining [325].

When machining a part forged to a porosity <2%, the cutting tool comes into contact not only with the surface layer but also with the homogeneous fully dense material, and therefore for the cutting tools and cutting conditions mainly the properties of the homogeneous fully dense base material should be considered.

7.11.2.1 Effect of porosity and inclusions

Effect of porosity. The porosity of a powder forged part is affected also by the forging temperature, *i.e.* lower forging temperature means higher porosity. With regard to machining, porosity levels must be considered. Low residual surface porosity (2.0–0.5%) deteriorates the tool life in surface turning at usual cutting speeds between 100–200 m/min and the roughness of the part. The surface roughness increases with increasing cutting speed and feed rate also in this case. The best surface finish can be reached by the use of superhard materials (ELBOR R, Hexanit).

Fig.7.119 shows sections of powder forged parts with oxidized surface pores which should be machined out. The thickness of a layer with residual porosity depends on material and mainly on the forging conditions. The material below this porous surface layer is fully dense. Under controlled conditions, pore-free surface layers

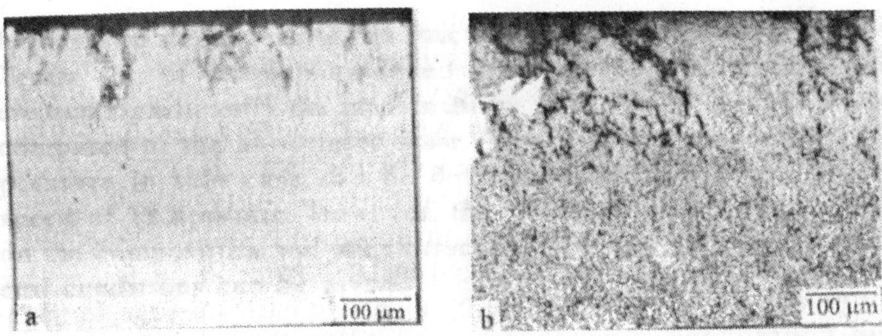

Fig.7.119 Section of a powder forged part with subsurface porosity: a – Astaloy A powder, 0.4% graphite, unetched; b – Sumiron 4100 S powder, 0.4% graphite. Density 7.83 g/cm^3. Forging temperature 1100°C. Optical micrograph. Nital etched.

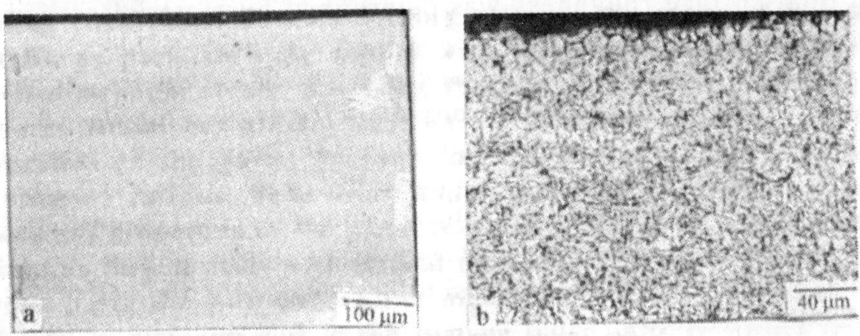

Fig.7.120 Sections of powder forged parts with pore-free subsurface layer (Astaloy A powder, 0.4% graphite), a – unetched; b – Astaloy A+2% Cu, 0.4% graphite. Optical micrograph. Nital etched.

of a part are also attainable by powder forging of components also attainable especially if, for example, 2% Cu was added to Astaloy A powder, Fig.7.120.

Secondly, if the part was forged to a lower density (2–5% porosity), the presence of pores in the whole volume deteriorates the machinability. The effect of the porosity in the range 0.5 to 2.5% was studied in powder forged Fe–2% Cu–2% Ni–1% Mo–0.35% graphite steel (mixed) with a hardness of ~200 HB (sintering 30 min at 1200°C, cracked ammonia, forging at 1100°C, annealing 30 min at 600°C) in turning with hardmetal and ceramic tools at cutting speed of 36 to 300 m/min and cutting depth of 0.5 mm. A fine-grained structure of the material consisting of ferrite and sorbite was the reason for a higher wear of the cutting tool compared with materials containing coarser pearlite [326].

The residual porosity extending to a larger depth of the part had a substantial effect on the extent and the type of the cutting tool wear. Higher performance was obtained with hardmetal cutting tools compared with HSS tools. The adhesive wear also of HM tools at higher cutting speeds was observed (built-up). The ceramic tools were not affected by adhesive wear. At cutting speeds > 200 m/min the ceramic cutting tools were the best. It was stated that attaining a maximum residual porosity of only ~0.5% makes it possible to increase tool life in turning of PF steels [326].

At machining of powder forged material with about 5% porosity, chip formation is accompanied by intensive built-up edge formation, and consequently, roughness increases. The roughness of the machined surface increased with residual porosity in the machined area. If a powder forged part was heat treated and then ground, the final surface roughness was determined by material properties and grinding conditions [327].

Effect of inclusions. Inclusions in a powder forged part can be introduced through the base iron powder particles (mainly sponge iron powder) and only to a minor extent through the atomised iron powder and prealloyed powder particles (mainly fine inclusions). Figure 7.121 shows the section of a powder forged sample Fe-0.35Mn–0.27Ni–0.30Mo–0.18Cr–0.4C (Astaloy D powder) with non-metallic inclusions and a similar image of a Fe–2Ni–0.5Mo–0.4C steel (Astaloy A powder), in which case inclusions are very rare. The chemical purity of a powder plays a very important role in the properties of PF (heavy duty) parts.

The presence of non-metallic inclusions in a powder forged part from any powder is more marked due to full density and relatively uniform distribution compared to wrought steel. If present, the non-

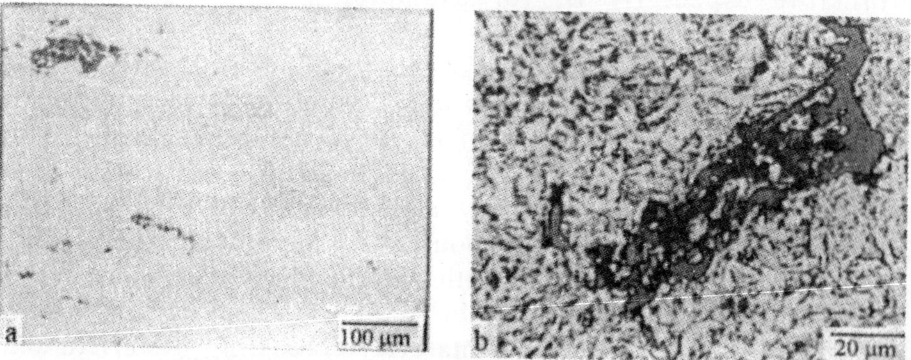

Fig.7.121 Sections of powder forged samples from prealloyed powders showing non-metallic inclusions. a – Astaloy D powder–0.4% graphite, unetched; b – Astaloy A powder–0.4% graphite. Sintering for 30 min at 1120°C, forging at 1100°C. Optical micrograph. Nital etched.

metallic inclusions do not form clusters, as, for example, in wrought steel since carriers of the inclusions in a PF material are singular powder paticles. Therefore, the machining behaviour of powder forged steel is affected substantially less by non-metallic inclusions (slag) in comparison to wrought steels, also due to the lower total amount of inclusions.

Larger non-metallic inclusions (>0.08 mm, rare) have a more detrimental effect on machining (and mechanical properties) of a powder forged material than would be expected from their total amount in a powder forged part [328]. The same holds for heterogeneously distributed carbides, which is also the reason for the higher toughness properties of powder forged steels compared with wrought steels.

It was shown that (deliberate) contamination of commercial powder grade with tundish slag can have a deteriorating influence on the machinability. This effect was observed under particular conditions in milling and also in turning tests, *e.g.* at relatively low cutting speeds and using high speed steel tools which are more sensitive to abrasive wear than carbide tips. The relation between the tool wear in turning and the cutting speed is shown in Fig.7.122 for some PM materials. It was supposed that for the reduced iron powder the matrix and the inclusions as well are much softer and less abrasive than in the case of the prealloyed powder. An increased amount of tundish slag in the atomised steel powders had a dramatically deteriorating effect on machinability. In contrast to the deteriorating influence of tundish slag additions, the relatively large amounts of inherent inclusions found in reduced iron powder seem to have rather a beneficial effect on machinability. This was found true in the milling test and also at higher cutting speeds in

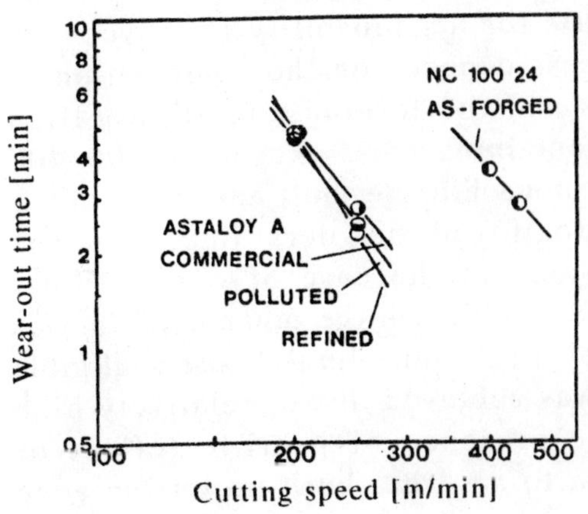

Fig.7.122 Influence of non-metallic inclusions on wear-out time of cutting edge in turning test for powder forged iron powder NC 100.24–0.39% C–0.15% O (135 HB) and prealloyed Astaloy A (commercial grade 0.38% C, polluted 0.39% C, refined 0.40% C)–0.10% O (200–210 HB) powder steel [328].

the turning test [326]. (It would be necessary to investigate this statement in greater detail).

It can also be stated that currently it is not possible to generalise the influence of non-metallic inclusions on the machinability of PF steels because their identification with regard to amount, size, and composition is both difficult and incomplete [162,163].

7.11.2.2 Influence of heat treatment

The following four types of heat treatment processes which can be commonly used for enhancing the machinability of powder forged parts where investigated for Astaloy A–0.4% C steel:

I – 'normal' cooling: from forging temperature, *e.g.* about 1000°C in nitrogen 10 min to about 400°C followed by air cooling; the consequence is relatively high hardness in dependence on the carbon content (bainite prevailing in the microstructure, minor amounts of pearlite, 203 HB),

II – soft annealing: after the procedure I 4 h at about 710°C, followed by furnace cooling to about 650°C and finally free air cooling, 175 HB,

III – step annealing/fast cooling: after the procedure as per I, austenitizing 1 h at 900°C, controlled cooling 45 min to 760°C, then controlled cooling in air 90 min to 500°C and then free air cooling (microstructure achieved – ferrite and pearlite, 210 HB),

IV – step annealing/slow cooling: after procedure as per I, austenitizing 1 h at 900°C, controlled cooling 45 min to 760°C, controlled furnace 10 h to 500°C, thereafter free cooling in air (microstructure achieved – ferrite and pearlite, 160 HB) [328].

These heat treatment processes are time and energy demanding, and, therefore, the spheroidizing annealing of high carbon steels seems to be more effective, Fig.7.116c, used also in annealing treatment of wrought forged parts for machinability improvement.

The optimum annealing process depends on the composition of the powder forged components and the forging conditions. High hardness of as-forged carbon containing parts exceeding 300–400 HV makes these steels virtually unmachinable. Soft annealing of as-forged materials based on prealloyed powders improved the machinability in the same way as in the case of conventional materials of the same composition [325]. On the other hand, it was also observed that by the performed soft annealing almost negligible improvement in machinability was achieved due to relatively high amounts of ferrite without the presence of pearlite (effect of carbon content), what gives rise to a severe built-up on the edge of the cutting tip mainly at low cutting speeds [328]. In large-scale production, soft annealing is realisable in practice so that the annealing furnace is a part of an automatic forging line and does not increase the labour cost [329,330]. This soft annealing mainly for mixed steels is not in all cases fully satisfactory if used in small-scale production due to the problems with the uniformity of processing also in regard to the size of the parts.

The bainitic structure achieved by normal cooling after forging, which yielded the most unfavourable milling results, yielded the best turning behaviour. It was also stated that bainitic structure was unfavourable at low cutting speeds but preferable to other microstructures at high cutting speeds. The pearlite–ferrite structure achieved by step annealing/slow cooling yielded a somewhat better machining behaviour than the soft annealed structure.

As stated in Ref. 328, the methods reportedly had only marginal effects on the machining behaviour of PF steels. It means that the conditions for the successful annealing of a material improving machinability must be very strictly adapted to the composition and to the homogeneity/heterogeneity of the microstructure.

Soft annealing of wrought steels with a carbon content above 0.3–0.5% is accompanied by spheroidisation of cementite. The prolongation of the critical time for soft annealing causes coarsening of the carbides. The most simple method of soft annealing is soaking below Ac_1. Annealing above Ac_1 is suitable mainly for the steels with higher carbon content (~0.9%) and for the steels alloyed with the elements which stabilise the cementite.

Annealing of carbon as well as of alloy steels is accompanied by coarseing of the carbides at a temperature of 450°C (size $\sim 10^{-5}$ mm). At 600–700°C, the crystallites are larger by about one order ($\sim 8–10 \cdot 10^{-4}$ mm). The carbide-forming elements (chromium, molybdenum, vanadium) retard high-temperature spheroidisation and nickel and cobalt accelerate the process of growth of carbides [331].

Usually, the spheroidising annealing of mixed powder forged steels has not been successful in changing carbide formation and hardness if it was not accompanied by a corresponding cooling rate. In contrast, it was fully successful for the Fe–Cr–Mo and Fe–Ni–Mo prealloyed steels with a carbon content of ~0.6%.

Final machining of heat treated powder forged parts (*e.g.* roller bearing races) is grinding up to super finishing and polishing, depending on the properties required. Grinding is carried out, for example, as oscillation or deep grinding. The roughness of ground

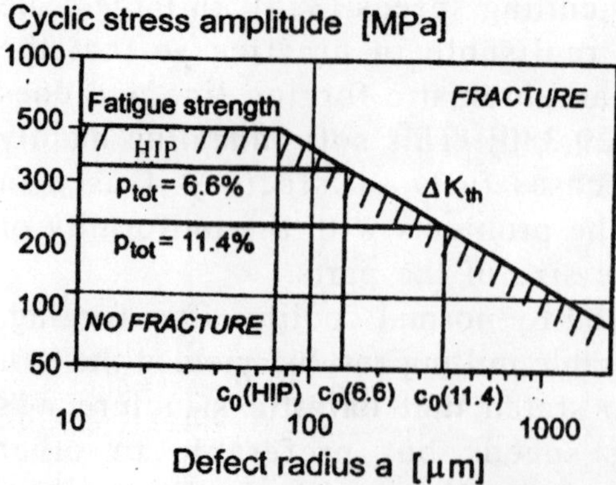

Fig.7.123 Kitagawa–Takahashi diagram for powder metallurgy of steels with varying density (Fe–1.5% Mo–0.7% C), quenched and tempered [321].

powder forged (Ni–Mo, Cr–Mo) bearing races was significantly lower compared to wrought bearing steel races, both heat treated and machined under equal conditions [23,332].

7.11.2.3 Effect of machining aids

When adding a machining aid to improve the machinability of powder forged steels, more factors affecting the machining conditions must be considered than for porous sintered steels. There is a marked difference in porosity between the surface residual porosity and the 'fully' dense core of the forgings. The thermo-stable machining aids in powder forged materials are incorporated in the matrix and can deteriorate the mechanical, esp. the fatigue, properties more than in porous materials in which the properties are affected mainly by the pores. This is clearly discernible from Kitagawa–Takahashi diagrams which show that with increasing fatigue endurance strength, as attained *e.g.* with HIPed or powder forged materials, the adverse effect of singular defects such as inclusions becomes progressively more pronounced, the critical defect size above which fatigue cracks are initiated being shifted to smaller diameters, Fig.7.123.

For fatigue loaded components such as *e.g.* conrods it is therefore of decisive importance to avoid larger inclusions of machining aids; in the case of MnS especially the tendency to agglomerate in humid environment has to be taken into account.

Sulphur prealloying. In testing, the effect of increased sulphur contents (0.13%), or rather manganese sulphide contents in PF steels (Astaloy A, Ancorsteel 4000), a beneficial influence on the machining behaviour was found [328]. Sulphur prealloying of powders not exceeding ~ 0.05% S seems to be a simple machining enhancement method in powder forged alloy steels which contain a certain amount of manganese (MnS formation). This sulphur content is also not detrimental for (contact) fatigue and fracture toughness of powder forged steels [23] and, therefore, can be a common addition for this purpose, as can be other sulphur containing compounds, *e.g.* manganese sulphide, in lesser amounts. Prealloyed powders (ATOMET 1001 = 0.45 to 0.63% Mn) containing 0.15% S exhibited in drilling (HSS 6.35 mm drill, 2220 rpm) better characteristics in machinability compared to FC-0205 steel containing 0.12% or 0.33% MnS [333].

Addition of *lead* in spite of environmental concerns was recommended as the chemical compound $(C_{17}H_3COO)_2Pb$ (lead stearate) in an amount of 0.1–0.45%. The machinability (turning, v_c = 180 m/min, feed 0.2 mm/rev., depth of cut 1 mm, criterion T 115 min→150 min) of a powder forged material containing this compound was increased by 50% without detrimental effect on mechanical properties [334]. Also, an addition of 0.5 up to 2.0% Pb at a carbon content not higher than 0.3% was recommended for increasing ductility (at a decrease of strength <5%) and machinability of powder forged steels. The lead reportedly coats the non-metallic inclusions [335,336].

Calcium containing compounds form a group of effective machining aids in powder forged steels (porosity 1–2%) as, for example, 0.01–0.05% calcium hydride [337], or 0.01–0.50% calcium carbide [338] or calcium difluoride (CaF_2) addition of 0.1–0.5% made possible to increase the cutting speed in turning by about 10% (from 160 to 180 m/min) and resulted in a decrease of the wear of the cutting tool by 50% compared with the state without this addition A slight decrease (~5%) in mechanical properties (elongation, impact strength) by this addition was observed [338]. It would be perhaps effective and interesting to investigate CaS as a machining aid due to its high thermostability (~2500°C) also for PF parts (see chapter 6.4.1.2).

Improved machinability of powder forged steels was demonstrated for addition of *bismuth* up to 0.2%, *tellurium* up to 0.25% and *selenium* ~0.05%. For example, addition of Te resulted in enhancement of machinability by 10 times (tool life time) and by Se addition improved surface finish was attained [262,339]. The addition of small amounts of tellurium (0.5%) and 0.5% copper to AISI 4600 powder forged steel improved the machinability without edge or surface cracking. The copper addition was required to prevent loss of tellurium in PM parts as mentioned before [340]. For the effect of bismuth on the behaviour of powder forged steel see also chapter 6.1.5.

7.11.3 Machining of Ni-and Cr-alloyed powder forged parts

The prealloyed powders Fe–2Ni–0.5Mo–0.2Mn (Astaloy A) and Fe–1Cr–0.7Mn–0.2Mo–0.05S (Sumiron 4100S) were used for experimental powder forging of roller bearing rings (races) in a pilot produ tion. Machining, which was a decisive factor for economical effectiveness of the production of the rings by powder forging from these powders, was compared to the machining of wrought bearing steel forgings under industrial conditions on automatic machine lines. The service life time of bearings assembled with PF rings was an essential part of a complex investigation. The production of roller bearing rings by powder forging is presented here in more detail since roller bearings are heavy duty parts. The properties of the outer surface layer, not exceeding ~0.8 mm and not the total cross-section of the part, are deciding for contact

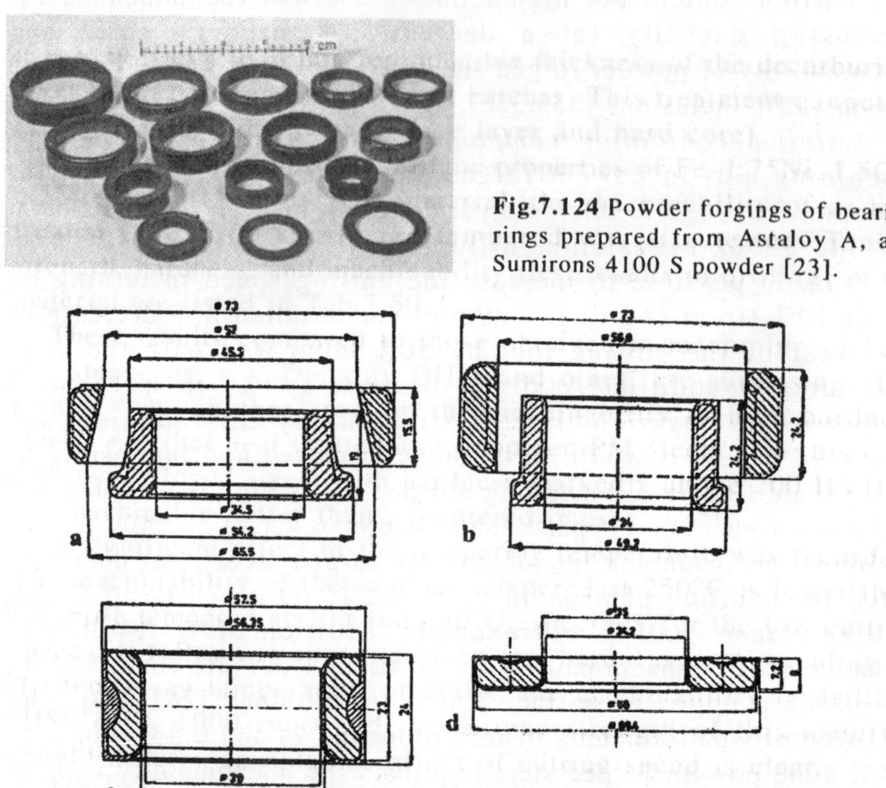

Fig.7.124 Powder forgings of bearing rings prepared from Astaloy A, and Sumirons 4100 S powder [23].

Fig.7.125 Powder forged roller bearing rings with as-forged dimensions for: a – tapered, b – cylindrical, c – inner ball, d – thrust rolling bearing.

fatigue endurance of a bearing. *i.e.* the area where failure through pitting occurs.

The production of outer and of inner rings for ball, roller, and tapered as well of the rings of thrust rolling bearings was investigated under laboratory and pilot-scale condition with the measurement of critical parameters of machining, Fig.7.124. The allowance for machining of powder forged compared with those of wrought bearing steel rings did not exceed 1 mm in diameter, as shown in Fig.7.125 and in more detail for some tested powder forged rings in Fig.7.126. Machinability was tested both for as-forged and as-annealed rings including grinding and super finishing after adequate heat treatment to fulfil the requirements for a roller bearing ring. The roller bearings with powder forged rings were tested for basic dynamic load rating and for fatigue life time. The

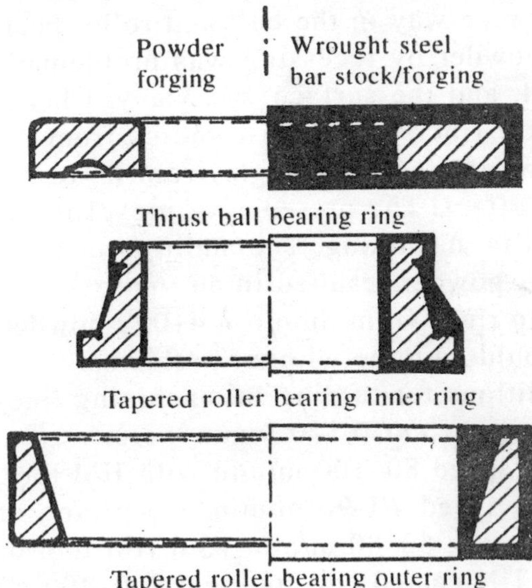

Fig.7.126 Intuitive illustration of the difference in material allowance for finishing operations by PF route and classical route in production of some roller bearing rings. Black – waste.

fatigue life time of bearings with PF rings from Astaloy A (AISI 4600) powder was on the same level as those for wrought bearing steels. The fatigue life time of bearings assembled with PF rings made from Fe–Cr–Mn–Mo prealloyed powder grade (Sumiron 4100 S) was 50 to 100% higher compared to standard ones made from wrought bearing steel (SAE 52100). The increased sulphur content in Sumirons 4100 S powder (0.05% S) for improvement the machinability did not adversely affected contact fatigue life of the bearing rings [23,329].

In comparison to machining of wrought bearing steel forgings in automatic machining lines, the following main results were attained

[314,332,341]:

- Power consumption in machining of PF rings was lower; the tangential component of the cutting force was lower by ~ 20% and the radial by 30–40%.
- In turning of a material with 0.2% C the formation of the locked layer on the chips was observed and regarded as a proof that the material was liable to built-up edge formation (presence of a larger proportion of ferrite).
- The chips were short and the need of the chip breaker in a machining line was thus eliminated, as shown in Fig.7.118.
- The optimum allowance in diameter of a ring for turning up to final grinding was estimated to be ~1 mm; at lower allowance the push off of the cutting tool from the cut was observed. In this case a soft grinding was effective.
- The machining of the inner race way in the ball and roller bearing rings from Astaloy A powder by recessing was problematic due to vibration of the tool, and the surface was wavy. Changing the cutting tool geometry (increase of nose radius from 0.5 to 1.0 mm and the decrease of clearance angle from 8° to 5°) eliminated this unwelcome effect. The presence of nickel in the steel caused some problems in machining. The addition of 0.2% MnS or 2% Cu to Astaloy A powder resulted in an improvement in this machining process. In rings from Sumiron 4100S powder, this turning by recessing could be done without problems.
- Recommended cutting conditions for turning in machining lines at feed 0.3 mm/rev. and depth of cut 0.1–0.3 mm were:

  outer surfaces – cutting speed 80–100 m/min with HM tool, inner surfaces – cutting speed 70–90 m/min, machining of inner race ways by recessing 60–80 m/min with HM tool or 120–200 m/min with a TiN-coated HM tool and 200–400 m/min with CBN tool.
- The machinability of the rings from Fe–Cr–Mo powder was better compared to those from Fe–Ni–Mo powder. The Cr-alloyed rings with 0.8% C as-annealed were machinable without any problems.
- Grinding up to super finish of case-hardened as well as of quenched and tempered PF rings was performed under conditions used for wrought steel in an automatic process.
- The roughness of as-super finished raceways in PF forged rings was lower compared to wrought bearing steel rings.

Tab.7.51 Characteristics of powder forged steels tested [10,162]. A – sponge iron powder (NC100.24)+0.4% graphite, D – prealloyed powder Fe–0.2Ni–0.3Mo–0.18Cr (Astaloy D)+0.4% graphite, A – prealloyed powder Fe–1.9Ni–0.55Mo–0.2Mn (Astaloy A)+0.4% graphite. Sintering and forging both at 1100°C

Steel designation	Density [g/cm^3]	C_c [mass %]	U.T.S. [MPa]	HB 62.5/2.5	Microstructure
N	7.73	0.39	428	135	Ferrite, pearlite
D	7.81	0.40	485	145	Ferrite, pearlite, secondary cementite at grain boundaries
A	7.80	0.37	675	215	Ferrite, bainite

Tab.7.52 Cutting speed for drilling and tapping (M10) for N, D and A powder forged steels. *Drilling condition: HSS 7.5 mm drill, emulsion, feed 0.1–0.2 mm/rev.; tapping conditions: HSS M 10 tap, cutting oil for steel N, emulsion for the steel D and A*

Steel	Drilling	Tapping	
		Through hole	Blind hole
	Cutting speed [m/min]		
N	35-45	20-25	10-12
D	20-25	10-15	5-7
A	15-20	8-12	4-6

Tab.7.53 Cutting speed for dry turning the steels N, D, and A with different hardmetal tools. Depth of cut 1.5 mm.

Cutting tool	P30			P25			P15		
Feed [mm/rev.]	0.60	0.40	0.20	0.60	0.40	0.20	0.60	0.40	0.20
	Cutting speed [m/min]								
N	130	200	250	160	280	250	200	350	450
D	55	110	140	125	225	280	140	250	320
A	45	85	110	100	180	240	130	235	300

Tab.7.54 Cutting speed for dry plane milling of powder forged steels N, D and A with different hardmetal tools. Depth of cut 1.5 mm

Cutting tool	K20		K10		K15	
Feed [mm/rev]	0.1	0.1-0.2	0.1	0.1-0.2	0.1	0.1-0.2
	Cutting speed [m/min]					
T	175	150	200	180	170	210
D	115	100	140	125	120	145
A	110	95	135	120	115	140

7.11.4 Some recommendations for machining of other powder forged steels

In the following the cutting data for three types of powder forged steels are given. The properties of these materials without a machining aid are listed in Tab.7.51 and the cutting data for drilling, threading, turning, and plane and end milling in Tabs.7.52–7.54.

The three investigated PF-materials designated N, D and A have two machining properties in common: in all studied machining operations, they render much shorter chips and appear to be much more consistent with regard to measured cutting forces and other machine parameters than comparable wrought steels, as was stated before for powder forged steels in general. Material N, which is based on sponge iron powder, has by far the best machinability in spite of its relatively larger content of oxide inclusions [162,342]. In fact, this material is notably better machinable than common wrought structural steel. The machinability of materials A and D, which are based on low alloyed atomised steel powders, is comparable to structural steel and markedly better than common quenched and tempered steel.

It has been proved to be possible to establish for these materials close to optimal cutting data applicable for turning, milling, drilling, tapping and reaming [10,152,228].

For drilling powder forged steels (density 7.7–7.8 g/cm^3) with the hardness of 136 to 215 HB using a HSS drill 7.5 mm in diameter in presence of emulsion at feed 0.1–0.2 mm/rev. a cutting (peripheral) speed in the range of 45 to 15 m/min was recommended. The recommended speed decreased with increasing hardness of the material.

7.11.5 Summary

- Machining of powder forged steels (porosity <5%, usually <2%) is characterized by significantly lower amount of removed material compared to drop forged components, since no or at least much less scale is present, and thus with lower power consumption. Regarding the present knowledge about powder forging materials and technique it is possible to produce the parts with the allowance only for final grinding without use the geometrically defined tools except some profiles.
- Machining of powder forged steels is 'short chipping' as with porous materials.
- Powder forged steels can be machined as well as or better than wrought steels. Machining of powder forged materials has some peculiarities compared to wrought steel of equal composition.
- Machining of nickel-containing low alloyed materials is more difficult compared to chromium low alloyed steels.
- Different machining aids tested improved machinability of powder forged steels. Sulphur prealloying (~0.05% S) of steels containing manganese seems to be very effective without deteriorating effect, *e.g.* on contact fatigue endurance which is a very specific form of loading.
- As-powder forged steels exhibit high hardness which makes them unmachinable without heat treatment or machinable only by

special hard turning operations.

- Machining of a powder forged material involves usually two types of material, *i.e.* at the surface a porous (oxidised) layer (residual porosity) and the fully dense core of a part, to which latter type the cutting conditions including tool material and geometry should be adapted.
- Machining of powder forged parts with higher porosity is affected by pores, but the effect of these pores on machining is lower compared to conventional porous sintered material.
- Addition of common machining aids increases the machinability of powder forged steels to a lesser degree compared to porous material. The use of sulphur prealloyed powders seems to be most effective for the machining. Use of other thermo-stable machining aids can deteriorate the properties of a fully dense material, the particles of the machining aid acting as non-metallic inclusions in the load bearing cross-section. This is of particular relevance for fatigue loaded components such as PF conrods.
- All machining methods including hard machining and all tool materials as in machining of wrought steels are applicable for machining of powder forged steels with some adaptation of the tool geometry. Depending on the type of cutting tool (hardmetal, coated hardmetal, CBN, ceramic) cutting speeds up to 400 m/min can be applied in turning.
- Due to the uniform microstructure with fine carbides, the surface finish after grinding of a heat treated PF part is better for a given grinding wheel quality compared to wrought steel.

8

Recommendations for Machining of PM Steels

The results presented in the previous chapters show the complexicity of machining processes in powder metallurgy parts production. Regarding this fact, in the following chapter generalized data are given which should, however, be taken only as cutting recommendations because they do not – and cannot – include all cutting variants which may occur in machining of PM steels. The results are based on very broad laboratory and industrial experience which makes them acceptable as starting basis for machining of the respective materials. On the basis of these results, relating tool material and geometry as well as cutting conditions to the material properties of the part to be machined can be a contribution for cost effective elimination of often unexpected problems in machining of PM steels.

As will be shown later, in the listed data for recommended machining parameters there are hardly any informations about the role of machining aids and friction in cutting. The reason is that there are no general rules that apply for larger groups of PM materials, *e.g.* a correlation to the hardness as common for reasonable clear machining of wrought steels, due to the much larger set of parameters affecting the machinability of sintered steels.

The recommendations for machining in reality involve the data regarding the interactions between the workpiece material properties, the cutting method, the cutting tool material and geometry, the cutting conditions, and the required final shape and quality of the part which must be obtained under acceptable economical conditions.

In machining of PM parts, the most frequently applied operations are drilling, tapping, turning and boring. Therefore, more data for them are available. Most drilling and turning operations on PM parts are performed usually with feeds and speeds approximately 20–25% lower than those used on wrought materials of similar composition, because of the relatively 'abusive' nature of this cutting process observed in some cases, outer and internal undercuts etc. [169]. It should however be noted that this general statement considering wrought steel as reference material is insufficient. Also materials described as, *e.g.* 'soft'; 'hard', 'low alloyed' and similarly in some recommendations or results in machining of PM parts cannot be correctly assessed based solely on these descriptions. As an illustration, Fig.8.1 shows some machined PM parts.

In this chapter recommendations from the literature for cutting PM parts in specific machining operation are given but also such more general recommendations for improving the machinability of the parts that are based on the results listed in previous chapters

a – Two side gear with 2 cold formed threads M8 (Metalsint, Dolný Kubín)

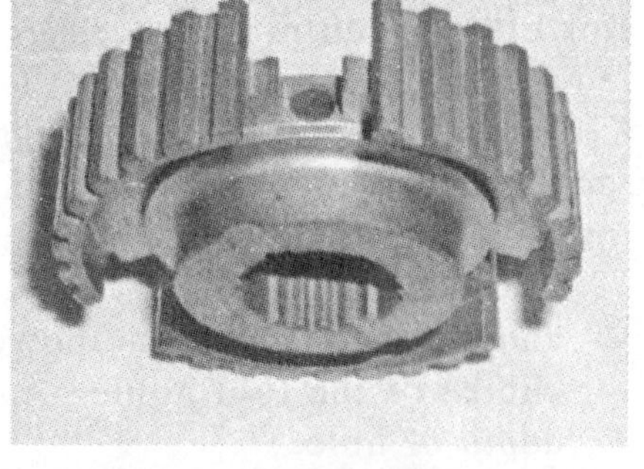

b – Synchronizer hub with 3 blind holes drilled Ø5.5 mm (MIBA, Vorchdorf)

c – Crankshaft sprocket for chain drive – turned to produce double row of teeth (MIBA, Vorchdorf)

d – Turned crown gear (GKN Sinter Metals, Krebsöge, Radevormwald)

Fig. 8.1 Examples of PM machined parts (Photos R. Bureš).

of this book. It was found that in many cases only the plain results were presented without detailed analysis about the relationship to material and/or specific cutting conditions. In spite of cutting recommendations being available, it must also be noted that for any part and machining operation, the machining variables need to be optimized. Therefore, available data may be used as a starting point, but optimization studies will mostly likely be required [150].

8.1 RECOMMENDATIONS FOR DRILLING

Drilling is the mostly widely used cutting process in production of PM parts which fact is also visible from the large number of published studies about this machining method, larger than, *e.g.* for turning of PM materials. Most drilling operations on PM parts are performed at 90° to the pressing direction (possibly higher hardness compared to those in the pressing direction). Considering the shape of the tool, a drill can be regarded as a very complicated cutting tool. From this point of view the summary of factors influencing the machinability of PM parts in the drilling process may be useful also for other machining processes with geometrically defined cutting edge.

When drilling a particular material at a given speed and feed rate, drill performance is governed by the drill quality and a range of other related factors influencing drilling performance as [118]:

- PM material, drilled,
- choice of tool machine power capacity, workpiece,
- choice of the tool holder,
- depth of hole,
- stability of workpiece hold,
- through or blind hole,
- horizontal or vertical drilling,
- dry cut or cutting fluid,
- stationary or revolving drill,
- condition of the machine,
- swarf control.

8.1.1 Selection of drill type

The selection of the drill type should be related only to the hardness of the material to be machined and the productivity. In the first approach for a PM material, hardness is used as an approximate criterion for machinability because there is no other more accurate property characteristic for the machining of PM material. With increasing hardness of the material firstly uncoated HSS drills are used, the coated HSS/HSCo-drills and finally coated solid carbide or carbide drill tips [118]. Figure 8.2 shows the A120, A520 and R520 drills with different flute angles and different cutting point geometry (see chapter 4.7.4).

8.1.1.1 Drill types

HSS standard drills. Standard HSS jobber drills (steam treated, uncoated and coated) for general applications and hole depths down to 4 × D are A100, and for depths to 2.5 x D, the A120. The shorter A120 has a special split point that ensures easy starting and accurate drill location. Both drills have a standard cylindrical shank.

Standard HSS drills are useful with a large variety of PM materials. They show best performance in soft materials where an open point geometry ensures that the cut material, *i.e.* the chips, cleanly leave the flutes.

HSS/HSCo high performance drills. The high performance family of HSS drills known as ADX drills with TiN and Futura Nano Top TiAlN coating (Balzers, high cutting edge stability and increased wear resistance), which include A150 and A520. The benefits include excellent swarf removal and accurate holes normally to the tolerance H9. The major advantage, however, is high productivity.

High performance solid carbide drills. Coated high performance solid carbide drills are known as the CDX family of

A120 – *stub drill for use in turret and automatic lathes, standard helix 118°, easy starting and accurate drill location*

A520 – *ADX stub length drill for NC and CNC machines, high productivity and accurate holes, special 130° thinned point convex cutting lip, TiN coating*

R520 – *CDX micrograin hardmetal stub drill, 130° special point convex cutting lip, TiN coating*

Fig.8.2 High performance drills A120, A520 and R520 (DORMER) [292].

drills. They include R510 and R520 drills. CDX and ADX drills can, under the right cutting conditions and with a stable set up, produce holes with a tolerance of H8 and H9, respectively. Standard drills reach a tolerance of H12.

Solid carbide (P40 and K10 most often used) drills are the only acceptable choice for machining of high carbon and alloy content as *e.g.* Fe–4Ni–1.5Cu–0.5Mo–C (Distaloy AE) and Fe–2Cu–4Ni–1.5Mo–C (Distaloy HP) and other materials with similar structure and mechanical properties [118].

8.1.1.2 Drill dimensions and geometry

Beside the type and the diameter of a drill, its geometry is defined by helix angle, point angle, and flute length. Complex tool point geometry should be developed for particular application cases, as *e.g.* in drilling stainless steel, to exploit the modified chisel point – the most efficient aspect of the drilling process [190]. Drill diameter and cutting speed have a greater effect on the cutting forces compared to the feed.

The drill length is considered as a primary factor for drill life improvements during PM machining. One of the demands for an approved operation is good stability. It is in some cases assumed that the presence of pores contributes to vibration. This can be the case if the cut occur interrupted through the pores (theory of the interrupted effect of pores on machinability) without deformation of a workpiece machined (as supported by deformation cutting theory). Surely, there are other undefined factors that can result from the microstructure of a material being machined and/or by the real technical practice of the drilling process which lead to vibration during drilling. Vibrations will be minimised if a short drill length is used [146,343].

The optimum short-series drills have a coated carbide substrate. Recently, the medium-temperature chemical vapour deposition (MTCVD) coating technique, applied for deposition of a TiCVN coating on a cobalt-enriched zone, has been shown to offer better wear and crater resistance than either physical vapour deposition (PVD) or the usual vapour deposition (CVD) process with the same TiCN coating. All of these coatings tend to reduce micro-fatigue on the cutting edge, an important factor when PM machining in general [344].

The helix angles of the drills are defined as low (15°), medium (normal) (28–30°) and high (40°). Low helix angle drills are not recommended for softer PM materials because of their poor chip ejection characteristics. Drills with 40° helix angles had twice the tool life of those with 28° helix angles in work performed with soft PM materials. Drills with a high helix angle are recommended therefore for all plain iron and soft low alloyed steel PM parts. For parts made of relatively hard and strong materials, especially those containing carbon, *e.g.* Ancoloy SA and SD, drills with a helix angle 28° are also recommended. For cemented carbide drills the helix angle 40° for soft steels and helix angle 28° for hard steels is recommended [345]. Carbide or HSS drills with a low right hand helix angle will help prevent the drill from digging in. The cutting edges should also be dubbed to reduce the axial rake [143,343].

Typical point angles of the drills are 135° point angles with a modified chisel point (Figs.3.14 and 3.15). The drill with a stronger point geometry than that of the conventional twist drill point and of 'short-series', rather than of jobber length, enhances drilling life and hole quality. This reduces the extruding effect of this point geometry and lowers the cutting forces, while improving its self-centering action [190]. The simplest technique is to grind a 135° point angle either with a web-thinned point or split-point onto short-series drills. If the point angle is ground to ~185° instead of ~135° better results can be reached, but a fixture must be applied to obtain the closest possible tolerances on the hole [143]. The drills with other point angles are used for improvement of the drill life or to eliminate the burrs in drilling a material with some defined properties. As stated in Ref. 191, hardmetal drills, especially short or extra-short ones, drills with helix angle 28°, relief angle 50–70° and point angles up to 190–225°, the average tool life can be increased five to eight times. Increasing cutting speed increases machining throughput, but also will increase tool wear and hole size variance.

A general rule when drilling relatively soft materials such as plain iron PM parts is that the point should be thinned so that the web thickness will be ~0.07 x drill diameter. For materials like Ancoloy SA and Ancoloy SD (Fe–1.25Mn–0.5Ni–0.6C) the corresponding web thickness should be 0.1 × drill diameter [143].

When through-drilling, breakout at the exit of the hole can be minimised if the point angle is specially designed so that the cutting edge corners are first to break through the underside of the hole being drilled. This can be achieved by grinding a 185° point angle on the drill, so that a negative plan approach angle occurs. If this introduces too high a wear at the drill corners, then a steeper drill point angle – about 100° – may solve the problem by gradually

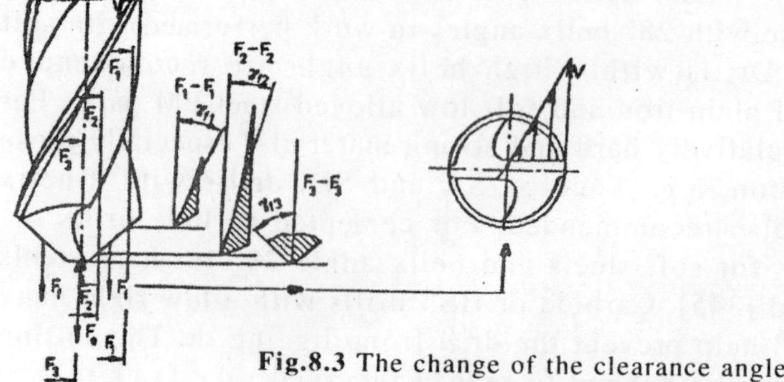

Fig.8.3 The change of the clearance angle in dependence on the distance from its axis and corresponding axis angles [346].

decreasing the thrust force on the transient surface at breakout [143].

In connection with the formation of the breakouts it is necessary to note that at classical spiral drills the cutting speed and the geometrical parameters of the drill change along the cutting edge, Fig.8.3. Beside these changes the existence of the cross edge markedly deteriorates the penetration of the drill from its first contact with the workpiece to achieve the contact along the total length of the cutting edges. Similar disadvantageous conditions are at the exit side of the drill from the hole when the cutting edges are failed. This can be eliminated by lower feed of the drill during entering and leaving the workpiece compared to the normal speed. This effect can be easily realised in automatic drilling.

Length of drill. There are also short (70 mm overall length) and extra short drills (39 mm) flute length (Fig.3.14). When drilling holes 12 mm deep into sintered iron (density 6.5 g/cm^3) with 3.5 mm drill diameter at 4000 rev./min (v_c = 44 m/min) at feed 0.08 mm (no coolant), the average number of holes drilled was 6 (short drill) and 48 (extra short drill), *e.g.* the drilled length per drill was 72 or 580 mm. It was generally stated that tool life in drilling can be improved considerably by using: a) a 40° helix angle for soft powder metallurgy materials; b) extra-short drills [143].

8.1.2 Drill recommendations and cutting data for structural steels

In general, for drilling sintered iron–carbon steel with porosity >10% it is recommended to use HSS drills and with the porosity <5% cemented carbides. For the span between them the choice depends on material, cutting conditions, and quality requirements

Tab.8.1 Machining data for drilling of sintered materials with HSS drills, helix angle of 28°
Group 1: plain-iron materials (sponge and atomised iron powder) of all densities and Fe–1.25Mn–0.5Ni–0.6 graphite (Ancoloy SD), group 2: Fe–1.75Ni–1.5Cu–0.5Mo–0.6 graphite, carbon steels, group 3: 0.5S or Te containing steels [10,143,348–350].

Material group	Cutting speed [m/min]	Drill diameter [mm]							
		2	4	6	8	10	12	15	20
		Feed [mm/rev.]							
1	10-15	0.05	0.08	0.12	0.15	0.20	0.25	0.30	0.30
2	10-20	0.04	0.06	0.08	0.10	0.12	0.14	0.18	0.22
3	30-60	0.04	0.06	0.08	0.10	0.12	0.14	0.18	0.22

[188]. It is also stated that the drilling of holes into iron compacts with a density of ~7.0 g/cm^3 can be performed with HSS drills at the cutting speed of about 20 m/min [347]. Giving only the porosity of a material, without other characteristic properties, cannot be sufficient for the selection of a of suitable drill material and then of the drilling conditions. In Tab.8.1, general machining recommendations are listed for three groups of materials.

These data can be accepted as a compromise between the technical and economical aspects of the drilling. According to these data, the effect of sulphur or tellurium as machining aids in the materials did not change the drilling conditions compared to those without them. According to these data, when drilling a relatively soft material with a drill with 28° helix angle the tool life can be increased by more than 100%. This is not consistent with the previous data concerning the helix angle. As stated before, when drilling soft materials such as plain iron PM parts the point should be thinned so that the *web* thickness will be ~0.07 × the drill diameter. For harder materials, the corresponding *web* thickness should be 0.1 × the drill diameter [143,345].

According to Ref. 351, the drilling of sintered iron base materials with the porosity 10–15% with hardmetal drills is not recommended due to the tendency to failure. Hardmetal drills are conveniet for materials with the porosity in the range ~5–10%. The cutting geometry of the drills should be other for the drills of smaller diameter as for drills of larger diameter (point angle 100°→116°).

The recommended cutting data for sintered iron listed in Tab. 8.1 and according this also for other materials cannot take into account all variables in material properties as shown in Fig. 8.4 for sintered iron. It follows from this that the differences in machinability

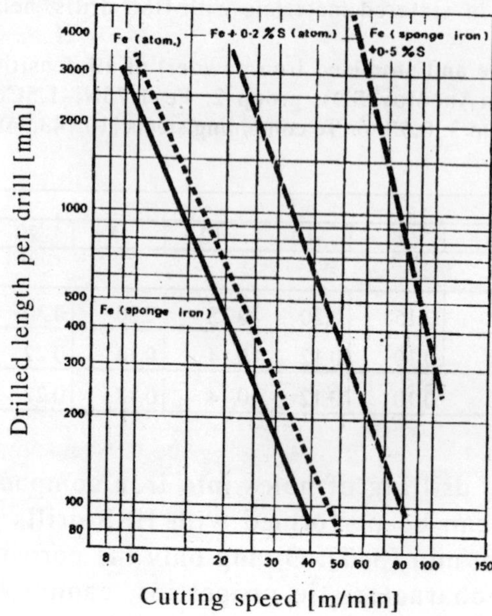

Fig.8.4 Drill life (drilled length per drill) vs. cutting speed for plain sintered iron based on atomised and sponge iron powder without and with 0.2% or 0.5% S addition. Density 6.5 g/cm³ [143].

between sponge and atomised iorn powder material are not negligible. Very good machinability was exhibited by Fe–0.5% S material based on sponge iron powder addition. The span in recommended cutting speed 10–15 m/min is optimum only for material without sulphur addition.

Tool life is nearly independent of the feed rate, which ranges from 0.05 to 0.22 mm/rev. but depends on cutting speed [146,343]. It is not possible to separate the effect of one factor on the drill life from the others.

Optimum drilling conditions for other sintered steels which include plain iron, Fe–P alloy and Distaloy SA and AB materials, some of them without and with 0.5% graphite addition and resin impregnated are listed in Tab.8.2. These alloys represent the materials with a wide range of mechanical properties. Nevertheless the recommended drilling conditions vary to a relatively small extent and can be attributed to the conditions listed in Tab.8.1, group 2. These optimum cutting speeds are in reality lower than those listed in Tab.8.1. An effect of drill point angle or resin impregnation is not stated in these recommendations. It shows that the recommendations have only a general validity.

In Tab.8.3 the recommendations are given for drills to match materials from the most common PM steel groups, in this case not modified with machining aids additions, and in Tab. 8.4 are listed

Tab.8.2 Optimum drilling conditions for sintered compacts at constant drill feed of 0.11 mm/s. Sintering 20 min at 1150°C, endothermic atmosphere [205]. ***Drilling test: HSS 9 mm drill, 29° helix angle, drill point angle α = 119°***

Steel No.	Steel	Density [g/cm^3]	Drill speed [rev./min]	Cutting speed [m/min]
1	Distaloy AB-0.5 % graphite	7.0	273	7.7
2	Distaloy AB-0.5 % graphite, IMP	7.0	273	7.7
3	Distaloy SA-0.5 % graphite	6.2	394	11.5
4	Fe (ASC100.29)	7.0	273	7.7
5	Fe-0.45P (PASC45)	6.2	394	11.5
6	Fe-0.45P (PASC45), IMP	6.2	394	11.5

Remark: IMP – workpiece resin impregnated; alloy No. 5, 6 – base powder sponge iron

the codes for recommended feeds for these materials. Each recommended drill application is rated as either excellent or acceptable. The speed and feed rates given which range from 0.04 to 0.43 mm/rev. are starting values for dry cutting of PM material with a density of approximately 7.0 g/cm^3. For each drill application, recommended cutting speed data are listed for a tool life of 15 min. Cutting data are given for each recommended application.

The drilling conditions recommended can be regarded to give the lower margin of the tool life band for minimum tool life because by using machining agents in the materials a reasonable improvement in machinability can be attained as shown before. Also when drilling with cutting fluid, the drilling speed can be increased by 10–50%.

Cutting speeds of up to 25 m/min and feed rates up to 0.25 mm/rev. in general are recommended for high speed steel drills. When hardmetal or hardmetal-tipped drills are used, cutting speed and feed rate could be as high as 120 m/min and 0.5 mm/rev., respectively. Hardmetal indexable drills are efficient because margins are eliminated. Abrasive margin wear and the welding of powder chips are also eliminated [129,152,343]. These general data can serve as a first view on possible drilling conditions.

The range for the use of HSS and/or cemented carbide drills is clearly shown. The effect of carbon content of a material on drilling conditions is related to drill type and mainly to workpiece material properties which were partly described in previous chapters.

For optimal drilling of sintered material, in addition to the mentioned minimum technical recommendations there are also the common practical hints for drilling which are universally known and

Tab.8.3 Recommended drill selection from eight drill types for a range of PM materials [118]. Comment: digits = cutting (peripheral) speed in m/min, mean range ± 10%; letters (T, U, V, W, X, Y) = feed rate code – Tab.8.4

Steel No.	Selection of drills	Standard drills		Application drills, HSCo			Application drills, hardmetal		
		A120	A100	R520	R522	R510	R520	R522	R510
			A001	R550	R522		R550	R522	
				A524	A504				
	Material group (base powder) – mix	Recommendations and cutting data: ■ excellent for application, ● acceptable for application							
		HSS 2.5 x D blue	HSS 2.5 x D blue	HSS/HSCo 2.5 x D TiN	HSS/HSCo 4 x D TiN	HSCo 5 x D TiN	K10 2.5 x D TiN	K10 2.5 x D AlTiN	K10 4 x D TiN
1	Fe (ASC100.29)	■ 50 V	■ 50 V						
2	Fe + 0.5C (ASC100.29)	■ 64 V	■ 64 V						
3	Fe + 0.4P (ASC100.29)	■ 38 V	■ 38 V						
4	Fe + 0.4P + 0.5 (ASC100.29)	■ 24 V	■ 24 V	■ 41 Y	■ 41 X	■ 41 V			
5	Fe + 2Cu (ASC100.29)	■ 24 V	■ 24 V						
6	Fe + 2Cu + 0.2C (ASC100.29)	■ 50 U	■ 50 U	■ 87 Y	■ 87 X	■ 87 V			
7	Fe + 2Cu + 0.5C (ASC100.29)	● 45 T	● 45 T	■ 78 Y	■ 78 X	■ 78 V			
8	Fe + 2Cu + 0.8C (ASC100.29)	● 25 T	● 25 T	■ 44 Y	■ 44 X	■ 44 V			
9	Fe - 4Ni - 1.5Cu - 0.5Mo (Distaloy)	■ 31 U	■ 31 U	● 45 X	● 45 W	● 45 U			
10	Fe - 4Ni - 1.5Cu - 0.5Mo + 0.25C (Distaloy)	■ 31 U	■ 31 U	■ 48 Y	■ 48 X	■ 48 V			
11	Fe - 4Ni - 1.5Cu - 0.5Mo + 0.5C (Distaloy)	● 20 T	● 20 T	■ 31 Y	■ 31 X	■ 31 V	■ 80 W	■ 80 W	■ 60 W
12	Fe - 4Ni - 1.5Cu - 0.5Mo + 0.8C (Distaloy)			■ 20 X	■ 20 W	■ 20 V	■ 80 W	■ 80 W	■ 60 W
13	Fe - 1.5Mo + 0.5C (Astaloy Mo)	■ 30 U	■ 30 U	■ 47 Y	■ 47 X	■ 47 V			
14	Astaloy Mo+ 4Ni + 2Cu + 0.5C)			● 8 X	● 8 W	● 8 U	■ 83 W	■ 83 W	■ 50 W

Tab.8.4 Recommended drill feeds in relation to the drill diameter of selected drills listed in Tab.8.3 [118]

Feed code	Drill diameter [mm]					
	3	5	8	10	12	16
	Feed [mm/rev.]					
T	0.04	0.06	0.09	0.11	0.13	0.17
U	0.07	0.09	0.14	0.17	0.20	0.23
V	0.10	0.13	0.20	0.25	0.28	0.32
W	0.13	0.17	0.26	0.33	0.38	0.43
X	0.15	0.21	0.33	0.42	0.48	0.55
Y	0.18	0.26	0.43	0.55	0.70	0.70

accepted. This relates to the selection of drill type for the appropriate application, rigidity of the workpiece clamping, stability of the tool holder etc.

In some high strength or hardened materials holes are best made by drilling when parts are in the presintered (partially sintered) state. If it is necessary to measure the position and size of holes after final sintering, a carbide reamer of carbide-tipped reamer, should be used [129].

The dominant tool for making holes in wrought metal components is the high speed twist drill. About 80% of all metal drilling and in PM drilling perhaps more, is done using high speed steel as cutting material. As well as being made from high speed steel, twist drills are also produced with brazed hardmetal tips and from hardmetal solid itself. TiN coating is also applied both to high speed steel drills and to solid hardmetal drills and so a variety of cutting materials are available. The popular so called 'short hole drill', used mainly in CNC machines, are fitted usually with specially shaped hardmetal indexable inserts. The advantage of short hole drills with hardmetal indexable inserts is high cutting speed, large feed, longer tool life, cutting edges that can be indexed (*i.e.* no regrinding of cutting edges necessary), high level of precision, quick changing of tools.

Twist drills and short hole drills are the most popular types of drilling tools used in PM machining.

For twist HSS drills, in drilling of wrought steels ranges for cutting speeds are suggested which can be used for comparison with those used in drilling of PM free machining and low carbon steels 25–40 m/min, alloy and medium to high carbon steels 10–15 m/min, and stainless steels 6–10 m/min. The feed of the drill depends on the workpiece material and on the diameter of the drill. A general guide of feed ranges for steps of drill diameter is given in Tab. 8.5.

The use of cutting fluids when drilling is recommended when appropriate. During drilling the chips can heat up to the point where they will weld and stick to the tool, and if this occurs the tool will fail very quickly.

Coating high speed drills with TiN significantly increases their

Tab.8.5 Recommendations for feed rates at different drill diameters in drilling wrought steel [204]

Drill diameter [mm]	up to 3	3 to 6	6 to 12	12 to 25
Feed range [mm/rev.]	0.02 to 0.05	0.05 to 0.1	0.1 to 0.2	0.2 to 0.4

life at higher cutting speed and increased feeds according to the instruction supplied by the manufacturers.

8.1.3 Recommendation for drilling stainless steel

In Tab.8.6 some results are summarised concerning the dry and wet drilling conditions for 430L and 316L grades of sintered stainless steels. These data can be accepted as recommendations and also as a view on the machining costs. In wet machining, it is advantageous to maximize the feed. This recommendation is not valid for dry machining of poorly machinable materials, *i.e.* the feed and the cutting conditions must be optimised.

Tab.8.6 Optimum cutting data in dry and fluid drilling of 430L and 316L steel (without and with 0.5% MnS addition) sintered 30 min at 1280°C in hydrogen and in $70N_2/30H_2$ atmosphere, respectively. Porosity 14%. *HSS 5 mm drill diameter* [205]

Material	Sintering atmosphere	Cutting speed [m/min]		Feed [mm/rev.]	
		Dry	Fluid	Dry	Fluid
430L, 0.5MnS	H_2	28.8	41.5	0.250	0.250
430L, 0MnS	H_2	16.4	30.6	0.250	0.250
430L, 0.5MnS	$70N_2/30H_2$	15.3	21.5	0.250	0.250
430L, 0MnS	$70N_2/30H_2$	11.3	16.2	0.160	0.250
316L, 0.5MnS	H_2	14.4	23.3	0.250	0.250
316L, 0MnS	H_2	10.1	19.9	0.206	0.250
316L, 0.5MnS	$70N_2/30H_2$	9.5	12.9	0.193	0.250
316L, 0MnS	$70N_2/30H_2$	8.6	14.0	0.096	0.175

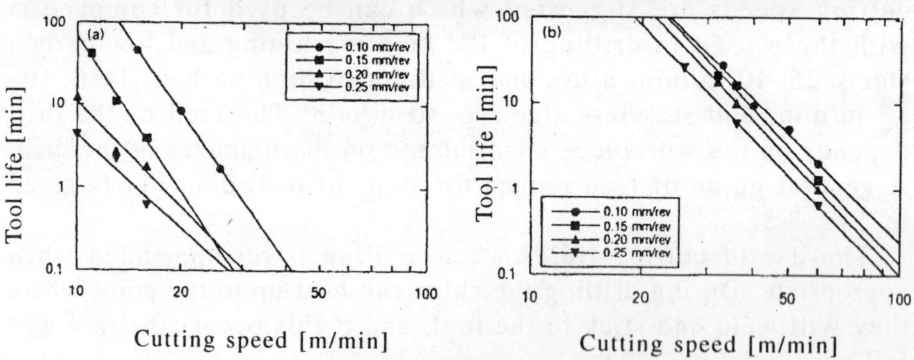

Fig.8.5 Influence of cutting speed and feed rate on tool life in drilling 430 LHC steel without MnS addition. Sintering in $70N_2$–$30H_2$ atmosphere. Drill 5 mm [290].
Fig.8.6 (right) As in Fig.8.5. Material with 0.5% MnS addition. Sintering in hydrogen [290].

Figures 8.5, 8.6 and 8.7 show the tool life in drilling of 430 LHC stainless steel without and with MnS addition as a function of feed rate and cutting speed and of drill diameter and cutting

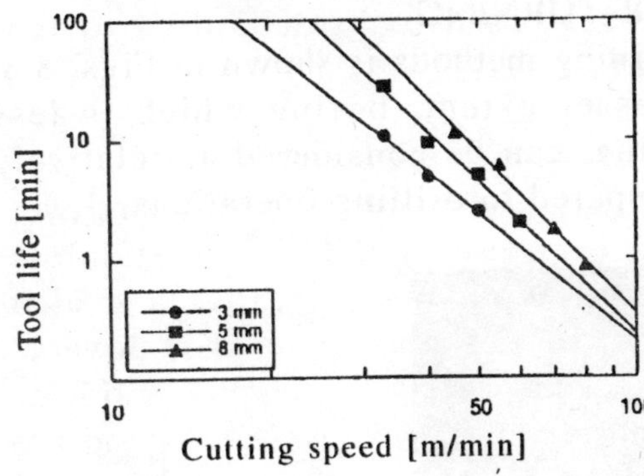

Fig.8.7 Influence of drill diameter and cutting speed on tool life in drilling 430 LHC steel witht 0.5% MnS addition. *Feed 0.1 mm/rev.* [290].

speed which complete the recommendation data.

8.1.4 Burrs

The burrs (frittering) occurring on the exit side of the drilled hole constitute a problem often arising in powder metallurgy but not only there. This can occur during machining, normally at the exit point from the cut. But it can be present, to a lesser degree, at entry and may be a cause of part rejection. For example, when drilling a hole through the wall of a bushing it may be difficult to eliminate the burrs of the hole. The formation of the burrs is closely connected to the drilling conditions and the material being machined.

The most important reason for the formation of burrs is high total thrust force. The chisel edge is responsible for 60–70% of the total thrust force and in order to decrease it, the chisel edge must be shortened. A small insert nose radius will exacerbate frittering, as a result of the axial force being magnified along a shorter cutting edge length [143].

As stated above, breakout at the exit of the hole can be minimised by a special design of the point angle so that the cutting edge corners (at the juncture of the margin and the lips) are first to break through the underside of the hole being drilled [190]. The other possibility would be a lower feed at the entry of the drill into the workpiece and at the exit of the hole realisable in an automatic process.

To decrease the amount of burrs the drills should be sharp, and therefore they must be reground more frequently. Burrs can be eliminated by shortening the chisel edge or increasing the point angle to 185° [143].

8.2 RECOMMENDATIONS FOR TURNING

An illustrative view on three turning methods is shown in Figs. 8.8, and 8.9. Turning and, to a lesser extent, boring which is less frequently used in PM machining, can be considered as relatively 'gentle' machining regimes compared to drilling operations.

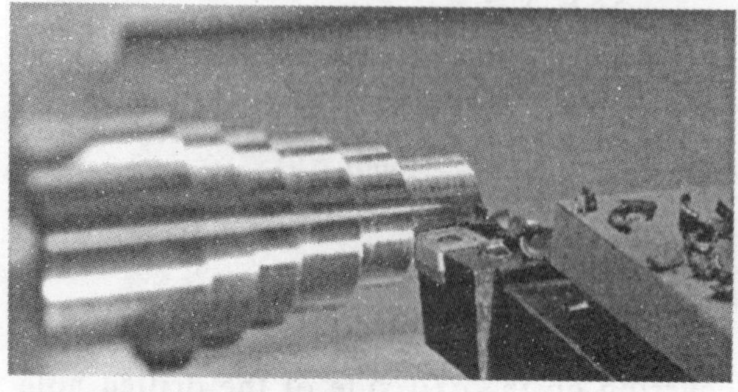

Fig.8.8 Illustrative view on straight turning of wrought steel bar.

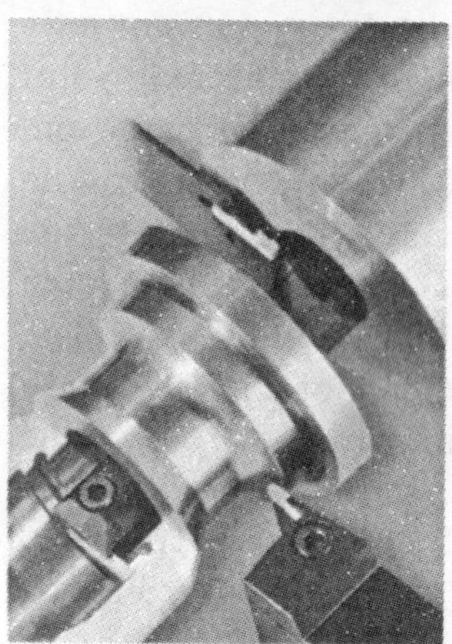

Fig.8.9 Illustrative view on curve and cut off turning of wrought steel bar.

8.2.1 Cutting tools and general cutting conditions

The application of the recommended cutting conditions depends also on the size of the parts being machined. For small parts some lathes are not able to perform corresponding revolutions [107]. For the selection of a tool for turning and boring there are many general recommendations. The type of cutting tool used should be selected carefully and it must be rigid, rigidity being also a requirement lathe. Furthermore, it must be considered that abrasive flank wear is a dominating wear mechanism in turning but the crater wear can be also massive. PVD–TiN coating of the hardmetal (HM) inserts reduces the wear rate, CVD coating (TiN, Al_2O_3) improves the performance even further. Oil impregnation and also suitable cutting fluids improve the machinability in general.

The (CVD and PVD) coated hardmetal as well as cermet and ceramic tools are generally used for turning and boring in PM. CBN should preferably be used, *e.g.* with the high alloyed Distaloy AE and Distaloy HP–1 grades also as-sinter hardened. It must be considered when machining hardened parts with CBN that the combination of negative rake and high cutting speeds generates heat and deforms the workpiece material in the shear zone [180,240]. Uncoated cemented carbide is seldom the best alternative unless a very low cutting speed is used.

For interrupted cuts, hardmetals are preferred to prevent breaking of the insert. In cases in which the preservation of surface porosity is vital, CBN inserts are used, especially with low porosity materials [129].

The tool nose shape, its radius, where applicable, its angle of attack (termed the 'plan approach angle'), and, in particular, the feed rate, will affect the resulting turned or bored surface topography. If a large nose radius is used in combination with a fine feed rate, this will impart a relatively even and consistent surface texture to the part. An alternative strategy however would be to use a small tool nose geometry with an insert having a trapezoid shape (*i.e.* the typical geometry for light turning and facing) in combination with a larger feed rate [190].

If an insert with large nose radius is used for turning or boring, it tends to blend-out the effect of feeding and is mechanically strong, but it is more prone to vibration problems. Conversely, a small nose radius is weaker but less influenced by vibrations. With boring bars, this general lack of rigid behaviour is pronounced, with rigidity decreasing with the cube of the bar length – as a result of its cantilever effect.

Insert geometry must be adapted to the machining conditions and the material being machined in order to meet the technical and economical requirements. All data concerning the tool geometry tested and recommended can be accepted only as general information. Such is the use of a geometry with positive rake and small edge rounding which also helps to preserve the surface porosity. From the chipping point of view, machining of PM materials is improved by the use of tools with sharp edge geometry with rake angles of up to 3° positive on top and side and frontal clearance of 3° and 5°.

Appropriate insert geometry is more important for interrupted turning and boring especially at smaller feed and when the selected grade has sufficient toughness [118].

For carbide tools (grade C-4 carbide) with a sharp nose point, accepted in many cases as a general turning tool, the following specifications are recommended [152]:

side cutting angle – 5°, end cutting angle – 10°, back rake – 10°, end relief – 10°, side relief – 10°, 0.8 mm nose radius for roughing cuts and sharp point for finishing cuts. Larger corner radius and lower feed can be accomplished by a lower depth of cut.

Hardmetal tips of ISO designation K10 with 0° top rake, 7° frontal clearance, and 5° side clearance (hardness of 93 HRA) give a satisfactory tool life and will accept some interruptions on the cutting surface especially in machining fully sintered parts (average hardness ~90 HRB, approx. 185 HV 30).

General turning conditions. Surface speeds of 105 to 120 m/min and feeds of 0.05 to 0.10 mm/rev. at a depth of cut of about 0.12 mm are satisfactory for turning, but surface speeds can be increased to 180 to 210 m/min in single-point turning. Feeds can be increased within the boundaries of economic tool life, the standard of accuracy, and the surface finish requirements. Tool life is nearly independent of the feed rate in a range of 0.05 to 0.2 mm/rev. [152]. According to Ref. 118, the general cutting recommendations depend on the grade of tool material used and the type of PM material being worked. Speeds can range from around 100 m/min for uncoated grade, up to 500 m/min with CBN tools.

Selection of the feed rate is of primary importance with PM materials, in fact more so than with most other workpiece materials. With the exception of high speed applications, in particular CBN, most operations are limited by flank wear. Obviously the highest feed rate possible should be selected; unnecessarily low rates should be avoided. An increase of the depth of cut (*i.e.* chip thickness) does not seem to have any negative effect on the flank wear.

There are now also inserts on the market which permit an increased feed rate while keeping surface roughness at low level.

8.2.2 Specific recommendations and cutting data

The recommendations of the cutting conditions for a given material are always connected with the tool life of the defined tool grade, so called 'economical tool life' which depends often on the 'economical cutting speed'. It may be impossible to reach the tool life corresponding to the economical cutting speed in machining of PM parts, which are relatively small, because of the limited number of revolutions of the lathe [143].

Tab.8.7 Cutting conditions and the tool geometry for turning porous sintered iron (density 6.2–6.9 g/cm³) according to different PM parts manufacturers [347]

Manufacturer		1	2	3	4	5
Tool geometry	Angle$^{(o)}$:					
	Back	5	6-8	8	10	8
	Wedge	65-80	62-6	62		72
	Cutting		70	70		
	True	5-9	20	20	15	10
	Setting		40	45		45
Rough turning	Feed [mm/rev.]	0.3-0.4	0.2-0.3	0.2-0.3	0.3-0.01	Larger than for finishing
	Depth of cut [mm]	0.25	max 1	max 2	max 2.5	
	Speed [m/min]	150	140-180	180-220	150-200	150-200
Finishing	Feed [mm/rev.]	0.05-0.2	max 0.1	Max 0.1	0.05-0.1	0.05-0.1
	Depth of cut [mm]	0.1-0.2	0.1-0.4	0.1-0.4	0.1-0.5	0.1-0.5
	Speed [m/min]	150-200	120-160	140-200	150-200	150-200

In Tab.8.7 cutting conditions and the tool geometry for turning the sintered iron materials are presented based on experiences of some PM part manufacturers.

Sintered iron parts with a porosity of ~6.0 g/cm³ are poorly machinable. It was supposed that the iron particles with insufficient interparticle bonds are torn out from the compact or are easily depressed by the tool. From this reason the cutting speed must be high (feed ~0.1 mm/rev., depth of cut 0.1–0.5 mm) and the hardmetal tool (H1 or H2) very sharp. Although not stated in Ref. 347, the formation of built-up edges can be expected to occur at low cutting speeds.

The general conditions for turning as stated of 'all sintered iron based materials' with cemented carbide tools with higher wear resistance are listed in Tab.8.8. These data can be accepted only as very informative due to very large differences in the properties of 'all' sintered materials.

An overview of data for turning of differently alloyed sintered materials (and thus with various base properties) with cemented carbide tools, given as cutting speeds for attaining three different economical tool live times (T_e), are listed in Tab.8.9.

As stated above, for PM parts which are relatively small, in many cases to reach economical cutting speeds it is necessary to use a lathe with high revolutions. At lower cutting speeds, much longer tool lives can be achieved.

Tab.8.8 Data for selecting tools and setting values for turning of sintered materials [10,348–350]

Material	Carbide Type ISO	Cutting Speed [m/min]	Feed [mm]		Depth of cut [mm]		Rake angle* (γ^o)		Flank angle (α^o)
			a	b	a	b	a	b	
All sintered materials	K01, K10, P10	min. 150	max. 0.3	max. 0.1	max. 0.3	max. 0.3	6-10	6-20	5-10

a – rough machining, b – finishing
Remark: *for soft materials larger rake angles in the given range must be used

Tab.8.9 Machining recommendations – cutting speed [m/min] for turning the various materials based on sponge iron powder [143]. *Depth of cut 1 mm, criterion: economical tool life (T_e)*

Carbide grade	ISO P10 (SIP)			ISO P40 (S4)		
Feed [mm/rev.]	0.3	0.2	0.1	0.3	0.2	0.1
Fe (density 6.0, 6.5, and 7.0 g/cm³)						
T_e = 7 min	335	390	480	290	345	430
T_e = 12 min	295	340	425	255	305	380
T_e = 15 min	280	325	400	245	290	360
Fe-0.5% S (density 6.5 g/cm³)						
T_e = 7 min	375	435	540	330	390	480
T_e = 12 min	325	375	465	285	335	415
T_e = 15 min	305	350	435	265	315	385
Fe-2% Cu (density 6.5 g/cm³)						
T_e = 7 min	315	365	450	280	325	405
T_e = 12 min	270	310	385	235	280	345
T_e = 15 min	255	295	365	220	265	325
Fe-1.75% Ni-1.5 %Cu-0.5% Mo-0.6% C (density 7.0 g/cm³)						
T_e = 7 min	260	300	370	225	270	330
T_e = 12 min	225	260	325	200	235	290
T_e = 15 min	210	245	300	185	220	270
Fe-1.25% Mn-0.5% Ni-0.6% C (density 7.0 g/cm³)						
T_e = 7 min	270	315	390	240	280	350
T_e = 12 min	235	275	340	295	245	300
T_e = 15 min	220	255	315	195	230	285

Data recommended for turning of three groups of sintered steels, *i.e.* iron and iron with admixed phosphorus or copper, Distaloy AE (Fe–4Ni–1.5Cu–0.5Mo) type, and prealloyed Astaloy Mo (Fe–1.5Mo) and Distaloy HP, all with different carbon levels added, are listed in Tab.8.10 with regard to four types of tool materials.

The previous data are augmented and refined by the recommendations listed in Tab.8.11 for the selection of inserts for

Tab.8.10 Recommendations – cutting speed [m/min] for dry turning of sintered steels with widely varying microstructure, mechanical properties and hardness, ranging from plain iron up to very hard Distaloy HP grade; density 6.7 to 7.1 g/cm³ [157]. *Dry cutting conditions: feed rate 0.1 mm/rev., depth of cut 0.5 mm, criterion: economical 15 min tool life (T_{15})*

Alloy No.	Base powder	Alloy element [%]	C [%]	Ferrite [%]	Hardness HV 10	Cutting speed [m/min] for tool			
						Coated HM	Uncoated HM	Cermet	CBN
1	Iron powder ASC100.29	-	0	100	50	245	120	120	
2		0.45P	0	100	100	175	90	90	
3		2Cu	0	100	70	220	110	110	
4		2Cu	0.25	80	120	260	130	225	
5		2Cu	0.5	30	130	250	125	220	
6		2Cu	0.8	1	150	230	115	200	
7	Distaloy AE		0	90	100	200	100	155	
8			0.25	35	160	180	90	140	465
9			0.5	8	200	160	80	120	430
10			0.8	2	230	140	70	110	375
11	Astaloy Mo		0.5		160	200	100		
12	Distaloy HP	4Ni +2Cu	0.5	1	270		55	300	300

turning of the PM steels. For Distaloy SA and AB materials with hardness of 160 to 250 HV10 and porosity of ~17.3% and ~13.5%, turning with cemented carbide inserts was recommended. A cutting speed of around 100 m/min for wet conditions gave the longest tool lives. Machining at a lower speed under dry conditions may give longer tool lives. At 200 m/min there was a large drop in insert lives when the depth of cut was increased to 0.5 mm. If lubrication was not used the tool lives could be very short. Machining at 300 m/min was not recommended [139,186].

A clear improvement in tool life can be attained in turning synchronising hubs from Distaloy AE material when adding MnS as free machining additive as shown in Tab.8.12. Furthermore it can be stressed that a relatively long tool life was attained in turning of Fe–Cu–C materials based on atomised iron powder ASC100.29 without addition of machining aid (machinability improving effect of copper).

As stated in Ref.351, the turning of sintered steels should be performed also with regards to surface roughness, and therefore the cutting speed should be higher than 60–100 m/min (at lower speed

Tab.8.11 Recommended start values for selected inserts (dry cutting) based on laboratory test, VB crit.= 0.30 mm, criterion: economical 15 min tool life (T_{15}) [118]. Material (alloy) designations as in Tab.8.10.

Alloy No.	Specific cutting force Kc=0.20 [N/mm²]	Tool grade					
		GC3005	GC3025	GC1025	CT5015	H13A	CB7050
		Feed [mm/rev.]					
		0.10-0.30	0.10-0.30	0.10-0.30	0.05-0.20	0.05-0.30	0.10
		Recommended cutting speed, start value [m/min]					
1	1900	220-230	195-200	195-200	115-125	115-120	
2	2150	130-165	115-145	115-145	80-100	80-100	
3	2050	175-210	155-185	155-185	100-120	100-200	
4	2100	200-260	175-230	175-230	195-250	110-145	
5	2150	220-250	195-220	195-220	205-235	120-130	
6	2300	170-235	150-205	150-205	170-230	100-130	
7	1900	200-210	175-185	175-185	150-150	95-100	
8	2100	165-170	145-150	145-150	1350140	85-90	465
9	2200	150-165	130-145	130-145	115-125	75-85	430
10	2200	120-135	105-120	105-120	105-115	65-75	375
11	2250	175-220	140-160	140-160	160-180	85-105	
12	2500	95-120	85-105	85-105	75-95	50-60	300

Tab.8.12. Tool life for turning synchronising hubs from different materials. Sintering 20 min at 1120°C in endothermic atmosphere for C containing materials, dissociated ammonia for the other materials [146]. *Turning conditions (facing): cutting speed 200 m/min, feed rate 0.1 mm/rev., depth of cut – 0.5 mm; dry; tool:PVD TiN coated hardmetal*

Distaloy AE -0.5%C	Distaloy AE-0.5% C+ 0.5% MnS	Distaloy HP- 0.5%C	ASC 100.29	ASC 100.29+2% C -0.5 % C
Tool life [min]				
3.36	31.62	2.79	5.54	26.63

increase in roughness), feed ~0.12 mm/rev. and depth of cut 0.2–0.3 mm.

For the PM materials tested, the dominating wear for the tool during continuos turning was abrasive flank wear. A depth of cut notch was typically formed which can limit the tool life. It was also observed that from the chipping point of view of a tool the machining of PM materials is improved by the use of sharp tool geometry [146].

The turning of sintered piston rings for diesel engines is one from the examples of machining operation. The piston rings were prepared from Fe–2% Cu–2% Ni–1% Cr–0.3% graphite mix by compaction at 620 MPa and high frequency sintered at 1150–1200°C (porosity 20±2%). The microstructure was formed by lamellar pearlite with precipitated carbides. The best turning of these parts was performed with hardmetal tool inserts with special geometry (not declared) having a positive effect also on surface finish. The highest surface quality was attained with the tools embedded with diamond inserts [352].

8.2.3 Surface finish

The requirement on surface finish is a question of the workpiece material properties and the cutting conditions on one side and on the other side of economy. From this reason, the criteria required for surface finish quality form a part of recommendations also for cutting conditions in turning. The best surface finish of a part with specific properties can be achieved by [107,160,345]:

- higher cutting speed,
- lower depth of cut,
- use of a larger nose radius and lower feed (Fig.4.25); (a larger nose radius considerably improves surface finish at higher feed rates since a larger nose radius gives smaller ridges at machining). Use of high feed rate is limited by the tool geometry,
- higher density and use of sulphur-alloyed powder resulting in the longest tool life,
- corner breakout is minimised by lower feed, lower depth of cut and higher density.

There are hardmetal inserts for turning of sintered steels that in general fulfil the current high requirements for the finish quality of a machined part and for productivity. The lower surface roughness demand can be achieved by some multilayer coated inserts [107].

The breakout at the corners of the machined part, especially at low densities, is a very critical factor in turning that affects the surface finish. These can be eliminated also by lower feed, smaller depth of cut, and higher density of the part [143].

8.2.4 Recommendation for turning with liquid nitrogen cooled cutting tool

A new and promising hard turning technology has been developed for wrought bearing steels, which can be useful also in PM machining. This method involves machining with Al_2O_3 ceramic tools cooled with small amounts of cryogenically cold, two-phase liquid nitrogen (LIN). Cryogenic liquid cooling was achieved by spraying the rake surface of the tool with two-phase jet (cold vapour and liquid droplets), Fig.8.10.

With regard to problems with hard turning of PM steels, this method was tested in hard turning for premixed sintered FN-0208

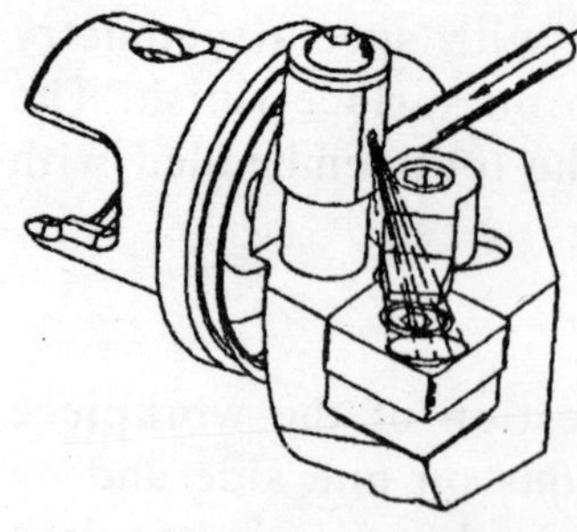

Fig.8.10 Mounted toolholder, liquid nitrogen nozzle and spray used for cooling cutting inserts [295].

(0.8Ni, 0.8–0.9C) composition of the density 6.67 g/cm^3 and 7.20 g/cm^3. The specimens (discs) were heat treated to hardness of at least 30 HRC in the lower density materials and at least 40 HRC in the high density material. For machining tests on the four materials a CBN tool combined with a conventional cutting fluid cooling and an alumina based, and Al_2O_3–TiC 'black ceramic' tool combined with cryogenic liquid nitrogen cooling were used. The 'low-content', TiN-bonded CBN tool (BN25) used was shown previously to exhibit predictable wear behaviour in the interrupted cutting of hard steels. The Al_2O_3 tool (ZC4-grade, PVD TiN coated) used was previously found to work extremely well with cryogenic cooling on hard, wrought steels, but failed by premature fracture if a conventional room-temperature cutting fluid was applied. The materials were tested as-sintered and as-heat treated.

From these tests, surface roughness values and tool life are given as results. As shown in Fig.8.11, a marked difference in surface roughness of the workpiece in dependence on cutting tool grades was attained.

As shown, cutting with the Al_2O_3–LIN tool significantly improved surface roughness in the case of 'soft' materials but not in the case of hard materials. It was also stated that the finish showed less dependence on cutting edge profile. Figure 8.12 shows the effect of the cutting method and base properties of PM workpiece material on tool edge life.

Significantly improved tool life was achieved in use of Al_2O_3–LIN tools compared to CBN tools, especially in 'soft' materials. Compared to CBN tool life in turning of 'hard' materials, the increase with ceramic tools is higher than 100%.

It was also determined that the cutting force does not correlate with tool life, with the change of material density, or with the use of lubricating cutting fluid. This shows according to the results attained the limited validity of force/tool life proportionality.

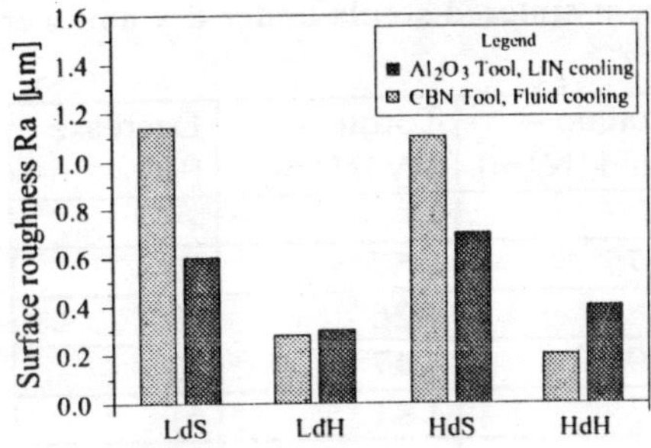

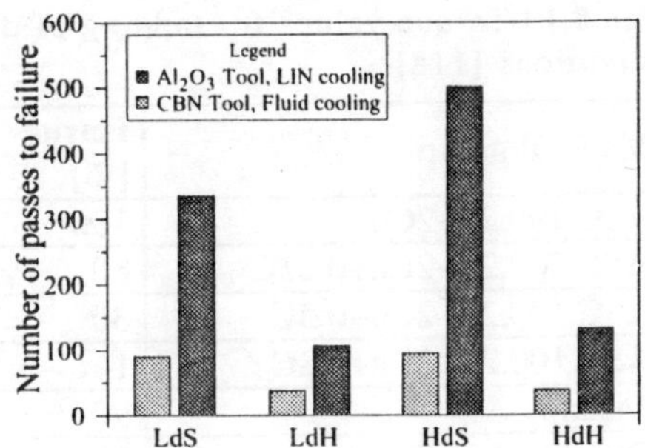

Fig. 8.11 Surface roughness in finish-turning operation of FN-0208 compact with CBN tool, fluid cooling and Al_2O_3–LIN tools. Material: LdS – low density as-sintered – soft (6.67 g/cm³), LdH – low density as heat treated sintered – hard (6.67 g/cm³, HdS – high density as-sintered – soft (7.2 g/cm³). HdH – high density as-heat treated – hard (7.2 g/cm³) [295]. *Cutting conditions: speed 152 m/min, feed 0.09 mm/rev., depth of cut 0.2 mm.*

Fig.8.12 (right) Effect of cutting method and properties of PM workpiece material on tool edge life (number of passes to failure). Material characteristics as in Fig.8.11 [295]. *Cutting conditions: speed 305 m/min, feed 0.18 mm/rev., depth of cut 0.23 mm, criterion 0.3 mm flank wear.*

8.3 TAPPING AND THREADING

8.3.1 Base characteristics for tapping

To classify PM material regarding tapping, evaluation of tool life of the tap like in other cutting methods is the best. A correct thread can only be made if the original hole is round, straight and of accurate diameter. To relate the workpiece properties and tapping, only the torque and apparent hardness were investigated. The correlation between the starting (as-sintered) hardness of the workpiece in the range of ~75 to 220 HV 10 and of torque in tapping (straight fluted tap, cutting speed 6.3 m/min, applied fluid) with emulsion in the range of 0.5 to 1.2 N·m was approximately linear [118]. Tapping is performed dry and fluid in dependence on workpiece properties and thread dimensions considering tap life.

Fluid versus dry tapping. Tapping with emulsion revealed no such effect of the amount of ferrite present in the microstructure of materials as was found at turning and drilling since tapping starts in the deformed surface of the hole drilled. When tapping dry, cutting speeds must be reduced.

In Tab.8.13, the torque values for tapping under dry and fluid conditions are listed for different PM materials according to the proportion of ferrite present in microstructures, which must be taken into account. Increasing ferrite amount in the microstructure results

Tab.8.13 Torque values for tapping of different sintered steels under dry and wet conditions [118]

Material group	Ferrite [%]	Torque – fluid [N·m]	Torque – dry [N·m]	Decrease [%]
ASC 100.29+2Cu	100	-	-	-
ASC 100.29+2Cu+0.25C	80	0.75	1.57	52
ASC 100.29+2Cu+0.5C	30	0.85	1.26	33
ASC 100.29+2Cu+0.8C	1	0.98	1.47	33
Distaloy AE	90	0.7	1.81	61
Distaloy AE+0.2C	35	1	1.19	16
Distaloy AE+0.5C	8	1.17	1.36	13
Distaloy AE+0.8C	2	-	-	-

Remark: Decrease = decrease of torque at fluid tapping vs. dry tapping

in a decrease of the torque in fluid tapping and increase in dry tapping.

The highest difference in torque between fluid and dry tapping was attained in plain Distaloy AE material. Copper and carbon content had some effect both on the absolute values of torque and also on its decrease when tapping with emulsion compared to dry tapping. For cooling and lubrication in tapping, an emulsion is used (12% concentration, thread cutting fluid), thread cutting oil, and thread cutting paste. Carbon addition decreases the performance.

Tapping under dry condition reveals problems with chip jamming. This can degrade the performance but it is remedied by the use of cutting fluid for each application what is one of basic hints for tapping.

Influence of additives. High strength material such as Distaloy AE-0.8C alloy could only be tapped with a machining aid addition. Both MnS and MnX additions and minimal lubrication decreased the torque. Materials with superior mechanical properties could not be tapped. Also the density has a strong influence on increase of the torque during tapping. Addition of 0.5% MnS revealed a large decrease in torque under dry conditions especially with increasing hardness of the material. In tapping with applied emulsion, there was a further small decrease in torque in the case of MnS added compared to dry machining. Addition of 0.3% MnX had no effect on tapping Distaloy AE-0.5C under dry condition. For Fe–2Cu–0.5C steel, in contrast, under equal conditions, the effect was significant [118].

For materials with high amounts of ferrite, MnX addition in combination with lubrication by either oil impregnation or minimal applied lubricant, is considered beneficial.

Generally internal lubrication (oil impregnation) or MnS/MnX addition is believed to be beneficial for the threading operation because it decreases the torque [118].

8.3.1.1 Selection and characterisation of the taps

According to the geometrical characteristics the taps are: spiral point taps, spiral flute taps, straight flute taps, and cold forming taps. The data concern the standardised threads in shape and in lead. Five main hole configurations were considered for tapping shown in Fig. 8.13.

Conventional tap drill charts should be followed to maintain 65 to 75% depth of thread. Two-flute taps are recommended for diameters up to 8 mm. Three-flute taps should be used for diameters of 8 to 12 mm [129,343]. By using fluted cemented carbide taps for holes these recommendations to less 12 mm can be changed for holes over diameter of 12 mm.

Spiral point taps. A tap with a straight, fairly shallow flute is often referred to as a gun nose or spiral point tap. This type of tap is recommended for threading through-holes and is used for hole types I and II. This type of taps is desirable because they throw the chip out. It may be used in blind hole applications where there is sufficient space to accommodate the swarf, hole type IV.

Spiral fluted taps. Taps with spiral flutes are intended primarily for threading in holes of type III, IV or V. The helical flute transports the swarf back away from the cutting edges and out of the hole and thus avoids packing of swarf in the flutes or at the bottom of the hole. The danger of breaking the tap is minimised.

Straight fluted taps. This type of taps is the most commonly used type of tap. Suitable for use on most materials, straight-fluted taps form the basis of most tapping operations. They are recommended for hole types I–IV.

Cold forming taps. No swarf is produced by cold forming and cleaning is eliminated. The application range are materials with good formability characterised by low hardness. It is also recommended that the elongation of the workpiece material should not be less than 10%. Cold forming taps without flutes are especially suitable when

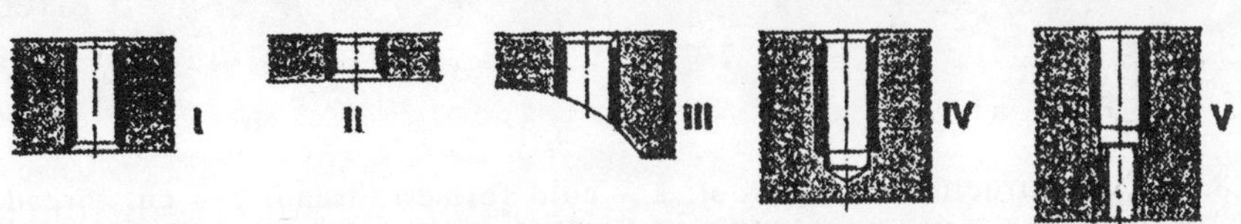

Fig.8.13 Five main types of holes in tapping operations.

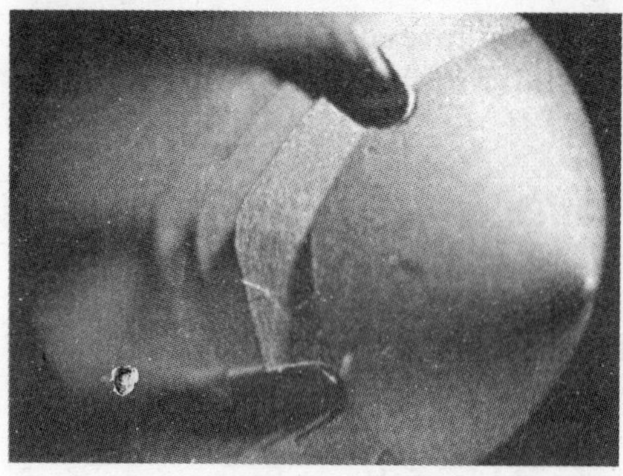

Fig.8.14 View on split point geometry of cold forming tap with flutes M8.

working with vertically tapped blind holes. Conversely, cold forming taps with flutes are especially suited for tapping in horizontal holes. The split point geometry of a cold forming tap is shown in Fig. 8.14.

As was determined, the cold forming tapping causes a significant cold deformation and by this an increase in hardness of the face and of the root thread. The marked deformation of the thread surface was confirmed by the microhardness measurement. The average microhardness of the surface of the cold formed thread was 294 HV 0.01 compared to the microhardness of the surface of the cut thread which was 206 HV 0.01 as shown in Fig.8.15. The microhardness of the core grains was 145 HV 0.01. The parts with two cold formed threads were finally nitrocarburised (Fig.8.1a). This confirms that in the case of tapping with a cold forming tap, the thread is formed by plastic deformation of the material. In the case of the predrilled hole with larger diameter 7.4 mm designated for cold formed thread M8 and 6.9 mm for cut thread tapped in the hole of larger diameter, at cold forming tapping the material flow

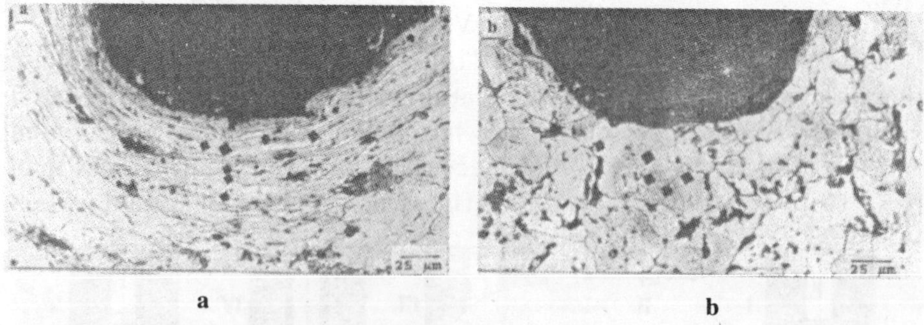

a b

Fig.8.15 Microstructure of the root: a – cold formed thread, b – cut thread by straight fluted tap, both thread M8 in Fe–0.5Mo–0.1C–0.3MnX steel. Iron powder SC100.26, density 6.4 g/cm^3, hardness 65 HV 10. Optical micrograph. Nital etched.

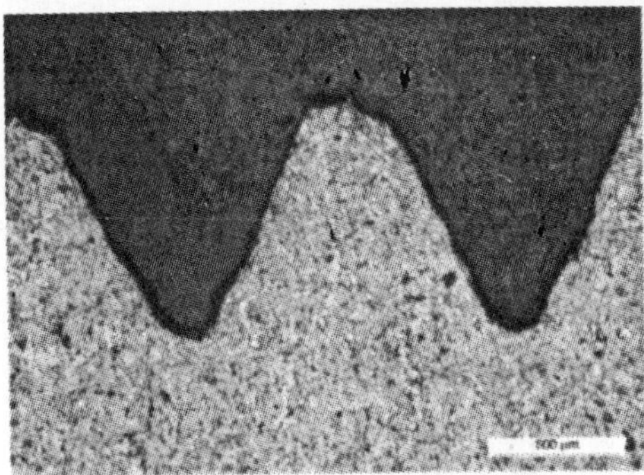

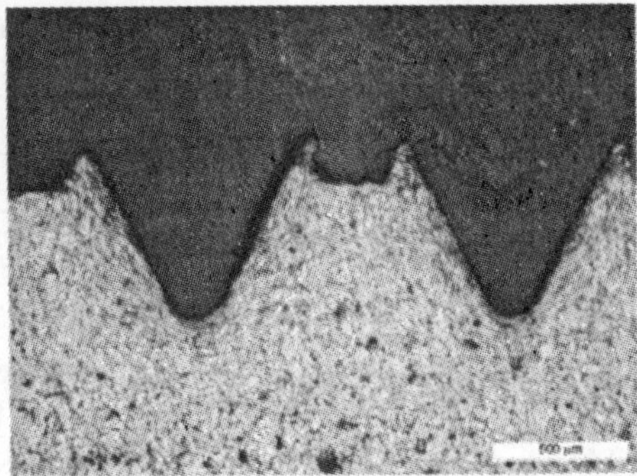

Fig.8.16 Section of a cold formed thread M8 in material as in Fig.8.15. Optical micrograph. Nital etched.
Fig.8.17 (right) Section of a cut thread M8 in material as in Fig.8.15. Optical micrograph. Nital etched.

filled the profile of the thread compared to the cut thread, Figs. 8.16, 8.17. Cold forming tapping is effective from the economical viewpoint.

In all cases tapping is performed in the hole the wall of which was deformed at drilling. The drill sizes differ between cutting and forming taps. The diameter of the hole drilled for cold forming taps is larger by ~5–8% compared to the diameter for cutting taps. The required base properties of the material being tapped by cold forming will depend on the diameter of the thread. The larger the diameter of the thread the softer the material must be due to larger deformation at tapping [118].

Tapping is easily accomplished by using two-spiral pointed taps which are most desirable because they throw chips ahead and prevent them from driving into the pores. If difficulties are encountered, the relief of the tap can be increased to nearly twice that used for conventional ferrous materials [152].

Fluid work with cutting lube to remove chips do not allow taps to bind. To increase tool life, it would be advisable to use spiral flutes since they will absorb more of the shock caused by interrupted cutting. Also, where possible, use 10% less of full thread in diameter (larger diameter of the hole) than with comparable wrought materials [345].

Different types of tool materials influence tap performance. Tapping is mostly carried out with high speed steel taps. The tool life can be improved by using Co-alloyed or nitrided taps. Within the selection of the right material the hardmetal and coated taps can also be used.

Tab.8.14 Data for tapping of sintered materials with high speed taps [10,344,348–350]

Sintered material	Cutting speed [m/min]	Flute		Cutting medium
		Through hole	Blind hole	
All	10-20	Straight	Helical, spiral	Cutting oil/air

Proper selection of tap geometry is considered to solve many problems in tapping PM materials but some experimenting in tapping PM parts may be required to determine which tap is best for a specific material [129,146,345]. Data for tapping of sintered materials with high speed steel taps are listed in Tab.8.14.

In regard to severity of the tapping process in materials with very large differences in hardness and in microstructure character, more detailed summarising data about the tapping in PM steels are lacking. As at other cutting methods the optimum tapping conditions must be adapted to the material properties. The correct cutting speed for tapping is specific and must be adapted to the general part properties (*e.g.* hardness) which are locally changed after drilling the hole.

8.3.2 Threading

Threading is not usual in PM parts machining. On the other side it is often applied on wrought steels. One of the most popular tools for threading consists of a solid piece of hardmetal with a dove tail cross section which was used to clamp the threading tool into the machine. The tool has to be inclined towards the workpiece so that clearance can be achieved. Usually three or four teeth are ground in with the leading tooth being topped and shorter than the rest with gradually increase in length up to the last tooth which then has the full thread profile.

Thread milling. Although it is strictly a milling operation, *i.e.* it is carried out by using a rotating cutting tool similar to a mill, it is a unique process and therefore is mentioned also in connection with PM machining. Thread milling cutters are used to mill both external and internal threads. They are particularly suitable for very large workpieces and for non-rotational, non-symmetrical parts [302].

Thread rolling operation is an operation in wrought steel machining, *e.g.* the production of the screws with threads formed by thread-rolling die head [134]. There are no data on the aplication of this method in PM machining.

8.4 RECOMMENDATIONS AND CUTTING DATA FOR MILLING, REAMING, BROACHING AND GREEN MACHINING

These cutting operations are used in PM machining in a relatively limited range but they complete a general view on machining and thus also on possible machining of PM steels. All these operations belong to the machining methods with geometrically defined cutting edge.

8.4.1 Milling

Milling is generally difficult because of the tendency of the softer materials to smear. To minimise the smearing, it is recommended that dead sharp helical tool cutters with an axial rake be used so that the chips are sheared on an angle. Most parts can be machined with the same cutters as are used on cast iron or alloys of low tensile strength and low hardness.

Slot and side milling cutters are often used for machining of PM materials. Speeds of 70 to 100 m/min, feed rates of 0.05 to 0.1 mm per tooth, and depths of cut of 0.13 to 0.4 mm are recommended in machining ferrous and non-ferrous alloys with uncoated carbide tools [129].

Speeds of 25 to 50 m/min are used in machining PM iron, steel, stainless steel, copper, and brass with high-speed end mills. Cutting speeds in the range of 100 to 200 m/min are recommended in machining soft iron, and low alloy steels with carbide tools. With harder steels, stainless steels, copper, and brass, speeds should be lowered to the range of 60 to 100 m/min [129,343].

In face milling with uncoated carbide tools, cutting speeds of 90 to 120 m/min, feed rates of 0.05 to 0.15 mm per tooth, and depths of cut of 0.12 to 0.4 mm are generally recommended for carbon and alloy steels and stainless steels. Cutting data recommendations for milling sintered materials under different conditions are listed in Tab.8.15.

The same geometry of the tools can be used for machining of steels with a strength up to 900 MPa. When surface or end milling, an emulsion is recommended. Air cooling is usually sufficient for slotting or sawing [344].

Using a double helix cutter is also recommended. A spiral or single helix cutter will not cut a straight or smooth line. Fully soft PM parts may be too ductile to mill and will require a change in processing to make them harder [108,345].

Tab.8.15 Cutting data for milling of sintered materials. a – coarse machining, b – light machining, max. 1 mm depth of cut [10,344,348–350]. Feed speed S' is the multiple of the feed per tooth s_z, the number of teeth z, and revolution n $S' = s_z z n$; HM – hardmetal

Sintered material	Tool	Cutting speed [m/min]		Feed s_z [mm/tooth]				Disc saw	
				Cylindrical cutter and shank end mill		Side-milling cutter		Cutting speed [m/min]	Feed S [mm/min]
		a	b	a	b	a	b		
All	HSS	17-20	18-30	0.12	0.07	0.06	0.04	35	50-70
	HM	70-90	70-100	0.12	0.07	0.06	0.04	300	180

8.4.2 Reaming

Reaming is definitely not recommended if porosity is to be maintained as, for example, in bearings because of the tendency to smear the surface. To control bore accuracy in PM parts, reaming is sometimes used instead of pin sizing, ball sizing, or burnishing. For structural parts standard reamers (HSS, HM) are satisfactory. The largest wear of the tool in reaming of the porous materials is formed in the junction place of the cutting to calibration part of the tool. A left-handed helical reamer with a right-hand periphery cut is recommended; left-hand spherical reamers have also proved successful. Allowances varying from 0.05 mm up to as much as 0.15 mm and recommended feeds, depending on diameter of the hole are listed in Tab.8.16.

If the surface finish is critical, reaming allowance should be doubled. Whenever possible, reamers should be used in floating holders and run at 7.5 to 15 m/min. The cutting speed 15–60 m/min, depth of cut 0.1–0.2 mm and feed 0.2–0.3 mm/rev. are recommended [351].

8.4.3 Broaching

Broaching is not advised if porosity is to be maintained. It is recommended that at least 0.38 mm of stock should be removed. Standard broaching is recommended to obtain best tolerances and finish. The best results are obtained when the broach has at least 16 or more cutting edges and minimum of six burnishing surfaces [152].

Tab.8.16 Allowances and feeds at reaming of holes [129,152,343]

Hole diameter [mm]	Allowance [mm]	Hole diameter [mm]	Feed [mm/rev.]
≤6.5	0.05	≤6.5	0.15
6.5-12.5	0.05-0.10	8-12.5	0.18
12.5-25	0.10-0.15	14-19	0.25

Normal broaching procedures can be used for material at 7.0 g cm^{-3} density and above. Below this density level, pores should be sealed with resin which can be burned out after broaching if necessary [345].

8.4.4 Green machining

For green machining, some requirements both regarding material properties and cutting conditions have to be met.

The following main recommendations for green machining are [115]:

- 'high' green strength >20 MPa by increasing green density, particle irregularities, addition of chemical adhesive acting by plastic deformation, mechanical interlocking and chemical adhesion,
- green density ~7.3 g/cm^3, or, if lower, using some mentioned methods to increase interparticle bond strength,
- high cutting speeds (100 up to 600 m/min) and low feeds (~0.025 mm/rev.), depth of cut (~0.01 mm) (more passes, the last with minimum depth),
- tools – sharp uncoated HM tools – the best. In turning cutting edge with higly positive rake angle,
- for improving machining performance the drill chisel angle should be increased from 118° to 135° (decreasing thrust force during drilling),
- breakout at the inlet and outlet side – cutting parameters must be addapted regarding workpiece nmaterial structure characteristics,
- surface finish is improved as cutting speed is increased,
- successful green machining is a function of both the green strength of the compact and the lubrication – binder type.

A summary of the view on the relative effect of singular factors taking part in green machining, including drilling conditions, drill type, density and representative base powders (Distaloy AE, Astaloy CrM) on thrust force is shown in Fig.8.18. It follows that the feed rate strongly influences the thrust force. A very important phenomenon in green machining is the breakout on the inlet, Fig.8.19, and on the oulet side. In this case, the strongest influence is found with feed rate and high speed drills on both sides. Surface finish is mostly affected by feed rate – low feed rate needed.

The drill type has a deciding effect on the axial (thrust) force and by this on drill life in green drilling, as shown in Fig.8.20. Hardmetal drills are the best also in regard to inlet and outlet breakouts.

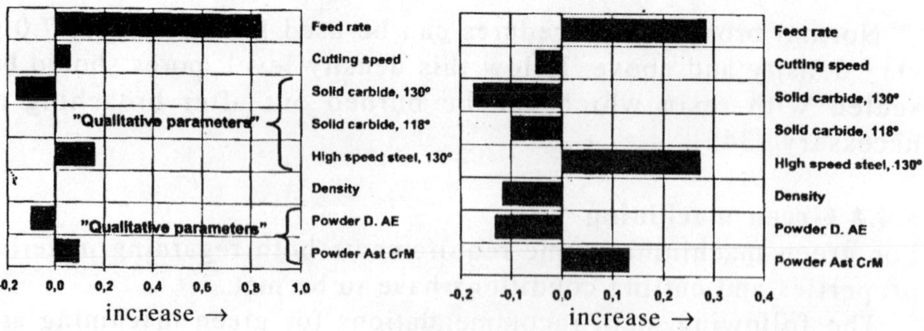

Fig.8.18 Values of significance for thrust force in green drilling, Fz [353].
Fig.8.19 (right) Values of significance for surface finish in green drilling in inlet hole face side [353].

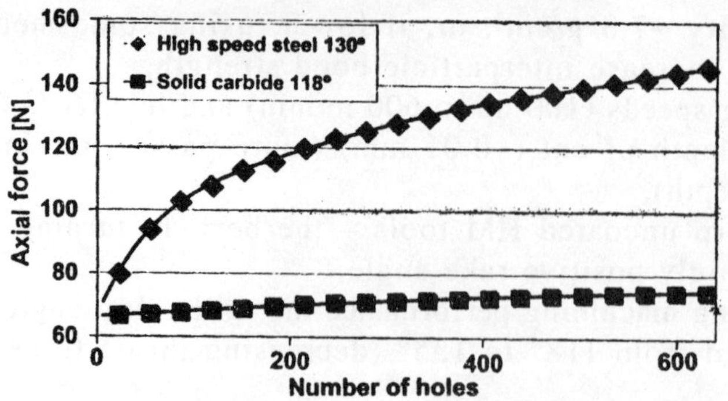

Fig.8.20 Axial force vs. number of holes green drilled in Ast.CrM+0.5% C, green density 7.15 g/cm³, green strength 30 MPa [353]. *Cutting conditions: HSS and HM 5 mm drill, cutting speed v_c = 80 m/min, feed = 0.05 mm/rev.*

8.5 RECOMMENDATIONS FOR MACHINING WITH GEOMETRICALLY NOT DEFINED CUTTING EDGE

8.5.1 Grinding

Grinding of PM parts can be very complex, especially with low density materials, because in many cases preservation of surface integrity is essential. Surface integrity is of special interest, especially when looking at highly loaded components. In many cases high shearing stresses (Hertzian stress) occur beneath the surface when the component is loaded, and here the presence of residual stresses after grinding has to be considered. The level of tensile residual stresses also depends on the grinding wheel material, speed and pressure.

It is necessary to note that in dependence on the cutting speed and wheel grains a large amount of the generated powder chips can be forced into the pores, and fine chips can be welded to the workpiece when high temperatures are attained at the wheel/workpiece interface which must not be the case in grinding PM steels. When grinding is necessary to achieve dimensional functionality of a part, and surface porosity needs to be preserved, special processes such ultrasonic or solvent cleaning are applied immediately after grinding. For rough applications, a downfeed of 0.025 to 0.075 mm is recommended, while for finish passes, a maximum of 0.013 mm should be used [129,152].

On the other side, it can be said that in grinding of a sintered part the surface layer is also densified, *i.e.* surface porosity decreases during grinding. This causes a change of the properties and by this way an improvement of the wear resistance is expected. The densification of the surface increases with increasing porosity. An increase of the grinding depth from 0.01 to 0.1 mm causes an increase in compressive microstresses by 15–30% [168,344]. The adaptation of machining parameters makes it possible to influence the grinding process. Higher cutting speed leads to a better surface quality and a reduction of the grinding forces. On the contrary, at very high cutting speeds grinding forces and wheel wear increase.

If the densification of the grinding surface is not acceptable, a larger transverse shift 1.7–3 mm/shift, the rate of the table 6–12 m/min and the depth of the cut 0.01–0.03 mm is recommended [351].

Stock removal rates should be either the same or less than those used in finish grinding of cast iron; wheels should be similar. It is important to keep a plentiful supply of coolant (containing an inhibitor) directed onto the wheel and the work to maintain a clean grinding wheel contact.

The surface finish of the components which, beside the dimensional accuracy is the main object of grinding, is influenced not only by the choice of the cutting parameters but also by the selection of the right grinding material [143]. The cutting effect of a grinding wheel depends beside the cutting conditions on the grain size and material and on the type and amount of the binder phase which affects also the porosity of a wheel. Particularly important is the bonding strength of the grains to one another, if it is 'soft' or 'hard' bonding, the former resulting in faster re-sharpening due to removal of worn grains but wear increases.

It is a tendency at grinding of wrought steel parts to limit the disadvantageous effect of heat formed in ground surface layer by the adaptation of the cutting conditions and by the arrangement of the grains of the grinding wheel and by this to attain the advantageous compressive stresses. Also deep grinding through one pass with porous wheels at low surface cutting speed with cooling is used more frequently instead of turning if the shape of the part corresponds to the process motion. The material removal can be up to 0.2 mm at one pass at the deviation of the alignment of the surfaces 2–4 μm [346].

The grinding tools can be subdivided into grinding wheels made from conventional grinding materials and diamond or CBN grinding wheels. Conventional grinding wheels, normally SiC or Al_2O_3, are particularly suitable for the machining of standard low alloy PM steels and can therefore be used for all short- or medium cutting alloys. For PM high speed steel they are not practical, given the hardness of the carbides and the extensive wear of the grinding wheels. CBN grinding wheels are also used frequently for grinding small holes. Cemented carbides are machined mostly by diamond grinding wheels. A basic requirement of the optimal application in grinding is that the topography of the grinding wheel must be adjusted to the grinding task. Conventional grinding wheels can be profiled and sharpened in one process step. Diamond and CBN grinding wheels are normally profiled and sharpened in different working processes with diamond profile and tracing rolls or at cylindrical profiles also by means of small cup wheels [108].

A very important factor in grinding is the coolant. Its main task is to reduce the friction at the point of contact of workpiece and tool and to carry off the heat from deformation and friction. For instance, the cutting fluid can also have an impact on results, as this can vary the cutting. A chemical coolant rather than oils or soluble oils is to be preferred [109].

8.5.2 Honing and lapping

Honing and lapping are usually finishing operations for heat treated materials, and only a small degree of pore closure occurs, however, if sintered non-heat treated parts are honed, the honing stock should be held less than 0.04 mm [129,152], because either of these processes will cause the pores to become filled with abrasive particles. For special applications, ultrasonic or solvent cleaning should be performed following grinding.

Sintered materials can be honed in the as-sintered and the as-heat treated condition. It is used when very tight tolerances and very good surface finish is required. The grain size and the hardness of the honing stones influence the surface finish. Finer grains give a better surface finish and softer honing stones give a coarser surface. A resin bonded grade Q/167 with a grain size of 150 is recommended for sintered material [143].

Holes requiring extreme accuracy can be honed or lapped by normal techniques if retention of porosity is required. However, size control of holes in PM parts can usually be obtained more economically by reaming or burnishing.

High-density ferrous metal parts, especially when hardened, have been successfully honed and lapped using conventional procedures. Honing of infiltrated parts is seldom practical, because the stones become clogged.

Diamond- and CBN-plated bore finishing tools are recommended for precise hole size control. These tools can be used on standard drilling or honing machines, as well as on multi-spindle or numerically controlled machines. The use of an adjustable sleeve attached to a mating tapered mandrel increases tool life. If the existing finish is 1.25 μm and the desired finish is 0.25 μm then 0.010 mm should be removed from the hole diameter.

In lapping, special attention should be paid to the influence of unit pressures and the type of abrasive mixture on the surface roughness. Maximum linear capacities have been obtained by using a paste with micrograins of 99C green silicon carbide (F500/13) (especially for hard sinters), whereas the lowest ones by using a 95A (F1200/3), electrocorundum paste (especially for soft sinters). The lower the size of abrasive micrograins, the less influence on lapping capacity is exerted by the type of grains.

8.5.3 Burnishing

Burnishing in PM is used mainly in sliding bearing production. When the clearance between a shaft and PM bearing is ±0.012 mm or less, burnishing the bearing bores after they have been installed in the housing is preferred for correcting the bore size. No more than 0.05 mm/mm of diameter should be displaced, and the smallest amount of displacement that will produce the true diameter is desirable. The type of burnishing tool recommended for this operation is illustrated in Fig.8.21.

The tool becomes progressively larger, and the bearing is expanded. If there were no spring-back, the operation would be stopped at the fourth or fifth tool land. However, the bearing would ordinarily be burnished to 0.010 mm oversize to allow spring back.

Roller burnishing is a cold working operation that compresses

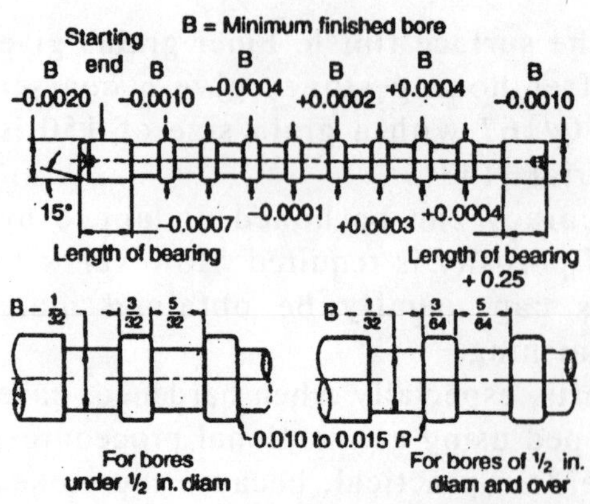

Fig.8.21 Ball broach for burnishing bores in PM parts (dimensions in inches) [129].

metal rather than removes it. The technique is suitable for sintered (not heat treated) powder metal materials for which maintenance of open surface porosity is not critical. Roller burnishing may be used to meet a significant improvement in accuracy and surface finish requirements. Any size change in the operation is related to the preburnished surface finish and the size of the hole.

In addition, the tool is adjustable to match individual product specifications as well as to compensate for wear on the rolls and mandrel. Both through holes and blind holes can be roller burnished. Hole size tolerance depends on the input tolerance of the hole; that is, a prepared tolerance of 0.050 mm can be reduced to 0.025 mm, or a ±0.0025 mm tolerance can be held if the input tolerance is 0.010 mm. A surface finish of 0.25 μm is common after roller burnishing. A lightweight, low-viscosity lubricating oil is recommended for PM materials. A change of 0.019 to 0.025 mm in diameter is representative for a hole about 25 mm in diameter [343].

Ball sizing or ball burnishing of holes maintains open pores, provided that proper sintered dimensions are held. Normally with carbon material, the sizing stock should not exceed 0.05 mm. To hold close tolerance, it must have a least three sizes of balls or burnishing punches having a difference of 0.005 mm in diameter. This is to allow for spring back from the sintered size [152]. The outer diameter on thin wall tubing can be distorted; however, automatic ball burnishing may be used on heavy walled or assembled parts [345].

8.6 COOL-LUBRICATION IN STEEL MACHINING

8.6.1 Cool-lubrication methods and matters

Use of coolant and lubricant during cutting is common with conventional steels. The general tendency is dry cutting of metals because the environmentally correct disposal of the spent cool-lubricant is in most cases more expensive than the cutting tool and other costs. The total cost of the use of fluids, workpiece cleaning, and the required changes to the environment should be compared with the cost, *e.g.* of modern drilling tools, especially coated HSS and hardmetals. Whenever possible the use of coolant and cutting fluids should be avoided. From a tool-life point of view, it is usually better to machine dry. This is in particular true with high performance tool grades such as CBN and CVD coated tools. Coated hardmetal tools with lower friction properties may be used so that they have an impressively long tool life, even with low-speed machines, if special attention is given to control of vibrations [118].

The cutting fluids in the cutting process have two tasks:

- to bring down the temperature of the cutting edge,
- increase the lubrication in the chip/tool interface

and so increase the life of the cutting tool and in many cases increase also the surface finish.

The use of cooling fluids basically increases the life time of the tools compared with dry machining. Its is caused particularly by lower thermal stress of the tool. The removal of the chips is also much easier. The application of lubrication to the cutting tool causes a reduction in the deformation in turning, drilling, tapping and in other procedures. It is necessary to decide if dry machining at lower cutting speed is better practice than wet machining at higher cutting speed, if it is basically acceptable [139,225].

The use of an appropriate cooling system contributes to the reduction of the frictional heat generated in cutting from the tool by the evaporation of the water contained in the coolant. The oily constituents of the coolant lubricate the tool, thus reducing the friction between the tool and work. There is the system based on a jet of coolant and another system based on the use of a *spray of coolant* as shown in Fig.8.22. The mist of coolant produced at the cutting edge by the spray cooling system has a higher cooling effect than a coolant jet.

The application of cutting fluids or liquid coolants generally improves machining performance. The prevention of possible oxidation when using cutting fluids, which is a very rare case, must be taken into account. If the oxidation of the component cannot be

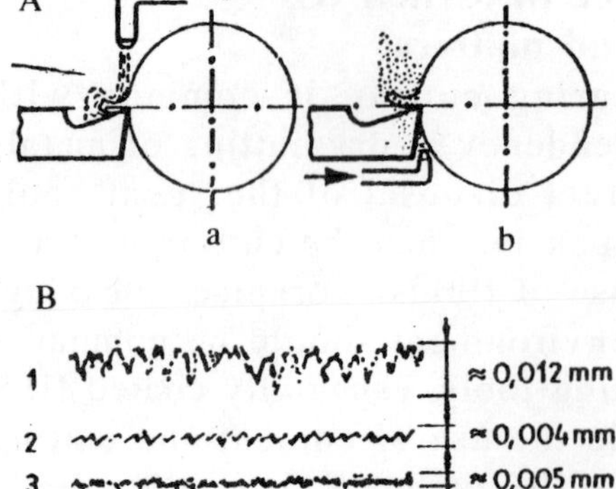

Fig.8.22 Cooling system – A: a – cooling system based on a jet of coolant, b – cooling system based on spray of coolant. B – microscopic representation of surfaces: 1 – turned surface, 2 – reamed surface, 3 – ground surface, 4 – lapped surface [106].

tolerated, dry cutting conditions are only possible which are also recommended, if not stated otherwise.

With some cutting materials, coolants cause thermal shock problems which result in cracking the tool. When using Al_2O_3 ceramics to turn steel, coolant should not be employed. Cermets are more sensitive to thermal shock than hardmetals. At rough turning of steel with cermets coolant is also not recommended. However, with very light finishing operations coolant can be beneficial to the surface finish of the workpiece. The cutting action of CBN is to raise the temperature of the workpiece to a point at which it softens and can then more easily be machined and so the use of coolant is for this reason not required [164,240].

Coolants are preferred in fast machining operations. Coolant pickup can be a problem. Ceramic and boron nitride (CBN) inserts are usually run dry; performance is typically better than or at least equal to that obtained with coolant. The coolant-lubricants used in machining are liquids and gases.

8.6.1.1 Liquid cool-lubricants

There are different cutting fluids available. Typically, cutting fluids are applied by flooding the workpiece with the liquid taking into account possible thermal shock of the tool, however, use of mist applicators is becoming more prevalent. Selection of a cutting fluid generally depends on the type of machining operation, the workpiece material being machined, and its final application, machining conditions, tool material, and environmental issue. In cases where air cooling is insufficient, liquid coolant containing an anti-corrosion agent is necessary [10,188,347].

Coolants can help to produce a good surface finish on the workpiece. This is particularly so with softer materials, *e.g.* low carbon or free machining steels. The most important effect of cutting fluids in the cutting zone is cooling. Its capacity depends on thermal conductivity, specific evaporation heat, and wettability of the cutting tool.

The lubrication effect decreases the wear on active faces of the cutting edge and by this decreases the cutting resistance and increases the finish of the machined surface.

The lubrication effect of water and of water solutions is low. The lubricating effect of oils is higher because they form a stronger film on the contact cutting surfaces. The formation of a film is increased by addition of some surface active agents to the fluid.

The main characteristics of effective cutting fluids are:

- they must not cause corrosion of the workpiece and of the tool,
- they must lubricate well and cool also at higher temperatures,
- they must be chemically and physically stable,
- they must have a small foaming effect and must be environmentally friendly.

The ageing of the cutting fluids used caused also by chemical reactions is a normal process at which the content of the impurities in them increases.

The cutting fluids are grouped as follows:

- water solution of electrolytes, *e.g.* sodium carbonate, sodium nitrate etc.,
- water soap solution (soft water), which have a good cooling effect and relatively good lubrication effect,
- oil emulsions which are a mix of water and of oil with addition of some emulsifying agent; they exhibit a good cooling and lubrication effect and are the most widely used cutting fluids,
- cutting oils, mineral and organic. Organic oils have good lubrication effect and by this contribute to the improvement in cutting properties. Their disadvantage is a poor chemical stability. The advantage of mineral oils is a great chemical stability but they do not form a sufficiently steady film on the tool. Therefore some surface active agents are added, *e.g.* MoS_2, MnS, S, graphite and others. By the use of emulsions or thin-bonded cutting oils the surface finish can be improved and the tool life can be increased [143,202].

For the selection of the appropriate cutting fluid the cutting method, properties of the workpiece material, the technological aspect (roughness, cutting forces etc.) and as in all cases the economy must be considered.

8.6.1.2 Gaseous cooling matter

The gas milieu has in some cases a similar effect on the durability of the cutting edge as a liquid. The gas more easily penetrates into the cutting zone than does a liquid.

It is supposed that the gases react (at elevated temperature) with the surface of the chip and of the tool and by this coatings on the tool are formed and, in dependence on the type of the gas, the intensity of the wearability of the tool is affected. The highest positive effect on WC–TiC hardmetal tool was attained with nitrogen compared to argon, air, CO_2 and oxygen, in that order. The more advantageous is the supply of the gas below the back surface of the tool. Increasing supply of the gas increases the durability of the cutting edge, which is also the result of a cooling effect [332].

8.6.1.3 Air cooling

Within the group of coolants-lubricants, compressed air serves as a cooling medium only. The use of other gases as cooling agent in PM machining probably was not tested.

Because of their porosity, sintered materials should be in general cooled with air. It is advantageous to use pressurized air of 0.15 to 0.2 MPa (1.5 to 2 bar), and by this as far as possible to prevent the penetration of liquids used into the pores. Compressed air is also used to cool the tool and maintain swarf clearances. Jets should be directed onto tool cutting edges and work surfaces. Air cooling is unavoidable when liquid coolants cannot be used because parts must be kept dry and clean. To avoid rusting, the components can be oiled after machining [344].

Also, the compressed air flowing through a discharge (ionised air, ozone) was tested in turning wrought steel and found to be effective in improvement of machinability as cooling and lubrication matter instead of cutting liquids or solely of plain air [354].

Effect of turning of wrought steel with an uncoated HM P20 insert under the jet of ionized air and without air jet on tool life is shown in Fig.8.23 and with an HSS 20 insert under equal conditions in Fig.8.24.

Use of the jet of ionized air in turning caused a decrease in flank wear of both cutting tools, relatively more of the HSS insert in spite of shorter cutting time but at higher cutting speed. It will be necessary to investigate the effect of highly oxidising ionized air in the cutting zone in contact with workpiece material and cutting tool material at elevated temperature. Very interesting and important would therefore be the analysis of the possible effect of ozone in

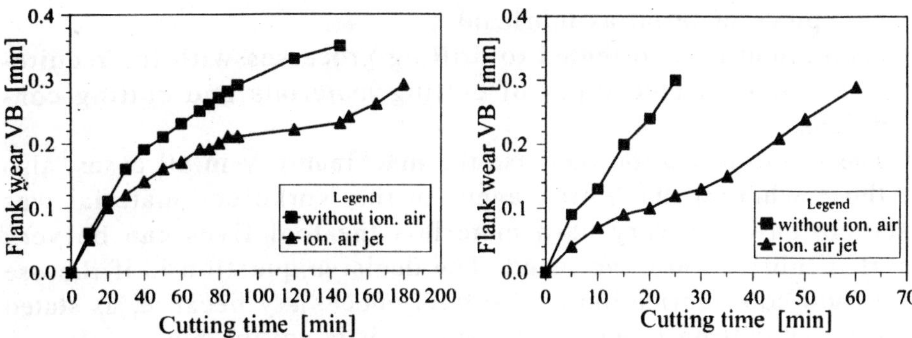

Fig.8.23 Flank wear of uncoated HM insert in turning of wrought steel (EN 10083-1; C50, 290 HB) in dependence on the cutting time (v_c = 12 m/min) without and with a jet of ionized air (average values from 5 measurements). (Data courtesy of K. Vasilko).

Fig.8.24 (right) Flank wear of uncoated HSS insert in turning of wrought steel (EN 10083-1; C50, 290 HB) in dependence on cutting time (v_c = 22 m/min) without and with a jet of ionized air (average values from 5 measurements). (Data courtesy of K. Vasilko).

machining of PM steels with addition of the machining aids.

8.6.2 Cool-lubrication in PM machining

The mentioned data concerning the use of cool-lubricants in machining of wrought steel are valid also in machining of powder metallurgy steels. In many cases, also in powder metallurgy machining, the use of a lubrication fluid is inevitable; it at least substantially facilitates the machining of a PM steel. In the case of PM machining, the cool matter takes out the heat from the cutting zone which is more important than in machining of wrought steel due to the lower thermal conductivity of the machined material. Selection of a cutting fluid generally depends on the type of machining operation, workpiece material, machining conditions, tool material, and environmental issues. If it is necessary to use a cutting fluid, a mist applicator may prove beneficial. The use of cooling matter has an especially positive effect on drilling of austenitic material [105].

The range of the use of some cool-lubricants in PM machining is limited. The main difference in use of the cool-lubricant in machining of sintered steels is due to porosity and microstructure compared to wrought steel and consists of:

- significantly needed lower removed material from a part (shorter cut time),
- addition of special free machining agents incorporated in the

workpiece material as lubricant,

- main attention is oriented to drilling processes with the requirement for restricted types of cutting materials and cutting conditions.

The fluid lubrication increase the machinability in all cases, also in those where a machining agent in the workpiece material was used. In cases of very hard materials the tool lives can be very short if lubrication is not used. The decisive question is if the use of a fluid cool-lubrication is absolutely necessary because, as stated above, dry cutting is most desirable, if it is acceptable.

The main disadvantage of the use of fluids in machining PM steels, in spite of their effect on machinability, is due to the porosity of the workpiece material. Porosity combined with the presence of water including water containing fluids can be detrimental to mechanical properties due to the resulting corrosion. Therefore, for machining sintered materials it is not possible to use common cooling and cool-lubricating fluids on water basis or fluids containing some acid. This is the reason why in most cases, PM materials are machined without using any cutting fluid [105]. Liquid coolants with inhibitors can be used in cases when the parts need not be kept dry and clean or when additional cleaning is acceptable. All cutting tools should be sharp when used a lubricant, and not only in this case.

The effect of a cutting fluid on the reduction of cutting forces is not uniform. For example, in one case when the use of a cutting fluid reduced the drilling force, in milling an oposite effect was observed [300]. Also, when using a cutting fluid in PM machining optimisation is needed with regard to the cutting methods, workpiece material and requirements on machined part properties.

Oil impregnation. The most simple and mostly used method is oil impregnation which invariably improves machinability. Internal lubrication on the cutting edge is possible with PM technology, when the material has been impregnated with oil. It is an advantage which can be offered only by PM compared to conventional steels. Porous PM materials can be readily oil impregnated to good effect. Oil impregnation of porous parts enhances surface quality, decreases the cutting force by 30 to 50% and increases the cutting rate up to 4 times. Due to corrosion protection, oil impregnated workpieces cut somewhat better than dry cutting. Oil impregnation is more effective with high strength materials. For example, with oil impregnation of Distaloy AE+0.5% C, machinability in dry cutting improved by a factor 1.5 and 2.2 at various feed rates. This can be explained by the reduction in cutting forces. The performance

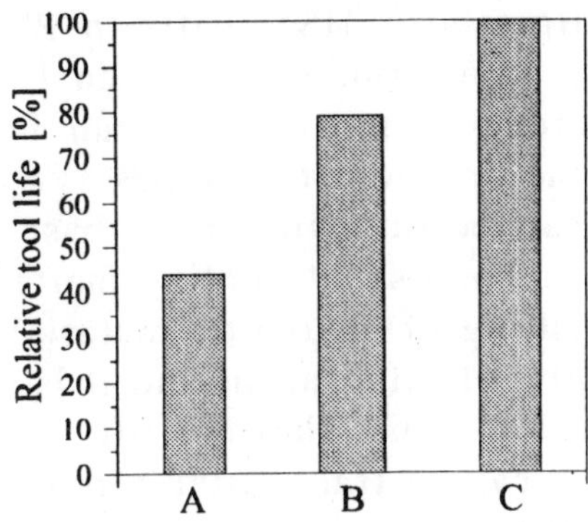

Fig.8.25 Cutting performance in terms of relative tool life for three conditions with Distaloy HP-1+0.5% C. A – dry cutting, feed 0.2 mm/rev., B – liquid coolant, feed 0.12 mm/rev., C – oil impregnation, feed 0.12 mm/rev. [118].

for oil impregnation compared to the cutting fluid of Distaloy HP-1+0.5% C is shown in Fig.8.25.

Oil impregnation enables internal lubrication on the rake face also throughout the turning operation. The effect of oil impregnation increases with feed rate [157]. Oil impregnation also increases the possibility of obtaining greater benefit from wear resistant coated CVD tools. Fundamentally, the lubrication on the cutting edge for high strength material contributes to its better performance.

The use of a mixture of 'lard' oil (5–10%) and of mineral oil or of chloroethylene (the last one is not acceptable from the environmental viewpoint) gave positive results in PM machining. It was used especially for filling pores in a sintered material, and therefore as was stated it can be excellent for light cutting in soft ferrous materials. A mixtures of 60% lard in a special cutting oil can be good also for hard ferrous materials. If parts are to be heat treated after machining, the oil is by this baked out [345].

PM Drilling. Large amounts of coolant are required in drilling medium- or low-density materials; the coolant should operate effectively at the drilling point to reduce abrasive wear caused by powder particles at the bottom of the hole (see Fig.7.79) (this is not the case in drilling wrought steel). A single-nozzle coolant system does not work properly because small or powdered chips do not easily exit through the drill flutes. A ring design system, however, is effective in eliminating the chip-clogging problem. Oil hole drills are the most effective means for removing chips from the cutting zone [129,343]. If a coolant can be used for hole drilling in the PM parts, then HM drills with integral 'oil-holes' to allow coolant access to the cutting edges are effective in improved drilling (Fig7.22) [190].

The effect of cutting fluids is strongly dependent on the cutting speed. Using cutting fluid when drilling with HSS-drills in PM materials increases tool life by factors approximately between 1.1 and 1.5 (10 and 50%). Large differences in tool life can be expected if an evaluation is made at the different tool types.

The drilling of stainless steels is carried out with water-based mineral oil containing a cool-lubricant. The use of cooling matter has an especially positive effect on drilling of austenitic material. The concentrate available contains ~40% of mineral oil and it is a 6% emulsion, or 2–4% emulsion in water. Thus, the tool life was markedly increased. The cooling matter can be transported through the drill helix into the cutting zone [225].

Treatments after machining. If liquid coolants have been used, it is essential to wash the components in hydrocarbons with a relatively high boiling point also from chips. An ultrasonic unit is highly efficient for washing. The components are subsequently dried above 100°C in order to remove all the liquids in the pores. The remaining liquid may cause rust or corrosion of the components. Therefore, the finished parts must be washed in oil and rinsed from the chips [340].

Liquid chlorinated hydrocarbons are not recommended for the cleaning of sintered components. On the other hand gaseous chlorinated hydrocarbons can be used [105,345].

Components containing oil for self-lubricating properties are re-impregnated with oil after machining. The best results are obtained by vacuum impregnation for 10 minutes at 70°C. This is in most cases not possible and the impregnation is done by more simple methods under ambient temperature, *i.e.* impregnation in an oil bath with a temperature of 70–80°C. The components should remain in the oil until bubbling action has ceased. This is usually achieved after 2–5 h, depending upon the cross-section of the component. To avoid losses in oil due to thermal contraction effects, the components should remain in the oil until the temperature was reduced back to room temperature.

8.7 PARAMETERS FOR OPTIMISING THE MACHINING OF PM PARTS

The special problems in machinability of PM steel parts compared to wrought steel parts are caused by the properties of the former. The complex property profile of the PM parts properties precludes attaining reasonable comparison with wrought steels of similar mechanical properties, especially hardness, which in frequently used in wrought steels as a comparative property to machinability.

Basically, the properties of PM steel parts produced through rather straightforward techniques, such as, for example, solid state sintering are affected by significantly more processing and material factors compared to wrought steel.

There are different manufacturing methods for base iron powders (especially reduction and atomisation) with corresponding characteristic properties of the powders. Different alloying method are used for the sintered steels (mixed, prealloyed, hybrid), and usually much higher amounts of alloy elements are added compared to wrought steels to achieve equal mechanical properties, *e.g.* solely copper contents of 2–8 (10.5)%, nickel 2–3%, phosphorus 0.6% and Ni, Cu and Mo in combinations totally up to 10% in a steel are used. In the last time, the carbide forming elements chromium and molybdenum (commonly as prealloyed powders) and more slowly also manganese accomplish the main group of the alloying elements.

The results of such alloying are in part heterogeneous microstructures, which do not have any equivalent in wrought steels. During cooling, each of the alloying elements used affects the transformation of austenite to different degrees and also the changes in the phases occurring during annealing when this is done for improving the machinability.

Furthermore, the porosity of the sintered parts to be machined is in the range of 20 down to 'zero' percentage (for powder forged parts). This significantly enlarges the complexity of machinability of PM steels.

Stainless steels containing porosity form a special group in terms of machinability as in the case with their ingot metallurgy counterparts, due to high alloying with nickel and chromium and higher porosity compared also to common sintered structural steels.

The data and the knowledge about machinability of PM steel parts attained in research and practice up to this time are very important since they involve the whole area of structural porous up to full density parts. They enable production of PM parts with very complicated shape. In spite of this, improving the machinability is a permanent research process due to existing problems in machining, and due to consistently increasing requirements towards quality and shape complexity of the parts as well as productivity and also due to development and application of new material systems in industrial production.

On the basis of all data about the machinability of PM steels that are presented in this book, it is attempted to show some for further investigations.

8.7.1 Material characteristics

Iron powder grade

It was recorded in all investigations that the base iron powder grades according to the manufacturing method only (sponge and atomised) differently affect the machinability of as-sintered materials. Furthermore, the effect of the iron powder grade was also determined in relation to the cutting method, *i.e.* if in drilling better machinability was attained with atomised iron powder, in turning it was with sponge iron powder. As an example, Distaloy SA (sponge) was more easily turned than Distaloy AB (atomised) but the opposite ranking was found in drilling (chapter 7.4.1.3). This effect was also observed at different cutting speeds and feeds (lower or higher). So far, it cannot be sufficiently explained by the known basic physical, chemical and technological properties of these powders how these characteristics affect the machining process in the sintered state.

Recommendation: The investigation of base iron powder grades and identification of those properties which adversely affect the machinability of sintered materials. The knowledge will facilitate the selection of the 'better', *i.e.* more suitable, iron powder grade for the production of alloy steel parts designed for machining by the required method.

Carbon (chapter 5.2.2.1)

The carbon content has a dominant effect on the machinability of wrought and also of PM steels. Plain carbon steels (ferrite – pearlite) nearly always have better machinability than alloy steels of comparable carbon content due to other microstructural type and usually higher hardness. Low carbon steels containing less than 0.15% C (few data about this C-content are available in PM steel machining) machine poorly because they are soft and ductile and adhere to the cutting tools (formation of built-up edges). PM steels in the 0.15 to 0.60% C range are machined satisfactorily. The steels with higher carbon content than 0.6% are poorly machinable up to non machinable in the sintered state.

These carbon contents show the ranges for an advantageous proportion of soft ferrite and hard pearlite with regard to the machinability in the material. In all cases higher carbon content (also fully pearlitic steel) is not recommended in terms of machinability. This is valid also for alloyed steel in which the formation of special carbides can also take place.

Recommendation: If carbon-free sintered iron is not essential for the function of a part, an addition of 0.2–0.3% C, tolerable according to the standards, *e.g.* SINT C 00, F-0000, can be recommended for improving the machinability.

Copper (chapter 5.2.2.2)
Copper addition, in addition to its strengthening effect, contributes to improvement of the machinability of PM steels not only in Fe–Cu–(0–0.6)% C materials (Fig.7.22, 7.23) but also in Distaloy type steels (machinability enhancer) (chapter 7.4.3.1). Especially in high copper (~10%) steel the improvement of machinability is marked (due to the presence of free copper). The improved machinability of sintered steels by copper is frequently explained by the postulated higher thermal conductivity of these materials, also of materials with dissolved copper, and from this lower temperatures in the cutting zone are deduced. However, adequate thermal conductivity measurements for Fe–Cu and Fe–Cu–C systems are lacking.

Recommendation: Firstly, the measurement of thermal conductivity of steels containing dissolved copper compared with plain iron with the aim to confirm that this is in fact that relevant property improving the machinability of the steels in the case of copper alloying. Secondly, the investigation of the microstructure characteristics of the Cu alloy steels in relation to the machinability through its friction – seizure behaviour. It could be a contribution to the knowledge of the copper enhancing effect in machining of sintered steels. (The authors suspect that thermal conductivity plays a minor role compared to the change in microstructure caused by Cu alloying).

Manganese (chapter 5.2.2.3)
Manganese, beside its role mainly in the formation of manganese sulphide, is also used without increased sulphur content in amounts of ~0.5% for improving the machinability of Fe–C and Fe–Cu–C steels (Figs.7.2 and 7.20). In such a case, machinability could be improved only through the microstructure being affected by manganese, in particular the special Mn distribution in the steel matrix (through the unique solid phase-gas phase sintering).

Recommendation: Detailed investigation of the microstructure of a material containing 0.5% Mn for improving the machinability (strengthened ferrite or a heterogeneous microstructure – ferrite, small proportion of bainite, also as a function of varying Mn distribution) could result in more general rules for improving the machinability of PM steels.

Nickel

Nickel alloyed steels of roughly the same mechanical properties compared to other steels, *e.g.* to copper alloy steels, exhibit poorer machinability. The same effect of nickel is found also in wrought steels. In PM steels this poor machinability also originates from considerably lower diffusivity of nickel in iron compared to other alloy elements, resulting in heterogeneous Ni distribution, and from the fact that the nickel martensite is only slightly harder than ferrite and thus also enhances adhesive wear which adversely affects the interaction of the chip and the cutting tool edge in the machining process (Fig.3.29). It must be noted that nickel as alloying element is mostly used in combination with copper, molybdenum and carbon, and only to a minor extent as Fe–Ni alloy (except for magnetic applications, in which case very high Ni contents and homogeneous microstructure are necessary). The chemical and microstructural heterogeneity of most Ni-alloy sintered steels significantly decreases their machinability. Also a soft annealing process (at rather low temperature) cannot generate a more uniform microstructure in a mixed or diffusion alloyed Ni-containing PM steel.

The deteriorating effect of nickel on the machinability of a steel is clearly shown, *e.g.* for Fe–P steel (Fig.7.60) compared to copper addition (Fig.7.59), or for Fe–0.85Mo steel (Tab.7.26), and stainless steels in general.

On the other hand, the prealloyed powder forged steel Fe–2Ni–0.5Mo–0.2Mn–C (base powder Astaloy A) was easily machined, due to its homogeneous microstructure, which indicates that at least for PM Ni-alloyed steels the microstructural heterogeneity is a greater obstacle for machining.

In connection with nickel, it must be noted that all alloying elements that increase the hardenability of the steel decrease the machinability; ferrite-strengthening elements, such as nickel and silicon, decrease the machinability more than equivalent amounts of carbide-forming elements, such as chromium and molybdenum.

Remark: In the case of nickel, with regard to machinability (without machining aid) a more homogeneous distribution of nickel can improve the machinability of a steel of the corresponding mechanical properties.

8.7.2 Machining aids

The addition of a machining aid is the simplest way for improving the machinability of a sintered material. The large number of machining aids available and of the methods for improving the machinability of PM steels (25 are presented in this book) generate a problem of choice in powder metallurgy. Each of them has advantages and disadvantages in dependence on material composition, processing, and cutting conditions. The mostly widely used aid is MnS in low purity and high purity quality (Tab.7.32) and the most recent development, the coated MnS grade, shows the way how to increase the beneficial effect of a machining aid.

The development of the 'new' machining aids should be oriented towards the stability of the lubrication effect in the cutting zone at elevated temperature, mainly in machining of high strength/hardness steels at higher cutting speeds. This is the area in which an appropriate machining aid is definitely lacking. The schemes proposed about the effect of a machining aid on the cutting tool edge are only illustrative, but are not supported by a detailed analysis of the chip and the cutting tool edge after the cut under the mentioned conditions.

The manufacturing of sulphurised atomised iron and prealloyed steel powders with a small addition of sulphur (~0.05%) is the simplest effective method for improving the machinability of PM steels.

In Ref. 355 it was also stated that the sulphide size and shape are important in free-machining steels and should be controlled. Such inclusions have a core-and-shell structure. The core consists of oxides ($CaO–Al_2O_3$) and the shell of sulphides (Ca, Mn)S. This improved the breakability of chips and reduced the size of sulphide inclusions. The change in the shape and size of calcium-containing shape-controlled sulphide inclusions was achieved by adding Ti at a rate of 0.003% max (max. 0.011%). The life of tools produced from this steel is four times longer than of tools made from lead-containing free-machining steel. These results may also be interesting for the machining of prealloyed PM steels.

Remark: The development of the machining aids should be concentrated on high strength/hardness steels with heterogeneous microstructure on the basis of an investigation of the real interaction in the cutting zone between the chip (considering the deformation cutting theory of the effect of the porosity) and the cutting tool edge. Basically, the evaluation of the effects of free machining agents and premix composition should begin at an early stage in part development.

8.7.3 Microstructure and heat treatment

The microstructure of a material, determining the mechanical and physical properties, is also responsible for machinability. All phases that are harder than ferrite solely deteriorate the machinability of a material which means that finally a compromise has to be found between the mechanical properties and dimensional stability required of a material and, on the other hand, the productivity in machining.

An appropriate annealing process can affect the microstructure of hard materials. The annealing processes used to improve the machinability of materials with the as-sintered heterogeneous microstructure were not sufficiently successful. This is caused by different transformation rates of the individual constituents in heterogeneous alloyed materials. At low cooling rates, the distribution of carbon in the Ni-rich area has also to be considered. The small changes of the Ni concentration which occurred during soft annealing, carried out around Ac_1 temperature (Fig.7.100), cannot sufficiently contribute to improving the machinability of such a steel (it must also be considered that heterogenous microstructures have benefits regarding mechanical properties; simply homogenizing the structure for better machinability would not be the solution).

For chemically homogeneous materials, the best microstructures for improved machinability are spheroidized ones with fine globular pearlite, which can be attained, if the starting microstructure is formed by martensite. Generally speaking, annealing treatments may be helpful in some cases but they cannot be regarded as the general solution to improve machining (also with regard to the effect on the mechanical properties).

The isothermal and continuous cooling transformation curves (TTT and CCT) for PM steels have to be fully established in order to define efficient annealing, sinter hardening, and other heat treatment processes. So far there are no detailed microstructural investigations showing the relationships between the heat treatments performed according to the TTT/CCT diagrams and changes of the (heterogeneous) microstructural constituents towards better machinability. Experience with the application of such curves, if existing, towards improved machinability of a given material are lacking.

A special case are steels hardened and tempered up to hardness levels > 300 HB. Under such conditions, alloy steels have superior machinability, exhibiting a rather regular microstructure of tempered martensite. This was also confirmed for high strength PM steel

(Tab.7.50).

Recommendation: To characterize the individual microstructural constituents regarding their respective effect of machinability with the aim to predict the optimum microstructure for machinability. This will also make it possible to define the cooling conditions of the parts from the sintering temperature (regarding the size, shape and mass of the parts in relation to the flowing atmosphere) to attain the 'desired' optimum microstructure for machinability which is often not sufficiently characterised, and also after annealing processes applied.

To establish the mentioned transformation curves accompanied by microanalysis of the final microstructure in relation to the machinability of a material. *(The best annealing process for a PM steel is that which need not be performed).*

8.7.4 Testing

The test methods used for drilling and turning, as presented in the previous chapters, represent the most widely applied cutting methods in PM machining. With regard to the different machinability test methods used it is very difficult to compare and generalize the results obtained.

Recommendation (chapter 3.4.2):

a – A short-time drilling test for a material, at least in the case of poor to moderate machinability – drilling with constant thrust force up to drill failure; the criterion – number of holes drilled or length of holes drilled. An additional criterion for this test method is the feed and the time to drill the first hole. The poorer the machinability of a material, the lower the feed, and the longer the time to drill the first hole. This criterion can also be applied for such materials for which the time to tool failure would be inconveniently long. This procedure can be used for a fast comparative testing of different materials or testing of HSS and HM drills with different geometry.

b – The turning tests – short time testing in face turning. At this test the turning (with constant workpiece revolutions) is done outward, from the surface of the center hole (r_0 radius), to the circumference of the ring shaped specimen, *e.g.* rollers, at constant feed and depth of the cut. Testing of different materials with one single type of cutting tool or testing of various types of cutting tools in machining of one material. The criterion – the length of the cut and/or amount of the removed material at the highest cutting speed.

c – The friction-seizure tests of the cutting pair, simulating the cutting process, and the microanalysis of the new surfaces formed will contribute to explaining the action of the machining aids on the tool wear. (A relationship was shown between the dry wear and the machinability of some structural steels, *i.e.* low wear – poor machinability, Fig.7.90).

8.7.6 Tools

In drilling only uncoated HSS, coated HSS, and hardmetal drills are used, in some cases with modified geometry. In turning there is a larger assortment of cutting tools, also for the most severe cutting conditions.

There are no accurate data indicating that in drilling of PM steels the PM HSS steel drills attained longer tool life compared to conventional (ingot metallurgy) HSS drills. The reason for this might be the larger size and angular shape of the carbides in a conventionally manufactured high speed steel as stated in the literature and by some PM part producers without detailed analysis. The carbides in PM HSS drills are possibly too fine and too rounded for drilling (although the PM HSS drills should be less prone to fracture). On the other hand, in turning PM HSS inserts exhibited significantly longer tool life compared to ingot metallurgy HSS inserts (Fig.4.10). There are no representative results comparing the drilling under equal cutting conditions of a PM steel with ingot metallurgy HSS drills and with PM HSS of the determined type, composition, and geometry. It is possible that the unsuffcient characterization of the drills in term of material and geometry is the reason for the scatter in some results attained in machining of PM steels.

Recommendation:

Drilling tests of some selected PM steels with conventional and PM high speed steel drills of identical (or at least similar) composition in relation to the differences in carbide size and shape. The knowledge will confirm/contradict the statement that conventional HSS drills are advantageous for drilling PM steels. The general statement that IM HSS drills are better than PM HSS drills without giving the specification is not sufficient. The results attained could then be a basis for a focused production of conventional or PM high speed drills for optimal drilling of sintered steels. The powder metallurgy route manufactures more types of high speed steels and, if necessary, enables increasing the alloy content in HSS and the addition of some other carbides of higher hardness and larger size (Fig.4.11), and by this should improve the drilling performance mainly in machining of high strength PM steels. This recommends a closer collaboration between the PM part producers (research in machining) and PM HSS manufacturer. *(The final goal could be: Manufacturing of powder metallurgy cutting material – for machining of powder metallurgy structural steels).*

However, it can also be expected that here the advantages of coatings can be brought to bear, both in the case when hard, abrasive sintered steels have to be machined and also in the case of softer workpiece materials when adhesive wear prevails; in the latter case the lower coefficient of friction obtained with coated tools should be a major advantage.

Cutting tool geometry plays an important role in machining and therefore its effect should be more closely investigated in the cutting process in relation to the specific PM workipece material and cutting condition.

8.7.6 Best results

The results listed in Tab.7.44 which exceed those usually attained in drilling of PM sintered structural steels show the unused potential in knowledge and practice of the drilling process of sintered steels. These results prove that the economically effective machining of sintered high strength steels, *e.g.* of FN-0205, and of Distaloy SE-0.5C steel is also possible without special measures (only 0.5% MnS), *i.e.* that PM steels can also exhibit superior machinability without additional special measures.

Recommendation: These results should be the basis for a more detailed investigation of the interaction between the workpiece material (with a machining aid), the chip, and the tool, and of the effect of cutting conditions to ensure their general reproducibility and by this to achieve a significant increase in the technical and economical effectiveness of PM machining.

9 Appendix

9.1 RELATIONSHIP BETWEEN HARDNESS VALUES DETERMINED BY VICKERS AND ROCKWELL METHODS

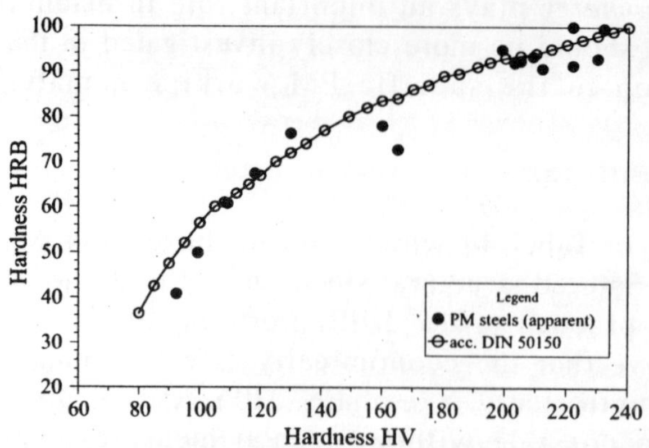

Fig.9.1 Relationship between HRB and HV hardness in wrought steels (according to standard DIN 50 150) and PM steels. (Apparent hardness HRB and HV values of PM steels of various densities and compositions as measured by the authors).

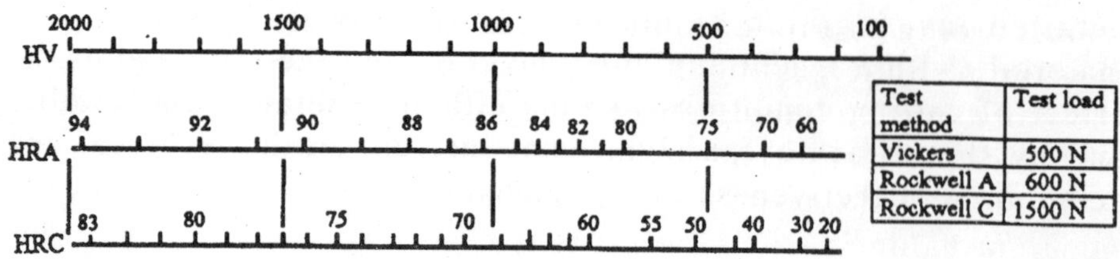

Fig.9.2 Relationship between HV, HRA and HRC hardness values for hard PM steels [356].

9.2 CHEMICAL COMPOSITION AND DESIGNATION OF POWDER METALLURGY STEELS

In the following the chemical composition and designation of powder metallurgy steels are listed according to MPIF Standard 35, 2000 Edition [296]. (In brackets the designations after DIN 30 910 are given where appropriate; 'X' represents the letter indicating the density, *i.e.* C, D, E).

Tab.9.1 Iron and Carbon Steel – Chemical Composition [mass %]

Material Designation	Fe	C	Element
F-0000	97.7	0.0	min.
(Sint-X 00)	100.0	0.3	max.
F-0005	97.4	0.3	min.
(Sint-X 01)	99.7	0.6	max.
F-0008	97.1	0.6	min.
	99.4	0.9	max.

Application: Unalloyed PM iron (F-0000) materials are typically used for lightly loaded structural applications and also for structural parts requiring self-lubrication when strength is not critical. At high densities, unalloyed iron is used for soft magnetic applications.

PM carbon (F-0005) materials are used primarily where moderate strength and hardness combined with machinability (drilling, tapping, lathe turning, milling, etc.) are desired.

PM steels with higher carbon content (F-0008) are used when loading is moderate. F-0008 is more difficulty to machine than F-0005. F-0008 and F-0005 materials may be heat treated to enhance strength and wear resistance. They may also be steam treated for improved shelf life, pore closure and to increase hardness.

All of the iron and carbon steel materials with densities of 7.0 g/cm^3 or less may be oil impregnated when self-lubricating properties are required.

Tab.9.2 Iron–Copper and Copper–Carbon Steel – Chemical Composition [mass %]

Material Designation	Fe	Cu	C	Element
FC-0200	93.8	1.5	0.0	min.
(Sint-X 10)	98.5	3.9	0.3	max.
FC-0205	93.5	1.5	0.3	min.
(Sint-X 11)	98.2	3.9	0.6	max.
FC-0208	93.2	1.5	0.6	min.
(Sint-X 11)	97.9	3.9	0.9	max.
FC-0505	91.4	4.9	0.3	min.
(Sint-X 21)	95.7	6.0	0.6	max.
FC-0508	91.1	4.0	0.6	min.
	95.4	6.0	0.9	max.
FC-0808	88.1	7.0	0.6	min.
	92.4	9.0	0.9	max.
FC-1000	87.2	9.5	0.0	min.
	90.5	10.5	0.3	max.

Application. PM iron-copper and copper-carbon steel materials find wide usage in medium strength structural applications. Copper contents of 2% are typical. When secondary machining is required, combined carbon contents of less than 0.5% should be specified. Materials in this category also can be heat treated to increase strength and wear resistance. Higher copper content materials (in the range of 5%) are recommended when maximum wear resistance is required and when heat treating is practical. Low density parts can be oil-impregnated for self-lubrication in use.

(At sintering the copper dissolves in the iron but does not penetrate to the center of the iron particles).

Tab.9.3 Iron-Nickel and Nickel Steel – Chemical Composition [mass %]

Material Designation	Fe	Ni	C	Cu	Element
FN-0200	92.2	1.0	0.0	0.0	min.
	99.0	3.0	0.3	2.5	max.
FN-0205	91.9	1.0	0.3	0.0	min.
	98.7	3.0	0.6	2.5	max.
FN-0208	91.6	1.0	0.6	0.0	min
	98.4	3.0	0.9	2.5	max.
FN-0405	89.9	3.9	0.3	0.0	min.
	96.7	5.5	0.6	2.0	max.
FN-0408	89.6	3.0	0.6	0.0	min.
	96.4	5.5	0.9	2.0	max.

This subsection covers PM materials manufactured from mixtures of elemental iron powder, elemental nickel powder, and graphite powder (carbon), if required. Nickel additions are typically in the range of 1 to 4%. If no carbon is present the product is designated PM iron-nickel.

Application. PM nickel steels are used typically for heat treatable structural parts requiring the combination of strength, wear resistance, and impact properties.

Microstructure. As-sintered nickel steels show lightly coloured austenitic nickel-rich islands with needles of martensite or bainite around their edges. In the heat treated condition, the nickel-rich islands are light coloured, austenitic at their center and with martensitic needles at the peripheries visible at high magnification.

Tab.9.4 Low Alloy Steel – Chemical Composition [mass %]

Material Designation	Fe	C	Ni	Mo	Element
FL-4205	95.90	0.4	0.35	0.50	min.
	98.75	0.7	0.55	0.85	max.
FL-4405	96.35	0.4	-	0.75	min.
	98.85	0.7	-	0.95	max.
FL-4605	94.20	0.4	1.70	0.40	min.
	97.50	0.7	2.00	1.10	max.
FLN-4205	93.95	0.4	1.35*	0.49	min.
	97.76	0.7	2.50*	0.85	max.
FLN2-4405	93.35	0.4	1.00	0.65	min.
	97.76	0.7	3.00	0.95	max.
FLN4-4405	91.35	0.4	3.0	0.65	min.
	95.95	0.7	5.00	0.95	max.
FLN6-4405	89.35	0.4	5.00	0.65	min.
	93.35	0.7	7.00	0.95	max.
FLNC-4405	90.35	0.4	1.00	0.65	min.
	96.95	0.7	3.00	0.95	max.

This subsection covers PM materials manufactured from prealloyed low alloy steel powders using nickel and molybdenum as the major alloying elements. Varying amounts of elemental powders may be also admixed.

These materials provide higher hardenability than is possible with admixed copper or nickel steels.

Application. Low alloy PM steel are used typically where high performance materials capable of being heat treated are required. These alloys will provide high strength and wear resistance after heat treatment.

Microstructure. In the as-sintered eutectoid product the carbide plates are coarser and spaced in such a way that the proeutectoid ferrrite is not clearly defined compared to plain iron-carbon system. Low alloy steel may contain up to 5% unalloyed iron.

*At least 1% of nickel is admixed as elemental powder. FLNC-4405 with 1.0 to 3.0% Cu.

Tab.9.5 Sinter Hardened Steel – Chemical composition [mass %]

Material Designation	Fe	C	Ni	Mo	Cu	Element
FLN2-4408	93.15	0.6	1.00	0.65	-	min.
	97.75	0.9	3.00	0.95	-	max.
FLN4-4408	01.15	0.6	3.00	0.65	-	min.
	95.75	0.9	5.00	0.95	-	max.
FLN6-4408	89.15	0.6	5.00	0.65	-	min.
	93.75	0.9	7.00	0.95	-	max.
FLN-4608	91.00	0.6	3.6*	0.39	-	min.
	95.41	0.9	5.0*	1.10	-	max.
FLC-4608	91.00	0.6	1.60	0.39	1.0	min.
	96.41	0.9	2.00	1.10	3.0	max.
FLC-4908	92.40	0.6	-	1.30	1.0	min.
	97.10	0.9	-	1.70	3.0	max.
FLNC-4408	90.15	0.6	1.00	0.65	1.0	min.
	96.75	0.9	3.00	0.95	3.0	max.

This subsection covers PM materials manufactured from prealloyed low alloy steel powders and elemental copper, and in some cases elemental nickel. Sinter hardenable materials are used in medium and high density applications. These materials provide sufficient hardenability to enable hardening during the cooling cycle following sintering.

Application. Sinter hardened PM steels are typically used where high strength and wear resistance are required.

Microstructure. The sinter hardened steels exhibit a bainite-martensite microstructure. Nickel-rich areas may be found in alloys containing admixed nickel.

*At least 2% of nickel is admixed as elemental powder

Tab.9.6 Diffusion Alloyed Steel – Chemical Composition [mass %]

Material Designation	Fe	C	Ni	Cu	Mo	Element
FD-0200	93.45	0.0	1.55	1.3	0.4	min.
(Sint-X 30)	96.75	0.3	1.95	1.7	0.6	max.
FD-0205	93.15	0.3	1.55	1.3	0.4	min.
(Sint-X 39)	96.45	0.6	1.95	1.7	0.6	max.
FD-0208	92.85	0.6	1.55	1.3	0.4	min.
	96.15	0.9	1.95	1.7	0.6	max.
FD-0405	90.70	0.3	3.60	1.3	0.4	min.
(Sint-X 39)	94.40	0.6	4.40	1.7	0.6	max.
FD-0408	90.40	0.8	3.60	1.3	0.4	min.
	94.10	0.9	4.40	1.7	0.6	max.

This subsection covers PM materials manufactured from diffusion alloyed steel powders. Varying amounts of elemental powders may be admixed also.
Application. Diffusion alloyed steels typically are used in medium to high density and medium to high strength structural products. These materials may be heat treated to increase strength and wear resistance.
Microstructure. These materials exhibit a heterogeneous microstructure. As-sintered diffusion alloyed steels show a micro-structure similar to the nickel steels with a higher proportion of bainite and martensite. After heat treatment these structures are similar to those in heat treated nickel steels.

Tab.9.7 Stainless Steel – 300 Series Alloy – Chemical Composition. (Mn – 0.0-2.0, Si – 0.0-1.0, Fe – balance) [mass %]

Material Designation	Cr	Ni	S	C	P	Mo	N	Element
SS-303N1, N2	17.0	8.0	0.15	0.00	0.000	-	0.20	min.
	19.0	13.0	0.30	0.15	0.200	-	0.60	max.
SS-303L	17.0	8.0	0.15	0.00	0.000	-	0.00	min.
	19.0	13.0	0.30	0.03	0.200	-	0.03	max.
SS-304N1, N2	18.0	8.0	0.00	0.00	0.000	-	0.20	min.
	20.0	12.0	0.03	0.08	0.045	-	0.60	max.
SS-304L	18.0	8.0	0.00	0.00	0.000	-	0.00	min.
	20.0	12.0	0.03	0.03	0.045	-	0.03	max.
SS-316N1, N2	16.0	10.0	0.00	0.00	0.000	2.0	0.20	min.
(Sint-X.40)	18.0	14.0	0.03	0.08	0.045	3.0	0.60	max.
SS-316L	16.0	10.0	0.00	0.00	0.000	2.0	0.00	min.
(Sint-X 40)	18.0	14.0	0.03	0.03	0.045	3.0	0.03	max.

This subsection covers PM materials manufactured from prealloyed austenitic stainless steel powders. With the use of prealloyed powder, a completely homogeneous steel microstructure is attained if following normal commercial sintering cycles.
Application. Stainless steel PM materials are generally selected because of their good mechanical properties, corrosion resistance, and surface quality:
SS-303 Austenitic Machining Grades – preferred for parts requiring extensive secondary machining. Strength and hardness and corrosion resistance are good; non-magnetic; SS-304 General Purpose Austenitic Grades – used in many applications; non-magnetic; SS-316 General Purpose Austenitic Grades – the best combination of properties in PM stainless alloy; first choice for general purpose applications; non-magnetic.

Tab.9.8 Stainless Steel – 400 Series Alloy – Chemical Composition. (S – 0.00–0.03, P – 0.00–0.04, Fe – balance) [mass %]

Material Designation	Cr	Ni	Mn	Si	C	Mo	N	Nb (Cb)	Element
SS-409L	10.50	-	0.0	0.0	0.00	-	0.00	8 x% C	min.
	11.75	-	1.0	1.0	0.03	-	0.03	0.80	max.
SS-409LE[(1)]	11.50	0.0	0.0	0.0	0.00	-	0.00	8 x% C	min.
	13.50	0.5	1.0	1.0	0.03	-	0.03	0.80	max.
SS-410	11.50	-	0.0	0.0	0.00	-	0.20	-	min.
	13.50	-	1.0	1.0	0.25	-	0.60	-	max.
SS-410L	11.50	-	0.0	0.0	0.00	-	0.00	-	min.
	13.50	-	1.0	1.0	0.03	-	0.03	-	max.
SS-430N2	16.00	-	0.0	0.0	0.00	-	0.20	-	min.
(Sint-X 42)	18.00	-	1.0	1.0	0.08	-	0.60	-	max.
SS-430L	16.00	-	0.0	0.0	0.00	-	0.00	-	min.
(Sint-X 42)	18.00	-	1.0	1.0	0.03	-	0.03	-	max.
SS-434N2	16.00	-	0.0	0.0	0.00	0.75	0.20	-	min.
	18.00	-	1.0	1.0	0.08	1.25	0.60	-	max.
SS-434L	16.00	-	0.0	0.0	0.00	0.75	0.00	-	min.
	18.00	-	1.0	1.0	0.03	1.25	0.03	-	max.

This subsection covers PM materials manufactured from prealloyed ferritic or martensitic stainless powders. When required, carbon content is controlled by admixed graphite powder additions. The corrosion properties are inferior to the 300 series stainless steel alloys when processed under the same conditions.

Application. The 400 series stainless steels normally are selected with regard to their magnetic properties or heat treat response: SS-409L – selected for welded applications; SS-410L – selected with regard to soft magnetic properties and for environments requiring the least corrosion resistance; SS-410-90HT – selected for the combination of strength, hardness and wear resistance; both corrosion resistance and machinability are poor; SS-430 – lower magnetic properties as compared with SS-410 but somewhat better corrosion resistance; SS-434 – similar magnetic properties to the SS-430- grade with slightly better corrosion resistance.

9.3 TRADE DESIGNATION AND BASE CHARACTERISTICS OF PLAIN IRON AND ALLOY POWDER PRODUCTS (FOR PM PARTS PRODUCTION) SUPPLIED BY HÖGANÄS AB, HOEGANAES CORP. AND QMP

9.3.1 HÖGANÄS AB [11]

Sponge Iron Powder Grades					
	Composition [mass %]		Apparent density [g/cm^3]	Compressibility at 600 MPa [g/cm^3]	*Green strength [MPa]
	C	H_2 loss			
NC100.24	<0.01	0.21	2.45	7.0	18
SC100.26	<0.01	0.12	2.65	7.1	13
MH100.23	0.08	0.35	2.30	6.75	24

*after compaction at 500 MPa

Atomised Iron Powder Grades					
	Composition [mass %]		Apparent density [g/cm^3]	Compressibility at 600 MPa [g/cm^3]	*Green strength [MPa]
	C	H_2 loss			
AHC100.29	<0.01	0.1	2.95	7.15	15
ASC100.29	<0.01	0.08	2.95	7.2	12
ABC100.30	0.002	0.05	3.00	7.25	10

*after compaction at 600 MPa

Distaloy Powders								
	Composition [mass %]					Apparent density [g/cm^3]	Compressibility at 600 MPa [g/cm^3]	***Green strength [MPa]
	C	H_2 loss	Cu	Ni	Mo			
Distaloy SA	<0.01	0.1	1.50	1.75	0.50	2.8	7.05	13*
Distaloy AB	<0.01	0.1	1.50	1.75	0.50	3.05	7.15	12**
Distaloy SE	<0.01	0.1	1.50	4.00	0.50	2.8	7.05	12*
Distaloy AE	<0.01	0.1	1.50	4.00	0.50	3.05	7.15	20**K, 35**LD
Distaloy AF	<0.01	0.15	2.00	5.00	1.00	3.05	7.14	22**K, 35**LD
Distaloy AE	<0.01	0.1	1.50	4.00	0.50	3.05	7.15	18
Distaloy DC-1	<0.01	0.1	-	2.10	1.47	3.15	7.1	10**
Distaloy DH-1	<0.01	0.1	-	2.00	1.47	3.1	7.1	10**
Distaloy HP-1	<0.01	0.1	2.00	4.00	1.41	3.15	7.1	12**

*after compaction at 700 MPa, ** after compaction at 600 MPa, K - lubricant Kenolube PH, LD - lubricated die

Phosphorus Alloyed Steel. PNC60 powder contains 0.6% phosphorus. To the base iron powder NC100.24, Phosphorus is added as fine Fe_3P powder. PASC60 powder is based on ASC 100.29 and contains 0.6% phosphorus.

Prealloyed Powder Grades

	Composition [mass %]						Apparent density [g/cm³]	Compressibility at 600 MPa [g/cm³]	*Green strength [MPa]
	C	O-tot	Ni	Mo	Mn	Cr			
Astaloy A	<0.01	0.13	1.90	0.55	0.20	-	3.1	7.02	17
Astaloy B	<0.01	0.13	0.45	0.60	0.30	-	3.1	7.09	14
Astaloy CrM	<0.01	<0.25	-	0.50	-	3.00	2.85	7.0	15
Astaloy CrL	<0.01	0.15	-	0.20	-	1.50	2.85	7.04	22
Astaloy 85Mo	<0.01	0.1	-	0.85	-	-	3.1	7.15	18
Astaloy Mo	<0.01	0.1	-	1.50	-	-	3.1	7.1	18

*after compaction at 600 MPa, Kenolube

Stainless Steel Powder Grades

	Composition [mass %]						Apparent density [g/cm³]	Compressibility at 600 MPa [g/cm³]	*Green strength [MPa]
	C	Si	Ni	Cr	Mo	Mn			
316 L	0.02	0.8	13.0	17.0	2.2	0.2	3.0	6.65	7
410L	0.02	0.8	-	12	-	0.2	3.0	6.53	10
430L	0.02	0.8	-	18	-	0.2	2.9	6.37	10

*after compaction at 600 MPa

9.3.2 HOEGANAES Corp. [357,358]

Sponge Iron Powder Grades

	Composition [mass %]			Apparent density [g/cm³]	Compressibility at 600 MPa [g/cm³]	*Green strength [MPa]
	SiO_2	C	H_2 loss			
Ancor MH-100	0.20	0.01	0.21	2.55	6.95	23.5
Ancormet 101	-	0.23	0.10	2.53	6.75	26.0

*after compaction at 600 MPa

Atomized Iron Powder Grades

	Composition [mass %]				Apparent density [g/cm³]	Compressibility at 600 MPa [g/cm³]	*Green strength [MPa]
	C	Mn	Cr+C+Ni	O			
Ancorsteel 1000	<0.01	0.20	0.25	0.14	2.94	7.1	18
Ancorsteel 1000B	<0.01	0.10	0.13	0.09	2.92	7.15	16
Ancorsteel 1000C	<0.01	0.07	0.09	0.07	2.92	7.25	16.8

*at the density of 7.1 g/cm³

Atomized Prealloyed Powder Grades								
	Composition [mass %]					Apparent density [g/cm^3]	Compressibility at 600 MPa [g/cm^3]	*Green strength [MPa]
	C	Mo	Ni	Mn	O			
Ancorsteel 2000	<0.01	0.61	0.46	0.25	0.13	2.98	7.0	11
Ancorsteel 4600 V	0.01	0.56	1.83	0.15	0.13	2.96	6.9	10
Ancorsteel 50 HP	<0.01	0.50	-	0.18	0.09	2.97	7.2	18
Ancorsteel 85 HP	<0.01	0.86	-	0.12	0.08	2.97	7.0	22
Ancorsteel 150 HP	<0.01	1.50	-	0.12	0.08	2.95	7.1	18
Ancorsteel 737 SH	<0.01	1.25	1.40	0.42	0.15	3.0	7.0	13

*after compaction at 600 MPa

Diffusion Alloyed Powder Grades								
	Composition [mass %]					Apparent density [g/cm^3]	Compressibility at 600 MPa [g/cm^3]	*Green strength [MPa]
	Ni	Mo	Cu	C	O			
Ancorsteel FD-4600A	1.75	0.50	1.50	<0.01	0.13	3.0	7.0	14
Ancorsteel FD-4800A	4.00	0.50	1.50	<0.01	0.13	3.0	7.0	18

*at the density of 7.1 g/cm^3

Ancorsteel Phosphorus Alloys. Ancorsteel phosphorus alloys are produced by adding phosphorus as Fe_3P to the highly compressible iron powders Ancorsteel 1000B and Ancorsteel 1000C. The products Ancorsteel 45P (conventional premix) and Ancorsteel 80P (binder treated premix) contain 0.45 and 0.80 mass % phosphorus, respectively.

Ancorloy Binder-Treated Powder Grades								
	Composition [mass %]						Apparent density [g/cm^3]	Compressibility at 410/600 MPa [g/cm^3]
	Ni	Cu	Mo	Si	C	Cr+Mn		
Ancorloy 2	1.75	1.5	0.55	-	-	-	3.24	6.85 (410)
Ancorloy 4	4.0	1.5	0.55	-	-	-	3.30	6.85 (410)
Ancorloy MDA	0.05	-	0.05	0.7	0.85	0.23	3.12	6.95 (600)
Ancorloy MDB	2.0	-	0.85	0.7	0.55	0.16	3.19	7.05 (600)
Ancorloy MDC	4.0	-	0.85	0.7	0.55	0.16	3.25	7.05
Ancorloy DH-1	-	2.0	1.45	-	-	0.15 (Mn)	3.10	7.10
Ancorloy HP-1	4.0	2.0	1.40	-	-	0.15 (Mn)	3.19	7.13
AncorMax D*	2.0	-	0.85	-	0.60	-	-	7.27 (690)

*Ancorloy 85 HP basic powder, 0.55% Lube + Binder, recommended processing temperature 93°C (145°F)

9.3.3 QMP (ATOMET Powder Metallurgy Products) [359]

Plain Powder Grades						
	Composition [mass %]			Apparent density [g/cm³]	Compressibility at 600 MPa [g/cm³]	*Green strength [MPa]
	C	O	Mn			
ATOMET 25	0.03	0.20	0.008	2.52	6.90	29
ATOMET 28	0.05	0.17	0.008	2.84	6.95	22
ATOMET 29	0.05	0.17	0.008	2.93	6.95	20
ATOMET 29M	0.05	0.16	0.008	2.95	7.00	14
ATOMET WPL200	0.01	0.15	0.20	2.65	7.00	20.9
ATOMET 1001HP	0.004	0.06	0.04	2.92	7.15	17
ATOMET 1001	0.003	0.08	0.20	2.95	7.10	16

*after compaction at 600 MPa

Prealloy Powder Grades									
	Composition [mass %]						Apparent density [g/cm³]	Compressibility at 600 MPa [g/cm³]	*Green strength [MPa]
	Mn	Ni	Mo	Cr	C	O			
ATOMET 4001	0.15	-	0.5	-	0.004	0.10	2.92	7.10	13.6
ATOMET 4201	0.28	0.45	0.6	-	0.004	0.10	2.92	7.05	14.5
ATOMET 4401	0.15	-	0.85	-	0.003	0.08	2.92	7.10	14.9
ATOMET 4901	0.20	-	1.5	-	0.01	0.15	3.00	7.00	10.3
ATOMET 4601	0.20	1.80	0.55	-	0.004	0.10	2.92	6.95	14.5
ATOMET 4701	0.45	0.90	1.0	0.45	0.01	0.25	2.92	6.90	9.0
ATOMET 4801	0.20	4.00	0.5	-	0.01	0.15	3.00	6.85	9.6

*after compaction at 600 MPa

9.3.4 An abbreviated list of comparable grades [360]

HÖGANÄS AB	HOEGANAES Corp.
NC100.24	Ancor MH-100
AHC100.29	Ancorsteel 1000
ASC100.29	Ancorsteel 1000B
ABC100.30	Ancorsteel 1000C
Distaloy AB	Distaloy 4600A
Distaloy AE	Distaloy 4800A
Distaloy SA	No equivalent
Distaloy SE	No equivalent
Distaloy DC-1	(See note below)
Distaloy DH-1	Ancorloy DH-1
Distaloy HP-1	Ancorloy HP-1
PNC60	Premix based on Ancor MH-100+Ancorsteel
PASC60	45P (60P on request)
Astaloy CrM, CrL	No equivalent
Astaloy Mo	Ancorsteel 150 HP
Astaloy 85Mo	Ancorsteel 85 HP
Astaloy A	Ancorsteel 4600 V

Note: Binder-treated (ANCORBOND) premixes based on Ancorsteel 85 HP are being used in USA rather than a Distaloy based on Ancorsteel 150 HP (Astaloy Mo).

9.4. CHARACTERISTIC TYPES OF WEAR OF HARDMETAL INSERTS

The characteristic types of wear of hardmetal inserts formed in turning and in milling which complete the data listed in chapter 3.3.7 and chapter 8 are presented in the following sketches with recommendations for its decrease. (Data from: *Exchangeable Cutting Tips*. PRAMET Diadur, Šumperk, VBD 0499-Cz, 1999, in Czech).

9.4.1 Types of insert wear in turning

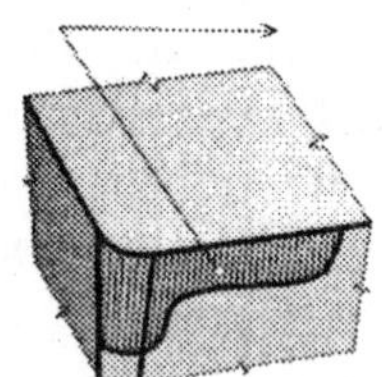

FLANK WEAR

Flank wear is one from the main criteria characterizing the lifetime of inserts. This is formed as a consequence of the effect of wear mechanisms on the tool. The intensity of this type of wear can only decrease.

Recommendation:-

- use more abrasive wear resistance HM type,
- use cooling emulsion resp. increase the cooling intensity,
- decrease the cutting speed,
- at feed <0.1 mm/rev. decrease the feed for CVD coated materials

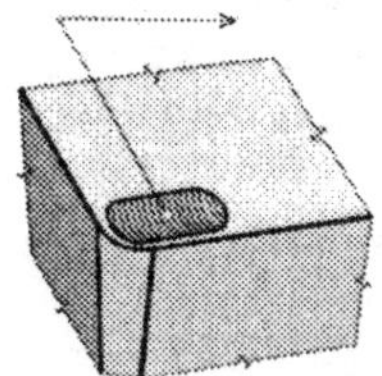

FACE CRATER

Face crater. This kind of wear occurs on inserts with a straight tool face and is not limited only to this type of insert.

Recommendation:

- use HM with higher abrasive wear resistance,
- use cooling emulsion or increase the cooling intensity,
- use other (more positive) type of cutting geometry

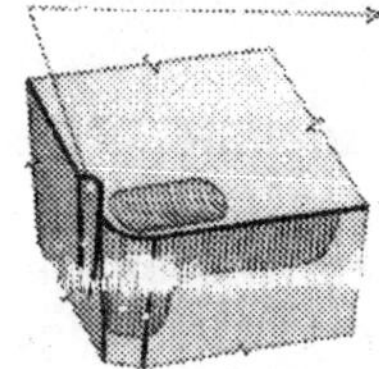

OXIDATION GROOVE ON FRONT CUTTING EDGE

Oxidation groove on front cutting edge – one of the most important criteria limiting the tool life of HM inserts. Interaction of the oxidized groove with the face crater causes increases the roughness of the machined surface. This feature is referred to as 'fuzziness'.

Recommendation:

- use a coated HM or one with higher abrasive wear resistance, if possible,
- use an insert with a coating containing Al_2O_3

(210K, 320P),
- use a cooling emulsion or increase the cooling intensity,
- decrease the cutting speed,
- use a cutting material with higher thermal stability and oxidation resistance

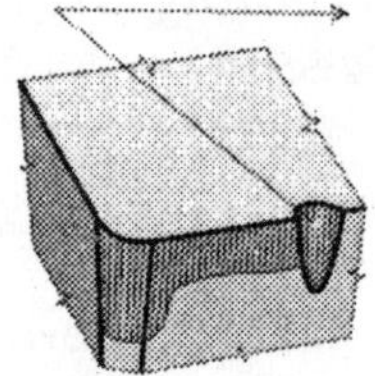

NOTCH WEAR ON SIDE CUTTING EDGE

Notch wear on side cutting edge forms in the area of contact of the cutting edge with the surface of the workpiece. It is caused mainly by strengthening of the surface layer of the workpiece and by burrs. This type of wear occurs mainly in stainless steels and at operations characteristics with variations of the cutting depth.

Recommendation:
- use the tool with lower angle,
- decrease the feed, if possible for applied operation,
- use an insert with the coating containing Al_2O_3 (210K, 320P)

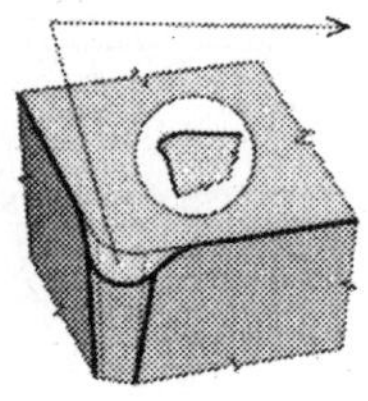

PLASTIC DEFORMATION OF THE TOOL POINT

Plastic deformation of the tool point. The reason of this kind of wear is overloading of the edge due to high cutting speeds and feeds.

Recommendation:
- use a more abrasive wear resistance HM type,
- decrease the cutting speed,
- decrease the feed,
- use a cooling emulsion, resp. increase cooling intensity,
- use an insert with larger chamfer radius,
- use an insert with larger point radius

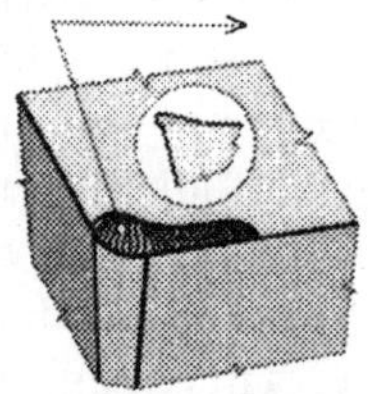

FORMATION OF BUILT-UP EDGE

Formation of built-up edge. This is sticking of the workpiece material on the cutting edge. The built-up edge has a character of a microweld on the cutting edge. At its breakaway the cutting edge can be failured. The deterioration of surface finish of the workpiece is a further consequence.

Recommendation:
- increase the cutting speed,
- increase the feed, use mainly PVD coated tools,
- use other (more positive or sharper) cutting

geometry,
- use an antibuilt-up fluid (if it is not available no cooling should be applied)

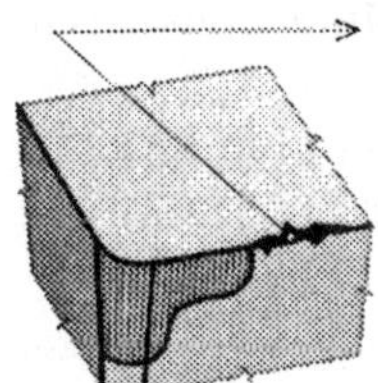

FAILURE OF THE CUTTING EDGE (out of engagement)

Failure of the cutting edge (out of engagement) is caused by unfavourable formation of the chip which by its flow away impacts on the cutting edge and by this it mechanically fails.

Recommendation:
- change the feed,
- use a tool with a different setting angle,
- use a tool with different cutting geometry (other chip former),
- use HM type with higher toughness

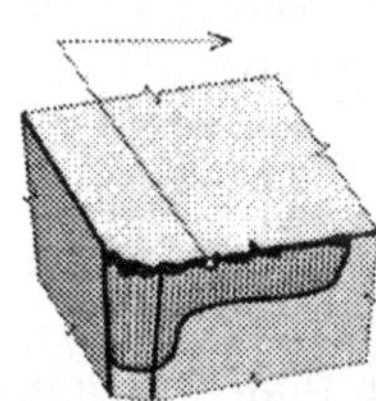

BRITTLE FAILURE OF THE CUTTING EDGE

Brittle failure of the cutting edge (microchipping) occurs with other types of wear. This type of wear is difficult to indentify separately.

Recommendation:
- use HM with higher toughness,
- choose less intensive cutting conditions,
- use different cutting geometry,
- decrease the feed at entry of the cutting edge into the cut

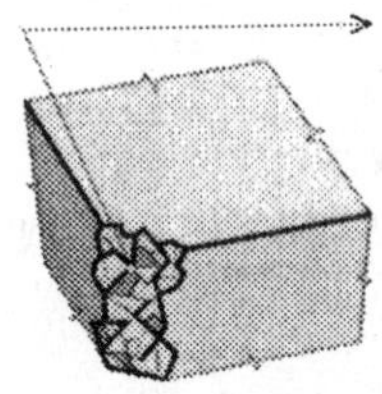

BREAKAGE OF THE CUTTING EDGE (of the tool point)

Breakage of the cutting edge (of the tool point). This failure can be caused by various factors such as the properties of the tool and workpiece material, the state and stiffness of the machine-tool-workpiece system, and also by the intensity of wear and engagement conditions.

Recommendation:
- use a HM tool with higher toughness,
- choose less intensive cutting conditions at the starting cut (decrease the feed and depth of cut),
- use an insert with a larger chamfer radius,
- use an insert with a larger point radius,
- use different cutting geometry (chip former),
- stabilize the cutting edge,
- decrease the feed at entry of the cutting edge

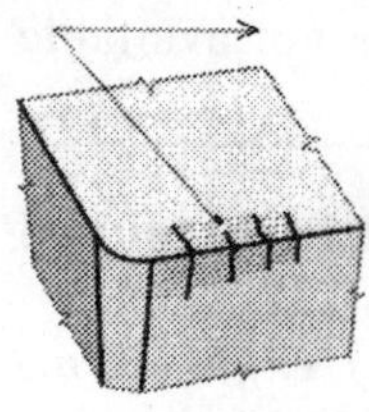

COMB CRACKS

Comb cracks are formed as a consequence of dynamic heat loading at interrupted cut. This type of wear is found mostly in milling operations.

Recommendation:

- use of a cooling fluid is not recommended; it is possible to use pressure air for removing the chips from the cutting zone

9.4.2 Types of insert wear in milling

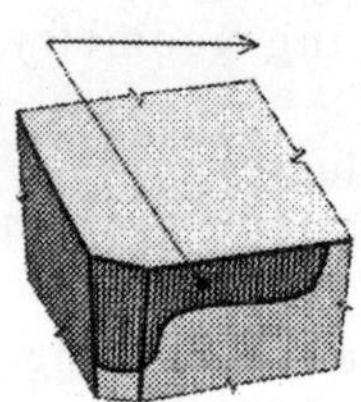

FLANK WEAR

Flank wear is one from the main criteria characterizing the lifetime of inserts. It occurs as a consequence of the effect of wear mechanisms on the tool. The intensity of this type of wear can only decrease.

Recommendation:

- use a HM type with wear resistance,
- decrease the cutting speed,
- increase the feed if it is lower than 0.1 mm/tooth

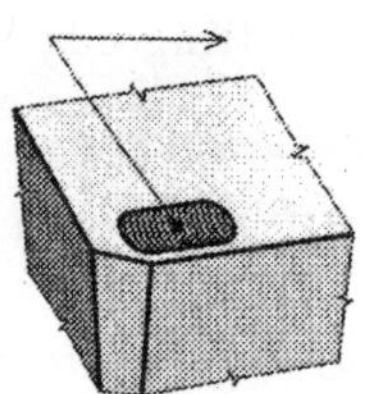

FACE CRATER

This kind of wear occurs mostly in inserts with a straight tool face and is not limited only to this type of insert.

Recommendation:

- use HM with higher abrasive wear resistance,
- use a cooling emulsion or increase cooling intensity,
- decrease the cutting speed,
- use other (more positive) type of cutting geometry

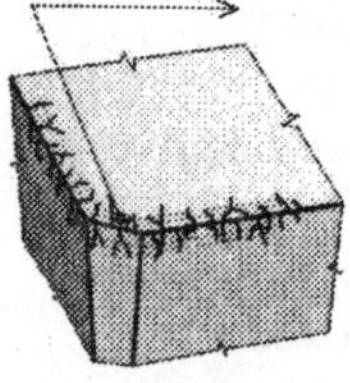

COMB CRACKS

Comb cracks are formed as a consequence of dynamic heat loading at interrupted cut. This type of wear is mostly found in milling operations.

Recommendation:

- use a cooling fluid not recommended; it is possible to use a pressure air for the taking off the chips from the cutting zone,
- use HM with higher heat impact resistance

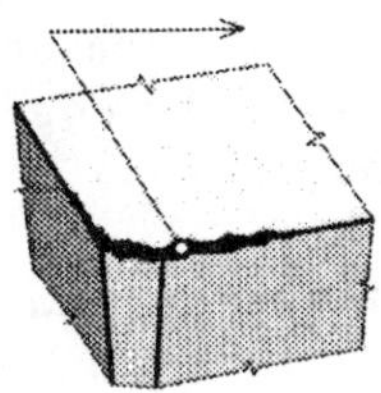

BRITTLE FAILURE OF THE CUTTING EDGE

Brittle failure of the cutting edge (microchipping) occurs with other types of wear. This type of wear is difficult to identify separately

Recommendation:

- use HM with higher toughness,
- choose less intensive cutting conditions,
- use different cutting geometry,
- decrease the feed at the start of the cut,
- change the method at entry and exit of the cutting edge

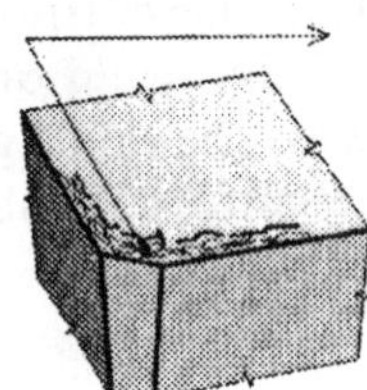

FATIGUE CRACKS ALONG FLANK

Fatigue cracks along flank form due to dynamic loading in the area closely behind the cutting edge.

Recommendation:

- use HM with higher toughness,
- change the method of the taking tool in and out,
- change the engagement conditions,
- use different type of cutting geometry or an insert with different cutting edge preparation (...T,...S),
- change the feed

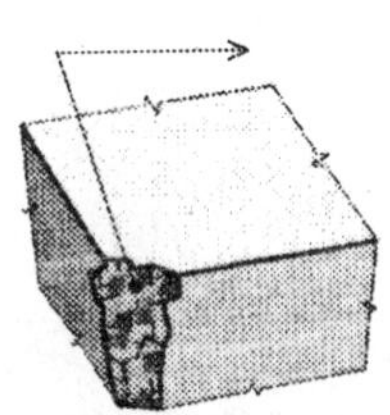

BREAKAGE OF CUTTING EDGE (point of the tool)

Breakage of the cutting edge can be caused by various factors such the properties of the tool and workpiece material, the state and stiffness of the machine-tool-workpiece system, and also by the intensity of wear and engagement conditions.

Recommendation:

- use a HM tool with higher toughness,
- choose less intensive cutting conditions at the starting cut (decrease the feed and depth of cut),
- use an insert with a larger chamfer radius,
- use an insert with a larger point radius,
- use different geometry (chip former),
- stabilize the cutting edge,
- decrease feed at entry of the cutting edge

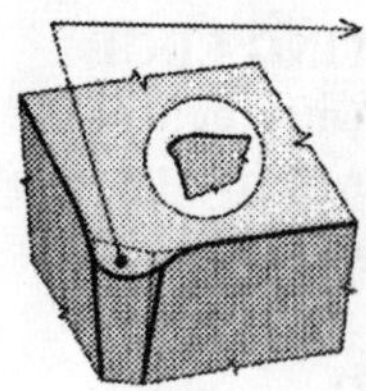

PLASTIC DEFORMATION OF THE TOOL POINT

Plastic deformation of the tool point. The reason of this kind of wear is overloading of the edge due to high cutting speeds and feeds.

Recommendation:

- use HM type with higher abrasive wear resistance,
- decrease cutting speed,
- decrease feed

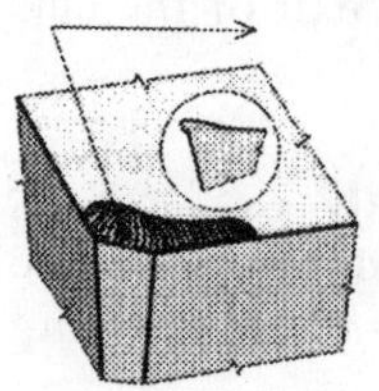

FORMATION OF BUILT-UP EDGE

Formation of built-up edge. This has the form of sticking of the workpiece material on the cutting edge. The built-up edge resembles a microweld on the cutting edge. When it breaks away, the cutting edge can be fail. Deterioration of the surface finish of the workpiece is a further consequence.

Recommendation:

- increase cutting speed,
- increase feed,
- use mainly PVD coated tools,
- use other (more positive or sharper) cutting geometry

CONSTANTS AND CONVERSION FACTORS
Fundamental and derived units

1 kg = 2.2 lb (pound)
1 lb = 0.454 kg
1 g = 0.035 oz (ounce)

Energy (joule = J)
1 J = 1 N·m = 0.737 ft·lb (foot pound)
1 J = 0.239 cal (calorie)
1 J = $9.48 \cdot 10^{-4}$ Btu (British thermal unit)
1 J = 4.83 hp·h (horsepower hour)
1 kWh (kilowatt hour) = $3.6 \cdot 10^{6}$ J
1 W = 0.737 ft·lb/s (foot pound per second)

Volume (litre = l) **and area**
1 l = 10^{-3} m^3 = 0.264 gallons = 1000 cm^3
1 gal = 3.79 l
1 cubic in. = 0.0164 l
1 cm^2 = 0.155 in^2
1 cm^3 = 0.06
1 in^3 (cubic inch)

Temperature
Celsius (Kelvin) (Fahrenheit) = °C (K) (°F)
Temp. °C = (Temp. °F–32) × 5/9
Temp. K = Temp. °C + 273
1120 °C = 1393 K = 2048 °F
1250 °C = 1523 K = 2282 °F

Force (newton = N) 1 N = 10^5 dyn
1 N = 0.225 lb force (pound force)
1 lbf = 4.44 N
1 kp = 9.807 N

Heating and Cooling Rate
1 K/s = 1 °C/s = 1.8 °F/s
1 K/min = 1 °C/min = 1.8 °F/min

Prefixes G (giga) = 10^9, M (mega) = 10^6k (kilo) = 10^3, c (centi) = 10^{-2}m (milli) = 10^{-3}, μ (micro) = 10^{-6}n (nano) = 10^{-9}, p (pico) = 10^{-12}

Pressure, Stress, and Strength (pascal = Pa)
1 Pa = 0.0075 torr (millimeters of mercury)
1 bar = 0.1 MPa
1 kPa = 0.145 psi (pounds per square inch)
1 MPa = 10 bar = 9.87 atmospheres
1 bar = 10^5 Pa
1 MPa = 145 psi (pounds per square inch)
1 MPa = 0.145 ksi (thousand pounds per square inch) = 13.79 tsi
1 ksi = 6.895 MPa
1 t cm^{-2} = 9.81 MPa
1 MPa $m^{1/2}$ = 0.91 kpsi $in^{1/2}$ (stress intensity conversion = fracture toughness)
1 Torr = 1000 microns Hg = 133.3 Pa

Length (meter = m)
1 m = 39.4 in (inch)
1 m = 3.28 ft (foot)
1 foot = 0.305 m
1 cm = 0.394 in (inch)
1 in = 25.4 mm
1 mm = 0.0394 in (inch)
1 μm = 39.4 μin (microinch)

Density 1 Mg/m^3 = 1 g/cm^3
1 g/cm^3 = 0.036 1 lb/in^3 (pound per cubic inch)
1 kg/m^3 = 10^{-3} g/cm^3

Mass (kilogram = kg)

References

[1] ŠALAK, A. and MIŠKOVIČ, V.: Properties of Sponge Iron Powder Produced by Reduction of Mill Scale. *Hutnické Listy*, 1972, vol. XXVII, no. 3, p. 159, in Slovak.

[2] KLIMENKO, VB., BULANOV, VJ., RUKIN, VV., MICHKOVA, ES. and ZAREROCHIN, LN.: *Iron Powders (Zheleznye poroshki)*. Ed. NAUKA, Moscow, 1982, in Russian.

[3] ŠALAK, A., SELECKÁ, M. and BUREŠ, R.: Manganese in Ferrous Powder Metallurgy. *Powder Metallurgy Progress*, 2001, vol. 1, no. 1, p. 41.

[4] ŠALAK, A.: Sintered Manganese Steeels, Part 1. Effect of Structure of Initial Iron Powders upon Mechanical Properties. *PMI*, 1980, vol.12, no. 1, p. 28.

[5] ŠALAK, A.: Evaluation of Structural Activity of Iron Powder by Tensile Strength of Manganese Steel Sintered in α-iron Region. *Poroshkovaya Metallurgiya*, 1984, no. 8, p. 58, in Russian.

[6] GERMAN, RM.: *Powder Metallurgy of Iron and Steel*. Wiley & Sons, Inc., New York, 1998.

[7] ŠALAK, A., SELECKÁ, M. and PARILÁK, Ľ.: Some Properties of Powder Forged Steel Based on Chromium Prealloyed Powders of High Oxygen Content. In AMPT'01. New Dev. on Powder Technology, J.M. Torralba ed., Univ. Carlos III, Madrid, vol. III, 2001, p. 1287.

[8] An Oil-Atomized Low-Alloy Steel Powder, UK patent GB 2 094 834 A; Sumitomo Metal Ltd, Technical information, 1985.

[9] ŠALAK, A.: Experimental Production of Metallic Powders by Dispersion by Diesel Oil and Gases. *Hutnické Listy*, 1990, vol. 45, no. 3, p. 185; Translation Riecansky Technical Translations, Cambridge, UK.

[10] ŠALAK, A.: *Ferrous Powder Metallurgy*. Cambridge International Science Publishing, Cambridge, UK, 1995.

[11] Höganäs Handbook. *Höganäs Iron and Steel Powders for Structural Parts*. Höganäs AB, 1996; 2002.

[12] *Höganäs Handbook for Sintered Components. 4 Warm Compaction*, Höganäs AB, 1998.

[13] SPOLJARIC, D., DANNINGER, H. and WEISS, B.: The Influence of the Surface Quality on Rotating Beam Fatigue Testing of PM Steels. In DF PM 1996. Deformation and Fracture in Structural PM Materials, Stará Lesná, Ľ. Parilák, H. Danninger, J. Dusza and B. Weiss eds., IMR SAS Košice, vol. 1, 1996, p. 159.

[14] ŠALAK, A.: Deformation Hardening of Iron Particles during Compacting Process. *Pokroky Práškové Metalurgie* (VUPM), 1975, no. 2-3, p. 45, in Slovak.

[15] Komage Extends Shaping Option for PM Compaction. *MPR*, 2001, no. 2, p. 20.

[16] VELTL, G. and PETZOLD, F.: Warm Flow Compaction for High Complexicity PM Parts. In 2002 World Congress on Powder Metall. & Part. Mat., Orlando, MPIF Princeton, Part. 4, 2002, p. 273.

[17] VELTL, G., OPPERT, A. and PETZOLD, F.: Flow Compaction Fasters more Complex PM Parts. *MPR,* 2001, no. 2, p. 26.

[18] SKOGLUND, P.: High Density PM Parts by High Velocity Compaction. *Powder Metallurgy*, 2001, vol. 44, no. 3, p. 199.

[19] DOREMUS, P., DUWA, F., FRANCOIS, P., PUENTE, G. and ALLIBERT, CH.: High Velocity Compaction. In 2002 World Congress on Powder Metall. & Part. Mat., Orlando, MPIF Princeton, Part 4, 2002, p. 96.

[20] STRÖMGREN, M. and LOCHON, M.: Development and Fatigue Testing of a Powder Forged Pinion Gear for a Passenger Car Gear Box. *Höganäs Iron Powder Information* PM 84-5.

[21] ARNHOLD, V.: Optimised Production Technics Opens New Applications for Powder Metallurgy Parts. *Werkstatt und Betrieb*, 1990, no. 11, p. 55, in German.

[22] ADAMS, VA. and GLOVER, D.: Powder Forging for Heavy Duty Parts. *MPR*, 1980, no.1, p. 10.

[23] ŠALAK, A. and BENKE, V.: Powder Forged Roller Bearing Races from Fe-Ni-Mo and Fe-Mn-Cr Powders. *PMI,* 1990, vol. 22, no. 5, p. 9.

[24] FERGUSON, BL.: Cold Forming of Metal Powder Preforms. *MPR*, 1987, no. 9, p. 583.

[25] SCHATT, W.: *Sintering Processes (Sintervorgänge)*, VDI Verlag, Düsseldorf, 1992, in German.

[26] DANNINGER, H., GIERL, Ch., KREMEL, S., LEITNER, G., JAENICKE-ROESSLER, K. and YU, Y.: Degassing and De-oxidation Processes during Sintering of Unalloyed and Alloyed PM Steels. *Powder Metallurgy Progress*, 2002, vol. 2, no.3, p. 125.

[27] KREMEL, S., RAAB, CH. and DANNINGER, H.: Contact Formation and Carbon Dissolution during Sintering of Steels Prepared from Astaloy CrM. In Euro PM2001. Eur. Congress and Exh. on Powder Metall., Nice, EPMA Shrewsbury, vol. 1, 2001, p. 52.

[28] SHIVANATH, R., JONES, PK. and LAWCOCK, R.: On the Synergies of High Temperature Sintering and Alloy Development for High Endurance of PM Powertrain Components. In World Congress on Powder Metall. & Part. Mat., Orlando, MPIF Princeton, Part 4, 1996, p.13.

[29] ŠALAK, A.: Sublimation and Condensation of Manganese Vapour. *Practical Metallography,* 1985, vol.22, no.1, p. 26.

[30] Höganäs Handbook for Sintered Components 2. *Production of Sintered Components*. Höganäs AB, 1997.

[31] BOCCHINI, GF.: Sintering of Carbon Steels: Controlled Atmospheres, Equipment, Practical Results. In Powder Metall. Summer School, Gothenburg, EPMA Shrewsbury, 1999, p. 185.

[32] BOCCHINI, GF.: Influence of Controlled Atmospheres on the Proper Sintering of Carbon Steels. *Powder Metallurgy Progress*, 2004, vol. 4, no. 1, p. 1.

[33] FAST, JD.: *Interaction of Metals and Gases*. Vol. 1, Phillips Technical Library, Eindhoven, 1965.

[34] ŠALAK, A., KUCHÁR, J., DUDROVÁ, E. and RUDNAYOVÁ, E.: Saturation of Iron Compacts with Nitrogen during Sintering in a Nitrogen Containing Atmosphere. *Kovové Materiály (Metallic Materials)*, 1971, vol. X, no. 4, p. 345, in Slovak; *Translation BISI* 9678/ 1972.

[35] SIBLEY, AT., MARSDEN, JG. and BUCK, DM.: Effect of Atmosphere Composition on Physical Properties in High Temperature Ferrous Sintering. In 1982 Nat. Powder Met. Conf., Montreal, MPIF Princeton, 1982, p. 48.

[36] TENGZELIUS, J., GREK, SE. and BLÄNDE, C-A.: High Strength Sintered Steels Containing Chromium and Manganese. In Int. PM Conf. Washington, MPIF Princeton, Part. 13, 1980, p. 159.

[37] CAMPOS-GOMEZ, M., KREMEL, S. and MARCU PUSCAS, T.: Sintered Behaviour and Mechanical Properties of Prealloyed Steel with Chromium. In Euro PM2000. Materials and Processing Trends for PM Components in Transportation, Munich, EPMA Shrewsbury, 2000, p. 47.

[38] DANNINGER, H., FRAUENDIENST, G., STREB, KD. and RATZI, R.: Sintered Steels Prepared Using Different Graphite Grades. In 2000 Powder Metall. World Congress, Kyoto, K. Kosuge and H. Nagai eds., JPMA, Part 2, 2000, p. 1010.

[39] ŠALAK, A. and SELECKÁ, M.: Mechanical and Friction Properties of Sintered (1-5)% Manganese Steel. In (CD) 3rd Int. Powder Metall. Conf., TTMD, Gazi Univ., Ankara, 2002, p. 230.

[40] DANNINGER, H.: Sintering of Mo Alloyed P/M Structural Steels. *PMI*, 1988, vol. 20, no. 4, p. 7. - Sintering of Mo Alloyed P/M Steels Prepared from Elemental Powders, I. Sintering Temperature and Mechanical Properties. *PMI*, 1992, vol. 24, no. 2, p. 73; II. Mo Homogenization and Dimensional Behaviour. *PMI*, 1992, vol. 24, no. 3, p. 163.

[41] ANCIFEROV, VB., AKIMENKO, VB. and GREBNOV, LM.: *Powder Alloy Steels (Poroshkovyje Legirovaniye Stali)*. Ed. METALLURGIYA, Moscow, 1991, in Russian.

[42] OGURA, K., TAKAJO, S., MAEDA, Y. and MORIOKA, Y.: Cr and Mn Containing Steel Powders Produced by Water Atomisation and Vacuum Annealing. *MPR*, 1987, no. 4, p. 292.

[43] DANNINGER, H., KREMEL, S., LEITNER, G., JAENICKE-RÖSSLER, K. and YU, Y.: Degassing during Sintering of Cr-Mo Steels Prepared from Prealloyed Powder. In 2002 World Congress on Powder Metall. & Part. Mat., Orlando, MPIF Princeton, Part 13, 2002, p. 291.

[44] DANNINGER, H., KREMEL, S., ALTENA, H. and YU, Y.: Sintering and Properties of PM Steels Prepared from Pre-alloyed Powder Astaloy

CrM. In 3rd Int. Conf. on Powder Metall. for Automotive Parts, Isfahan, A. Arvand and H. Danninger eds., MPM Co, Tehran, 2002, p. 26.

[45] LINDNER, KH.: High Density P/M Gears by Enhanced Solid-State Sintering. In Adv. Powder Metall. & Part. Mat., comp. by H. Cadle and KS. Narasimhan, MPIF Princeton, Part 11, 1996, p. 221.

[46] DANNINGER, H., WOLFSGRUBER, E. and RATZI, R.: Sintering of Different PM Ferrous Materials at Moderate Temperatures. In 1998 PM World Congress & Exh., Granada, EPMA Shrewsbury, vol. 2, 1998, p. 290.

[47] ŠALAK, A.: Development Trends in Ferrous Powder Metallurgy. In DF PM 1996. Deformation and Fracture in Structural PM Materials, Stará Lesná, Ľ. Parilák, H. Danninger, J. Dusza and B. Weiss eds., IMR SAS Košice, vol. 1, 1996, p. 205.

[48] TORRALBA, JM. and CAMPOS, M.: Low Alloyed Cr-Mo Sintered Steels - Update. *Powder Metallurgy Progress*, 2002, vol. 2, no. 3, p. 177.

[49] ŠALAK, A., SELECKÁ, M. and PARILÁK, Ľ.: Properties of Hybrid Fe-Mn-Cr-C Sintered Steels. In Euro PM2001. Eur. Congress & Exh. on Powder Metall., Nice, EPMA Shrewsbury, vol. 1, 2001, p. 251.

[50] DANNINGER, H., STROBL, S., GÜRTENHOFER, R. and DUDROVÁ, E.: PM Structural and Tool Steels Prepared from Coated Powders. In 2000 Powder Metall. World Congress, Kyoto, K. Kosuge and H. Nagai eds., JPMA, Part 1, 2000, p. 394.

[51] MENAPACE, C., MOLINARI, A. and COSTA, P.: Effect of Matrix Characteristics on the Properties of Infiltrated Steels. In Euro PM2003. Eur. Congress & Exh. on Powder Metall., Valencia, EPMA Shrewsbury, vol. 1, 2003, p. 364.

[52] MOLNÁR, F., LEŽANSKI, J., DUDROVÁ, E. and FRYDRYCHOVA, H.: Properties and Structure of Sintered Fe-Cu Systems of Iron Particles Coated with Copper. *Pokroky Práškové Metalurgie* (VUPM), 1989, no. 3, p. 67, in Slovak.

[53] DANNINGER, H.: Pore Formation during Sintering of Fe-Cu and its Effect on Mechanical Properties. *PMI*, vol. 19, 1987, no. 1, p. 19.

[54] *Machinability of Steels*. Metals Handbook, Tenth Ed., vol. 1, ASM Metals Park.

[55] PROCHÁZKA, V.: Sintered Phosphorus Alloyed Steel. In Problems in Powder Metallurgy (Problémy práškovej metalurgie), Ed. Vydavateľstvo Slovenskej akadémie vied, Bratislava, 1963, p. 86, in Slovak.

[56] JANGG, G., DROZDA, M., DANNINGER, H. and EDER, G.: Magnetic Properties of Fe-P Materials. *PMI*, 1984, vol. 16, no. 6, p. 264.

[57] DANNINGER, H., WOLFSGRUBER, E. and DALAL, K.: Sintering and Microstructure of PM Valve Seat Inserts Materials. In PM'94. Powder Metall. World Congress & Exh., Paris, EPMA Shrewsbury, vol. II, 1994, p. 875.

[58] ŠALAK, A. and SELECKÁ, M.: Properties of Sintered Fe-1.65Cr-XMe-C Steels Alloyed with a Master Alloy Forming a Liquid Phase and with Boron. *Trans. of the Tech. Univ. of Košice*, 1995, vol. 5, no. 2, p. 102.

[59] OGURA, K., TAKAJO, S., YAMATO, N. and MAEDA, Y.: Low Oxygen Alloy Steel Powder by Vacuum Annealing. In Horizons of Powder Metallurgy (1986 Int. PM Conf. & Exh., Düsseldorf), WA. Kayser and WJ. Huppmann eds., Verlag Schmid, Freiburg, Part I, 1986, p. 37.

[60] SANDEROW, HI. and RODRIGUES, H.: High Strength 4100 Alloy PM Steels. *MPR*, 1986, June, p. 423.

[61] OGURA, K., TAKAJO, S., YAMATO, N., MAEDA, Y. and MORIOKA, Y.: Cr and Mn Containing Low Oxygen Steel Powder Produced by Water Atomisation and Vacuum-Annealing. *MPR*, 1987, no. 4, p. 292.

[62] KREMEL, S., DANNINGER, H. and YU, Y.: Effect of Carbothermic Reduction of Surface Oxides on the Mechanical Behaviour of Sintered Steels Prepared from Cr-prealloyed Powders. In DF PM 2002. Deformation and Fracture in Structural PM Materials, Stará Lesná, Ľ. Parilák and H. Danninger eds., IMR SAS Košice, vol. 1, 2002, p. 132.

[63] DANNINGER, H. and LEITNER, G.: De-gassing and Reduction Processes at Sintering of Iron Materials. *Pulvermetallurgie in Wissenschaft und Praxis*, Bd. 19, H. Kolaska ed., Fachverband Pulvermetallurgie, Haagen, 2003, p. 211, in German.

[64] ORTIZ, P. and CASTRO, F.: Influence of Carbon Activity and Oxygen Potential of Sintering Atmospheres on Microstructure Characteristics of Low Alloy PM Steels. *Material Science Forum*, 2003, vols. 426-432, p. 4337.

[65] BENNER, A. and BEISS, P.: Wear Progress during Green Drilling of High Strength PM Steels. In Euro PM2003. Eur. Congress & Exh. on Powder Metall., Valencia, EPMA Shrewsbury, vol. 1, 2003, p. 354.

[66] CASTRO, F. and ORTIZ, P.: Study of Gas Solid Interactions during Sintering of Cr-containing PM Steels. In Euro PM2003. Eur. Congress & Exh. on Powder Metall., Valencia, EPMA Shrewsbury, vol. 1, 2003, p. 243.

[67] ŠALAK, A.: Manganese Vapor-Protection of Premixed Manganese Steels against Oxidation during Sintering, *PMI*, vol. 1, 1986, no. 4, p. 266.

[68] ŠALAK, A., SELECKÁ, M., ČAJKA, A. and FRANEK, Š.: Tribological Properties of Sintered Manganese Steel. In DF PM 2002. Deformation and Fracture in Structural PM Materials, Ľ. Parilák and H. Danninger eds., Stará Lesná, IMR SAS Košice, vol. 1, 2002, p. 284.

[69] ŠALAK, A., SELECKÁ, M.: Wear Behaviour of Carbon and Low Alloyed Sintered Steels. *Powder Metallurgy Progress*, 2002, vol. 2, no. 4, p. 231.

[70] ŠALAK, A.: Sintered Manganese Steels. Part 2. Manganese Evaporation during Sintering. *PMI*, 1980, vol. 12, no. 2, p. 72.

[71] ŠALAK, A.: Effect of Compacting Pressure and 0.5% Mo on Properties of Sintered Iron Powder and Manganese Steels. *Powder Metallurgy*, 1981, no. 2, p. 64.

[72] UNAMI, S., OGURA, K. and UENOSONO, S.: Effect of V on the Strength of Sintered Cr-Mo Alloyed Steel Powder Compacts. In 1998 PM World Congress & Exh., Granada, EPMA Shrewsbury, vol. 3, 1998, p. 173.

[73] ORBAN, M.: Vanadium Using as Microalloying Element in Sintered Structural Steels. In Euro PM2003. Eur. Congress & Exh. on Powder Metall., Valencia, EPMA Shrewsbury, vol. 1, 2003, p. 399.

[74] KLEIN, AN., OBERACKER, R. and THÜMMLER, F.: High Strength Si-Mn-alloyed Sintered Steels. *PMI*, 1985, vol. 17, no. 1, p. 13.

[75] ŠALAK, A., DUDROVÁ, E. and SELECKÁ, M.: Magnetic Properties of Fe-Si Alloys Sintered in High Vacuum. *Kovové Materiály (Metallic Materials)*, 1995, vol. 33, no. 5, p. 289.

[76] SELECKÁ, M., ŠALAK, A. and DANNINGER, H.: The Effect of Boron Liquid Phase Sintering on Properties of Ni, Mo and Cr-alloyed Structural Steels. *J. of Materials Processing Technology*, 2003, vol. 141, p. 379.

[77] COLE, ChJ.: Gear Wheels Rolled from Powder Metal Blanks. Patent WO 92/05897, 1992 (GB).

[78] BENGTSSON, S., FORDÉN, L., DIZDAR, S. and JOHANSSON. P.: Surface Densified P/M Transmission Gear. In 3rd Int. Conf. on Powder Metall. for Automotive Parts, Isfahan, A. Arvand and H. Danninger eds., MPM Co, Tehran, 2002, p. 125.

[79] FORDÉN, L., BENGTSSON, S. and KUYLENSTIERNA, C.: Performance and Properties of Surface-densified PM Trasmission Gear. In 2002 World Congress on Powder Metall. & Part. Mat., Orlando, MPIF Princeton, Part 2, 2002, p. 50.

[80] JONES, P.K., BUCKLEY-GOLDER, K. and SARAFINCHAN, D.: Developing P/M Gear Tooth and Bearing Surfaces for High Stress Application. *The Int. J. of Powder Metall.*, 1998, vol. 24, no. 1, p. 26.

[81] SELECKÁ, M. and ŠALAK, A.: Durability and Failure of Powder Forged Rolling Bearing Rings. *WEAR*, vol. 236, 1999, p. 47.

[82] ŠALAK, A.: Gears for Hydrogenerators Produced from Powder Manganese Steel. *Strojírenská Výroba*, 1990, vol. 38, no. 12, p. 45, in Slovak.

[83] GURYEVICH, JuG. and RACHMANOV, VI.: *Heat Treatment of Powder Steels (Termoobrabotka poroshkovykh staley)*. Ed. METALLURGIYA, Moscow, 1985, in Russian.

[84] MATTY, MT.: Heat Treatment Processes Applicable to PM Parts. *MPR*, 1982, no. 10, p. 511.

[85] DUCHESNÉ, E., L'ESPERANCE and De REGE, A.: Sinter-hardening and Hardenability. *The Int. J. of Powder Metall.*, 2000, vol. 36, no. 1, p. 49.

[86] YU, Y.: Characterization of Sinter Hardening Effect of Astaloy CrM. In Euro PM2001. Eur. Congress & Exh. on Powder Metall., Nice, EPMA Shrewsbury, vol. 1, 2001, p. 52.

[87] CIAS, A., SULOWSKI, M., MITCHELL, SC. and WRONSKI, AS.: Sinter Hardening of Fe-Mn-C Steels. In Euro PM2001. Eur. Congress & Exh. on Powder Metall., Nice, EPMA Shrewsbury, vol. 4, 2001, p. 246.

[88] SUZUKI, H., SATOH, M., YOSHIDA, M. and SEKI, Y.: Sinter Hardening of Prealloyed Powder. In 2000 Powder Metall. World Congress, Kyoto, K. Kosuge and H. Nagai eds., JPMA, Part 2, 2000, p. 931.

[89] ROSTOKER, W. and CLEMENS, DR.: A Hardenable PM Steel through Infiltration. *The Int. J. of Powder Metall. & Powder Tech.*, 1981, vol. 17, no. 4, p. 277.

[90] DANNINGER, H., KREMEL, S., MOLINARI, A., MARCU PUSKAS, T., TORRALBA, JM., CAMPOS-GOMEZ, M. and YU, Y.: Heat Treatment of Cr-Mo Sintered Steels Based on Astaloy CrM. In Euro PM2001. Eur. Congress & Exh. on Powder Metall., Nice, EPMA Shrewsbury, vol. 1, 2001, p. 28.

[91] ALTENA, H. and SCHRANK, F.: Low Pressure-Carburizing with High Pressure-Gas Quenching. *Berg- und Hüttenmännische Monatshefte*, 2003, vol. 148, no. 3, p. 89, in German.

[92] SEYRKAMMER, J., BLAIMSCHEIN, F., DELARCHE, C. and POURPRIX, Y.: Bimetal Sintered Rocker Arm. In Adv. in Powder Metall. & Part. Mat., MPIF Princeton, vol. 5, 1992, p. 541.

[93] ROSSO, M., GUELFO, S. and ACTIS GRANDE, M.: Induction Hardening of Sintered Steels: Novel Potentiality. In Euro PM2003. Eur. Congress & Exh. on Powder Metall., Valencia, EPMA Shrewsbury, vol. 1, 2003, p. 345.

[94] ŠALAK, A. SELECKÁ, M. and DANNINGER, H.: Manganese Assisted Pack Boriding of Sintered Steels. *The Int. J. of Powder Metall.*, 2001, vol. 37, no. 4, p. 53.

[95] ŠALAK, A. and SELECKÁ, M.: New Aspects for Sinter Boriding PM Steels. *Powder Metallurgy Progress*, 2002, vol. 2, no. 3, p. 161.

[96] DANNINGER, H., SPOLJARIC, D. and JANGG, G.: Characterization of Pressed and Sintered Ferrous Materials by Quantitative Fractography. *Pract. Metallography*, 1994, vol. 31, no. 2, p. 56.

[97] DANNINGER, H., SONNTAG, U., KUHNERT, B. and RATZI, R.: Automatic Measurement of the Effective Load Bearing Cross Section Ac in Sintered Steel. *Pract. Metallography*, 2002, vol. 39, no. 8, p. 414.

[98] DANNINGER, H., JANGG, G., WEISS, B. and STICKLER, R.: Microstructure and Mechanical Properties of Sintered Iron. Part I. *PMI*, 1993, vol. 25, no. 3, p. 111; Microstructure and Mechanical Properties of Sintered Iron. Part II. *PMI*, 1993, vol. 25, no. 4, p. 170.

[99] ŠALAK, A., MIŠKOVIČ, V., DUDROVÁ, E. and RUDNAYOVÁ, E.: The Dependence of Mechanical Properties of Sintered Iron Compacts upon Porosity. *PMI*, 1974, vol. 6, no. 3, p. 128.

[100] ŠALAK, A.: Relation between Ultimate Tensile Strength, Elongation and Hardness of Sintered Porous Iron Materials. *Strojírenství*, 1974, vol. 24, no. 7, p. 413, in Slovak.

[101] ŠALAK, A.: Effect of Iron Powder Grades and Processing Conditions upon Variational Range of Mechanical Properties Values of Sintered Compacts. *Kovové materiály (Metallic Materials)*, 1977, vol. 15, no. 1, p. 95, in Slovak.

[102] BOULGER, FW: *Machinability of Steels*. Metals Handbook, vol. 1, 10th Ed., Selections: Iron, Steels & High Performance Alloys, ASM Metals Park, 1990.

[103] BLACK, JT.: Introduction to Machining Processes. *Metals Handbook*, 9th Ed., vol. 16, Machining, ASM Metals Park, 1987, p. 1.

[104] SANDEROW, HI. and PRUCHER, T.: The Effect of Manganese Sulfide on the Mechanical Properties of P/M Steels. In Adv. in Powder Metall. & Part. Mat., APMI Princeton, vol. 4, 1993, p. 97.

[105] MADAN, DS.: The Importance of Machinability in the Processing of P/M Parts. In Adv. in Powder Metall. & Part. Mat., MPIF Princeton, vol. 8, 1996, p. 55.

[106] GENSKY, R.: *Metal Cutting Operations - Technical Fundamentals*. Edition Leipzig, 1966.

[107] GEIJER, E. and JAMISON, RB.: Machinability of Sintered Iron. In Source Book on Powder Metallurgy, comp. by S. Bradbury, ASM Metals Park, 1979, p. 256.

[108] TÖNSHOFF, HK., WOBKER, HG. and FRITSCH, A.: Machining of Powder Metallurgy Materials. In Adv. in Powder Metal. & Part. Mat., MPIF Princeton, vol. 8, 1995, p. 105.

[109] PIGNON, AP.: Secondary Machining Operations for P/M Products. In Mod. Dev. in Powder Metall., MPIF Princeton, H. Hausner, W. and E. Smith eds., vol. 8, 1974, p. 163.

[110] WEBER, H. and LOLADZE, TN.: *Fundamentals of Cutting (Grundlagen des Spanens)*. VEB Verlag Technik Berlin, 1986, in German.

[111] BARYLSKI, A.: Surfaces Formability of PM Materials by Lapping. In Euro PM2003. Eur. Congress & Exh. on Powder Metall., EPMA Shrewsbury, vol. 1, 2003, p. 182.

[112] VASILKO, K., HRUBÝ, L. and LIPTÁK, J.: *Technology of Cutting and Mounting* (*Technológia obrábania a montáže*). Ed. ALFA Bratislava, 1991, in Slovak.

[113] TRENT, EM.: Metal Cutting and the Tribology of Seizure: I. Seizure in Metal Cutting, *Wear*, vol. 128, 1988, p. 29.

[114] MONKOVÁ, K. and MONKA, P.: Optimization of Geometrical Parameters of Cutting Tools. In Int. Conf. 'Strojárska technológia a valivé ložiská 97', Žilina University, Žilina, p. 258, in Slovak.

[115] CIMINO, TM. and LUK, SH.: Machinability Evaluation of Selected High Green Strength P/M Materials. In Adv. in Powder Metall. & Part. Mat., MPIF Princeton, vol. 8, 1995, p. 129.

[116] SMITH, GT: Drilled Hole Quality Assesment in Ferrous PM Components Using Surface Integrity Techniques. *Powder Metallurgy*, 1988, vol. 31, no. 2, p. 117.

[117] TRENT, EM.: *Metal Cutting*. BUTTERWORTHS & Co, London, 1977.

[118] *Machining Guidelines. Höganäs Handbook for Sintered Components* 5. Höganäs AB, 1998.

[119] POIRIER, JP. and GAGNÉ, M.: A new Approach to Improve Machinability of High Strength PM Parts. *MPR*, 1990, vol. 45, no. 3, p. 184.

[120] SOVKOV, AF and GOGOLEV, AJa.: Experimental Determination of Cutting Force at Cutting with Cutting-off Tools. In Isl. v oblasti

poroshkovoy metallurgyi. Novoch. Politech. Inst., Novocherkask, 1972, p. 7, in Russian.

[121] DUBOVSKÁ, R.: Physico-chemical Features in Cutting Zone and Conditions for their Stabilisation. In Int. Conf. 'Strojárska technológia a valivé ložiská 97', Žilina University, Žilina, 1997, p.183, in Slovak.

[122] ÁBEL, M. and IŽO, P.: Effect of Energy on the Cutting Process. In Int. Conf. 'Strojárska technológia a valivé ložiská 97', Žilina University, Žilina, 1997, p. 186, in Slovak.

[123] ŠLESÁR, M. and VASILKO, K.: Physico-netallurgical Views on the Process of the Formation of a Chip. *Strojírenství,* 1988, vol. 38, no. 7, p. 426, in Slovak.

[124] FLOREK, R.: Wear Behaviour of C-free Carbide HSS at Cutting of Iron Based Materials. PhD Thesis, Vienna University of Technology, Vienna, 2003, in German.

[125] KENDALL, LA: Tool Wear and Tool Life. *Metals Handbook*, Ninth Ed., vol. 16, Machining, ASM Metals Park, 1987, p. 38.

[126] AGAPIOU, JS., HALLDIN, GW. and De VRIES, MF.: Effect of Porosity on the Machinability of P/M 304L Stainless Steel. *The Int. J. of Powder Metall.,* 1989, vol. 25, no. 2, p. 127.

[127] PAUER, G.: Use of Cemented Carbides in Automobile Manufacturing. In 2nd Int. Conf. on Powder Metall. for Automotive Parts, Isfahan, A. Arvand and H. Danninger eds., MPM Co, Tehran, 1999, p. 65.

[128] *Modern Metal Cutting - a Practical Handbook.* AB Sandvik Coromant, 1994.

[129] CHANDLER, E.: Machining of Powder Metallurgy Materials. In ASM Handbook, ASM Metals Park, 1990, p. 799.

[130] ROMERO, A., VIVANCOS, J., SIERRA, C. and PRADO, J.M.: Rapid Characterisation of Machinability on PM Steel Parts. In 1998 PM World Congress and Exh., Granada, EPMA Shrewsbury, vol. 2, 1998, p. 364.

[131] FIELD, M., KOHLES, JF. and KOSTER, WP.: Surface Finish and Surface Integrity. *Metals Handbook,* Ninth Ed., vol. 16, Machining, ASM Metals Park, 1987, p. 19.

[132] KAHLES, JF.: Elements of the Machining Process. *Metals Handbook*, Desk Ed., ASM Metals Park, 1985, p.27.

[133] Machining Performance of Ferrous Metals Using an Automatic Screwbar Machine. E 618-81-03.01, *Annual Book of ASTM Standards.* Am. Soc. for Testing and Materials, ASM Metals Park.

[134] REEN, OW.: The Machinability of P/M Materials. In Mod. Dev. in Powder Metall., HH. Hausner and PV. Taubenblatt eds., APMI Princeton, vol. 10, 1976, p. 431.

[135] CHOPRA, KS.: Manganese Sulfide in Machining Grade Ferrous P/M Alloys. In World PM Congress & Exh., Orlando, APMI Princeton, vol. 21, 1988, p. 361.

[136] CAUSTON, RJ and CIMINO, T.: Machinability of P/M Steels. *ASM Handbook*, Powder Metal Technologies and Applications, ASM Metals Park, vol. 7, 1998, p. 671.

[137] CAUSTON, RJ: Machinability of P/M Steels. In Adv. in Powder Metall. & Part. Mat., MPIF Princeton, vol. 8, 1995, p. 149.

[138] LESKOVAR, P. and GRUM, J.: The Metallurgical Aspects of Machining. *Annals of CIRP,* 1986, vol. 36, no. 2, p.537.

[139] CHADWICK, GA. and KEHOE, FP.: Machinability of Sintered Ferrous Powder Metallurgy Compacts. *Report Chandlers Ford,* Eastleigh.

[140] HAZRA, J., TARAMAN, K., KENNEDY, FM. and BADIA, F.: Machinability of Powder Metallurgy Components. A Look at Surface Finish. In Mod. Dev. in Powder Metall., HH. Hausner and P. Taubenblatt eds., MPIF Princeton, vol. 10, 1977, p. 385.

[141] ANDERSEN, PJ. and HIRSCHORN, JS.: Use of Additives to Improve the Machinability of Sintered Steels. In Mod. Dev. in Powder Metall., HH. Hausner and P. Taubenblatt eds., MPIF Princeton, vol. 10, 1977, p. 477.

[142] AGAPIOU, JS. and DeVRIES, MV.: Machinability of Powder Metallurgy Materials. *The Int. J. of Powder Metall.*, 1988 vol. 24, no. 1, p. 47

[143] KVIST, S.: Turning and Drilling of Some Typical Sintered Steels. *Powder Metallurgy,* 1969, vol. 12, no. 24, p. 538.

[144] CAPUS, JM. and FURNEL, C.: Tool Wear Measurements in Machining of Sintered Ferrous Alloys. In Progress in Powder Metallurgy. Nat. Powder Metall. Conf., Philadelphia, JM. Capus and DL. Dyke eds., MPIF Princeton, vol. 37, 1981, p.165.

[145] SANDEROW, HI., SPIRKO, JR. and CERRENTE, R.: The Machinability of P/M Materials as Determined by Drilling Tests. In Adv. in Powder Metall. & Part. Mat., (PM Conf. Chicago), comp. by RM. McKotch and R. Webb, MPIF Princeton, vol. 2, 1997, p. 15.

[146] BERG, S., THOORS, H. and STEEN, B.: Machinability of Sintered Steels - Guidelines for Turning, Drilling and Tapping. In Euro PM97. Adv. in Structural PM Component Production, Munich, EPMA Shrewsbury, 1997, p. 166.

[147] CHOPRA, KS.: Improvement of Machinability in PM Parts Using Manganese Sulfide. In Progress in Powder Metall. 1987 Annual Powder Metall. Conf., Dallas, comp. By CL. Froeby and H. Hjort, APMI Princeton, vol. 21, 1987, p.501.

[148] ENGSTRÖM, U.: Machinability of Sintered Steels. In Progress in Powder Metall. 1982. Nat. Powder Metall. Conf., Quebec, JG. Bewley and SW. McGoee eds., APMI Princeton, vol. 38, p. 417.

[149] HAMMIUDIN, M. and MURTAZA, Q.: Machinability of Sintered and Heat Treated Steels. In 7th Int. Conf. on Sintering, New Delhi, 2000, p. 2.

[150] *Machining*. ASM Metals Handbook, Desk Ed., ASM Metals Park, 1985.

[151] SHAREEF, I. and BOSWELL, KH.: Production Models for Drilling Sintered Steel. In Adv. in Powder Metall. & Part. Mat., comp. by Ch. Lall and AJ. Neupauer, APMI Princeton, vol. 7, 1994, p. 153.

[152] PEARSON, GL: Machining P/M Components. In Source Book on Powder Metallurgy, comp. by S. Bradbury, ASM Metals Park, 1979, p. 273.

[153] ŠALAK, A., VASILKO, K. and SELECKÁ, M.: Machining of PM Materials by Drilling with Constant Thrust Force. In DF PM 2002. Deformation and Fracture of PM Materials, Stará Lesná, Ľ. Parilák and H. Danninger eds., IMR SAS Košice, vol. 1, 2002, p. 143.

[154] ANDERSSON, O.: A Machinability Enhancing Additive for High Strength Materials. In 1993 Powder Metall. World Congress, Tokyo, JSPM, 1993, p. 1101.

[155] CAPUS, JM.: Choosing PM Steels for Improved Machinability. In Productivity and Powder Metall., Metal Powder Conf., Frankfurt a. M., EPMA Shrewsbury, 1982, p. 39.

[156] HRNCIAR, J: Cutting of Powder Forged Steels in Turning. In Int. Conf. 'Strojárska technológia a valivé ložiská 97', Žilina University, Žilina, 1997, p.138, in Slovak.

[157] BERG, S. and THORNDENBERG, H.: Cutting Data Recommendations for Turning of Various PM Materials Using Different Tool Grade and Geometry Sections. In 1998 PM World PM Congress & Exh., Granada, EPMA Shrewsbury, vol. 2, 1998, p. 390.

[158] BÉKÉS, J.: *Engineering Technology of Machining of Metals (Strojárska technológia obrábania kovov)*. Ed. ALFA Bratislava, 1981, in Slovak.

[159] JAKUBÉCZYOVÁ, D. and BLACH, J.: Microstructure of PM High Speed Steels and Effect of Heat Treatment on Distribution of Carbides. *Hutnické Listy*, 2001, no. 4-5, p. 28, in Slovak.

[160] PÁLKA, V., JANÁC, A. and LIPA, Z.: Contribution to Testing of Cutting Tools Life. In Int. Seminary 'Nástrojové Materiály' (Tool Materials), Ostravice. Technical University Ostrava, 1994, p. 238, in Slovak.

[161] KUTSCH, W. and BEISS, P.: Machinability of Sintered Stainless Steel 316L. In Adv. Powder Metall. & Partic. Mat., APMI Princeton, vol. 5, 1996, p. 17.

[162] KOOS, R., BOCKSTIEGEL, G. and MUHREN, C.: Machining Studies of PM Forged Materials. In SAE Technical Papers 790192. In Congress and Exh. of Aut. Eng. Soc. Warrendale, Detroit, 1979, *MPR*, vol. 35, no. 10, p. 449.

[163] KÖNIG, W., SALÉ, G., RÖBER, G., STRÖMGREN, M., DORWEILER, HJ., LINDNER, KH., SCHILLER, H. and SCHRIEFER, K.: Powder Forging of Cylindrical Gears with Multiphase Structure (PM World Congress, Orlando, 1988); *Höganäs PM Iron Powder Information* PM 88-6.

[164] EDWARDS, R.: *Cutting Tools*. The Institute of Materials, London, 1993.

[165] KÖNIG, W. and NEISES, A.: Comparison of the Cutting Stage - Tendencies and Potentions in Machining. In Symp. 'Pulvermetallurgie in Wettbewerb', Haagen, 1994, Nov., p. 143, in German.

[166] KOURIL, K.: The Application of Super Hard Cutting Materials. In Int. Conf. 'Strojárska technológia a valivé ložiská 97', Žilina University, Žilina, 1997, p. 116, in Czech.

[167] PASTOR, H.: Materials for Cutting Tools, Mining Tools and Wear Parts. In (CD) 3rd Int. Powder Met. Conf., Ankara, TTMD, Gazi Univ., 2002, p. 68.

[168] EDWARDS, G., KRAUS, G. and KENNEDY, R.: *Tool Steels*. ASM Metals Park, 1998.

[169] JAKUBÉCZYOVÁ, D. and FÁBEROVÁ, M.: Mechanical Properties and Surface Treatment of PM Cobalt High Speed Steel. *Powder Metallurgy Progress*, 2002, vol. 2, no. 3, p.188.

[170] FREMUNT, P., KREJCÍK, J. and PODRÁBSKÝ, T.: *Tool Steels (Nástrojové oceli)*, Ed. Dom techniky Brno, 1994, in Czech.

[171] KARAGÖZ, S. and FISCHMEISTER, HF.: Cutting Performance and Microstructure of High Speed Steels: Contribution of Matrix Strengthening and Undissolved Carbides. *Met. Trans.*, 1998, vol. 29A, p. 205.

[172] HELLMAN, P., LARKER, H., PFEFFER, JB. and STROMBLAD, I.: The ASEA-Stora-Process: A New Process for the Manufacture of Tool Steel and Other Alloy from Powder. In Mod. Dev. in Powder Metall., HH. Hausner ed., Plenum Press, New York-London, Part 4, 1971, p. 573.

[173] RABITSCH, R., LIEBFAHRT, W. and MAKOVEC, H.: Properties of PM High Speed and Tool Steels Produced with the Newest PM-Technology. In Euro PM2001. Eur. Congress & Exh. on Powder Metall., Nice, EPMA Shrewsbury, vol. 1, 2001, p. 219.

[174] SPIEGELHAUER, C. and SANDBERG, O.: Production of Tool Steels Using the Spray Forming Technology. In Adv. Powder Metall. & Part. Mat. 2002, comp. by V. Arnhold, CL. Chu, WF. Jandeska and H. Sanderow, MPIF Princeton, vol. 4, 2002, p. 84.

[175] BEISS, P., WÄHLING, R. and DUDA, D.: Toughness of Vacuum Sintered P/M High Speed Steels. In Mod. Dev. in Powder Metall., EN. Aqua and CI. Whitman eds., MPIF Princeton, vol. 17, 1985, p. 331.

[176] SMART, J., REED, BW. and BREWIN, PR.: Pressing and Sintering Methods to Produce High Grade Tool Steels. *MPR*, 1980, vol. 35, no. 6, p. 241.

[177] HILLSKOG, T.: Advantages of Powder Metallurgical Tool Steels. In Int. Seminary 'Nástrojové Materiály' (Tool Materials), Ostravice, Technical University Ostrava, 1994, p.17.

[178] MICROCLEAN. High Performace Steels Produced by Powder Metallurgy Methods. Böhler Kapfenberg - prospectus, 2001.

[179] JAKUBÉCZYOVÁ, D., FÁBEROVÁ, M. and PARILÁK, Ľ.: The Effect of Co-, Nb- and Ti-Addition on the Structure and Properties of PM High Speed Steel. In Euro PM2001. Eur. Congress & Exh on Powder Metall., Nice, EPMA Shrewsbury, vol. 1, 2001, p. 297.

[180] CAPUS, JM.: Machinability Studies with Sintered Ferrous Alloys, SAE Technical Paper 810246, SAE Int. Congress and Exh., Feb., 1981, p. 55.

[181] TORRALBA, JM. and GORDO, E.: PM High Speed Steel Matrix Composites. State of the Art. *Powder Metallurgy Progress*, 2002, vol. 2, no. 1, p. 1.

[182] SCHEDLER, W.: *Hardmetal for the Practician (Hartmetall für den Praktiker).* VDI-Verlag, Düsseldorf, 1988, in German.

[183] BROOKES, JAK.: *Hardmetals and other Hard Materials.* Inter. Carbide Data, East Barnet, UK, 1992.

[184] LASSNER, E. and SCHUBERT, WD.: *Tungsten.* Kluwer Academic/Plenum Publishers, New York, 1999.

[185] NEUSTUPA, Y. and SITTA, Š.: Present State and Production of Cemented Carbides. In Int. Conf. 'Spekané karbidy', Dum techniky, Ostrava, 1983, p. 2, in Czech.

[186] KIPARISOV, SS. and LIBENSON, GA.: *Powder Metallurgy (Poroshkovaya Metallurgiya).* Ed. METALLURGIYA, Moscow, 1991, p. 421, in Russian.

[187] Böhler Kapfenberg; personal communication, 2004

[188] *Machining of Sintered Steels (Zerspanung von Sinterstählen).* Protokol des Auschusses für Pulvermetallurgie, DGM, Oberursel, June, 1995.

[189] KRISHNAMURTHY, R., SENTHILKUMAR, SB. and GOKULARATHAM, CV.: Dispersion Strengthening of Ceramic Cutting Tools for the Improvement of Tool Life Performance. In Adv. in Powder Metall. & Part. Mater., comp. by M. Phillips and J. Porter, MPIF Princeton, vol. 2, 1995, p. 8-91.

[190] SMITH, G.: Getting the Measure of PM Machinability. *MPR,* 1998, no. 5, p. 31.

[191] FURUKUWA, N.: The Application of CBN Tools to Machine PM Parts. *MPR,* 1985, no. 2, p. 99.

[192] NARUTAKI, N. and USUKI, H.: Machinability Sintered Steels. In Horizons of Powder Metall.- 1986. Int. PM Conf. & Exh., Düsseldorf, Verlag Schmid, Freiburg, 1986, p. 393.

[193] BRANDT, G.: Advanced Tool Materials. In Euro PM2001. Eur. Congress and Exh. on Powder Metall., Nice, EPMA Shrewsbury, vol. 1, 2001, p. 90.

[194] SCHUBERT, WD. and HAUBNER, R., Vienna University of Technology, Inst. for Chemical Technologies and Analytics, Vienna; private communication, 2004.

[195] ŠTEKLÁC, D., STANCEKOVÁ, D. and NESLUŠAN, M.: Comparison of Coated and Uncoated HM in Turning and Milling. In Int. Conf. 'Strojárska technológia a valivé ložiská 97'; Žilina University, Žilina, 1997, p. 248, in Slovak.

[196] BROOKS, KJA.: Chicago's IMTS Takes Cutting-Tool Honours. *MPR,* 2001, Jan., p. 16.

[197] JAKUBÉCZYOVÁ, D. and FÁBEROVÁ, M.: Analysis of Structure of PM High Speed Coatings. In Int. Conf. 'Fractography', Stará Lesná, IMR SAS Košice, Ľ. Parilák ed., 2003, p. 410.

[198] VILAR, R., COLACO, R. and AlMEIDA, A.: Laser Surface Treatment of Tool Steels. *Optical and Quantum Electronics,* 1995, vol. 27, p. 1273.

[199] HEBEL, T.: A Better Way to Relieve the Stress. *American Machinist,* 1986, Dec., p. 65.

[200] JABLONOWSKI, J.: Pulsed Magnetism Boots for Tool Life. *American Machinist*, 1987, Jan., p. 80.

[201] ŠALAK, A., DUFEK, V., GAJDUŠEK, J. and ŠOŠKA, J.: Contribution to Improvement the Cutting Tool Life through Magnetic Field. *Strojírenství*, 1995, vol. 41, no. 5, p. 75, in Slovak.

[202] DOROFEYEV, JuG.: *Dynamic Hot Pressing of Porous Powder Preforms (Dinamicheskoye pressovaniye poristykh poroshkovykh zagotovok).* Ed. METALLURGIYA, Moscow, 1977, in Russian.

[203] USUKI, N.: Machinability of Sintered Steels. *MPR* , 1986, no. 11, p. 839.

[204] ABOTT, AG., BALL, AW., WARTMAN, W. and AKPAN, E.: Machining - a Discussion of Some Variables. In Adv. in Powder Metall. & Part. Mater., comp. by M. Phillips and J. Porter, MPIF Princeton, vol. 2, 1995, p. 8-69.

[205] JAMES, PJ.: Factors Affecting Quality of Drilled Holes in Sintered Steels. *Powder Metallurgy*, 1994, vol. 37, no. 2, p.133.

[206] OGATA, Y., SHIRASI, J. and NAKAI, T.: Various PM Parts Machinability using Cutting Tools. In 2000 Powder Met. World Congress, Kyoto, K. Kosuge and H. Nagai eds., JPMA, Part 2, 2000, p. 1578.

[207] LIERSCH, A.: Efffect of Machining Aids on Wear Behaviour and Machinability of Sintered Steel. PhD Thesis, Vienna University of Technology, 1994, in German.

[208] ISHII, K.: Development of High Machinability Sintered Material for P/M Machine Parts. In 2000 Powder Metall. World Congress; 9th Case Studies on New Product Development, Kyoto, JPMA, 2000, p. 1.

[209] BENNER, A.: Machining of PM Steels: Cost Aspects. Höganäs Chair Seminar 2002, University of Trento, Trento, 2002.

[210] SHAREEF, I. and BOSWELL, KH.: Prediction and Modelling of Surface Finish in Drilling of P/M Parts. In Adv. in Powder Metall. & Part. Mat., comp. by RA. McKotch, and R. Webb, MPIF Princeton, vol. 2, 1997, p. 15.

[211] OGATA, Y., SHIRASHI, J. and NAKAI, T.: Various PM Parts Machinability Using Cutting Tools. In 2000 Powder Metall. World Congress, Kyoto, K. Kosuge and H. Nahai, eds., JPMA, Part 2, 2000, p. 1578.

[212] EISENMANN, MR. and NAYAR, HS.: The Influence of Sintering Atmosphere and Powder Composition on Machining Grade Ferrous Components. In Mod. Dev. in Powder Metall. (Powder Met. Conf., Toronto), MPIF Princeton, vol. 15, 1984, p. 785.

[213] FROHBERG, MG.: *Thermodynamics for Metallurgists and Material Engineers (Thermodynamik für Metallurgen und Werkstofftechniker).* VEB DVG, Leipzig, 1980, in German.

[214] BEISS, P., Inst. of Materials Science, RWTH Aachen University, Aachen; personal communication, 2003.

[215] BOCCHINI, GF. and De CATALDIS, C.: The Base Thermodynamics Aspects of Sintering Chromium Steels. *La Metalurgia Italiana*, 1986, no. 6, p. 457, in Italian.

[216] GERMAN, RM.: *Sintering Theory and Practice,* John Wiley & Sons, Inc., New York, 1996.

[217] DANNINGER, H. and XU, Ch.: De-gassing and Reduction during Sintering of High Density PM Steels. In Euro PM2003. Eur. Congress & Exh. on Powder Metall., Valencia, EPMA Shrewsbury, vol.1, 2003, p. 250.

[218] KAUFMANN, SM.: Carbon Control during Sintering of Iron. *PMI,* 1973, vol. 5, no. 1, p. 11.

[219] KONONENKO, WI.: Some Laws Governing Tool Wear during Machining of Porous Material. *PMI,* vol. 2, 1976, p. 24.

[220] ROY, LG., De REWGE, AF. and PEASE, LF.: Relationship between Machinability and Strength in Prealloyed Manganese Sulfide Sintered Material. In World PM Congress & Exh. on Powder Metall., Orlando, APMI Princeton, vol. 21, 1988, p. 327.

[221] BOLCHKOV, AN.: Effect of Porosity of Powder Materials and Cutting Conditions on Densification of Chips. In 'Isl. v oblasti dinamicheskoy shtampovki poroshkovykh materialov', Novoch. Politech. Inst., Novocherkask, 1984, p. 45, in Russian.

[222] GAGNÉ, M., GAUMONT, M. and OLSCHEWSKI, G.: Effect of Machinability Enhancing Additives on the Machining Behaviour of PM Materials. In Euro PM2000. Conf. on Materials and Processing Trends for PM Components in Transportation, Munich, EPMA Shrewsbury, 2000, p. 25.

[223] LOEB, AE.: Thermal Conductivity VIII, A Theory of Thermal Conductivity of Porous Materials, *J. of Am. Ceram. Soc.*, 1954, vol. 37, no. 2, p. 96.

[224] MADAN, DS.: An Update on the Use of Manganese Sulfide (MnS) Powder in Powder Metallurgy Application. In Adv. in Powder Metall., APMI Princeton, vol. 3, 1991, p. 101.

[225] RUTHARDT, R.: *Optimisation of Production. Powder Metallurgy Science and Practice* (*Fertigungsoptimierung. Pulvermetallurgie in Wissenschaft und Praxis),* vol. 12, DGM Verlag, 1996, in German.

[226] SIMCHI, A. and DANNINGER, H.: Electrical Conductivity and Microstructure of Sintered Iron. *Powder Metallurgy*, 2000, vol. 43, no. 3, p. 209.

[227] KRAVCHENKO, VF. and USTIMENKO, VI.: Effect of Porosity of Powder Metallurgy Materials on Cutting Temperature. In Isl. v oblasti poroškovoj metallurgiji, Novoch. Politech. Inst., Novocherkask, 1972, p. 34, in Russian.

[228] KASIJAN, MV. and ZAJMTSIJAN, GM.: Effect of Porosity on Cutting Forces. *Poroshkovaya Metallurgiya*, 1973, no. 6, p. 58, in Russian.

[229] DOROFEYEV, JuG. and USTIMENKO, VI.: Machining of Powder Metallurgy Materials Prepared by Hot Dynamic Pressing. *Poroshkovaya Metallurgiya,* 1971, no. 5, p. 76, in Russian.

[230] CAPUS, JM. and FOURNEL, C.: The Influence of Mix Composition on the Machinability of Sintered Iron. In Mod. Dev. in Powder Metall., MPIF Princeton, vol.13,_1981, p.137.

[231] HOUDREMONT, E.: *Manual of Special Steel Science* (*Handbuch der Sonderstahlkunde*). Verlag Stahleisen Düsseldorf, vol. I, 1956, in German.

[232] BERG, S.: Machinability of Sintered Steels - Guidelines for Turning, Drilling and Tapping. In Adv. in Powder Metall. & Part. Mat.-1997, comp. by RA. McKotch and R. Webb, MPIF Princeton, vol. 2, 1997, p. 15.

[233] ANDERSSON, O.: A Machinability Enhancing Additive for High Strength Materials. In 1993 Powder Metall. World Congress & Exh., Tokyo, JPMA, 1993, p. 1101.

[234] NAVARA, E.: Investigation of Friction Properties of Structure Constituents in Sintered Steel. In Problems in Powder Metallurgy (Problémy v práškovej metalurgii), SAV Bratislava ed., 1964, p. 127, in Slovak.

[235] *Atlas of Microstructures.* Ferrous Powder Metallurgy. Höganäs AB PM 86-6.

[236] HUPPMANN, WJ. and DALAL, K.: *Metallographic Atlas of Powder Metallurgy.* Verlag Schmid, Freiburg i. H., 1986.

[237] *METALLOGRAPHY.* Höganäs Handbook for Sintered Components. Höganäs AB, 1999.

[238] BEISS, P., DALAL, K. and PETERS, R.: *International Atlas of Powder Metallurgical Microstructures.* MPIF Princeton, 2002.

[239] IACOCCA, RG., SMITH, DC. and GERMAN, RM.: A New Technique for the Evaluation of Die Compaction Lubricants. *The Int. J. of Powder Metall.*, 1999, vol. 35, no. 5, p. 54.

[240] CAPUS, JM.: Machinability Studies with Sintered Ferrous Alloys. *MPR,* 1982, no. 5, p. 247.

[241] CLAEYS, SF. and CHOPRA, KS.: Enhanced Machinability and Oxidation Resistance of P/M Steels Using Modified MnS Additions. *The Int. J. of Powder Metall.,* 1998, vol. 34, no. 3, p. 29.

[242] CAUSTON, RJ., CIMINO, TM. and SCANLON, HM.: Machinability Improvement of P/M Steels. In Adv. in Powder Metall. & Part. Mat.-1994 (PM Conf. Toronto), MPIF Princeton, vol. 7, p. 169.

[243] TRUDEL, Y., CILOGLU, C. and TREMBLAY, S.: Selecting Additives to Improve Machinability of Ferrous PM Parts. In Mod. Dev. in Powder Metall., MPIF Princeton, vol. 15, 1984, p. 775; *MPR,* 1985, no. 11, p. 639.

[244] MADAN, DS.: Effect of Manganese Sulfide (MnS) on Properties of High Performance P/M Alloys and Applications. In Adv. in Powder Metall. & Part. Mat. (PM World Congress, San Francisco), comp. by JM. Capus and RM. German, APMI Princeton, vol. 4, 1992, p. 245.

[245] DUDROVÁ, E. and ŠALAK, A.: Effect of Sulphur Addition on Some Properties of Sintered Steel. *Pokroky Práškové Metalurgie* (VUPM), 1971, no. 1, p. 3, in Slovak.

[246] EXNER, HE. and FISCHMEISTER, HF.: Theories of Sintering, Part III: Liquid Phase Sintering. *Metall*, 1965, vol. 19, no. 9, p. 941, in German.

[247] OKAMOTO, H.: *Monograph Series on Alloy Phase Diagrams 9. 'Phase Diagrams of Binary Iron Alloys'*. ASM Metals Park, 1993.

[248] ROY, LG., ESPÉRANCE, GL. and LAMBERT, P.: Prealloyed MnS Powders for Improved Machinability in PM Parts. *MPR*, 1989, no. 2, p. 116.

[249] KOLESNICHENKO, LF., TRUSKO, PV., PUGINA, LI., AGEYEVA, VS. and PUSHKAREV, VV.: Thermal Stability of Sulphides of Some Metals in Sintered Materials on the Iron Basis. *Poroshkovaya Metallurgiya*, 1972, no. 7, p. 56, in Russian.

[250] RAGHAVAN, G.: *Phase Diagrams of Ternary Iron Alloys Part 2: Ternary System Containing Iron and Sulfur*. The Ind. Inst. of Metals, Calcutta, 1988.

[251] DANNINGER, H., LIERSCH, A. and RATZI, R.: Influence of Additives in PM Steels on the Quality of Machined Surfaces. In 1998 Powder Metall. World Congress & Exh., Granada, EPMA Shrewsbury, vol. 2, 1998, p. 474.

[252] EISENMANN, MR., NAYAR, HS. and HILL, M.: Sintering of Machining Grade Ferrous Compacts for Improved Machinability. *MPR*, 1985, no. 2, p. 96.

[253] LEE, KY., PARK, DK. and KIM, HB.: Sintered Stability and Machinability of PM Steels Using Modified MnS Additions. In Adv. 2002 World Congress on Powder Metall. & Part. Mat., Orlando, MPIF Princeton, Part 12, 2002, p. 72.

[254] CLAYES, SF. and CHOPRA, KS.: Enhanced Machinability and Oxidation Resistance with MnS Containing Additions of Iron. In Adv. in Powder Metall. & Part. Mat. - 1997 (Int. PM Conf. Chicago), comp. by RM. McKotch and R. Webb, MPIF Princeton, 1997, p. 15-111.

[255] PATEL, BK. and MADAN, DS.: Machining Unleaded Brass PM Parts. In Adv. in Powder Metall. & Part. Mat. - 1992 (PM World Congress San Francisco), comp. by JM. Capus and RM. German, MPIF Princeton, vol. 4, 1992, p. 297.

[256] MADAN, DS.: Shelf Life of MnS Powder and MnS Containing Mixes. In Adv. in Powder Metall. & Part. Mat., MPIF Princeton, vol. 2, 1995, p. 177.

[257] Chemetall GmbH, Arnoldstein, Austria, unpublished report; private communication, 2000.

[258] PUGINA, LI., SINAVSKAYA, MD. and MAKSIMCHUK, IM.: *Molybdenum Disulphide*. Ed. NAUKOVA DUMKA, Kiev, 1968, in Russian.

[259] SIMILEJSKIJ, BM., MIROSHINKOV, VI. and LOZOVOJ, BI.: Effect of Heating Modes on Thermal Stability of Molybdenum Disulphide in the

Fe-MoS_2 Composite. *Poroshkovaya Metallurgiya,* 1985, no. 6, p. 86, in Russian.

[260] ŠALAK, A. and HAVALDA, A.: Properties of Powder Forged and HIPed Composite Materials - Steel - MoS_2. In Sixth Int. Symp. on Comp. Mat., Stará Lesná, Inst. of Metals SAS Bratislava, vol. II, 1986, p. 333, in Russian.

[261] BROWNLIE, HA. and PUMELL, CG.: Reaction Sintering to Improve Machinability of Ferrous PM Materials. In PM94. Powder Metall. World Congress, Paris, EPMA Shrewsbury, vol. II, 1994, p. 891.

[262] VEROPACHA, DN.: Investigation of the Additions of Bismuth, Tellurium and Selenium on Properties of Steels. *Izv. VUZ,* Sev. Kavk. Region, Techn. Nauki, 1994, p. 1, in Russian.

[263] TAUBENBLATT, PW., SMITH, EE. and BLADT, FA.: Iron-Copper-Tellurium - a New Machinable Iron-base P/M Alloy. In Mod. Dev. in Powder Metall., MPIF Princeton, vol. 10, 1977, p. 467.

[264] DANNINGER, H., YANGG, G. and GIAHI, M.: Production of Stainless Steel Filters by Liquid Phase Activated Gravity Sintering. *Mat.-wiss. und Werkstofftechnik,* 1988, vol. 19, no. 6, p. 205, in German.

[265] MUISENER, ChM.: Resin Impregnaton of Powder Metal Parts. In ASM Handbook, ASM Metals Park, vol. 7, 1998, p. 688.

[266] FULMER, JJ. and BLANTON, JM.: The Effect of Microstructure on the Machinability of an MPIF FC-0208 Copper Steel. In Adv. in Powder Metall. & Part. Mat. -1992 (PM World Congress San Franciso), comp. by JM. Capus and RM. German, MPIF Princeton, vol. 4, 1992, p. 283.

[267] BENNER, A. and BEISS, P.: Green Machining of Warm Compacted PM Steels. In Euro PM2000 Conf. on Materials and Processing Trends for PM Components in Transportation, Munich, EPMA Shrewsbury, 2000, p. 101.

[268] KUCHUGURENKO, RV. and GOGOLEV, AYa.: Effect of Solid Grease Coating on Wear and Life Time of Cutting Tools at Turning of Hard Machinable Steels. In Isl. v oblasti poroshkovoy metallurgiyi, Novoch. Politekh. Inst., Novocherkask, 1972, p. 73, in Russian.

[269] ŠALAK, A., RUDNAYOVÁ, E. and SELECKÁ, M.: Internal Nitriding of Sintered Iron in Nitrogen. *Powder Metallurgy,* 1999, vol. 45, no. 1, p. 45.

[270] ALAM, AM., CHIERAGATTI, R., AKAMATSU, M. and CAUMEL, Z.: Effect of Machining Procedure on the Surface State of Low Carbon Steel. *Scripta Materialia,* 1997, vol. 37, no. 10, p. 1559.

[271] PLAMONDON, P., L'ESPÉRANCE, G. and BLAIS, C.: Optimization of Prealloyed MnS Steel Powders for Improved Machinability. In World Congress in Powder Metall. & Part. Mat., New Orleans, APMI Princeton, Part 6, 2001, p. 29.

[272] POIRIER, J., ST-LAURENT, S., ILLIA, E., O'NEILL, M. and LEE, J.: A Success Story on the Use of Hexagonal BN for Demanding Machinability Applications. In Euro PM2003: Eur. Congress & Exh. on Powder Metall., Valencia, EPMA Shrewsbury, vol. 1, 2003, p. 166.

[273] CAUSTON, RJ.: Role of Additives in PM Machining. In Adv. 2002 World Congress on Powder Metall. & Part. Mat., Orlando, MPIF Princeton, Part 12, 2002, p. 47.

[274] UENOSONO, S., UNAMI, S. and OGURA, K.: A New Improvement in the Machinability of P/M Steel due to Retained Graphite Particles. In Adv. in Powder Metal. & Part. Mat., MPIF Princeton, 1995, p. 8-171.

[275] YAMANISHI, Y., TSUTSI, T. and ISHII, K.: The Development of a High Machinability Sintered Material for Bearing Caps by Inhibiting the Diffusion of the Graphite. In 2000 Powder Metall. World Congress, Kyoto, K. Kosuge and H. Nahai eds., JPMA, Part 2, 2000, p. 955.

[276] SHAREEF, I. and BOSWELL, KH.: Residual Stress Measurements in Machined P/M Steels. In Adv. in Powder Metall. & Part. Mat., MPIF Princeton, 1995, p. 8-183.

[277] SANDEROW, HI., SPIRKO, JR. and CORRENTE, R.: Machinability of P/M Materials as Determined by Drilling Tests. *The Int. J. of Powder Metall.,* 1998, vol. 34, no. 3, p. 37.

[278] DAVIES, TJ.: Dual-phase Sintered Steels: To Achieve Good Strength and Machinability. In (CD) 3rd Int. Powder Met. Conf., TTMD, Gazi Univ., Ankara, 2002, p. 39.

[279] JAMES, PJ.: The Machinability of Sintered Steels. *PMI*, vol. 22, 1990, no. 6, p. 19.

[280] BERG., S. and MÅRS, O.: Investigating the Relationship Between Machinability Additives and Machining Parameters. In World Congress in Powder Metall. & Part. Mat., New Orleans, MPIF Princeton, Part 6, 2001, p. 50.

[281] KREMEL, S., DANNINGER, H. and YU, Y.: Effect of Sintering Conditions on Particle Contacts and Mechanical Properties of PM Steels Prepared from 3% Cr Prealloyed Powder. *Powder Metallurgy Progress,* 2002, vol. 2, no. 4, p. 211.

[282] ŠALAK, A., DANNINGER, H., SELECKÁ, M., OGURA, K. and UNAMI, S.: Properties of Sintered Fe-Cr-V Steels. In 2nd Int. Conf. on Powder Metallurgy for Automotive Parts, Isfahan, A. Arvand and H. Danninger eds., MPM Co, Tehran, 1999, p. 177.

[283] SUZUKI, S., HAMAZAKI, A., KARASUNO, I., KOSHIRO, K. and UMINO, M.: Machinability of 4100 Sintered Steel Containing Sulphur. In Progress in Powder Metall.- 1987, comp. by CL. Freeby and H. Hjort, APMI Princeton, vol. 43, 1987, p. 511.

[284] KAWATA, H. and HAYASHI, K.: Valve Seat Insert Material with Good Machinability. In 2002 World Congress on Powder Metall. & Part. Mat., Orlando, APMI Princeton, Part 2, 2002, p. 78.

[285] ŠALAK, A. and DUDROVÁ, E.: Magnetic and Mechanical Properties of Fe-Si Materials Sintered in Vacuum. *Powder Met. Science & Technology*, 1993, vol. 4, no. 3, p. 27.

[286] ENGSTRÖM, U.: Machinability of Sintered Steels. *Powder Metallurgy*, 1983, vol. 26, no. 3, p.137.

[287] LYULKO, VG., MARTIROSIYAN, RA., and HLEBUNOV, SA.: Machinability of Magnetic Soft Materials Containing Phosphorus. In All Russian Conf. 'Powder magnetic materials', Penya, 1992, p. 19, in Russian.

[288] ROSSO, M., ACTIS GRANDE, M. and ORNATO, D.: Sintering of Duplex Stainless Steels and their Properties. *Powder Metallurgy Progress*, 2002, vol. 2, no. 1, p. 10.

[289] AMBS, HD.: Machinability Studies on Stainless Steel. In Adv. in Powder Metall., APMI Princeton, vol. 3, 1991, p. 89.

[290] BEISS, P. and KUTSCH, U.: Machinability of Stainless Steel 430LHC. *Powder Metallurgy*, 1996, vol. 29, no. 1, p. 66.

[291] YAMAGUCHI, K. and NAKAMOTO, T.: Development of Free Machining Sintered Austenitic Stainless Steel (304L Stainless Steel) including Non-metallic Materials. In 1993 Powder Metall. World Congress, Tokyo, JPMA, Part 4, 1993, p. 1104.

[292] DORMER, Technical Sheet.

[293] CAUSTON, RJ., CIMINO, TM. and SCANLON, HM.: Machinability Improvement of P/M Steels. In Adv. in Powder Metall. & Part. Mat., APMI Princeton, vol. 7, 1994, p. 169.

[294] BOCCHINI, JF., RIVOLTA, B. and SILVA, G.: An Application the Universal Hardness Test to P/M Materials. In 2002 World Congress on Powder Metall. & Part. Mat., Orlando, MPIF Princeton, Part 13, 2002, p. 99.

[295] ZURECKI, Z., GHOSH, R. and FREY, JH.: Finish-turning of Hardened Powder Metallurgy Steel Using Cryogenic Cooling. *The Int. J. of Powder Metall.*, 2004, vol. 40, no. 1, p. 19.

[296] MPIF Standard 35. Materials Standards for P/M Structural Parts. 2000 ed., MPIF Princeton.

[297] BENNER, A. and BEISS, P.: Green Turning of Warm Compacted PM Steels. In World Congress on Powder Metall. & Part. Mat., New Orleans, MPIF Princeton, Part 6, 2001, p. 1.

[298] ANDERSSON, O.: Machining of Warm Compacted PM Parts in Green State. In World Congress on Powder Metall. & Part. Mat., New Orleans, MPIF Princeton, Part 6, 2001, p. 16.

[299] BENNER, A. and BEISS, P.: Green Turning of Warm Compacted PM Steels. In PM Euro2001. Eur. Congress & Exh. on Powder Metall., Nice, EPMA Shrewsbury, vol. 3, 2001, p. 59.

[300] GAGNÉ, M. and POIRIER, J.: Key Parameters for Machining PM Materials. In Symp. Recent Processes in Powder Metallurgy, Tohoku Univ., Sendai, 1992, p. 1-17

[301] ASAMI, J., HIROSE, N., SAWA, J. and UMEDA, A.: Cutting of Green Compacts. *J. of Jpn. Soc. Powder Metall.*, 2001, vol. 48, no. 11, p. 1082.

[302] DANAHER, J., TREMBLAY, L., ARMSTRONG, R. and ATTEWELL, L.: Green Machining for PM Applications. In Euro PM2003. Eur. Congress & Exh. on Powder Metall., Valencia, EPMA Shrewsbury, vol. 2, 2003, p. 134.

[303] RAMSTEDT, M. and ANDERSSON, O.: Improved Green Strength Opens Up Possibilities for Green Machining. In Euro PM2003. Eur. Congress & Exh. on Powder Metall., Valencia, EPMA Shrewsbury, vol. 3, 2003, p. 304.

[304] JOHANSSON, R.: Different Methods to Improve Machinability of High Strength Sintered Steels. In World PM Congress in Powder Metall & Part. Mat., Orlando, APMI Princeton, vol. 21, 1988, p. 381; *Höganas Iron Powder Information* PM 88-3.

[305] CHAGNON, F., POIRIER, PJ. and TRUDEL, Y.: Enhancing Machinability of PM Materials. Int. Congress & Exh., Detroit, SAE Inc., 940359, 1994, p. 1.

[306] CHAGNON, F. and GAGNÉ, M.: Machining Sinter Hardenable PM Materials. In 1998 PM World Congress & Exh., Granada, EPMA Shrewsbury, vol. 2, 1998, p. 384.

[307] DANNINGER, H. and GIERL, Ch.: Processes in PM Steel Compacts during the Initial Stages of Sintering. *Mater. Chemistry & Physics*, 2001, vol. 67, p. 49.

[308] BLAIS, C. and L'ESPÉRANCE, G.: Turning and Drilling of Parts Made from Sinter Hardenable Steel Powders. *Powder Metallurgy*, 2002, vol. 45, no. 1, p. 39.

[309] BENNER, A. and BEISS, P.: Drilling and Tapping of Warm Compacted PM Steels after Sintering. In Euro PM2000 Conf. on Materials and Processing Trends for PM Components in Transportation, Munich, EPMA Shrewsbury, 2000, p. 159.

[310] BENNER, A. and BEISS, P.: Turning of Warm Compacted High Density PM Steels in the As-sintered Condition. In 2002 World Congress on Powder Metall. & Part. Mat., Orlando, MPIF Princeton, Part 12, 2002, p. 16.

[311] BLAIS, C., YOUNG, B. and L'ESPÉRANCE, G.: Hard Turning of Parts Made from Sinter Hardenable Powders. In 2002 World Congress in Powder Met. & Part. Mat., Orlando, MPIF Princeton, part 12, 2002, p. 1.

[312] LIPP, K., STRAFELLINI, G. and SONSINO, CM.: Fatigue Resistance and Machinability of High-strength Sintered Steel in Dependence on Tempering Temperature. *PMI*, 1993, vol. 25, no. 6, p. 261.

[313] ASHOK, S., THAGGARSE, Y., SARATHY, ZD. and MADHUSUDAN, P.: Effect of Machining Aids on the Machinability of Steam Treated PM Parts. In 2000 Powder Metall. World Congress, Kyoto, K. Kosuge and H. Nagai eds., JPMA, Part 2, 2000, p. 1548.

[314] ŠALAK, A. and BENKE, V.: Automatic Forging Lines for Production of Parts by Powder Forging. *Strojírenská Výroba*, 1991, no. 6-7, p. 26, in Slovak.

[315] ARVAND, A.: MPM Co, Tehran; personal communication, 2003, 2004.

[316] MITTRACH, J.: Development of a Connecting Rod from Powder up to Finished Part. In Pulvermetallurgie in Wissenschaft und Praxis, Bd. 12, R. Ruthardt ed., DGM, Frankfurt a. M., 1996, p. 119, in German.

[317] BODE, U.: Powder Forging of Engine Transmission Parts. In Euro PM2000, Munich, EPMA Shrewsbury, 2000, p. 117.

[318] SKOGLUND, P. and BENGTSSON, S.: Materials for Highly Loaded Parts such as Connecting Rods. In 3rd Int. Conf. on Powder Metallurgy for Automotive Parts, Isfahan, A. Arvand and H. Danninger eds., MPM Co, Tehran, 2002, p. 185.

[319] WILLIAMS, VA.: Powder Metal + Forge = Net Shape. *Production*, 1977, no. 7, p. 10.

[320] ŠALAK, A. and SELECKÁ, M.: Mechanical and Toughness Properties of Powder Forged Low Alloyed Cr-steel. In 3rd Int. Powder Metall. Conf., Ankara, TTMD, Gazi Univ., 2002, p. 408.

[321] RUSNÁK, Z.: Fracture Toughness of Bearing Steel and Life Time of Rolling Bearings. *Technický Spravodaj*, ZVL Považská Bystrica, 1982, no. 2, p. 6, in Czech.

[322] HAIMERL, M.: Powder Forged Conecting Rod for BMW-V-8-cylinder. In Pulvermetallurgie in Wissenschaft und Praxis, H. Kolaska ed., VDI Düsseldorf, Bd. 9, 1993, p. 76, in German.

[323] MEHLHOSE, J., KUTSCHERA, H., TEICHERT, B., WÄHRIG, A. and JOST, H.: Additional Processing of Powder Forged Parts. In Int. Conf. 'Neue Technologien für Wälzlagerindustrie', Ingenieurhochschule, Chemnitz, 1985, in German.

[324] MEHLHOSE, J., LORENZ, K., TEILCHERT, B. and MICHEL, P.: Processing, Cold Work Hardening and Machining Behaviour of Powder Forged Materials. In VII. Int. PM Tagung in der DDR, Dresden, 1981, vol. 2, p. 61, in German.

[325] STRÖMGREN, M. and KOOS, R.: Powder Forging at Höganäs AB. *Höganäs PM Iron Powder Information PM 83-7.*

[326] DOROFEYEV, YuG., USTIMENKO, VI. and SKORIKOV, AV.: Some Specifics of the Wear of a Cutting Tool in Finishing of Low Porosity Alloyed Steels. *Poroshkovaya Metallurgiya,* 1984, no. 2, p.74, in Russian.

[327] KRAVCHENKO, VF., LUTKEVICH, EG. and USTIMENKO, VI.: Effect of Built-up in Turning of Sintered Materials. *Poroshkovaya Metallurgiya,* 1973, no. 10, p. 93, in Russian.

[328] KOOS, R. and BOCKSTIEGEL, G.: The Influence of Heat-Treatment, Inclusions and Porosity on the Machinability of Powder-forged Steel. In 1981 Nat. Powder Metall. Conf. & Exh., Philadelphia; *Höganäs PM Iron Powder Information* PM 81-4.

[329] ŠALAK, A. and SELECKÁ, M.: Manufacturing and Properties of Powder Forged Rolling Bearing Rings. In PM Auto '99. Int. Conf. on Powder Metall. for Automotive Parts, Isfahan, MPM Co, Tehran, A. Arvand and H. Danninger eds., 1999, p. 35.

[330] FEY, K.: Powder Forging on Fully Automatic 'Lasco' Press Lines. In 1998 Powder Metall. World Congress & Exh., Granada, EPMA Shrewsbury, vol. 2, 1998, p. 227.

[331] BLANTER, ME.: *Phase Transformations at Heat Treatment of Steel (Fazovyye prevrashcheniya pri termoobrabotke staley).* Ed. METALLURIZDAT, Moscow, 1962, in Russian.

[332] ŠALAK, A. and BENKE, V.: Machinability of Powder Forged Structural Steels. In Conf. 'Konštrukčné materiály na báze práškových kovov a ich využitie', IMR SAS Košice, Oct., 1988, p. 71, in Slovak.

[333] St-LAURENT, S., GAGNÉ, M. and CHAGNON, F.: Effect of Sulfur Content on Machinability of Forged Specimens made by Water Atomized Steel Powders Prealloyed with MnS. In Adv. in Powder Metall. & Part. Mat. - 1997 (PM^2TEC '97 Chicago), comp. by RA. McKotch and R. Webb, MPIF Princeton, Part 15, 1997, p. 95.

[334] KUBICKI, B., CHEN, N. and SELVARAJAH, S.: Cutting Forces Investigation by Tapping and Drilling of the PM Materials. In 2000 Powder Metallurgy World Congress, Kyoto, K. Kosuge and H. Nagai eds., JPMA, Part 2, 2002, p. 1544.

[335] DOROFEYEV, JuG., KOZLOV, EV. and SHISHKA, VG.: Technological Aspects of Preparation of High Density Steels Containing Lead. In Antifriktsionyye materialy spetsialnogo naznacheniya, Mezhvuz. Sb., Novocherkask, 1991, p. 118, in Russian.

[336] MALEVANNYJ, VI., SKORIKOV, AV., KOZLOV, EV. and YEREMEYEVA, ZhB.: Effect of Lead on Structure and Properties of High Density Powder Ferrous Materials. *Izv. Sev. Kavk. Nauchn. Centra,* 1990, no. 4, p. 110, in Russian.

[337] ARTAMONOV, AJa. and KONONENKO, VI.: Wear of Cutting Tools at Machining of Porous Materials. *Poroshkovaya Metallurgiya,* 1967, no. 4, p. 78, in Russian.

[338] DOROFEYEV, YuV., SKORIKOV, AV., SHISHKA, VG., VEROPACHA, DN. and YAYCKIY, DL.: Effect of Calcium Carbide on Machinability of Powder Forged Structural Steel. *Izv. VUZ, Sev. Kavk. Region. Tekhn. Nauki,* 1994, no. 1-2, p. 245, in Russian.

[339] SKORIKOV, AV., KOZLOV, EV., SHISHKA, VG, ERESHEYEVA, ZhV. and VEROPACHA, DN.: Effect of Calcium Fluoride on Machinability of Powder Forged Steel. *Izv. VUZ, Sev. Kavk. Region. Tekhn. Nauki,* 1994, no. 1-2, p. 9, in Russian.

[340] TAUBENBLATT, PW., SMITH, WE. and BLADT, FA.: Forging of P/M Alloys with Improved Machinability. In P/M 78 SEMP 5, Stockholm, EPMA Shrewsbury, vol. 1, 1978, p. 143.

[341] KARLSSON, B., LARSSON, L-E. and STRÖMGREN, M.: How to Aspect the Surface Temperatures during Powder Forging. In Horizons of Powder Metallurgy (PM86 Int. Conf Düsseldorf), WA. Kaysser and WJ. Huppmann eds., Verlag Schmid, Freiburg, 1986, p. 252.

[342] KOOS, R. and BOCKSTIEGEL, G.: Machining Studies on Powder Forged Steels. *MPR*, vol. 35, 1980, no. 10, p. 116.

[343] BERG, S.: Machining of Powder Metallurgy Materials. ASM Handbook, vol. 7. Powder Metal Technologies and Applications. ASM Metals Park, 1998.

[344] *General Guidelines for the Machining of Sintered Materials.* Höganäs PM Iron Powder Information PM 77-14.

[345] FEIR, M.: Post Treatment of PM Parts. *MPR*, 1980, vol. 30, no. 1, p. 28.

[346] VASILKO, K. et al.: *Rolling Bearings (Valivé ložiská).* Ed. ALFA Bratislava, SNTL Praha, 1988, in Slovak.

[347] MINISTR, Z.: *Machine Parts from Powder Metals (Strojní součástí z kovových prášku).* Ed. SNTL Praha, 1963, in Czech.

[348] *Sintered Metal Materials - Sint-guide-lines* - Part 1: Machining of Sintered Parts, Standard DIN V 30912, in German.

[349] *Sintered Metal Precision Parts (Gesinterte Präzisionsteile).* Literature of MIBA Vorchdorf, in German.

[350] ZAPF, G.: *Handbook of Processing Technique. Powder Metallurgy (Handbuch der Fertigungstechnik. Pulvermetallurgie).* Carl Hanser Verlag, Munich, 1981, in German.

[351] ROMAN, OV.: *Overview in Powder Metallurgy.* Part V. Pecularities of Cutting of Powder Steels. Minsk, Ed. BELARUS, 1988, p. 114, in Russian.

[352] DOMOGACKIY, VI. and KOSTYUKOV, AV.: Pecularities of Turning of Sintered Piston Rings. In Isl. tech. metall. poroškov i spechennych materialov, AN SSSR, Ural Center, Sverdlovsk, 1980, p. 120, in Russian.

[353] ANDERSSON, O.: Green Machining of PM Steels. Höganäs Chair Seminary, Madrid, June, 2004.

[354] VASILKO, K., Faculty of Production Technologies in Prešov, TU Košice; personal communication, 2004.

[355] SHIIKI, K., YAMADA, N., KANO, T. and TSUGUI, K.: Development of Shape-Controlled Sulfide-Free Machining Steel for Application in Automobile Parts. *Advanced Mater. & Processes*, 2004, vol. 162, no. 7, p. 33.

[356] SCHATT, W.: *Powder Metallurgy. Sintered and Composite Materials (Pulvermetallurgie. Sinter- und Verbundwerkstoffe).* VEB Deutscher Verlag für Grundstoffindustrie, Leipzig, 1979, in German.

[357] HOEGANAES. CD - Product Data Sheet International.

[358] WOJTASIK, K. and MISSOL, W.: Technology Pushes Green Strength up in Single Press and Sinter Cycle. *MPR*, 2004, no 7, July/August, p. 27.

[359] ATOMET Powder Metallurgy (P/M) Products, QMP - prospectus.

[360] JAMES, B., Hoeganaes Corp.; e-mail information, 2000.

Index